D1272479

Naval Warfare in the Age of Sail

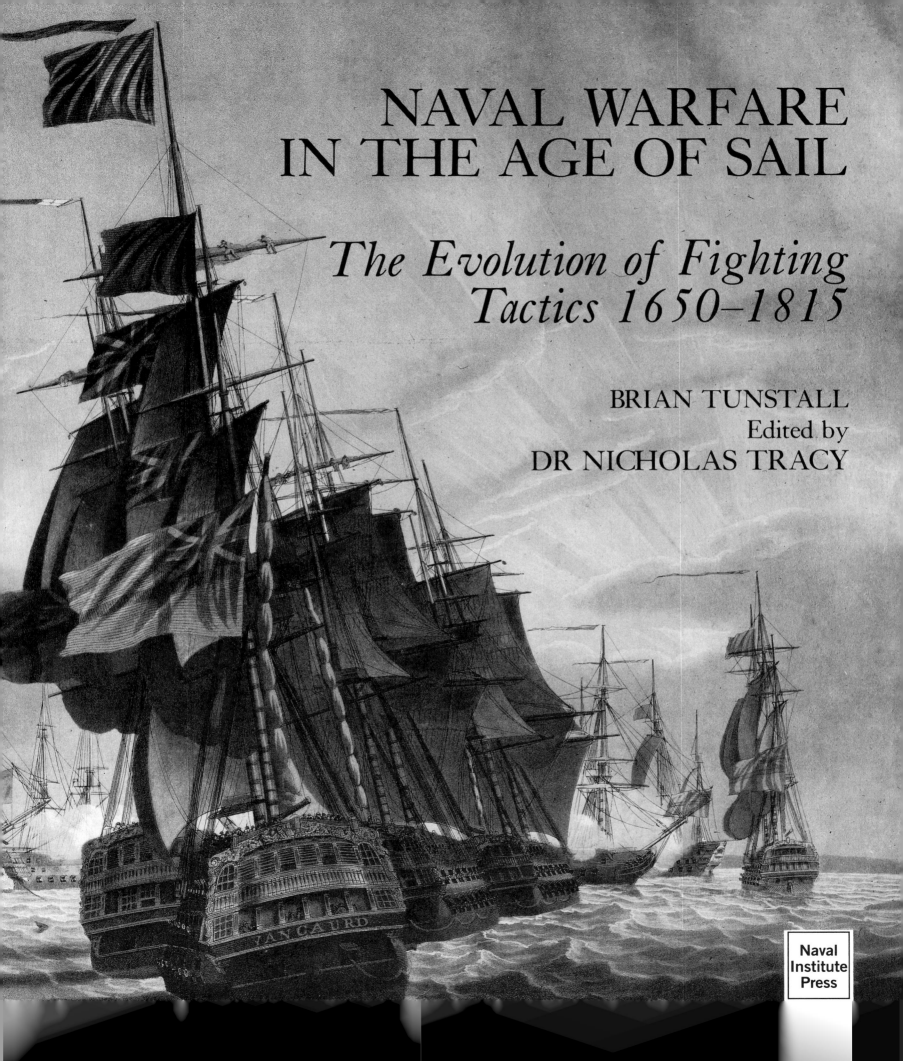

NAVAL WARFARE IN THE AGE OF SAIL

The Evolution of Fighting Tactics 1650–1815

BRIAN TUNSTALL

Edited by

DR NICHOLAS TRACY

Naval Institute Press

Frontispiece: *Nelson's fleet sailing into action at the Battle of the Nile.*
NMM

© 1990 Brian Tunstall
Additional material © 1990 Nicholas Tracy 1990

First published in Great Britain 1990 by
Conway Maritime Press Limited
24 Bride Lane, Fleet Street
London EC4Y 8DR

Published and distributed in the United States
of America and Canada by the Naval Institute
Press, Annapolis, Maryland 21402

Library of Congress Catalog Card number
90–62558

ISBN 1–55750–601–9

Manufactured in Great Britain

Contents

William Cuthbert Brian Tunstall FSA

This book represents the most sustained contribution to Naval History of a considerable scholar. After coming down from St John's College, Cambridge, Tunstall taught at Oundle School, and at the Royal Naval College, Greenwich. During the war he served as a naval expert in the BBC. Ultimately he became Senior Lecturer in International Relations at the London School of Economics.

He made his scholarly reputation as a naval historian. Not only was he Sir Julian Corbett's son-in-law, but he also gathered, preserved and catalogued Corbett's papers. He wrote a strong, realistic account of Admiral George Byng while he was editing the Byng Papers (three volumes) for the Navy Records Society. Of the commercial edition of the biography, a reviewer observed, 'Mr Tunstall moves easily under the burden of heavy research' - happy sentence. Tunstall also wrote a life of the Great Commoner, the Elder Pitt, which represented one of the first attempts to use psychiatric insights in a responsible way in biography. For much of the 1930s he was Honorary Secretary of the Navy Records Society, and he blasted Sir Geoffrey Callendar's bland patriotic pap with that little gem *The Realities of Naval History*. The book startled the naval establishment, along with Callendar, and doubtless contributed to Tunstall's departure from the College in the late 1930s.

After the war he wrote three celebrated chapters in the *Cambridge History of the British Empire*. He taught not only international relations, but also national policy in relation to sea power, to a generation of graduate students. He was a superb teacher, invariably advancing new knowledge or fresh approaches under the guise of tentative suggestions. His mind was as powerful as his deployment of its products was unorthodox.

Tunstall's work on signals derived directly from the work of Corbett. He carefully and systematically organised and added to his father-in-law's collection of signal books. He traced the whereabouts of complimentary collections, and in his latter years had come to know where all were to be found. But the signal books, Corbett's work, and Tunstall's general knowledge of naval history came together as he began to examine the changes in tactics that had taken place since 1654, a reassessment based on a detailed appraisal of the signal books themselves. He did not neglect the framework that Corbett had pioneered. Nevertheless, he looked at every change in usage pragmatically, and observed the struggle of the practitioners of the signaller's art as a part of fleet policy with ultimate national purposes. It is not too much to say that the book he has produced has irrevocably changed the world. His view was that the relationships of naval strategy (including tactics) at sea to state policy cannot be appreciated without a thorough knowledge of the significance for fleet tactics of fleet signalling.

This is, therefore, an important work and Dr Tracy's service to naval historical scholarship is, consequently, hard to overestimate.

DON SCHURMAN

Author's Preface

THE FIRST move towards a systematic study of British naval tactics was made by Sir Julian Corbett in 1905. In that year, the centenary of Trafalgar, he edited a collection of documents for the Navy Records Society, entitled *Fighting Instructions, 1530–1816*. These documents showed for the first time something of the origins, development and continuity of tactical methods as pursued by a long line of naval commanders from the early Tudor ship captains to the authors of the Signal Book of 1816. Until that moment neither historians nor naval officers had appreciated that naval tactics had a history and lineage which could be traced and studied in serious terms.

As a result of Corbett's *Fighting Instructions*, more documents were brought to light, and in 1908 he edited a second volume, entitled *Signals and Instructions, 1776–1794*. This volume included selections from a valuable collection of manuscript and printed sailing and fighting instructions, orders of battle, tactical memoranda, signal books and projects for new systems of signalling, made by Admiral Sir Thomas Graves during the War of American Independence. Graves, who was a member of the famous naval family, was a comparatively young man at the time of the American War. His collection helps to trace the course of this period of important change in British tactics. He was afterwards second-in-command to Nelson at the Battle of Copenhagen (1801); he died in 1814. 'His representatives patriotically presented his collection to the [Royal] United Service Institution. A few of the documents found their way to the shelves of the Library, but the bulk remained together and uncatalogued.' They were discovered again by Sir John Laughton, whose original researches in British and French naval history, begun many years previously, had inspired Captain A T Mahan, USN, to begin writing *The Influence of Sea Power upon History*.

Corbett's two volumes of documents, together with his substantial editorial comments, provided a corpus of source material linking Blake, James Duke of York, Monck, Prince Rupert, Rooke, Vernon, Anson, Hawke and Rodney with Hood, Howe, St Vincent and Nelson. He drew distinctions between the ideas embodied in critical treatises, the instructions issued by admirals and the way tactics were employed in battle. Relationships were shown between traditional ideas and innovations and the reader glimpses something of the intellectual ferment which inspired the work of Howe, Hood and Kempenfelt. Nevertheless, these volumes were not intended to do more than provide a corpus of source material. They do not, and were never intended to constitute a history of British naval tactics. Attempts to isolate the editorial matter, and use it in the manner of a textbook, while at the same time ignoring the huge accretions of new source material, can only produce misleading and unscholarly results.

Before his death in 1922, Corbett had accumulated enough new material for a completely revised edition of both volumes. Since then, the flow of newly discovered documents has been very considerable and, as a result Corbett's work, can be seen in better perspective. Many of the distinctions and conclusions he drew have been confirmed. Several of his more speculative conclusions, based on brilliant guesswork, have been triumphantly vindicated. Documents have now appeared which he had assumed to exist but had never seen. In other cases, his views can be shown to be in need of serious revision, especially where his assumptions about the authorship and dating of documents were in error. It is a tribute to his great work as a naval historian that so many writers can still permit themselves to make completely uncritical use of conclusions based on a corpus of original source material amounting to less than a quarter of what is available today.

Apart from some important items acquired by the Admiralty Library and the Royal United Service Institution since 1908, the bulk of the new material now available for study is in the National Maritime Museum, Greenwich.[1] This great collection of documents is rich in French items, including the instructions and signals issued by the Comte de Tourville in 1690–3. It is not only the most important collection, in both quality and quantity, but it is also the best administered. The combined Corbett and Tunstall collection is now considerable and includes French, Spanish, Dutch and Italian items.[2] The collection formed by the late Commander Hilary Mead is specially important for its Howe, Kempenfelt and St Vincent items. Altogether it may be said that the amount of documentary material available today is more than three times the amount available to Corbett in 1908.

After 1908 the most important contributor to the history of British tactics was Admiral Sir Herbert Richmond. Though primarily concerned with the broader strategic, political and economic aspects of sea warfare, he dealt very fully with tactics in *The Navy in the War of 1739–48* (1920) and *The Navy in India 1763–1783* (1931). A few years later Lt-Commander J H Owen made a number of invaluable studies of British and French tactics in the American War using the original logs and journals of British ships. Rear-Admiral A H Taylor has recently made important and scholarly reconstructions of a number of fleet actions, including Trafalgar.[3] All students of the Dutch Wars are indebted to R C Anderson for his masterly elucidations of Dutch and English texts of fighting instruction, order of battle, signals and tactical narratives. *A History of Naval Tactics 1530 to 1930* by Rear-Admiral S S Robinson, USN and Mary L Robinson (US Naval Institute, 1942) is the only work in existence of its kind. Unfortunately, it is not an original history at all but a series of summaries and excerpts taken from a wide range of printed secondary sources.

In all his writings on naval history, of which the volumes of tactical documents formed only a very small part, Sir

---

1 The Royal United Services Institute collection has now been broken up and divided between the Ministry of Defence Library, Whitehall (Naval Library), and the National Maritime Museum, Greenwich.

2 The Corbett, Tunstall and Mead collections are now all in the National Maritime Museum.

3 *The Mariner's Mirror*, 36 (1950), no 4

Julian Corbett emphasised the importance of signalling at sea. He was indeed the first historian to draw attention to the fact that in addition to showing how battle orders were actually transmitted, signals often provide evidence of tactical ideas and plans otherwise undiscoverable. In dealing with signalling techniques, he relied very considerably on the resources of William Gordon Perrin, the Admiralty librarian, whose immense erudition was always at the disposal of even the most unsophisticated enquirer. Perrin's *British Flags* (1922) includes a valuable section on naval signalling systems and signal books but he did not live to write the full-dress work he intended on this subject. Vice-Admiral Lancelot Holland accumulated a valuable collection of instructions and signals, including those of the Comte de Tourville now in the National Maritime Museum. He was killed in HMS *Hood* in action with the *Bismarck* in 1941, and left only a very short study for publication.[4] Commander Hilary Mead was amongst the leading world authorities on the subject of flags generally, including all methods of visual signalling. His large collection of books and manuscripts includes many naval items but he did not make naval signalling his sole specialisation.

Of the French naval historians, G Lacour-Gayet and Charles de la Roncière have each made considerable use of original source material in their analyses of tactics. The first detailed study of particular tactical experience was made by Colonel Edouard Desbrière, whose *La Campagne Maritime de 1805 – Trafalgar*, published in 1907, is a monument of able and devoted scholarship. Nevertheless, it covers only the battles of Ferrol, Trafalgar and Cape Ortegal. In 1933 Constance Eastwick not only translated the whole work but re-examined the original French, Spanish and British documents and was thus able to add some useful corrections to Desbrière's work. The most original, stimulating and irritating book on naval tactics by a French author is *Les Idées Militaires de la Marine du XVIIIme Siècle: de Ruyter à Suffren* (1911) by Captain R Castex. It is based on research and written in brilliantly critical terms. Almost all that he says about the difference between tactical theory and tactical practice, and the doctrine or ethos underlying official pronouncements, is excellent. Unfortunately, much of his work is marred by violent exaggerations and a number of sweeping generalisations unsupported by evidence.

Such are the writers who, in company with Corbett, have contributed towards a better understanding of the tactical theories, and tactical and signalling techniques during the sea wars of the seventeenth, eighteenth and early nineteenth centuries. The present work has a comparative basis and is the first in which any attempt has been made to evaluate the new evidence, as well as to re-examine the old.

---

4 'The Development of Signalling in the Royal Navy', *The Mariner's Mirror*, 39 (1953), no 1

Editor's Preface

BRIAN Tunstall's great work, completing the analysis of naval tactics under sail which had been begun by his father-in-law, Sir Julian Corbett, was not published during his lifetime. The reasons will be apparent to anyone who cares to read his manuscript, which is part of the collection at the National Maritime Museum. Perhaps the convoluted style of Admiral Lord Howe, who did more than anyone to reform British naval tactics, laid a curse on the subject. In the posthumous preparation of this book for publication, editorial intervention in those sections describing tactical ideas and signalling systems has had to be so extensive that it virtually amounts to co-authorship. The objective has been to bring out as clearly as possible Tunstall's research and perceptive analysis. Since his death important books have appeared on various aspects of the naval history of the period covered by Tunstall's book, but the value of his work has not been materially affected. Because it is inevitable that scholarly attitudes will vary over the years, it has been the policy of this editor to present Tunstall's case without significant alteration, so that the text is a true statement of Tunstall's ideas, expressed as far as possible in his own words.

In contrast to the sections describing the efforts of tactical reformers, the sketches Tunstall has written of naval battles have required little if any substantive intervention by the editor. Their value is all the greater because they are tightly focused on the important tactical developments which occurred during these actions.

Over half the manuscripts to which Tunstall referred were in private collections, including his own and that of the Royal United Services Institute. These collections are no longer intact. It has been necessary to track down the manuscripts in their new depositories. In this task I have been helped by Roger Morriss of the National Maritime Museum, and David Brown of the Admiralty Library. Unfortunately, not all the manuscripts became part of public collections, and in those cases the references given in the following pages are to the original collections, which are no longer in existence.

Notes on the drawings

The drawings which accompany the text of this book have
been drawn mainly from the information given by the
author, but also with reference to other relevant sources, such
as battle plans in standard works, or the original books in the
case of theoretical tactics. A standard set of symbols has been
used throughout. All battle plans are oriented to the north,
except where a compass rose is used to indicate north. A
version of the tricolour has been used to indicate French
flagships, though of course that flag was not in use until the
revolution of 1789.

Key

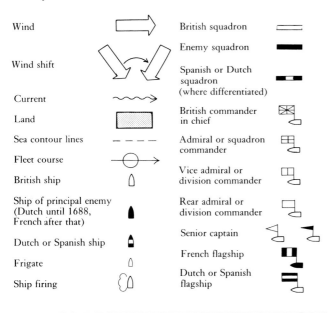

Wind	British squadron
Wind shift	Enemy squadron
	Spanish or Dutch squadron (where differentiated)
Current	
Land	British commander in chief
Sea contour lines	Admiral or squadron commander
Fleet course	Vice admiral or division commander
British ship	Rear admiral or division commander
Ship of principal enemy (Dutch until 1688, French after that)	
	Senior captain
Dutch or Spanish ship	French flagship
Frigate	Dutch or Spanish flagship
Ship firing	

Abbreviations

BM   British Museum
MM   Musée de la Marine, Paris
MOD   Ministry of Defence Library, Whitehall (Naval
        Library)
NMM   National Maritime Museum, Greenwich
PRO   Public Record Office, Kew Gardens
RUSI Royal United Services Institute

The Art of the Admiral

NAVAL tactics are an admiral's art, and in a fleet propelled by wind and sails he exercised his art with difficulty. At no time could he make use of more than five-eighths of the sea around him, and when a calm prevailed he could not move his fleet at all. As he stood on the poop or quarterdeck of his flagship with the wind blowing, the remaining three-eighths of the sea were barred to him because his fleet could not sail directly into the wind. When close-hauled, that is to say when sailing as close as possible to the wind, his ship could not keep closer than six points of the compass on either the starboard or the larboard (port) tack. Hence progress directly into the eye of the wind could only be made by tacking across from one side to the other, and never closer than six points of the compass (67.5°). Even when running before the wind, it was generally best to go quartering to and fro rather than have the wind directly astern, since, in square-rigged ships, the sternmost sails would blanket the sails ahead and prevent them catching a stern wind.

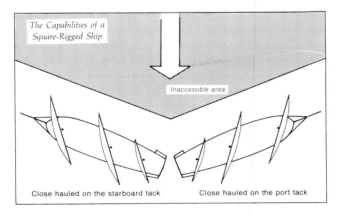

The Capabilities of a Square-Rigged Ship

Inaccessible area

Close hauled on the starboard tack   Close hauled on the port tack

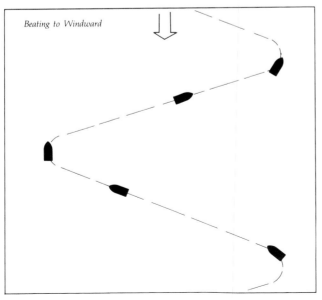

Beating to Windward

The enemy were of course subject to exactly the same limitations. When two fleets sighted each other, one of them was usually to windward, and so held what was termed the weather gage. Since the weather gage automatically conferred some measure of tactical advantage, especially in the choice of whether, when and how to attack, the manoeuvres preliminary to a battle were often concerned with keeping, gaining or disputing the windward position.

In battle each admiral had to make a double set of calculations relating his fleet to that of the enemy in terms of distance, speed and course. Not only were there the wind and sea to consider, as they were at the moment, but also possible change in the force and direction of the wind and the influence, when near land, of tides and currents. If the wind dropped altogether the rival fleets might be left immobilised, half engaged, fully engaged or with the engagement over and still in sight of one another. Everything depended on the wind and the weather and, unless these were favourable, no battle could take place, nor could either fleet be formed in battle order.

The warships of the sailing period 1650–1815 were not built primarily for speed or manoeuvrability. They were wooden fortresses, propelled by wind and sails, furnished with two large batteries of guns, one on each broadside, and manned by a huge garrison, that is compared with their size and tonnage. They were heavily timbered, both for defence and to withstand the strain caused by the discharge of their own guns. They had stowage space for large quantities of food, stores, equipment and ammunition. They could stay at sea for many weeks, and they could withstand heavy weather. Each ship was a self-contained fighting unit of considerable power, capable of developing a heavy 'fire face' on each broadside, though with only a very few guns mounted to fire directly ahead and astern.

Success in war depended on an admiral's ability to organise a body of ships into a disciplined fleet, capable of obeying his instructions and signals. Only when his fleet was properly organised was he in a position to execute such tactics against the enemy as he deemed both possible and remunerative. Rapid changes of formation in the face of the enemy were useless and even dangerous if the captains were not sufficiently practised to carry them out. Nor could the admiral expect wholehearted support for a form of attack of which the captains did not approve or which they did not thoroughly understand. Much depended on the time available for exercising the whole fleet together at sea. If its composition was continually changing or if detachments had to be made for convoy work, watching enemy ports or minor operations, efficient organisation was not possible. This was especially the case at the beginning of a war. By the end, a common experience had generally been built up, thus allowing interchanges to take place between ships of different fleets and squadrons without any severe diminution of tactical efficiency. An anonymous admiral wrote of the 'Glorious First of June' (1794) that it 'was the *first* general action fought in the course of the war, and led to many glorious results; had it been the *last*, not one of the French ships would have been allowed to return to port'.

At sea each captain had to be capable of keeping his ship on the course set by the admiral and at the required distance – and, if necessary, on the required bearing from the next ship ahead or astern, or to starboard or larboard. Changes in sail carried by the admiral had to be promptly copied, with such adjustments as were necessary for correlating the known sailing powers of the individual ships of the fleet with that of the flagship. The larger the fleet the more difficult it was to control, and the longer it took to carry out evolutions. Tacking through the wind, and wearing or gybing (changing course with the wind astern), whether in succession or together, were movements which tested the seamanship and discipline of every single ship. If the fleet was in close order there was always danger of collision if the instructions were not properly observed. Station keeping was all-important, and correct alignment had always to be regained after tacking and wearing. Losing station by falling to leeward was the commonest kind of lapse, and British fleets always steered seven points from the wind when close-hauled, even though six points was possible, so as to enable ships which had fallen to leeward to work up again to their station. Well-trained French fleets – and before the Revolution most of them were – could manage six points off. A very high degree of corporate skill was needed to keep this alignment. Great responsibility rested on captains leading the fleet on either tack in line of battle ahead, that is, with the ships in file on ahead of another. This also applied to the leading captains if the ships were disposed in two or more parallel columns in line-ahead. Leading the line or lines and station keeping in general were particularly important and especially difficult when the ships were in line of bearing and in echelon.

All this skill in controlling the individual ships composing a fleet was required for one purpose only, the defeat of the enemy fleet. Evolutions could never be an end in themselves, though this is how they sometimes seemed to have appeared in the French service. Yet, without a properly maintained order of sailing, a fleet could be taken unawares, especially if it became dispersed during the night. Admiral Thomas Mathews failed to beat Admiral La Bruyere de Court off Toulon in 1744 largely because his fleet failed to get into station on the previous evening. Vice Admiral John Byron was beaten off Granada in 1779 by the Comte d'Estaing largely because his fleet had become dispersed in the night. Admiral Augustus Keppel was much praised after the Battle of Ushant (1788) for engaging the Comte d'Orvilliers without wasting time in making the signal for 'form line of battle.' This was certainly creditable but if he had had his fleet better in hand at day-break he would not have had to commit himself to a ragged attack. Lord Nelson did not have the whole of his fleet as close to him as he might have wished on the morning of Trafalgar (21 October 1805), especially when he decided to attack immediately. This disability was never put right and it can be argued that he might have done better to wait until his fleet was more collected before launching his attack.

Night action itself was shunned until Nelson risked it at the Nile (1798). There were signals for it in both navies, since it was recognised that admirals might have night action forced on them through chance encounters. Strong fleets shunned night fighting, however, because they could not use their strength effectively. Weak fleets equally shunned it because they feared that they might become heavily engaged and be unable to withdraw.

Failure to maintain a proper system of reconnaissance seems to have been one of the worst faults of the sailing fleet admirals. The Comte de Grasse risked complete defeat because he was surprised at the Chesapeake in 1781. Admiral François Brueys d'Aiguïlliers failed to use his frigates to keep a look-out for Nelson before the Battle of the Nile. The inadequacy of reconnaissance was to some considerable extent a result of the lowly status of private ships, which were unable to signal back to the admiral in the same terms in which he could signal to them. No reconnaissance system was of any great use unless the ships conducting it could make full and detailed reports as to the composition, strength and disposition of the enemy, their present course and seeming intentions and the amount of sail they carried. In the French service more advanced ideas prevailed. Nelson at Trafalgar had an excellent system which was able to employ Sir Home Popham's 'telegraphic' system of signalling.

The most perfect signalling system for look-out frigates was useless unless the frigates were sufficiently bold and cunning to get up with the enemy without being winged or captured. This meant being able to avoid or beat off the enemy's frigates. Hence the importance attached to the qualities of frigate captains, one or more of whom might even be senior to some of the captains of the ships-of-the-line.

Discussions often arose as to whether the admiral ought not to be in a frigate instead of in one of the most powerful ships in his fleet, which must by its very character become heavily involved in the pell-mell. Lord Howe actually tried it and so did Admiral George Rodney. Audibert Ramatuelle discussed it at length in his *Cours Elementaire de Tactique Navale*, 1802. Another advantage was that the admiral could carry out advance reconnaissance in person instead of having to rely on signals sent back to him, as Napoleon pointed out to the Comte de Ganteaume. In battle, however, his position would be equivocal, since he could hardly be judge of tactical movement in which he himself was not participating. If his fleet was engaging from windward he would have to be to windward of his own line and if from leeward, then to leeward, with the additional disadvantage of his view being blocked by smoke. In a light breeze the smoke from the guns hung over the battle area and prevented the admiral from seeing anything beyond the ships immediately around him. His fleet meanwhile might be strung out over a wide sea area, under conditions which made signalling difficult even if there was no smoke at all.

When attacking a fleet already formed in line of battle, it was customary to make a slow and stately approach. This enabled the attacking ships to maintain their alignment and not mask each other's fire by running too far ahead or accidentally steering across the line of approach of another ship. Even with a stiff breeze in their favour, they might well be advancing at only about four knots. It is difficult for modern readers to visualise the agonising slowness of most sea battles under sail. Hours and even days might be spent in trying to get into the desired order of battle before launching the attack. It is easy to see that under these circumstances the strain on the admiral was terrific and why it was that both Rodney and Howe were incapable of pursuit after their respective victories of the Saints (1782) and the First of June.

Diagrams printed in the tactical text-books and official instructions of the sailing-ship period naturally show ships carrying out evolutions in perfect alignment, though this was not always attainable in practice. For this reason diagrams of sea battles under sail are misleading; the reader is tempted to project the modern capacity for station keeping into the past, simply on the analogy of what modern machine-driven warships can do with the aid of radio and radar.

Even under sail, relative speed was always of great tactical importance, and this was often determined less by the ship's design than by the amount of weed and barnacles fouling her bottom. Prior to the introduction of copper sheathing, a fleet of newly cleaned ships, fresh out from dock, could outsail by a significant margin a fleet of ships exactly similar in build, but in need of scraping after many weeks at sea. The introduction of built-in speed differentials as part of steamship design was one of the most important influences on the development of naval tactics in the iron-clad period.

No one ever really knew the exact time at which things occurred in a sea battle under sail. Despite the efforts of the clockmaker John Harrison, chronometers did not appear until towards the end of the period, and even then they were not supplied by the Admiralty. Some officers bought their own at a cost of £100, a very large sum in comparable terms today; assuming that he was at sea continuously, the pay of a captain commanding a line of battle ship would vary between about £246 and £365 a year, depending on the rating of the ship. Although ships' logs may record the times of battle movements and of signals sent and received, these times are only of value in so far as they relate to the progression of time in a particular ship. Instruments were used for estimating the distance between one ship and another when formed in line of battle but there were no range finders for estimating the distance of the enemy's ships.

The sailing ship-of-the-line experienced the problem of bringing its armament into action far more than did the warships of the machine age. Even when weather conditions were ideal no ship-of-the-line could advance directly on its enemy without disabling its own batteries, which could fire only in a narrow arc on either beam. Systems of naval tactics were thus inevitably designed to bring fleets into action as rapidly as possible while exposing them to as little hostile fire as possible. This could be done by approaching the enemy's line end on, so that the enemy was prevented from firing his batteries; by an oblique approach which allowed the ships to begin firing on part of the enemy line even if they had to shift targets as they advanced; or by a bold head-on charge at the enemy line, as at Trafalgar, which reduced the time of exposure to long-range enemy fire.

Once fleets were engaged, there was more to gunnery than firing and reloading rapidly, although that was indispensably important. Effective gunnery depends on morale and training, either of which alone is useless. This was the lesson of the famous single-ship action between the *Chesapeake* and the *Shannon* in 1813. The latter frigate's crew not only had excellent morale but were the best trained body of gunners in the whole Royal Navy at that time. Her captain, Philip Broke, had been preparing them for a fight such as this for over six years.

It is frequently asserted that, whereas the British fired at the French hulls, the French fired at the masts, rigging and sails of the British ships. There is some truth in this, though Admirals Edward Vernon and Richard Kempenfelt were all for imitating the French. The hull was more difficult to hit but was the more decisive target. The piercing of the ship's sides for gunports made them comparatively vulnerable and a series of well-aimed shots would quickly damage and dismount guns, and kill and wound the gunners. Besides this, hull shots cut deep into the ship and might well crack or damage the base of the masts in addition to disorganising the control of the ship and making it difficult to bring up fresh powder and cartridges from the magazine. Waterline and below waterline shots started leaks which the carpenter and his mates might find difficult to plug. British ships were often so crank that their seams leaked through the strain of their own guns being fired, so a few well placed enemy shots would soon have them hard at work at the pumps. One of the most damaging methods of attack at close quarters was alternative depressing and elevating the guns. Under such conditions the guns could be fired with reduced charges so as to cause the maximum amount of splintering in the enemy ship.

Seen at a distance, the whole ship, rather than the hull alone, presented an easier target; in this case, damage aloft could stop a ship's advance or send her reeling out of the line. Although there were standard methods of strengthening the rigging and chaining the yards to prevent their fall, everything aloft was vulnerable. Sails could be slashed to ribbons, important rigging such as the stays could be cut, yards brought down, together with topmasts. Sometimes the whole mast itself splintered, cracked and finally broke at deck level. This generally meant a huge cluster of sail, woodwork and rigging over the side of the ship, entirely masking the guns in front of which it fell.

With anything like a sea running, it was difficult to aim at the enemy's hull at any distance as it kept partially disappearing from view. There was also the difficulty of timing the fire; the French generally fired on the upward roll of the firing ship.

Battle damage and battle casualties were often disproportionate. In the Battle of St Vincent (1797) there were only 73 men killed in the whole British fleet, whereas the *San Nicolas* alone had 144 killed. In the Battle of the First of June

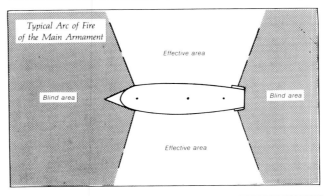

Typical Arc of Fire of the Main Armament

Effective area

Blind area

Blind area

Effective area

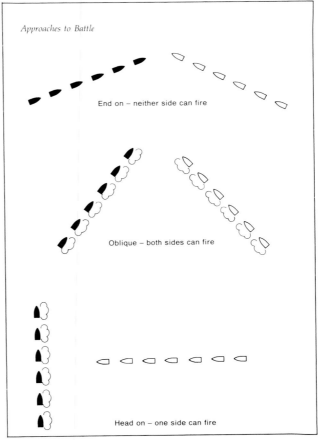

Approaches to Battle

End on – neither side can fire

Oblique – both sides can fire

Head on – one side can fire

the number of French killed in the six ships captured was more than three times that of the total killed in the whole British fleet. Once a ship with good gunners had the upper hand over another, the result was decisive, as is shown by an account of the sinking of the *Vengeur* at the First of June written by a lieutenant of the *Brunswick:*

As the *Vengeur* pass[ed] we split her rudder, stove her stern post into shivers; we saw the water pouring into her counter by tons. The *Ramillies* was yet waiting for her to settle more from us. Looking out of the stern ports, I could well perceive every precaution taken by them to train their muzzles clear of our stern, and point them to the Frenchman's quarter. At length went their broadside, and such a tremendous fire, in my life, I never saw or could have believed; I verily think every shot struck her, and as she was within pistol distance, the effect must have been terribly severe. . . .

We do not remember the mizen mase of *Le Vengeur* falling, when near the *Ramillies.* All her masts went soon after, but from the weight of shot thrown into her, viz. 3 rounds shot of 32 lb., i.e. 96 lb. of ball from each gun, driving home our coins at one time, watching their rising, to fire beneath the water line, next withdrawing the coins to elevate the muzzles and rip up her decks, this alternative mode of firing for two hours and a half alongside, we rather think it may probably happen her masts were cut by our shot from within board. The silence of their fire for the last two hours, shews the people were employed at their pumps, but all their efforts could not preserve their ship from sinking.[1]

Although British admirals paid lip-service to the idea that battles were decided by individual ships and squadrons pressing home their attacks – the notion of the 'pell-mell' – only a few attempted to fight a major battle without exercising close tactical control. Still fewer were the successes resulting from such pell-mell actions. Many British admirals honestly believed that a good cannonade in line ahead, parallel with the enemy's line at fairly close range, was a satisfactory enough form of engagement. Advantage would then rest with the fleet having the most effective gunnery and the best-formed line from which to apply it. Ships disabled aloft or heavily leaking from wind and water shots would try to withdraw from their respective lines; those in the line to leeward would manage this with greater ease. In the end the less effective fleet would withdraw altogether.

British admirals seem to have preferred these linear battles because they could see no other way of dealing with a well-drilled French fleet in a well-formed line. Splitting up the fleet, and attacking with separate squadrons, seemed less effective because it led to dispersion, risking piecemeal defeat; in any case it was not the best way of developing the maximum fire face against the enemy. The main reason, however, for not adopting a more elaborate form of attack was the difficulty of organising and executing anything requiring sophistication and finesse, especially in view of the limited British signalling system. The problem became greater once ships were damaged aloft; when any of the leading ships were thus disabled, and unable to keep station, the whole attack became disorganised. But, as John Clerk of Eldin so cogently argued in his *Essay on Naval Tactics* (1790–97), it was the slanting attack on the enemy's van, as so often practised, which proved unsatisfactory, rather than the linear engagement itself.[2]

When two fleets of unequal strength, or with sharp differences in gunpower, met in battle, there were various

devices by which the weaker fleet might space out its ships so as to leave some of the enemy ships without any target to fire at; since broadside guns had a very restricted arc of fire, this happened frequently. Nor was it easy for the ships left unengaged to swing themselves into position from which to open fire. By doing so they risked upsetting the forward movement of their own line of battle, which necessarily carried sufficient sail to maintain steerage way.

The least likely form of engagement to produce any substantial results was when the fleets passed each other on opposite tacks, as at the Battle of Ushant. This kind of fighting seems often to have taken place in the Anglo-Dutch Wars, but by the end of the eighteenth century it had become discredited, at any rate in the Royal Navy. A more promising scheme was for the stronger fleet to use its excess of ships to harry the van or rear of the weaker fleet by raking their bows or sterns and, if feasible, doubling on them. This involved getting right round onto the disengaged side of the enemy's van or rear, so putting them between two fires and forcing them to engage with both broadsides at once, a difficult task requiring high efficiency and morale. On these occasions the doubling ships had to be careful not to hit each other when firing at the enemy ships sandwiched between them.

A more sophisticated form of attack was by concentrating the bulk of the fleet against part of the enemy's line, and leaving the remainder practically unengaged. In this way the attackers might hope to bring a crushing superiority of fire against a smaller number of enemy ships and thus achieve decisive results before the unengaged ships could rally to the support of that part of their fleet already under attack. In such cases concentration on the enemy's rear seemed to offer the best results. Much depended, however, on the extent to which the attackers could get into sufficiently close order to bring their superior weight of fire to bear and yet avoid obstructing each other.

Still more sophisticated and ambitious was the manoeuvre of breaking the enemy's line. In theory this form of attack had many possible variations. It could be executed from windward or leeward, by all or part of the fleet, and at any point in the enemy's line. Lord Howe's special variation, which he actually attempted at the First of June with some measure of success, was for his whole fleet to cut through the gaps between all the enemy ships. In doing so, his ships could rake the bows and sterns of the ships respectively to starboard and larboard of them as they passed through and then indulge in a general pell-mell. It is, of course, obvious that if part of an attacking fleet cut through the enemy's line, both were liable to become separated into two parts; in other words, there were risks for the cutters as well as for the cut. The assumption was, however, that the cutters had superior fighting efficiency and morale, and that in the general confusion which ensued they would benefit either by retaining their tactical cohesion or in the pell-mell fighting. The assumption was also made that by forcing the ships in the line being cut to give way, the cutters would pass through intact as far as the rearmost ship, even though the cutting movement may have been initiated by a ship only part of the way along the line.

Despite the frequent issue of orders not to open fire too soon, or until signalled to do so, battles were often begun and sustained at a range as great as 1000 yards. This might happen when a ship in an attacking fleet was damaged aloft, and, swinging broadside on, opened fire more or less in self defence against the ship engaging her. This in turn might set off other ships and so prevent the rest of the attacking fleet getting much nearer, as otherwise they would be in danger of masking each other's fire. This emphasises the overriding importance of fire discipline. Long range cannonading also

1 Source unknown
2 *An Essay on Naval Tactics*, London, 1782 (a rare copy is in the Royal Naval College library, Greenwich).

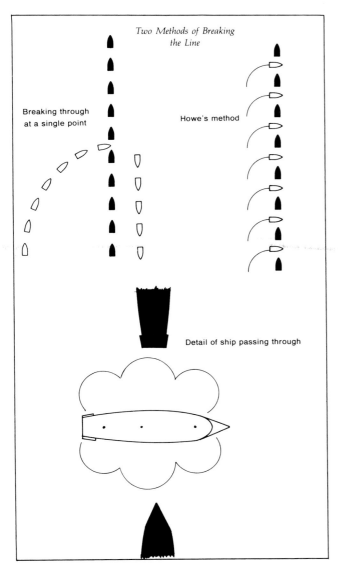

Two Methods of Breaking the Line

Breaking through at a single point

Howe's method

Detail of ship passing through

fire along all the gundecks, if necessary on both sides of the ship. As the action continued, guns were put out of action by enemy shots and gun crews killed or wounded. Further casualties resulted when overheated guns recoiled so hard that they broke the double breeching ropes specially fitted for battle but possibly cut by shot. Marines and seamen exchanged musket fire from the forecastle, quarterdeck and poop, and also from the tops. Swivel guns, and later carronades mounted on the upper works and loaded with canister, could be used to sweep an enemy's deck clear of boarders forming up for an attack. Boarding was only possible when two ships were locked together; such a position was dangerous if either ship caught fire. Mutual bombardment might sometimes go on for several hours until casualties amongst gun crews in one ship reduced their rate of fire and finally left them unable to resist further.

Seamanship in a pell-mell meant, first, trying to rake an enemy ship by bringing the broadside of the attacking ship against the enemy's bow or quarter. It also meant using such movement as was possible with spars and sails shot away to get into a more favourable position or to engage a new opponent; this was also what was meant by mutual support, – one ship coming up to relieve another, either by drawing the enemy's fire or by interposing the ship between the enemy and the hard-pressed ship.

Ships-of-the-line were seldom sunk in battle as a result of gunfire alone. The fate of the *Vengeur* at the First of June was exceptional. Ships which caught fire might blow up if the fire reached the magazine. Badly damaged ships might sink eventually as a combined result of shots below the waterline and heavy weather. Really crank ships were in danger of falling to pieces simply as a result of firing their own guns.

When, in 1689, the navies of England and France began that long series of contests which ended in 1815, their respective systems of tactics were very similar. Both navies used the 'line-ahead' formation, with each ship directly astern of the one ahead. In battle the opposing fleets were ranged roughly parallel so as to develop the maximum degree of broadside fire. French tactical doctrine, like the English, favoured the close-hauled line-ahead ('au plus près du vent') as a preliminary tactical formation. But, whereas the English system had been gradually developed through the experience of the three Dutch Wars, the French system had come into being as part of the naval expansion directed by Jean Baptiste Colbert (1619–83), statesman and virtual founder of the navy of Louis xiv. When the new French navy began to emerge, with a strong corps of well-built and well-gunned broadside battleships, it became possible, and indeed necessary for the fleet to adopt the line-ahead formation in battle though the formation's origins were acknowledged: 'Cet arrangement a été inventé par les Hollandois et les Anglois.'[3] So quick were the French to adopt what seemed the best tactics for their new broadside battleships that, by the Third Anglo-Dutch War of 1672–3, they were able to contribute a well-drilled and powerful squadron to the allied Anglo-French fleet.

As Colbert's reforms took root, French tactics came under the influence of the mathematicians and scientists, who were playing such an important part in French naval development as a whole. The navy of Louis XIV was a direct expression of the royal will. The new warships were works of art and science; art in the form of carvings and decorations designed by the leading sculptors, architects and painters of the day, science through the advice of the newly founded Académie Royale des Sciences. No longer could the construction of

occurred when one fleet was trying to keep another from carrying out some manoeuvre or was trying to preserve its own freedom of manoeuvre. 'Close range' is an elastic term, and seems to have meant anything under a quarter of a mile. 'Musket shot' and 'pistol shot', used in both the French and British services, are also inexact terms. Musket shot presumably meant the range at which the marines would usually open fire at an enemy ship, but this might differ greatly according to circumstances. It might well have been between 150 yards or more, even if the musket was not regarded as accurate beyond 80 yards. Pistol shot again might mean anything between duelling range, say 25 yards or less, and long-distance pistol shot up to 50 yards. On the whole it would seem that in the majority of battles the ranges may possibly have been slightly greater than had hitherto been supposed. We know of course that at the First of June, St Vincent and Trafalgar, some ships were actually touching their opponents, but this seems to have been exceptional. Though preferring longer ranges than the British, the French, once in a pell-mell, favoured boarding, believing that this was something at which they themselves excelled and the British feared.

In a real pell-mell, with the ships lying close together in no particular order, everything depended on efficient gunnery and seamanship and on what the British called 'mutual support'. In terms of gunnery this meant keeping up a steady

3 G Daniel (de la Compagnie de Jésus) *Histoire de la Milice Françoise*, Paris, 1721, II, p735

ships be left to the traditional empiricism of untutored shipwrights. Much theoretical work was being done on the resistance offered by the sea to various shapes of hull, to problems of stability and buoyancy, and to the effect of wind pressure on sails of various shapes and sizes, set at varying heights from the deck and at varying angles to the centreline of the ship. From this study of the movement of the individual ship, as governed by wind and water, it was an easy step to the study of ships moving together in squadrons or fleets involving angular measurements of the ships from one another and the mathematical problems raised by changes of formation in relation to the wind.

In navigation, as in shipbuilding, the best scientific advice was obtained through the new Service Hydrographique and the new Paris Observatoire. The later editions of Georges Fournier's *Hydrographie,* really a work on pilotage, first published in 1643, became a fine example of contemporary science applied directly to maritime needs.[4] When it came to setting out in textbook form the various orders of sailing and orders of battle which the fleet should assume or the methods of changing from one to another, the mathematicians were heard with respect. Indeed it was only in France that the need for such a textbook was as yet recognised. The need produced the book, *L'Art des Armées Navales,* by Père Paul Hoste, a mathematician.[5] It proved however to be prejudicial to the French service in that it tended to substitute elaborate evolutions for actual battle tactics and, in so far as it dealt with battle at all, placed undue emphasis on defensive fighting and defensive needs. Hoste was a protégé of the great Comte de Tourville, and he wrote in the shadow of French defeats. No doubt his mathematically based love of complex patterns and his overwhelmingly defensive attitude were suited to the mood of the age in which he wrote.

Amongst the new discoveries of exceptional interest which have, however, been available for study for many years are the printed sailing and fighting instructions, signals, and orders of sailing, issued by Tourville during 1690–1 and 1693. They show an immense superiority over contemporary English methods of fleet organisation and signalling, though the actual battle instructions and battle signals show no advance on the ideas of James Duke of York or of Prince Rupert.

Eventually a doctrine emerged, this time strategic in intent, but fitting in only too well with defensive tactics: the extension and control of the world-wide French empire was to take precedence over winning battles. Battle tactics and battle opportunities, therefore, because subordinated to more far-reaching strategic enterprises. It is no accident that naval commanders such as the Marquis de La Jonquière, the Marquis de La Galissonière, the Comte de Bougainville, the Marquis de Vaudreuil and the Comte d'Estaing played a leading part in French imperial development.

These are some of the elements leading towards the development of a French view of naval tactics, a view which seems to have hardened as time went on and did not change until the Revolution. In England the situation was different. The hard fighting experience of the Anglo-Dutch Wars had led both to the consolidation of the line-ahead formation in battle and to various special methods of attack. As in the case of the Dutch, the greatest obstacle to tactical efficiency was the size of the fleets employed. With seventy, eighty, ninety and even more ships included in the line of battle, supported by twenty or thirty fireships and a host of auxiliaries, it was impossible to obtain tactical cohesion, especially as more than

10 per cent of the line ships were hired merchantmen. Nor was it possible to do much in the way of tactical training, as the fleets were based in ports comparatively close to the Dutch ports and anchorages. Broadly speaking, they set out with the intention of fighting and immediately returning to base for repairs. Although this did not always happen, the situation was very different from that on a distant station, where there was time and opportunity to develop a small fleet into a well-drilled tactical combination. In the Anglo-Dutch wars proper tactical control was impossible: ships followed their own squadron commanders or even the commanders of their squadron sub-divisions, who often acted quite independently, so that the fleet tended to become dispersed in a series of desultory piecemeal engagements. Tactical sophistications were unattainable because the preliminaries of tactical control had not yet been achieved. Nevertheless, it is perfectly clear that the aim of English commanders was to defeat the enemy, and that tactics were no more than a means to that end.

The so-called 'fighting instructions' (a term abused by many writers) never existed as a separate work. They were merely one chapter in what became the permanent 'Sailing and Fighting Instructions for Her [later His] Majesty's Fleet'. This book includes signals by day for when at anchor, weighing anchor, sailing and anchoring; a similar series for the night; distinguishing lights for the flags; instructions for sailing in a fog; instructions about ship precedence between 'younger' and 'elder' captains; signals for calling officers; instructions for masters, pilots, ketches, hoys and smacks, attending the fleet; the 'encouragement' for fireships and instructions for fireships.

The sailing instructions were initially very rudimentary, though they provide the only evidence we have of the signals used for controlling the fleet when not in the presence of the enemy. Gradually the sailing instructions came to include more details and, eventually, to provide a whole set of fleet cruising formations, with the methods of changing from one to another or into order of battle. From these we gain a picture of what a great fleet must have looked like when searching for the enemy and ready to deploy into line of battle according to the various methods available. They thus form a parallel set of documents to the fighting instructions.

It has become fashionable to attribute British tactical failures between 1704 and 1781 to a pedantic, formalistic and docile adherence to the order of battle in line-ahead formation. The chief reason for the line's disrepute was its association with the notions of defensive fighting arising from the experience of the Battle of Malaga in 1704. In the hands of General at Sea Robert Blake, Prince Rupert and James Duke of York, the line-ahead formation became the best means of bringing the maximum number of guns to bear on the enemy so as to defeat them. When Admiral Sir George Rooke and his contemporaries and successors began to emphasise the line's defensive virtues, those responsible for the earlier development of the line could be dubbed formalists. By contrast, those who put less stress on warlike order and still favoured the independent squadron movements of an earlier age came to be regarded as the truly offensive fighters. It is important to appreciate the changes which took place in the structure of the line of battle between 1672 and 1744 and the difference made by the increased proportion of three-deckers. Much of the criticism levelled against docile British admirals who failed to force a pell-mell, in contrast with dashing men such as Lord Anson, Admiral Edward Hawke and Admiral Edward Boscawen, takes no account of the fact that it needs two to make a battle. As Admiral John Jerius, Lord St Vincent, so aptly put it, 'I have often told you that two fleets of equal strength can never produce decisive events, unless

---

4 Georges Fournier, *Hydrographie,* Paris, 1643
5 See below, p59

they are equally determined to fight it out or the Commander-in-Chief of one of them so bitches it as to misconduct his line.'

General chase, the action so popular with writers of stirring tales, was not in itself a battle formation and, unless carefully controlled, could only invite defeat in detail. Nor was it quite such a rapid business as these robust writers would have us suppose. In a moderate breeze an attacking fleet might approach the enemy at a speed of between 2 and 3 knots, so as to make sure of keeping their alignment and not masking each other's fire. How many knots would general chase produce with a similar breeze? Hardly more than 6 or 7 – not a very breath-taking speed at which to 'hurl' a fleet at the enemy.

The American War of Independence was the highest competitive point in French and British tactics. Except when led by Admiral Pierre André de Suffren, French fleets never attacked, as indeed they had never attempted to do since 1704. Their tactics when opposed by fleets of equal strength were mainly defensive. Nevertheless, they outpointed the British again and again, Rodney's great victory at the Saints coming too late to alter the strategical issues of the war. The fleets depended on the new, unwise and unnecessary elaboration of French signalling and tactical formations which accompanied the naval revival following the defeats of the Seven Years War. In 1763 the Vicomte de Morogues produced in *Tactique Navale ou Traité des Évolutions et des Signaux*. a voluminous work whose sheer extent and complexity have defeated many critics. His influence on French tactics was disastrous. As late as 1806, the French navy rejected the proposed adoption of a true numerical system of signalling, by a vote taken at each of the naval bases. Instead, they voted to continue using the highly complicated system introduced by Captain du Pavillon in 1776. They retained this system until after the end of the Napoleonic Wars.

During the American War, a differentiation between British tactical traditionalists and reformers can be seen in both the form and the content of the additional fighting instructions and signals issued by the Admiralty (though not in title) and by particular admirals for particular fleets. It can also be seen in the examples provided by actual battle. In many respects the form governed the content. Signals as mere words coming at the end of what were often long-winded instructions could never be as effective as signals directing this or that immediate and direct manoeuvre divorced from any instructional element. What the British most needed was a pure signal book with signals for attacking and fighting the enemy in various ways. Instructions were absolutely necessary for fixing how certain battle manoeuvres should be executed, but there was no need to keep new battle signals chained to the instructions, especially when an instructional element was involved.

Rodney never issued a signal book in the whole of his career. His signals still lay embedded in the printed text of his instructions. The term 'Rodney's signals' refers to private signal books made by officers of his fleet for their personal use. By contrast, Lord Howe was of pre-eminent importance in the reform of British tactics and signalling. He was the first to separate signals and tactics, and he gave his fleet an official, printed signal book, the accompanying book of instructions being 'explanatory of and relative to the Signal Book herewith delivered'. This was as early as 1776, when he commanded on the North American station. In 1782 he applied his new system to the Channel Fleet. In 1790 he gave the Royal Navy a true numerical system of signalling. A lucky discovery of a set of printed additional sailing and fighting instructions, unsigned, undated and unissued, though clearly composed by Howe, provides an important link between his activities in the Seven Years War and the American War.[6]

Howe also has the unique distinction of being the executor of his own themes. In the Battle of the First of June he attempted with some success his long planned cutting of the enemy's line at all points. He was then sixty-eight, the greatest age at which any British admiral has ever won a major battle. Howe, unlike Kempenfelt, refused to be dazzled by the elaborate complexities of the French tactical and signalling methods or the French theoretical treatises. Nevertheless, he could never have won the battle if the French had been unwilling to fight it out and it is to the credit of his opponent, the Comte de Villaret de Joyeuse that, with a comparatively untrained fleet, he unflinchingly accepted the full hazards of a real pell-mell.

Howe's triple success as tactician, signaller and commander-in-chief in battle is all the more remarkable in view of the appallingly long-winded obscurantism of several of his sailing and fighting instructions. When he gave oral instructions the result was sometimes as bad or even worse.[7] Despite this handicap, he succeeded mainly because he had only one great object in view: the defeat of the enemy.

It is sad to record that Kempenfelt fails to emerge from detailed investigation as anything more than a vigorous exponent of Howe's earlier ideas. The evidence is prolific and its implications unmistakable. Vice-Admiral Sir Roger Curtis, Howe's chief of staff, has been revealed, by a lucky discovery of manuscript working papers, as a complete traditionalist, so far as signalling is concerned. Their outlooks were entirely different.[8]

Admiral Sir Charles Henry Knowles was a great experimenter in tactics and signalling. His Set of Signals, printed in 1777, was an astonishing achievement for an officer aged only twenty-three.[9]

After the Revolution the tactical efficiency of French fleets declined, though unlike the fleets of the *ancien régime* they did not invariably try to avoid close action. By contrast, the need for a more offensive system of tactics was now proclaimed by the writers of theoretical treatises. Beginning with the Vicomte de Grenier in 1777, the demand for offensive tactics increased with the Comte d'Amblimont in 1788, Ramatuelle in 1802 and Le Chevalier Delarouvraye in 1815, who had *Delenda est Britannia* printed on the title page of his book. Inside the service itself, however, no changes took place. There is no evidence that the work of these men influenced service thinking – and, after 1802, it was in any case too late.

The final developments of British tactics and signalling, after the issue of the first official Admiralty books of 1799, were gathered together in the Admiralty Signal Book of 1816. In this book, Howe's system is abandoned in favour of that invented by Popham, whose *Telegraphic Signals or Marine Vocabulary* had already proved so useful to Nelson. This work was now adopted by the Admiralty and issued as a companion volume to the Signal Book.

Naval tactics are normally the expression of the tactical doctrine prevailing at the time in the particular service concerned. The doctrine is given official and systematic form in the manuals and books of instruction under which the officers are trained, and in the light of which they fight. We know, however, from experience in other fields that the actual working of a system cannot be understood solely by reference to a corpus of official regulations. The picture they provide is never complete, their interpretation depending on

6 Tunstall Collection, *NMM*, TUN/111 (formerly FI/5)
7 Sir John Barrow, *Life of Richard, Earl Howe*, London, 1838, pp118–9
8 Corbett Collection, *NMM*, TUN/7 (formerly S/MS/Am 7)
9 Sir Charles Henry Knowles, *NMM* S(H)25. There is also a mutilated copy in the Mead Collection, *NMM*, TUN/136 (formerly Mead/S/1).

professional practice, the 'custom of the service'. Beyond this again, there must always be a philosophy, a sentiment, a tradition, a prejudice or an ethos which both inspires the users of the system and governs their interpretation of it. Broadly speaking, the correlation between official thought and fighting practice is closest when morale and fighting efficiency are high. It is when defeat or internal disintegration has brought the spirit low that the official regulations take on the character of theoretical propositions divorced from reality. It is only too easy for a navy in decline to be issued or re-issued with fighting instructions and signals, detailing methods of attack which experienced officers recognise as unlikely even to be attempted.

In practice, no admiral could afford to ignore the strategic implications of his battle tactics. The instructions he carried with him and under title of which he held his command, specified, often in detailed terms, the strategic objects he was bound to pursue, and to some extent the methods by which he must attain them. Princes, ministers, admirals and generals of allied states must be persuaded and placated, and governors of neutral ports must be cajoled; while over all loomed the constant need to protect trade with frigates and to supply escorts for convoys. To fight or not to fight, and if to fight, when and in what manner, were not matters which could invariably be settled simply by noting the wind and watching the movements of the enemy.

British and French signals

Standard methods of signalling had become established in the English, Dutch and French navies well before the end of the seventeenth century and, apart from developments in codes and techniques, they remained unchanged until after 1815. Signals were originally conceived as a means by which the admiral communicated his immediate orders and future intentions to his subordinate flag officers and captains. They themselves could acknowledge or repeat his signals but not much more. Private ships had a limited number of signals which they could make direct to the admiral or repeat to him, mainly in terms of 'land discovered', 'strange sail sighted', signals for distress or for springing a leak. They could, of course, make such signals to each other and were indeed expected to do so. The full range of signals, however, belonged to the admiral alone. In the Royal Navy, in early days, he alone possessed a full set of flags. French private ships-of-the line possessed all the flags and repeated the admiral's signals. The admiral might appoint one or more repeating frigates, with flags, to stay on his disengaged side in battle, to repeat his signals so that the fleet could see them free from ships ahead or astern and the smoke overhanging the line of battle.

The admiral also had other means of communication. Where tactics were concerned and a special situation envisaged, he might call a council of war, often as a means of obtaining sanction for not obeying his instructions to the full. He might also call his flag officers and captains on board to explain his intentions, though not all admirals were prepared to risk this kind of conference, with the embarrassment of being outfaced in argument by able and aggressive subordinates. Alternatively, he might distribute special tactical instructions by boat; this seems generally to have been done by signalling each ship to send a boat to receive a copy.

By day, in port, he could signal by sail movements and guns, since flags might not blow out enough to be seen. At sea he had various choices. For communicating with the fleet as a whole or with squadrons or even divisions, flag signalling was best, the signal generally being accompanied by the firing of one or possibly two guns to draw attention to it. Signals to individual ships were made by first hoisting the recognition pendant for that particular ship. Orders could also be sent to individual ships by boat, though this was an extremely wasteful method as it generally required an officer of the rank of lieutenant to take it either in writing or by word of mouth. Nevertheless, orders were often sent by boat in battle, especially if the fleets had only steerage way on, or it was dead calm so that a single boat might visit several ships in succession. Ships-of-the-line, frigates, fireships, fleet auxiliaries and boats, if near enough, could be given direct orders by speaking trumpet. Orders could also be sent by hailing a frigate, attendant cutter or ketch and ordering her captain or commander to deliver a spoken message to a flag officer or captain of a private ship.

In the battles of the Dutch Wars, when large fleets were spread out too far for signals to be easily recognised or understood, especially with smoke from the guns obscuring the view, the dispatch of orders by boat or some larger vessel seems to have been the most effective method. In fact it was only during the eighteenth century, when fleets on the whole were smaller and the technique of repeating frigates better developed, that the admiral could do much with flag signals once the battle was joined. Throughout the Dutch Wars, it seems doubtful if the respective commanders-in-chief ever exercised or expected to exercise much control over the fleet as a whole during the battle. This, of course, had an important bearing on the extent to which fighting instructions with signals attached, purporting to deal in different ways with various tactical situations, ever gained practical recognition.

Night signalling was always difficult and, in any case, fewer fleet movements could be attempted. In the early days signals for running into danger, strange sail sighted and recognition signals for 'losing company and meeting again', cutting and slipping, weighing anchor or anchoring, were the main ones. The basis of night signalling was with lights (lanterns), generally up to four, displayed in various places and patterns, together with guns (always fired from the same side of the position). 'False fires' were also used in addition to or as substitutes for lights. These were manufactured from gunpowder and wheat flour, in the proportion of 1lb to 6oz (8 to 3), contained in a tube made by rolling paper thickly round a stick, which was then withdrawn. The composition was then hammered home. A quick-match was used for firing. Later in the eighteenth century, blue lights were used, composed of a mixture of 7lb of saltpetre, 1lb 12oz of sulphur, and 8oz of blue orpiment.[10]

Sky rockets were also in use at least as early as the Third Dutch War. In the French navy they became divided into *fusées en étoiles*, *fusées en pluie* and *fusées en sermentaux*. They were especially useful towards the end of an action when fresh gunfire would attract little attention.

One of the chief difficulties of night signalling was that of not being able to see the signals made by the admiral when blanketed by his own sails or the sails of intervening ships. Much also depended on the lights displayed in each ship, by his orders, to guide the fleet on its immediate course, regardless of any attempt to change formation. The tactical disadvantage of night signalling was that it gave warning to the enemy of the presence of another fleet. Vernon realised this, and cut down the number of guns used for night signalling in the West Indies.

Fog signalling was even more limited in scope, guns being the chief method, with later sophistications embracing quick

---

10 The formulae are from the Admiralty Night & Fog Signals of the Victorian period.

and slow firing with measured intervals between individual guns, and with sequences of guns. For exchanging recognition at close quarters it was possible to give the respective pass-words by hailing with a speaking trumpet, just as in day time or at night. In the French service, auxiliary fog signalling methods included bugle calls, drums, bells, trumpets, fifes, muskets and swivel-guns, alone or in combination with signal guns.

In both the English and French navies, as well as in the Dutch, the early system of signalling was that each flag was known by its description and had no number or letter attached to it. A signal flag, therefore, only meant something when it was hoisted in a particular position. The English worked with very few flags. The French were more enterprising, largely because they had already developed a more elaborate series of instructions which, in turn, required more signals. In the English navy the positions for hoisting signal flags were the fore, main and mizzen topmast heads; the fore yardarm, the fore topsail yardarm and the fore shrouds; the main backstay, the main yardarm, the main topsail yardarm, the main shrouds and the main topmast shrouds; the mizzen peak, mizzen shrouds, mizzen topmast shrouds, the crossjack yardarm and the mizzen backstay; and the ensign staff. The French used only about eleven of these positions.

The English and French used different flags for signalling. The English flags were: Royal Standard, Union Flag, Red Ensign, red, white, blue, yellow, red and white stripes horizontal, red, white and blue stripes horizontal, yellow and white stripes horizontal, yellow and white stripes oblique, blue and white stripes horizontal, red and white stripes horizontal, white with a red cross, blue with a red cross, and red, white, blue and yellow pendants. The French flags were: red, white, blue, white with a red cross, red-white-red horizontal stripes, white and blue horizontal stripes, white-red-white horizontal stripes, white and red horizontal stripes, blue-white-blue horizontal stripes, white-blue-white horizontal stripes, white and blue horizontal stripes, white and red horizontal stripes, white with a blue border, Dutch Jack, white with a red border, white with a blue cross, blue with a white cross, white and red chequered, white and blue chequered, White, Red and Blue Ensigns, and red, white, blue, white and red, white and blue pendants.

Apart from flags, there was a sharp difference in the way in which signals were treated by the two navies. Tourville had already established the principle of issuing printed *signaux généraux* in addition to instructions. Thumb-indexed signal books with painted flags could be made from these for personal use. In the British service a somewhat similar system had prevailed until Rooke removed the signal indices from the Sailing & Fighting Instructions, leaving the details of the signals embedded in the words of the instructions. As a result, personal signal books made by extracting the actual signals from the instructions together with a few words of 'signification' became practically a necessity, whereas in the French service they were more of a luxury. Luckily for the British, there were few flags and far fewer signals, so that the construction of the books from the thumb-indexing standpoint was not difficult.

Captain of the Fleet

The chief of staff system developed less easily at sea than on land. From quite early times, however, it was recognised that a flag-captain to a commander-in-chief held a very responsible and difficult position. He had to assist the admiral as well as assume responsibility for the ship, which was generally the largest and strongest in the fleet and therefore needed the most supervision. During the Restoration period it

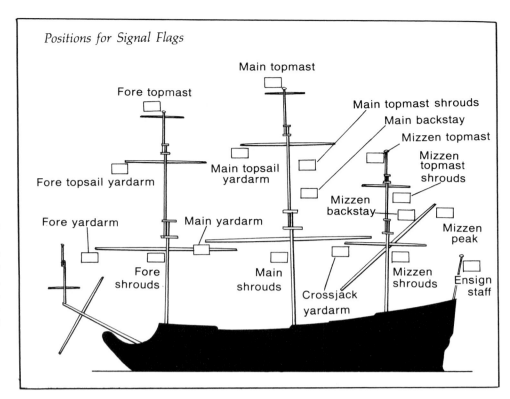

Positions for Signal Flags

Main topmast

Fore topmast

Main topmast shrouds

Main backstay

Mizzen topmast

Main topsail yardarm

Mizzen topmast shrouds

Fore topsail yardarm

Mizzen backstay

Fore yardarm

Main yardarm

Mizzen peak

Fore shrouds

Main shrouds

Crossjack yardarm

Mizzen shrouds

Ensign staff

became customary to appoint an additional captain to the commander-in-chief's ship, to act solely as his chief of staff and without any responsibility for the ship itself. He would usually be a senior man, with the same pay as a rear-admiral while he held the appointment.[11]

By 1731 it was established that this so-called First Captain of the Fleet should 'be esteemed as a Rear-Admiral, and take place at all Councils of War'.

Not all commanders-in-chief took advantage of this facility. Some, perhaps, felt no need for it; others did not qualify because of the size of their fleets. Nevertheless, there must have been many fruitful collaborations between commanders-in-chief and their respective chiefs of staff, of which the Curtis-Howe and Douglas-Rodney combinations are amongst the best known. Kempenfelt's letters to Admiral Sir Charles Middleton give a vivid description of the duties and trials of being a First Captain to Admiral Sir Charles Hardy and Admiral Sir Francis Geary, good officers in their time but worn out by age and unaccustomed responsibility.

In 1747 it was laid down that a command of fifteen British ships-of-the-line, or twenty partly British, was the minimum qualification for the appointment of a First Captain.[12] In 1795 this figure of fifteen British ships was repeated, nothing being said about combined fleets with allies.[13]

The main tactical duty of a First Captain seems to have been to help the commander-in-chief draw up orders of sailing and battle, and draft operational instructions. Sir Robert Calder seems to have supplied Sir John Jervis with important tactical memoranda but the full extent of their collaboration is not clear. The First Captain was also concerned with signals, though in special cases, such as that of

11 This arrangement dates from 1668 at the latest. It was confirmed by Order in Council, 26 June 1674. Sir Clowdisley Shovell was given power to appoint a second captain (to take charge of the ship) in case he had a prospect of engaging the enemy. Papers of Thomas Corbett, Admiralty; notes by Admiral Sir Edmond Slade for Sir Julian Corbett.
12 Order in Council, 22 March 1747
13 Order in Council, 6 November 1795

John McArthur, the organisation of the signal book and the drafting of the new signals might be done by the admiral's secretary. Nelson never had a First Captain; until Trafalgar, he did not qualify for one as he was not a commander-in-chief. At Copenhagen (1801) he used Captain Foley, to whose ship he had just shifted his flag, and also Captain Edward Riou, chosen, apparently, on the spur of the moment, as he and Nelson had not met before. On the morning of Trafalgar, Nelson gave certain staff duties to Captain Blackwood of the frigate *Euryalus*, including the power to signal to ships in the rearmost line of battle in the admiral's name.

With the growing complexity of naval operations, the position of First Captain became increasingly important, and his title was changed to that of Captain of the Fleet. By a regulation of 1806, the minimum size a fleet required to qualify was reduced to ten of the line, the officer appointed to be a rear-admiral or senior captain. He was empowered to give orders in the name of the commander-in-chief, though under his direction, to all officers, including flag-officers senior to himself. But he was not 'to alter, without his permission, the temporary position of the fleet, by signal or otherwise unless some evident necessity require it'.

The development of the chief of staff system in the French navy seems to have been on similar lines, and possibly started sooner. By the early eighteenth century it seems to have been customary to appoint a major to all fleet commanders, whether or not they had the titular style of a commander-in-chief. The major, and such aide-majors as might be appointed, were concerned both with guards and with signals, no doubt because of the security link. An aide-major's duties at sea were similar to those of a junior staff officer in the army.

At the beginning of the American War, staff officers were particularly important because of the increased size of the French fleets and the new developments in signalling. Pavillon's appointment as *major d'escadre* to d'Orvilliers in 1778 seems to have been made chiefly for signal purposes. One would like to know exactly what part de Vaugirauld played as *major d'escadre* to be Grasse in influencing French tactics during 1781–2, and what part Buer de La Charulière played as *major d'escadre* to de Guichen in 1780.

CHAPTER ONE

The Development of Line Tactics in England

LITTLE is known about the tactical systems of the sailing fleets of the leading maritime states between the years 1500 and 1650, though there is useful evidence for the galleys. On the whole, broadside tactics lagged behind contemporary developments in shipbuilding and gun mounting. Despite the obvious implications of broadside fire, fleets still seem to have been organised to fight in line abreast, as if their main offensive power lay in the bows of their ships, and their main object to promote single-ship encounters in which grappling and boarding could still be decisive. The English, however, developed a practice of avoiding close quarters where Spanish infantry could get at them. To do so, they had inevitably to develop tactics to make best use of their artillery.

Philip II of Spain seems to have been well informed about English tactics. He wrote to Medina Sidonia, commander in chief of the Armada of 1588:

You should take special note, however, that the enemy's aim

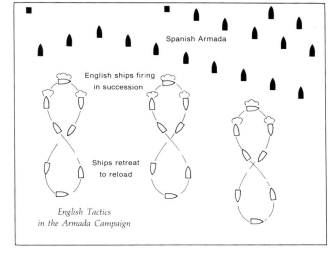

English Tactics
in the Armada Campaign

will be to fight from a distance, since he has the advantage of superior artillery and of the large number of fireworks with which he will come provided; while ours must be to attack, and to come to grips with the enemy at close quarters, and to succeed in doing this you will need to exert every effort. That you might be forewarned, you will receive a detailed report of the way in which the enemy arranges his artillery so as to be able to aim the broadsides low in the hull and so sink his opponents' ships.[1]

According to two sets of orders issued, respectively, by or under the names of Sir Walter Raleigh and his cousin, Sir William Gorges, and most probably reflecting Armada or even earlier practice, English ships were to attempt to get to windward of the enemy and to follow one of the flagships 'or other leading ship within musket shot of the enemy'.[2] The leading ship, having fired her broadside into the 'windermost ship or ships of the enemy', was then to tack, followed in succession by the rest of the squadron or *ad hoc* group. As the sternmost English ship tacked, the leading ship would appear from the windward on the opposite tack to batter the enemy once more, forcing them 'to bear up and so entangle them and drive them foul one of another to their utter confusion'. This evolution would then be repeated. Accounts of the Armada operations, vague and piecemeal as they are, suggest that this is what happened on several occasions, the English squadron attacks being in rudimentary line-ahead formation, moving in a succession of figures of eight.

The only other tactical lesson learnt from the Armada operation was the use of eight fireships by the English. True, they were completely unsuccessful as a means of directly damaging the Spanish ships, but they certainly provoked confusion and even panic in the Spanish fleet and led to a precipitate flight from Calais and a loss of cohesion and control which was never properly recovered.

In 1625 Sir Edward Cecil, Viscount Wimbledon, commander-in-chief of the English fleet sent to capture Cadiz, issued sets of instructions with a strongly professional

Sir Edward Cecil, Viscount Wimbledon (1572–1638); engraving by Simon Pasaeus. BM

1 Christopher Lloyd, editor, *The Naval Miscellany*, vol IV, London, 1952, p16

2 Julian S Corbett, editor, *Fighting Instructions*, London 1905 (hereafter cited: *Fighting Instructions*) p42

flavour. He and all his flag and squadron commanders were land officers, mostly drawn from the regiments serving in the Low Countries against Spain. He repeated the instructions of Gorges and Raleigh for the attack from windward, but combined these with orders for the whole fleet to 'follow in their several places, the admirals with the head of the enemy, the vice-admirals with the body, and the rear-admirals with the sternmost ships of the chase' (Article 35). This put the admiral in the van and not the centre. Another set of instructions prescribed detailed tactical arrangements. The fleet was to be organised in three squadrons each of nine ships, each squadron being again sub-divided into three divisions, each of three ships. Three extra ships were to be in reserve. The allied Dutch squadron also taking part in the expedition was to be posted 'on the starboard side of the admiral [the post of honour] and observe their own order and method of fighting' (Article 2). The vice-admiral's squadron was to be to larboard of the admiral, the post of lesser honour, while the rear-admiral's squadron was to act as a general fleet reserve. Nothing was specifically stated about

English ships, mostly hired merchantmen, on the expedition to the Ile de Ré in 1627. NMM

Wimbledon's Fleet Disposition, 1625

Rear-admiral	Commander-in-chief	Vice-admiral

Reserve

the method of fighting except that the nine ships of each squadron 'should discharge and fall off three and three, as they are filed in this list' (Article 1). Should the enemy fall into confusion 'our rear-admiral and his squadron with all his divisions should lay hold thereof and prosecute it to effect' (Article 8).

More interesting for the future is Article 10:

That if any ship or ships of the enemy should break out or fly, the admiral of any squadron which should happen to be in the next and most convenient place for that purpose should send out a competent number of the fittest ships of his squadron to chase, assult, or take such ship or ships breaking out; but no ship should undertake such a chase without the command of the admiral, or at leastwise the admiral of his squadron.

This article was clearly intended to take advantage of any weakness shown by the enemy while, at the same time, curbing irregular pursuit by individual captains seeking plunder.[3]

When these instructions were discussed by Lord Wimbledon's council of war it was contended that the squadron organisation was too difficult to comply with, and it was only accepted on the understanding that it was to be regarded as perfectionist:

It was observed that it intended to enjoin our fleet to advance and fight at sea, much after the manner of an army at land, assigning every ship to a particular division, rank, file, and station: which order and regularity was not only improbable but almost impossible to be observed by so great a fleet in so uncertain a place as the sea. Hereupon some little doubt arose whether or not . . .[4]

In the autumn of 1628, after the murder of the Duke of Buckingham had removed the expedition's most obvious

commander, an attempt was made under the command of Robert Bertie, first Earl of Lindsey, to relieve the blockaded Huguenots in La Rochelle. His instructions repeat Article 10 of Lord Wimbledon's third set, about pursuit of the enemy, and add another: 'If the enemy be entangled among themselves, or forced to bear up, that an advantage may be had, then shall the rear-admiral with his squadron lay hold thereof and prosecute it with effect.'[5] The rear-admiral's squadron, being anyhow regarded as a reserve, is thus given the more positive task of exploiting weakness in the enemy's fleet (as Vernon provided for later) rather than that of assisting disabled or hard-pressed ships in his own.[6]

A further set of instructions issued at about this time and certainly after the Ré and La Rochelle expeditions which it mentions, provided for the vice-admiral's squadron to attack the enemy from windward, followed by the admiral's squadron[7] The rear-admiral's squadron was to act in support of any ships which were hard pressed or disabled. Ships were to maintain sufficient distance from each other as the situation required. If the rear-admiral happened to be nearest to the enemy and to windward of them he might begin the fight, unless ordered not to by the admiral.

If any ship or ships of the enemy shall break out and stand away then the Admiral of any squadron which shall be next them, is to give chase with all his force if it be needful and send out a sufficient number of his fittest ships to take them, but never to undertake such chase without command of the Admiral or at the least the Admiral of his squadron.

Lord Lindsey's instructions for the Ship Money Fleet in 1635 lay down an important though not very progressive principle, namely that the admiral, vice-admiral and rear-admiral should, if possible, engage the enemy flagships opposed to them, private ships being forbidden to do so. Instead they were 'to match themselves accordingly as they can, and to secure one another as cause shall require.'[8] This is interesting, first, because it seems to drop the whole idea of a

Algernon Percy (1602–68), tenth Earl of Northumberland and Lord High Admiral, artist unknown, in the style of Van Dyck. NMM

3 *Fighting Instructions*, pp52–64

4 *Fighting Instructions*, pp63–4

5 Source unknown

6 The great series of engravings by Jacques Callot illustrating the operations give a chaotic impression of contemporary tactics.

7 Unpublished notes from the Leconfield MSS by Sir Julian Corbett

8 Article 18; M Oppenheim, *The Naval Tactics of Sir William Monson in Six Books*, London, 1902–14 (hereafter cited *Monson*), vol iv, p8

The Battle of the Gabbard, 2 June 1653, by Heerman Witmont. NMM

force in reserve and, second, because it suggests the notion of fleets fighting each other in two coterminous lines.

Next year, 1636, Sir Algernon Percy, Earl of Northumberland, commanded the Ship Money Fleet and issued fleet instructions similar to the above but with additional 'Directions for the Admiral's Squadron'. Were the other flag officers left to make their own squadron arrangements? We do not know. The last article has a much more practical ring about it than anything recorded before.

Article 10. The uncertainty of a sea fight is such that no certain Instructions can be given by reason; until we come to it we know not how the enemy will work, and then (as often befalls) one ship will becalm another and some not possible to luff or bear up as they would because of ships that are near them; and many other accidents which must be left to every captain to govern by his own discretion and valour.

This article neatly resolves the conflict between pursuit of a disabled enemy and support of a disabled ship of the flag officer's fleet. At the same time, pursuit of a fleeing enemy is checked in the interest of crushing the enemy's main core of resistance, and the importance of the fleet takes precedence over the fortunes of individual ships. Here are perhaps the first true set of English fighting instructions: every one of the ten instructions is concerned with tactical arrangements and the conduct in battle (see the summary of these instructions given below).

Northumberland's Instructions of 1636

Articles 1–3. Northumberland's own squadron of nine ships was equally divided between himself and his squadron vice-admiral and rear-admiral. Each squadron commander was to be supported by his two other private ships.

Article 4.   Ships were to observe a proper distance from each other to leave room for firing.
Article 5. If the admiral was attacked by a fireship, his second was to cover him from the enemy to give him time to clear himself and put out the fire.

Article 6.   Any ship in danger of capture was to be relieved by his second even if it meant letting a hard-pressed enemy ship escape him.

Article 7.   Ships hit below the waterline or having lost a mast or yard or whose guns had grown too hot to fire were to be relieved by their seconds, who were to rake the enemy fore and aft.

Article 8.   The vice-admiral of the fleet was to begin the fight, supported by his seconds.

Article 9.   No ship was to chase an enemy ship which fled but to 'engage yourself where you shall see the most resistance to be made by the enemy, to weaken the force of theirs that shall most strongly oppose us'.

Article 10. 'The uncertainty of a sea fight is such that no certain instructions can be given by reason; till we come to it we know not how the enemy will work, and then (as often befalls) one ship will becalm another and some not possible to luff or bear up as they would because of ships

that are near them; and many other accidents which must be left to every captain to govern by his own discretion and valour'.[9]

Nevertheless, there was little progress generally in naval tactics. Captain Nathaniel Butler, author of *Boteler's Dialogues*, written between 1634 and 1643 and first published in 1685, emphasised the obscurity surrounding tactics immediately before the Civil War. He wrote that in all the sea fighting between the Spaniards, Portuguese, French, Dutch and English since Lepanto, 'there is not so much as a sentence or word to be found, either of the manner of fight, or form of battle practised by any of them'. However, he was quite prepared to state his own views. A fleet, he said, should be divided into three squadrons unless it contained a hundred ships or more, when the number should be increased to five, the extra two to act as 'Wings to the van, Battle [centre], and Rear of the rest of the fleet'. Each squadron should be subdivided into three equal divisions plus a reserve equal in number to one-third of the total squadron. To avoid confusion and collision, 'they are in the fight to charge and fall off by threes and fives' with the reserve groups ready to relieve disabled ships. He thought that a small fleet 'should be brought up to the battle in one only front; with the Admirals in the midst of them, and the chief admiral in the midst of these; and on each side of him the strongest and best provided ships'. These ships would 'have an especial regard in the fight to all the weaker ships of the side, and to relieve and succour them upon all occasions'. A large fleet may follow the same tactics, provided there is sea room enough in which case the strongest ships should be 'in the windwardmost' so as to be able to relieve the weaker ships to leeward of them. For small fleets he seemed to favour an old-fashioned mass action and, for large fleets, group action by fives and threes, presumably in line-ahead. He placed little confidence in fireships, though acknowledging their importance in 1588, chiefly because they required a combination of favourable circumstances to be effective.[10]

Tactical development during the First Anglo-Dutch War

The English navy apparently plunged into war against the Dutch with no real tactical system at all beyond a fleet organisation of three squadrons, the admirals leading their squadrons, the squadrons and individual ships under orders to assist each other, and all attacks to be made from windward if possible. Little is known, however, about what really happened in the first four battles. In 1648, just about the time that the bulk of the royalist ships went over to Parliament, the Committee of the Lords and Commons for the Admiralty and Cinque Ports announced that they would issue 'more particular directions' for use against any fleet 'which may probably be conjectured and have a purpose to encounter, oppose, or affront the fleet in the Parliament's service'.

But for the present . . . you are to leave it to the vice-admiral to assail the enemy's admiral, and to match yourself as equally as you can, to succour the rest of the fleet as cause shall require, not wasting your powder nor shooting afar off, nor till you come side by side.[11]

There is nothing new in this. It is vague, unrealistic and a mere repetition of post-Elizabethan forms. It was a new departure, however, that the Lords and Commons had appointed a committee to sit in London to give tactical instructions to the fleet at sea. During the course of the First Anglo-Dutch War the English conception of naval forces and naval warfare was to be transformed. The Civil War had

9 Leconfield MSS (33); notes made by R C Anderson and sent to Corbett 7 April 1914; part printed in *Historical MSS Commission Report*, vi, p304

10 W G Perrin, editor, *Boteler's Dialogues*, London, 1929, pp279 and 308–12

11 *Fighting Instructions*, pp87–8

ensured that fleets were no longer scraped together by ship money. Instead the navy became the most important charge on the Commonwealth budget. The appointment by Parliament of experienced soldiers as generals-at-sea transformed disciplinary and tactical control over the merchantmen included in the fleet. A fighting organisation was created which became the Royal Navy of the English Restoration.

In February 1649 the command of Parliament's fleet at sea was given to Colonel Robert Blake, aged fifty, Colonel Richard Deane, aged thirty-nine, and Colonel Edward Popham, aged fifty. In modern phraseology, they were seconded from the New Model Army for sea service and each had the title of general-at-sea. Blake and Popham had been Members of Parliament. Deane later fought at Worcester as a major-general and commanded the Commonwealth army in Scotland before returning to sea to meet his death in the Battle of the Gabbard. All three were men of exceptional ability and between them they disposed of considerable experience in commerce, politics, revolutionary government and land fighting.

The first battle of the First Dutch War, fought off Dover on 19 May 1653 (Old System), pitting Blake and Colonel Nehemiah Bourne against Admiral Maarten Tromp, could not possibly have been based on planned tactics, as it arose more or less fortuitously over 'the honour of the flag'. On 16 August (OS) Sir George Ayscue, a quasi-professional sea commander, and Vice-Admiral Michiel Adriaenszoon de Ruyter fought an indecisive action in the Channel. There was repeated mention of 'charging' in the eye-witness accounts. If this way from to windward it probably meant that the ships which charged through the enemy soon found themselves to leeward and vulnerable in turn to an identical charge. Popham died in 1651, and William Penn, a real seaman by trade and upbringing who had already done distinguished service for Parliament against Prince Rupert in the Mediterranean, was appointed to be Blake's vice-admiral. Descriptions of the Battle of the Kentish Knock, in which Blake and Penn fought against Admiral Witte Corneliszoon de With and de Ruyter on 28 September (OS), read rather like a maritime version of the *Iliad* with attention fixed on the flagships as personal fortunes of the leaders swayed this way and that. On 30 November (OS) Tromp, compelled by the States-General to try to force a convoy through the Channel in mid-winter, succeeded in defeating Blake in a short action, begun in the afternoon and known as the Battle of Dungeness. In trying, however, to bring home a convoy early the following year, he suffered a severe defeat which, but for his own skill and leadership, might have been a complete disaster. Meeting Blake and Deane off Portland on 18 February (OS) 1652, he at first held the advantage but during the next two days, while the fleets fought their way up the Channel to Cap Gris Nez, the English gradually gained the upper hand until they were at least able to break through the Dutch battle fleet and attack the convoy. Père Hoste, writing more than forty years later, says that Tromp formed his fleet in an 'order of retreat' to cover the convoy in the space between the two sides. Otherwise, no tactical forms are discernible, despite the large amount of descriptive material in both English and Dutch. Nevertheless the battle is important in the history of naval tactics because George Monck, lieutenant-general of the ordnance and later com-

General at Sea Robert Blake (1599–1657), artist unknown (English school, seventeenth century). NMM

mander-in-chief in Scotland, appeared in it as a co-general with Blake and Deane.

The supposition that no form of line-ahead tactics had yet been adopted by either side is strengthened by a set of fighting instructions issued to Penn, then vice-admiral of the fleet, for issue to his own squadron. They are dated 10 February (OS) 1652, the day the English fleet left the Thames and only eight days before the battle.[12] The instruc-

Luitenant-Admiraal Marten Harpertszoon Tromp (1597–1653), by Jan Lievensz. NMM

12 Samuel Rawson Gardiner and C T Atkinson, editors, *Letters and Papers Relating to the First Dutch War*, London, 1899–1930, (hereafter cited: *The First Dutch War*), vol iv, pp34–8, including Penn's appointment. The instructions are almost word for word the same as those printed in *Fighting Instructions* pp88–90 from an unsigned draft in the Harleian MS in the British Museum.

General at Sea Richard Deane (1610–53), by Robert Walker. NMM

tions, consisting of nine articles, are signed by Blake, Deane and Monck. With one exception, they contain nothing new. The vice-admiral's and rear-admiral's squadrons were 'to make what sail they can to come up with the admiral on each wing', thus suggesting a line-abreast formation. Ships disabled or in danger of sinking or capture were to be relieved, for which 'the flagships are to have a special care'. It would seem then that on the very eve of the so-called Battle of Portland, Blake, Deane and Monck, 'Admirals and Generals of the Fleet appointed by Parliament for the Expedition',

were instructing Penn to fight in a manner no different from that of a quarter of a century earlier.

The new feature of the fighting instructions has nothing to do with the adoption of line tactics. Article 7 states that 'Commanders and masters of all the small frigates [fighting ships], ketches, smacks, etc' were to be stationed to windward of the fleet and to observe the enemy's fireships and be prepared to intercept and capture or burn them. If unable to stop them striking English ships, they were to tow them clear with grapnels and destroy them, 'which if honourably done, according to its merits, shall be rewarded, and the neglect thereof strictly and generally called to account'. The English fleet's own fireships were instructed in Article 8 to keep to windward of the fleet and both they and the anti-fireship flotilla were 'to be as near the great ships as they can – to attend the signal from the commander-in-chief and to act accordingly'. The anti-fireship flotilla was reckoned as a special service involving special dangers to be rewarded, when successfully faced, by the payment of danger money, something quite outside the notion of seamen's wages. It may be questioned, however, whether 'backwardness' in this specially dangerous service, even if officially punishable, could ever be conclusively proven, especially as armed boats from the bigger ships would probably be intermixed with the flotilla in the general scrimmage to ward off the dreaded fireships.

On 29 March (OS), 1653, while the fleet was at Portsmouth refitting after the Battle of Portland and about to proceed to sea once more, Generals Blake, Deane and Monck issued two new sets of instructions, one 'for the better ordering of the Fleet in Fighting'. For the first time, cruising or sailing instructions had been brought together with operational or fighting instructions and issued simultaneously in two separate but companion codes, thus laying the foundations of the later Sailing and Fighting Instructions issued as separate codes but paged and bound as one volume. These instructions, moreover, included nothing but what they purported to include, the whole clutter of administrative, disciplinary and standing orders having been omitted. No doubt their sudden disappearance was connected with Parliament's promulgation in the previous December of the earliest Articles of War. These 39 Articles covered in broad terms almost everything previously included in the usual hotch-potch of admiral's instructions.[13]

The order and form of these Sailing Instructions were simple and objective. Of the twenty-one articles, twelve were wholly or partly concerned with night cruising and sixteen took the form of executive signals. The articles were primarily concerned with keeping the fleet going both by day and by night, and had little to say about how the fleet was to be organised for sailing and nothing at all about any order of sailing. Nor was there any mention of signals for tacking or wearing or for the direction of a particular squadron. Nevertheless, the appearance of an entirely separate issue of Sailing Instructions, however rudimentary in form, marks the beginning of a new era. From these stemmed the development of the future cruising formations so closely connected with tactical deployment in the presence of the enemy.[14]

The Fighting Instructions were shorter. Of the fourteen articles only one was concerned with night action, while

Admiral Sir William Penn (1621–70), by Sir Peter Lely. NMM

13 'Laws of War and Ordinances of the Sea ordained and established by the Commonwealth of England', *The First Dutch War*, vol iii, pp293–301

14 Printed in *The First Dutch War*, vol iv, pp266–73. They were reissued by Blake, Disbrowe and Penn in 1653 much in the same form; *NMM*, WYN/6/1 and WYN/7/14.

eight include one or more signals. In neither set was any signal given for the ordinary handling of the fleet at sea, even in the rudimentary form 'the van to fill and stand on' or 'the fleet to tack the rear to begin'. How did the admiral control the fleet as a whole when sailing? Did each squadron follow its own flagship and the flagships the admiral's flagship? If so, what happened when he wished to detach one of his squadrons from the fleet? Presumably orders were often sent by boat, or by hailing between flagships and between flagships and private ships.[15]

Despite their limitations, the two sets of instructions were quite revolutionary, not for their form and content alone, but because they were issued under realistic conditions. Each flag officer and each captain, as he read those two documents, knew that Penn with his squadron was about to sail under orders from the generals, of the same date, and that the remainder of the fleet would sail very shortly. This meant that within a matter of days practically the whole English navy would be cruising under the new sailing instructions and that Penn and the rest might at any moment have to fight under the new fighting instructions. The Dutch were known to be at sea again, though their main fleet was still refitting. Inevitably, there would be yet another big battle before the summer was gone. The whole situation was one of action and urgency. It was nothing whatever like the peregrinations of the Ship Money Fleet.[16]

As for the authorship of the new instructions, there seems good reason to suggest the influence of Monck, now at sea for the first time, and, next to Cromwell, the leading soldier in England, though far more experienced. Monck was a real professional, having fought in the Low Countries, taken part in the Cadiz expedition of 1625, fought for Charles I in Ireland and been captured on the Royalist side in the Civil War. Blake had been severely wounded in the Battle of Portland. Deane appears to have been more interested in naval administration than tactics, but this is only conjecture. To Monck, then, may be due the credit for the initiative in trying to give the fleet, with its many ex-merchantmen and ex-merchant officers, a greater measure of warlike order. Monck, moreover, with his long experience of artillery work, would be quick to realise the importance of tactical order as a means of developing full and unhindered fire-power.

The inevitable next battle, the Battle of the Gabbard, 2 and 3 June (OS) 1653, again proved a victory for the English. Despite the usual exasperating reticence, there are enough tactical references in the sources to suggest that English tactics were different from and better than those used before. 'Our fleet did work in better order than heretofore, and seconded one another,' wrote Richard Lyons, flag captain to the Generals Monck and Deane in the *Resolution*.[17] An anonymous intelligence agent writing from the Hague on June 9 or 19, says of the first day, that

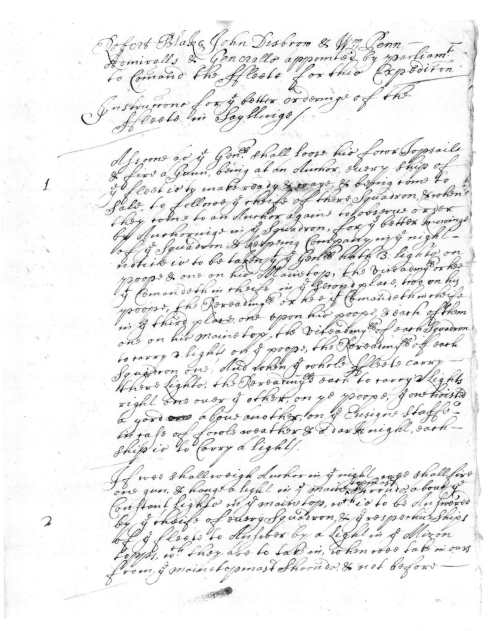

Instructions for Sailing, issued in 1653 by Robert Blake, John Disbrowe and William Penn. NMM

General at Sea George Monck, (1608–70), first Duke of Albemarle, by Sir Peter Lely. NMM

---

15 Printed in *The First Dutch War*, vol IV, pp262–6; and *Fighting Instructions*, pp99–104. Another copy is in the *NMM*, WYN/7/13.

16 Sir Julian Corbett was of the opinion that the fighting instructions amounted to establishing the line of battle ahead as the standard fighting formation. If this is a correct interpretation of them, then the formation could not have been thrust on the fleet *de novo*. As he says, it must have been 'empirically practised', perhaps for many years, 'a practice which had long been familiar though not universal in the service'. These are cautious words, befitting so intriguing a problem. 'How far the new orders were carried out during the rest of the war it is difficult to say. In both official and unofficial reports of the actions of this time an almost superstitious reverence is shown in avoiding tactical details.' *Fighting Instructions*, p95–7.

17 R Lyons to President of COS, June 4–14 1653, *The First Dutch War*, vol v, pp82–5

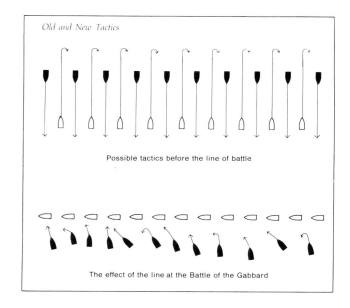

Old and New Tactics

Possible tactics before the line of battle

The effect of the line at the Battle of the Gabbard

The English, having the wind and more and greater guns, made use of these advantages, playing on the Dutch only with their ordnance. And when the Dutch, finding the great disadvantage they were at, endeavoured to get the wind that they might come nearer, the English by favour of the wind, still prevented them . . .[18]

The implication here is that the English exploited their superior firepower and refused to let the Dutch close for boarding. If so, how could they have done this without adopting some kind of line-ahead formation to exploit broadside fire?

The best evidence that a line formation was attempted is from another letter from the Hague 'to my Lord Went-worth', in exile in Copenhagen:

[The English] stayed upon a tack, having the wind, within twice cannon shot about half an hour, to put themselves in their order they intended to fight in, which was in file at half cannon shot, from whence they battered the Hollanders

The Battle of Scheveningen, 31 July 1653, by Willem Van de Velde the Elder (dated 1655). NMM

furiously all that day, the success whereof was the sinking two Holland ships . . . The second day the English still battered them in file, and refusing to board them upon equal terms kept them at bay but half cannon distance, until they found some of them disordered and foul one against another, whom they presently boarded with their frigates (appointed to watch that opportunity) and took; and this they continued to do until the Holland fleet approached the Wielings, when they left them (by reason of these sands) upon Saturday night.[19]

This account seems clear enough. The English ships, being on the whole of deeper draught than the Dutch and more heavily gunned, exploited their advantage and refused to come close enough for the Dutch to board them. When, at last, the Dutch began to fall into confusion, with some of their ships disabled, the smaller English fighting ships which had been kept in reserve were able to push forward and capture those most heavily damaged. It reads like a battle actually fought according to plan – and what plan could it have been other than one involving at least some uses of the line-ahead formation? The English fleet, deployed either by squadrons or as a whole, held the Dutch in check by the fire delivered from their well-formed line. The second day they did the same, in file, until their success enabled them to take prizes.

On the Dutch side every effort seems to have been made to come to close quarters at once and board; so to begin with, at any rate, the line-ahead appears to have been associated with avoiding the old time pell-mell battle of the pre-cast gun age in which boarding was the only way of forcing surrender. When Howe and Nelson and their contemporaries in the late eighteenth century revived pell-mell battle tactics, they did so simply as a means of breaking up the enemy's formation and stopping their escape. Boarding by then was no longer regarded as the standard method of capture even though it had been brilliantly executed by Nelson at St Vincent.

The last battle of the war was fought on 31 July (OS) 1653, off Scheveningen. Deane had been killed at the Battle

18 *The First Dutch War*, vol v, p100
19 *The First Dutch War*, vol v, p109

of the Gabbard, and though Blake had joined the fleet during its later stages, he was ill ashore three weeks later. Monck, therefore, was the only general on the English side; the Dutch were commanded by Tromp supported by de Ruyter and de With. The battle was an overwhelming victory for the English, and Tromp was killed in it. There are several descriptions of the fighting but none contains any real tactical information. Captain Cubitt's account, which is a fine piece for anthologies, is full of 'chargings':

We tacked upon them and went through their whole fleet, leaving part on one side and part on the other side of us . . . as soon as we had passed them we tacked again upon them and they on us, passed by each other very near . . . As soon as we had passed each other both tacked, the Hollander having still the wind and we keeping close by . . . We cut off this bout some of his fleet which would not weather us . . . We tacked again upon them and they upon us and this bout was most desperately fought by either almost at push of pike. . . . Our General must needs gall them very much this bout and so did all our ships, being constantly very near specially this last charge . . .

Theophilus Sacheverell, another eye-witness, writes:

We engaged their whole fleet and charged through them and cut off many of their ships; they fired three or four of their fireships, but, (blessed by our God), they did us no hurt. After that we tacked about again and charged through them a second time, and then a third time . . .[20]

In none of this is there the slightest suggestion of line-ahead. On the contrary, there seems to be a reversion to the older tactic of fleets passing through each other on opposite tacks, apparently in open order, possibly in line-abreast or lasking, with the leeward fleet quickly tacking to make another pass and the windward fleet bearing down again and quite prepared to fall temporarily to leeward. Père Hoste, writing over forty years later, quotes a French eye-witness as saying that on 28 July (7 August) he saw the Dutch fleet

renagée en trois Escadrons, et elle faisoit vent-arrière pour aller tomber sur les Anglois, qu'elle recontra le même jour à peu près en pareil nombre, rangez[21] sur une ligne qui tenoit de plus de quartre lieues Nord-Nord-Est et Sud-Sud-Oeust, le vent étant Nord-Oeust.[22]

The difficulty with this account, however, is that it would seem to be describing cruising rather than actual battle formations.

The was no mention made of line tactics when in March 1654 new fighting instructions were re-issued by Blake and Monck, together with Penn and Major-General John Disbrowe, the two new generals-at-sea. They contain fifteen very minor changes in wording, none of which is in any way tactically significant.[23] On 26 December 1654 Penn issued a shortened form of the sailing instructions.[24] However, despite the absence of any mention of line tactics in these instructions, and the evidence that English experiments with the line of battle were not pursued in the Battle of Scheveningen, the idea must have been fairly firmly established. By the outbreak of the Second Dutch War, line tactics had apparently been accepted as necessary, and the first indications are available that a prescribed order of battle was being drawn up before contact with the enemy.

English tactics in the Second Dutch War

The Restoration of 1660 brought England a wealth of naval capacity, unparalleled in any country at that time. Prince Rupert, the king's first cousin, aged only forty-one, besides fighting on land in Germany, Flanders and England, had

Edward Montague (1625–1672), first Earl of Sandwich, by Sir Peter Lely. NMM

spent four years at sea, on the Irish coast, in Portuguese waters, in the Mediterranean and in the West Indies. General Monck, aged fifty-two, now Duke of Albemarle and captain-general of the Land Forces, one of the most considerable men in the realm, had the full experience of the First Dutch War behind him. Colonel Edward Montagu, aged thirty-five, now a Knight of the Garter and Earl of Sandwich, had been co-general-at-sea with Blake in 1656. Sir William Penn, aged thirty-nine, vice-admiral of the Commonwealth Fleet and co-general at the conquest of Jamaica, was amongst the most experienced sea officers in the country. Only the Lord High Admiral, James Duke of York, the king's brother, aged twenty-seven, was totally lacking in sea experience. Nevertheless his personal insight into the high command of the armies both of France and Spain and his personal experience of many battles, skirmishes and sieges seemed to mark him out as a man well-fitted to learn the business of a naval commander and administrator.[25]

---

20 *The First Dutch War*, vol v, pp367–8 and 372

21 The final 'z' added to the unaccented 'e' appears slightly out of alignment in the text.

22 Père Hoste, p78. Sir Julian Corbett deduced from this, assuming that the account is correct (which is a little doubtful), that 'this is the first known instance of the Dutch fleet forming in a single line, and, so far as it goes, would tend to show that they adopted it in imitation of the English formation. *Fighting Instructions*, p97–8.

23 *Fighting Instructions*, pp99n and 100–3nn

24 *NMM*, WYN/9/6; the articles and their numbers correspond with the shortened form numbering shown in *The First Dutch War*, vol iv, pp266–73.

25 The full details of James II's hitherto unknown military experiences have been revealed by the recently published: A Lytton Sells, translator, *The Memoirs of James II . . . 1652–60*, Bloomington, 1962.

James, Duke of York (1696–
1746), by Nicolas de Largillière.
NMM

Throughout the autumn and winter of 1664 the English fleet was fitted out at Portsmouth and, on 9 November, James arrived to take command. Two days later he divided the fleet into three squadrons,[26] and on 16 and 22 November he issued new sailing and fighting instructions. The sailing instructions were the same as those issued under the Commonwealth with slightly different numbering and the substitution of 'Admiral' for 'General' throughout. There were additional signals for cutting and slipping by day or night.[27] The 'Instructions for the better ordering of His Majesty's fleet in fighting', consist of sixteen articles very like those issued under the Commonwealth.[28] The differences were that Articles 2 and 3, equating to no 3 in the Commonwealth instructions, instead of merely ordering captains 'to keep in the line' and 'to get in a line with the Admiral', require the captains to 'put themselves into the place and order which shall have been directed them before in the order of battle' (Article 2), and 'to engage with the enemy according to the

order prescribed' (Article 3). Clearly it had been decided that the order in which ships took their place in the line of battle should be determined before contact with the enemy. It had also been decided that even the restrictions placed by the Commonwealth upon prize taking during battle were too few to prevent disruption of the tactical organisation of the fleet at what might be critical times. Commonwealth Article 10 originally stated that prizes were to be sunk or burnt immediately, after saving the crew, so 'that our own ships be not disabled or any work interrupted by departing of men or boats from the ships, and that we require all commanders to be more than ordinarily careful of'. The new fighting instructions now stated that 'ships in condition of pursuing the enemy are not during fight to stay, take, possess or burn any of them [disabled enemy ships], lest by so doing the opportunity of more important service be lost, but shall expect command from the flag officers for doing thereof when they shall see fit to command it'.

The last two articles (15 and 16) of James's fighting instructions were new. They were for forming line-ahead on the starboard or larboard tack, and for commanders not to fire 'until the ship be within distance to do good execution' under penalty of severe punishment by court-martial.[29]

This issue of instructions, combined with assigning ships to squadrons, is an early example of fleet organisation. Although Sandwich only mentions squadrons, it is reasonable to assume that the organisation went further, each squadron being again subdivided under a total of nine flag officers, and that an actual order of battle was also issued, giving each ship her place in the line. The evidence for this is that on 1 February 1665, Sandwich, putting to sea from the Downs with a squadron of fifteen ships, issued what amounts to four additional fighting instructions.[30] These consisted of five unnumbered signals and articles. The first was for forming a line abreast on each side of the admiral 'in such berth as opportunity shall present most convenient, but if there be time they are to sail in the foresaid posture', that is, in the order of sailing or battle already issued to the squadron. This was a separate order from that issued by James to the fleet as a whole from Portsmouth. The second and third signals were for forming line-ahead from line-abreast, either to starboard or larboard. The fourth was for forming in 'this posture, every ship in the place and order here assigned', again referring to some order of sailing or battle previously issued. The fifth was an instruction for the lesser ships of the fleet to keep on the unengaged side of the fleet and to come up under the admiral's stern for orders when signalled.

The implication of the first signal seems to be that captains unable to reach their correct station in the line were to take station as best they could. It was no more than an escape clause in case of difficulty and was not the equivalent of saying that captains were free 'to take station for their mutual support' in the highly sophisticated sense in which Howe, St Vincent and Nelson understood this phrase. In Sandwich's time the order of battle was a novelty, its object being to give every ship her own place in the line. Captains who found they could not reach their correct station were to get into line as best they could. It was only much later that captains were trusted to use their professional skill to act to the best advantage.

The English fleet was fitting out at the Gun Fleet anchorage in the Thames on 23 March 1665 when James arrived to take command. During April the whole fighting organisation of the fleet was again overhauled. Not only was the fleet reorganised in three squadrons under James, Prince Rupert and Sandwich, but an order of battle was issued.[31] From this it appears that the idea of each flagship being supported by two powerful seconds had not yet been accepted. On the contrary, five out of the six seconds of the three flagships

26 R C Anderson, editor, *The Journal of Edward Montagu, First Earl of Sandwich*, London, 1929 (thereafter cited: *Mountagu*), p157, entries for 9 and 11 November 1664

27 There are four copies of these instructions in the *NMM*, Wynne Collection, which includes the papers of Sir William Penn: WYN/12/1 (where they are wrongly headed 'fighting'), WYN/12/5, WYN/12/8 and WYN/13/5. There are copies of both the sailing and the fighting instructions in the Bodleian Library, Rawlinson MS 919; letter from Sir Charles Firth to Corbett, 1908 (no day or month).

28 Penn's two dated copies in the National Maritime Museum, Wynne Collection, WYN/12/6, WYN/13/4 (the first signed by James and countersigned by Sir William Coventry) prove conclusively that these instructions were issued in 1664; printed in *Fighting Instructions*, pp122–6 from the reissue of 10 April 1665.

29 Both copies include the word 'do' (good execution) omitted in the 1665 issue: WYN/13/4 has only fifteen numbered articles, but includes the whole of the text.

30 *Fighting Instructions*, pp108–9. The actual document was issued to Hugh Seymour, Captain of the frigate *Pearl*, and is the Duke of Somerset's MSS, printed by HMC Rep XV, part vii, p100.

31 *Mountagu*, pp195–8; for guns see pp174–7

were ships of very inferior force, stationed next to the big ships as if to receive protection rather than supply it, and possibly to act as auxiliaries. The same may be said of the ships seconding the ships of the divisional vice-admirals and rear-admirals of the three squadrons, though in these cases the disparity of force was not so marked. In Rupert's White squadron his vice-admirals and rear-admirals were stationed next but two from the outer ends of their respective divisions. In Sandwich's Blue squadron his vice-admirals and rear-admirals were next but one to their divisional extremities. This undoubtedly reflects both a desire to strengthen the fighting power of the extremities and to have men posted there better able to give a lead in executing fleet tactics than captains of private ships. This was emphasised by Admiral Sir Edward Spragge eight years later.[32]

On 1 April 1665 James issued sailing instructions to the fleet very similar to those of the Commonwealth,[33] and on 10 April he issued 'Instructions for the better ordering of his Majesty's Fleet in time of fighting'.[34] The instructions themselves are simply a re-issue of those of the previous 22 November, but on 18 April they were reinforced by ten 'Additional Instructions for fighting.'[35] The general purpose seems to have been to give the fleet a better sense of warlike order and at the same time to try to establish some fairly simple tactical principles. Commanders were explicitly instructed in Article 1 'to endeavour to keep the fleet in one line, and as much as may be to preserve the order of battle which shall be directed before the time of the fight', and Article 4 directed the ships to fight at half a cable's distance 'in reasonable weather', Articles 2 and 3 established that if the fleet should engage the enemy from windward the leading squadron was to 'steer for the headmost of the enemy's ships', and if from leeward 'commanders . . . shall endeavour to put themselves in one line close upon the wind'. Articles 6 to 8 were concerned with ensuring that the tactical cohesion of the fleet was not destroyed by battle damage or by the chance of capturing a disabled enemy ship:

Article 6. 'None of the ships of his Majesty's fleet shall pursue any small number of ships of the enemy before the main [body] of the enemy fleet shall be disabled or run.'
Article 7. No ships to chase 'beyond sight of the flag' and all chasing ships to return at night.
Article 8. Disabled ships not in danger of sinking or capture to be relieved by 'the sternmost of our ships' and the rest of the fleet to press on: '. . . nothing but beating the enemy can effectually secure the lame ships'. This article superseded Articles 4 and 5 of the main instructions enjoining the nearest ships to give relief.

Two days later James added to his instructions 'Encouragement for the Captains and Companies of fireships, small frigates and ketches'.[36] This was a revised form of the instruction issued to Penn by the generals on 29 March 1652 (OS) but in much more detailed and persuasive terms. The crew of a fireship which carried her against a 40-gun ship of the enemy and burned her were each to be paid £10 and the captain was also to receive a gold medal. If the enemy flagship were destroyed the reward was doubled. Those in the anti-fireship flotilla who preserved one of the king's ships of fifth rate or above from being burnt were to receive 40s. Private ships hired for fireship or anti-fireship service and destroyed in action would be paid for by the treasurer of the navy. Whatever may have been the real, as opposed to paper, value of these promises they clearly show the need for a greater measure of encouragement than was forthcoming under the Commonwealth. The fireship, and to a lesser extent the anti-fireship, service could only be staffed effectively by the offer of danger money.

On 27 April and 30 May, only four days before the Battle of Lowestoft, James issued further additional instructions establishing signals requiring 'all the ships to fall into the order of Battailia [sic] prescribed', for 'the other squadrons to make more sail, though he himself [the admiral] shorten sail', for the different squadrons to chase the enemy, and finally for the larger ships, fourth rates and upwards, to 'make what sail they can to come up with the Admiral and so get into a line, for the better doing whereof', with other squadron commanders to repeat the signals.[37]

Sir Julian Corbett, and later Admiral Sir Herbert Richmond, characterised the orders against pursuit of individual enemy ships, which eventually became incorporated in the so-called permanent Sailing and Fighting Instructions as Article 21, as a regrettable example of formalism and regimentation, since it deprived captains of individual initiative.[38] Admiral Richmond pushed the argument still further, showing that because the offending article had been accepted by later admirals

the original conception of keeping the fleet in hand in order to effect a complete victory merged into a formalism of the worst kind . . . in which the fleet was kept together in order to secure it against attack. This was the view which tended to become prominent in a prolonged period of peace. Defence took a higher place than attack.[39]

However, this is no reason for condemning the issue of the article in the first instance. It is one thing to regiment a well-ordered fleet of comparatively modest size but quite another to try to keep in hand a very large fleet that had never yet attained to the tactical status of a well-ordered body. Words such as 'formalism' can only be applied in cases where some degree of formal order already exists. There is no clear evidence that the offending instructions were the work of James advised by Penn. Accordingly, Corbett's identification of these two as the pedants, and Rupert and Albemarle as a rival school favouring individual initiative, seems based on doubtful premises.

It is necessary to differentiate between fighting instructions written on paper and the actual events in sea battles. The fighting instructions issued by the Restoration admirals were counsels of perfection, recognised by those who drafted them as unlikely to be attained in any full measure in the heat of battle. Even the idea of a set order of battle was only just obtaining acceptance and this in itself is sufficient evidence of the primitive tactical methods then accepted. To say, as Sir Herbert Richmond did, that Wimbledon, Lindsey and Northumberland had shown a wise restraint in not seeking to impose a complete ban on ships turning aside to pursue small groups or individual ships of the enemy is merely another

32 *Vide* Spragge's journal, R C Anderson, editor, *Journals & Narratives of the Third Dutch War*, London, 1946 (hereafter cited: *Third Dutch War*), p323–4

33 Bodleian Library, Rawlinson MSS 919 D

34 *NMM*, Dartmouth MS DAR/2; the phrase 'in time of fighting' is unusual.

35 *NMM*, DAR/2, WYN/12/7,WYN/13/1, and WYN/13/4; copy also in *BM*, Harleian MS 1247, with slightly different arrangement; printed in *Fighting Instructions*, pp126–8

36 *BM*, Harleian MSS 1247 f. 53; printed in *Fighting Instructions*, pp149–51

37 *NMM*,WYN/12/8; WYN/13/4 (two instructions only dated 24 April from the *Royal Charles*); WYN/13/2 and /5 (signed 'James' and countersigned by Sir William Coventry, *Royal Charles*, 30 May). The first is printed in *Fighting Instructions*, pp128–9.

38 *Fighting Instructions*, pp115 and 134–5, and letter from Richmond to Corbett, 14 March 1914 (Corbett Papers)

39 *The Navy in the War of 1739–48*, vol iii, p259

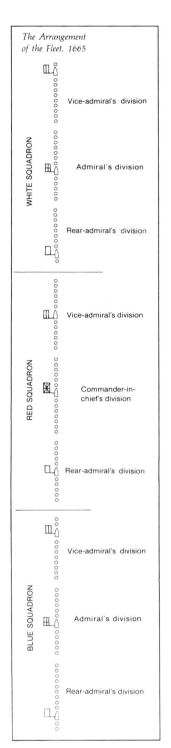

The Arrangement of the Fleet, 1665

WHITE SQUADRON

Vice-admiral's division

Admiral's division

Rear-admiral's division

RED SQUADRON

Vice-admiral's division

Commander-in-chief's division

Rear-admiral's division

BLUE SQUADRON

Vice-admiral's division

Admiral's division

Rear-admiral's division

way of saying that they saw no hope of enforcing such a ban.

The Restoration fleets, under the leadership of highly intelligent and progressive commanders, were just beginning to emerge from the first stages of tactical development imposed, or at least attempted, by the Commonwealth generals. The fixed order of battle was the most important new development and had to be imposed on conservative captains even at some cost. The English admirals knew the evil of allowing the battle to degenerate into a series of private duels and detached skirmishes. They knew the temptation to acquire glory by boarding and taking an enemy ship, just as they knew the danger of allowing cavalry to make an undisciplined outflanking manoeuvre in a land battle, or to ignore their proper tactical role in favour of attacking the enemy's baggage train. More particularly they knew the unreliable nature of some of their gentlemen captains and captains of hired merchantmen, only too anxious to make private captures. In the English fleet of 1665, out of a total of one hundred and six ships in the line, nineteen were hired merchantmen. Besides, with so many ships involved on either side it was extremely difficult to maintain any cohesion at all. If the enemy's fleet began to shed small groups here and there, either through disablement, inefficiency, faint-heartedness or as part of a *ruse de guerre*, common sense demanded complete concentration on the main force so as to take, sink or disable as many as possible, including the flagship of the

commander-in-chief. There can be no helpful comparison here with the far smaller and more highly trained fleets of Nelson's time.

What were the sea battles of the Second and Third Dutch Wars really like? Despite narratives and observations in flag-officers' journals and elsewhere, the evidence for effective tactical cohesion is slight. It was the squadron commanders of the Red, White and Blue who dictated the form of the battle, though even they could not wholly prevent their own vice-admirals and rear-admirals from fighting independent small divisional actions. In view, moreover, of the individual differences in gunpower of the so-called ships-of-the-line, it is scarcely surprising that the 'flagmen' in the big ships tended to become centres of local engagement. There was of course a great awareness of the need to co-operate, and particularly of the need to relieve ships in distress. There was also a keen sense of the need to anticipate the next move of the nearest enemy and to take action in advance to cover or relieve ships of another squadron or division rather than merely to safeguard one's own. All this showed an admirable sense of what a hundred and thirty years later could be characterised as mutual support. It stemmed, however, from tactical anarchy rather than from tactical formalism, which was wisely ignored, when deemed necessary, by sophisticated professional commanders.

Tactical anarchy naturally reduced the effectiveness of

The Battle of Lowestoft, 3 June 1665, showing HMS Royal Charles and the Eendracht, by Hendrik van Minderhout. NMM

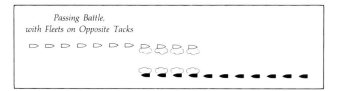

Passing Battle,
with Fleets on Opposite Tacks

Prince Rupert (1619–82), first Duke of Cumberland and Count Palatine of the Rhine, by Sir Peter Lely. NMM

naval gunfire. Rear-Admiral A H Taylor observes that in the Four Days Battle, which was fought the subsequent year,

The fleets engaged for the most part by passing on opposite courses at a range of 200 yards or more. There were at least ten such passes, each taking about 3 hours; the English ships carried 40 rounds per gun and no ship had less than 10 remaining; this gives an average expenditure of 3 rounds per gun per pass or, overall, about one round per gun per hour.[40]

In addition to these passes along the enemy line, groups of ships, divisions and even whole squadrons passed or charged through the enemy's fleet and back again without necessarily gaining or losing any great tactical advantage. Cutting, breaking, or dividing, charging or passing through the enemy's line was regarded as incidental to the fight. Inevitably, gunfire was not concentrated. Ships' hulls were strong, but continuous waterline hits could start leaks and flood the hold, thus causing the ship to lose speed and manoeuvrability, so becoming an easier target for guns and fireships. Provided, however, that it could withdraw from the line, a badly damaged ship might well get home.

On 31 May 1665, the day before the Dutch fleet was first sighted, Sandwich proposed at a council of war that the merchant ships, now totalling twenty-four, should be taken out of the fleet and put all together in a rear squadron with three more flag officers in command of them,

By which means our ships of force of the King's would have had their strength contracted into a lesser room (by near a league) . . . They would have been much stronger to make an impression on the enemy in any part, or to resist any combined force of the enemy attempting us. They would have had no impediment by bad sailors. And the commanders of the King's ships more entire and resolved to aid one another than it is to be feared the others are.

Sir John L[a]wson was for this the day before and others seemed to like it, but now nobody was forward to speak, and so agreed to continue our former order of battle.[41]

This idea of disembarrassing the line of the hired merchantmen, and at the same time concentrating the hitting power of the king's ships, was thoroughly progressive and seamanlike, though it is not surprising that on the eve of battle it failed to gain acceptance.

This first battle, fought on 3 June 1665 off Lowestoft, showed most of the tactical shortcomings characteristic of both wars. Sandwich recorded his observations,

Whereas our order of battle (as is before showed in this book) was a line, that so every ship might have his part in fighting and be clear of his friends from doing them damage, yet many of our ships did not (even in this first pass) observe it, but luffed up to windward, that we were in ranks of 3, 4 or 5 broad, and divers out of reach of the enemy fired over us and several into us and did us hurt.[42]

The mere recital of these facts shows how far removed were the actualities of battle from the tactical assumptions underlying the issue of the fighting instructions.

Three months later, on 29 August 1665, Sandwich, in command of the English fleet lying in Southwold Bay, called 'a General Council of War of all the Captains' (an unusual procedure) at which he 'did admonish the commanders of some ships for our advantage in fighting & sailing.' The

lessons he felt the fleet had to learn as a result of the battle of Lowestoft are eloquent about the condition of tactical discipline then prevailing.

In order of fighting:

1. To be in their place according to the order of battle, at the first if possible.
2. If they were hindered of that by any accident, then to be sure to put themselves in a line anywhere, to have their broadsides to the enemy.
3. In tacking and sailing in time of fight to have special care of fouling one of another, which is the great occasion of destruction.
4. If by accident they be out of the line, take heed not to fire at the enemy through our friends, but catch an opportunity to have the enemy clear.
5. We meeting now with a mixed fleet of men-of-war, East Indiamen, etc., no man to seize a merchant till victory be obtained for certain.

In order to sailing:
1. Give good berth, to avoid disabling our ships by tacking or falling foul.
2. Take special heed not to lose company of the fleet, which whoso does shall justify himself at a Court-Martial.
3. None to chase but by order of the Flag; the contrary to be examined at a Court-Martial.
4. Sail in such order as you may most readily fall into the posture of battle.[43]

Sandwich was clearly a realist and knew the futility of grandiose tactical projects unsupported by a proper order of

40 Private publication

41 *Mountagu*, p222

42 *Mountagu*, p224. Four English earls and a viscount were killed in the battle.

43 *Mountagu*, pp269–70

Engraved for the Univerfal Magazine

Sr.EDWARD SPRAGGE.

sailing; hence particularly sailing instruction no 4. With an armada of ninety-five ships-of-the-line under his command it would be impossible to form the line in the face of the enemy unless the fleet was already in tolerable order.

The campaigns of 1666

The removal of James Duke of York and Lord Sandwich as naval commanders left Prince Rupert the sole superior commander of the English fleet. For the campaign of 1666, however, George Monck, Duke of Albemarle, was appointed to command with him as co-general. Rupert reissued James's 'Encouragement for the captains and companies of fireships', probably on first taking command. While the fleet was fitting out for the season's campaign, he and Albemarle also issued 'Instructions for the better ordering of his Majesty's Fleet in Sailing, Given under our hands aboard the *Royal Charles* at the Boy [sic] of the Nore, the 1st day of May 1666.'[44] These are almost a repeat of the Commonwealth sailing instructions of 1653, with extra articles for the sternmost ships to tack first, and for cutting or slipping by day or night.

At some unknown date, Prince Rupert issued 'Additional Instructions for fighting' to Spragge as vice-admiral of the Blue squadron, a post he received on the fourth day of the Four Days Battle (June 1–4), following the death of Sir William Berkeley.[45] The second article was a signal that 'all the best sailing ships are to make what way they can to engage the enemy, that so the rear of our fleet may better come up; and so soon as the enemy makes a stand then they are to endeavour to fall into the best order they can'. The instructions certainly have about them an air of recent fighting experience, but Sir Julian Corbett was probably wrong to believe that Article 2 anticipated 'by a century the favourite English signals of the Nelson period for bringing an unwilling enemy into action, ie for general chase, and for ships to take suitable action for mutual support and engage as they got up'.[46] The analogy is misleading, for the circumstances were very different. Rupert's fleet was not homogeneous; some of his line ships were still only hired merchantmen while others were notably slow sailers through poor or outmoded design. His object was to force a battle, presumably making use of the new line technique. In the Nelson era the object was more to force a pell-mell against a less effective fleet tactically weakened by upheavals in its command during the French Revolution. Slow sailers might be out-distanced but the whole notion implied a sophisticated appreciation by highly skilled captains of the tactical options best open to each. Not even Nelson would have risked a scrambling attack against de Ruyter.

The circumstances of the Four Days Battle were unusual. For the first three days of the battle Albemarle, with some fifty-four ships in his line, fought a very unequal action off the Thames estuary against a Dutch fleet of eighty-four ships commanded by de Ruyter, with Cornelis Evertsen (the Elder) and Cornelis Tromp as his squadron admirals. Meanwhile Prince Rupert had taken twenty of the best ships down Channel on what turned out to be a very wild goose chase againse the French, and did not return to take part in the battle until the fourth and last day.

44 *NMM*, WYN/14/10 (no date or signatures but endorsed 1666) and WYN/14/11 (Sm/1 is another copy).

45 *NMM*, DAR/2; printed in *Fighting Instructions*, pp129–30. The implication seems to be that they were issued at that time, as they were signed by Rupert alone. Rear-Admiral A H Taylor, however, suggests that they were issued at a council of flag officers held on 29 June 1666, while the fleet was refitting *after* the Four Days' Battle.

46 *Fighting Instructions*, p130n

The second day began with the usual passing movements, each fleet tacking in succession van first as soon as its rear had passed the rear of the enemy. In an unusually heavy concentration of forces after Tromp had been cut off and de Ruyter had returned to relieve him, it seemed as if the English might gain a clear advantage. Having received considerable damage, however, and being outnumbered anyhow, Albemarle chose to draw off. He spent the third day returning slowly into the Thames estuary and covering his damaged ships in what later came to be called the 'order of retreat'. The *Royal Prince*, originally built by Phineas Pett for James I, ran on the Galloper, surrendered and was burnt by the Dutch. That evening Rupert made contact with Albemarle and it was agreed that next day he should take the van. The opposing fleets were then about equal in strength and after some twelve hours fighting, an evening fog obliged them to separate.

Despite de Ruyter's skill and the superiority which the Dutch had held for the first three days, they had by no means won a devastating victory. English losses were ten ships to the Dutch four, and the fleets expended six and five fireships respectively.

While the fleet was refitting, the English co-generals must have been considering their future tactics somewhat anxiously. Slackness and inefficiency in station keeping were problems they could hardly hope to cure when handling so large and heterogeneous a fleet, especially as they were never able to carry out any tactical training. Apart from this, there was the semi-independence of action adopted by, and allowed to, squadron and even division flag officers. This was valuable when it arose out of firm, yet sophisticated disciplinary control, but it was merely anarchic when it was a response to an ideal, postulated on paper but never realised in practice.

On 18 July 1666, only a week before the next action, known as the Battle of St James's Day and fought on 25 July, Rupert and Albemarle issued their 'Further Instructions for Fighting'.[47] The face of these is headed, 'To keep the enemy to leeward':

In case we have the wind of the enemy, and that the enemy stands towards us and we towards them, then the van of our fleet shall keep the wind, and when they are come to a convenient distance of the enemy's rear shall stay until our whole line is come up within the same distance of the enemy's van, and then our whole line is to stand along with them the same tacks on board, still keeping the enemy to leeward, and

47 There are three copies of these Further Instructions in the Dartmouth MS, *NMM*, DAR/2, 3, 13; printed in *Fighting Instructions*, pp129–30

Admiraal *Cornelis Evertsen the Elder (1600–66)*, engraving by Arnold de Jude after a painting by P Borselaer. NMM

not suffering them to tack in the van, and in case the enemy tacks in the rear first, then he that leads the van of our fleet is to tack first, and the whole line is to follow, standing all along with the same tacks on board as the enemy does.

Viewed in terms of recent fighting, this article seems to aim at getting rid of continuous passing by bringing about a

Luitenant-Admiraal *Cornelis van Tromp (1629–91), by Sir Peter Lely*. NMM

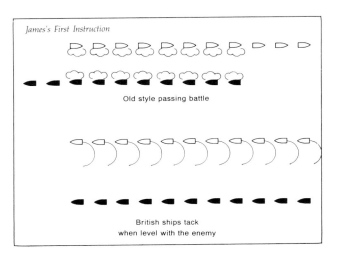

James's First Instruction

Old style passing battle

British ships tack when level with the enemy

general engagement as soon as the two fleets are level and roughly parallel, assuming them to be roughly equal in number. James had tried to execute this manoeuvre at the Battle of Lowestoft the year before, but there was some delay in making the required signal and the opportunity was lost.[48] No signal ordering this manoeuvre is attached to Rupert's and Albemarle's instruction.

Although the general intention of the article is clear enough, there does seem some difficulty in the phrase 'in case the enemy tacks in the rear first, then he that leads the van of our fleet is to tack first'. What does this mean, especially in view of the fact that nothing is specifically said about the van of the English fleet tacking at all? In James's instruction of 1672 the wording is precisely the opposite: 'and in case the enemy tack in the rear first, he who is the rear of His Majesty's is to tack first'.

To the historian of naval tactics, this article is by far the most important to be issued to the English fleet in the whole of the seventeenth century, apart from that for the line-ahead. By continuous repetition and subsequent incorporation as Article XVII of the fighting instructions in the permanent Sailing & Fighting Instructions, it had currency for over a hundred years as the only positive, as well as mandatory, fighting instruction actually specifying to the admiral how to fight the battle, given certain circumstances. On a strict reading, it came to mean that given the prescribed circumstances, the admiral had no tactical choice, except as regards the exact moment when he should signal the fleet to tack so as to get on the same course as the enemy.

The second article is headed 'To divide the enemy's fleet', and reads:

The Battle of St James's Day, 25 July 1666, from the Instructions for the better Ordering of the Fleet, c1688. MOD

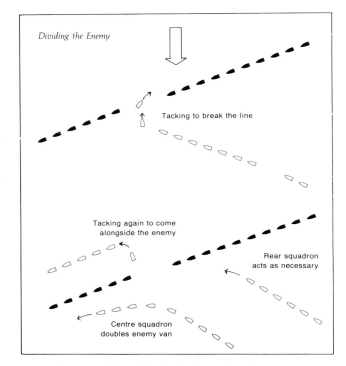

Dividing the Enemy

Tacking to break the line

Tacking again to come alongside the enemy

Rear squadron acts as necessary

Centre squadron doubles enemy van

In case the enemy have the wind of us and we have sea room enough, then we are to keep the wind as close as we can lie until such time as we see an opportunity by gaining their

48 *Mountagu*, pp1ii–1iii

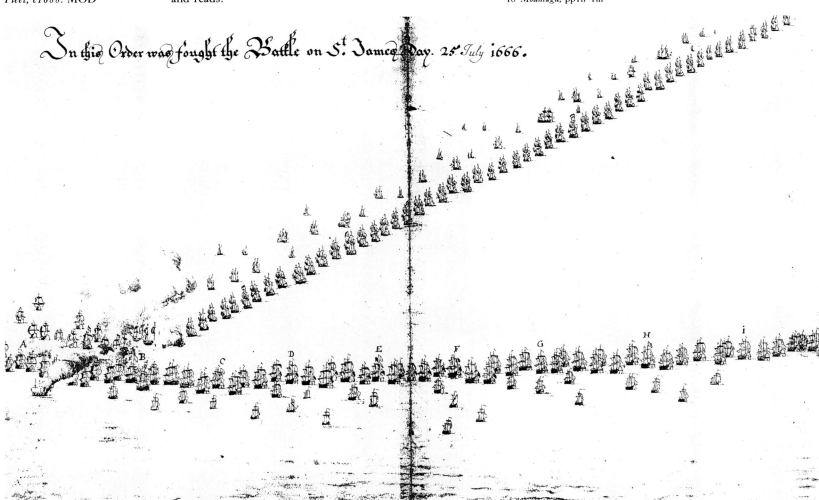

In this Order was fought the Battle on St. James Day. 25 July 1666.

wakes to divide their fleet; and if the van of our fleet find that they have the wake of any part of them, they are to tack and to stand in, and strive to divide the enemy's body, and that squadron which shall pass first being come to the other side [of the enemy] is to tack again, [ie, get on its original course and parallel with the enemy] and the middle squadron is to bear up upon that part of the enemy so divided, which the last [the rear] is to second, either by bearing down to the enemy or by endeavouring to keep off those that are to windward, as shall be best for service.

The origin of this instruction is not hard to find. Prince Rupert had provided a brilliant example of this manoeuvre in the Four Days Battle. Nevertheless, it seems to be reading rather too much into the evidence even to suggest, as does Corbett, that this 'is the first time, so far as is known, that the principle of containing was ever enunciated. In this, it compares favourably with everything we know of until Nelson's famous Memorandum'.[49] In Nelson's terms, 'containing' meant using a part of the British fleet to hold off the remainder of the enemy's fleet, while the other part cut off and destroyed the fighting power of that part of the enemy's fleet chosen to be attacked. As such the manoeuvre depended on very close action by the holding force to make it as difficult as possible for the remainder of the enemy (assumed to be the van) to work back by tacking or wearing to 'succour their rear'. In fact, as Corbett says, the new instruction 'seems designed rather as a method of gaining the wind than as a method of concentration, and that the initiative of the manoeuvre is left to the discretion of the leading flag officer, and cannot be signalled by the commander-in-chief'. Naturally, because under existing conditions the 'general'

had little hope of controlling his immense fleet in which divisional flag officers, acting in the manner of Homeric warriors, fought their own battles with the commander-in-chief's flag out of sight five or six miles away. We should also note, as Corbett implies, that the whole idea of dividing the enemy was concerned with free movement and not with coming together in close action and the pell-mell, in the sense of the term in a later age.

The third article is quite simple: 'To keep the line: The several commanders of the fleet are to take special care that they keep their line, and upon pain of death that they fire not over any of our own ships.' This was bravely written indeed, but still not strictly enforceable, with the penalty no doubt little more than an accepted formula.[50]

---

49 *Fighting Instructions,* pp135–40

50 Some explanation is required of the date of these three articles. The letter copy printed by Sir Julian Corbett bears the copies signature 'James, By command of his R Highness M Wren.' Wren became the Lord Admiral's Secretary in 1677 and died on 14 June 1672. Corbett dated the articles tentatively as 1672, but conceded that they might date back to 1666. They were in fact first issued on 18 July 1666, as stated. They therefore immediately precede the Battle of St James's Day and belong to the Second and not the Third Dutch War; *Fighting Instructions,* pp133–9 and 148–9. Corbett used the copy in Spragge's order book, forming part of the Dartmouth MS now in the National Maritime Museum and known as DAR/2. They come immediately *after* a copy of instructions dated 1666 and immediately *before* other instructions dated 1665. Nothing as regards date can, therefore, be concluded from their position in the order book. The instructions are here headed 'to be observed in the next engagement'. But in DAR/3 they are included in the entries for 18 July 1666 and are initialed 'R & A'. This date is proved by Basil Lubbock's discovery of a copy in the *(continued over)*

The later stages of the Battle of St James's Day, 25 July 1666, from the Instructions for the better Ordering of the Fleet, c1688. MOD

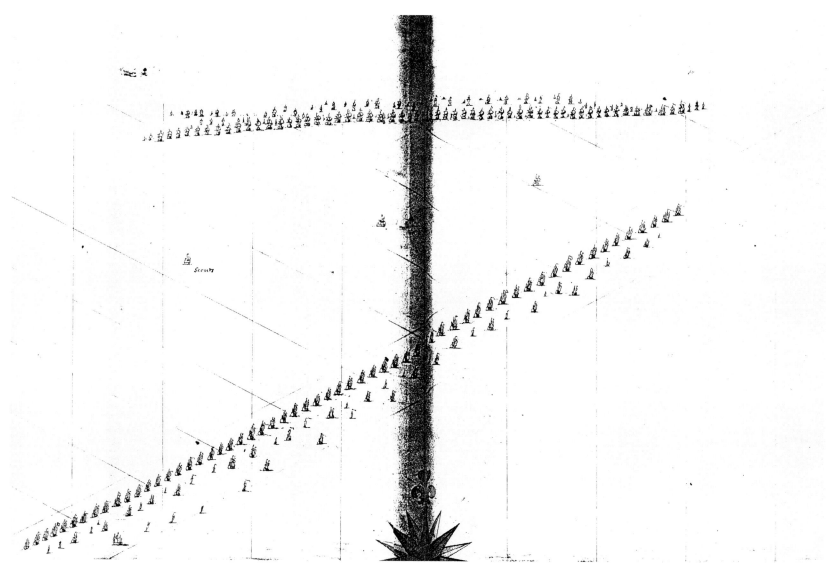

The English to windward at the beginning of the St James's Day battle, 25 July 1666, from the Instructions for the better Ordering of the Fleet, c1688.
MOD

Admiral Sir Thomas Allin (1612–85), by Sir Godfrey Kneller. NMM

It is extremely difficult to gain any coherent picture of the tactics of the Battle of St James's Day, 25 July 1666. Quite early, the van and centre divisions of the English van (White) squadron began to gain the advantage in this exchange of fire over the Dutch van, and this advantage was never lost. The English kept on pressing the Dutch and by evening had won a complete victory, the two fleets being roughly equal at the start. Admiral Sir Thomas Allin, commanding the English van squadron, recalled the action in classical style:

We having the Van our people were very awk [reluctant] to get into line and some never did, as Day, Sackler [captains] and some others, but those shot through several of our ships contrary to a strict order. We fell to fighting between 9 and ten. There were two [Dutch] Vice Admirals with about 5 ships more, then an Admiral, then 6 ships and a Rear Admiral, then another Admiral. Sir Tho Teddeman [Vice-Admiral of the White and van squadron] fought bravely upon his party, although the *St George* and *Anne* did him no service and the *Old James* did us a little. The *Rich[ard] and Martha* went away

Calendar of State Papers Domestic, vol CLXIV, no 72, with the same date and endorsed in a covering letter to Charles II from Rupert and Albemarle dated 23 July 1666 (letter of 1 April and another undated to Sir John Laughton in the Corbett Papers). V Vale discovered yet another copy, also dated 16 July 1666 in Sir William Coventry's papers at Longleat (*The Mariner's Mirror*, 38 (1952) no 3).

from us. The Rear-Admiral's division [of the van] did us little help.[51]

After the campaigning season was over and the main fleet laid up for the winter, James issued Sailing Instructions, dated Portsmouth, 16 November 1666.[52] They repeat, almost word for word (substituting 'Admiral' for 'General') the twenty-one Articles of the Commonwealth Instructions, but add four new articles dealing with manoeuvre, and with slipping cables. These instructions were reissued with Mathew Wren's countersignature at some point between 1667 and the spring of 1672, most likely during the preparation for the Third Dutch War early in 1672. They were accompanied by eight 'Additional Sailing Instructions', together with 'Instructions for Masters, Pilots, Ketches, Hoys and Smacks', signed James and countersigned by Wren. There is a signal for tacking in fog, five signals detailing squadrons or individual ships to chase a strange sail,

---

51 R C Anderson, editor, *Journals of Sir Thomas Allin 1660–1768*, London, 1939, vol i, pp277–8. 'Naval Operations in the Latter part of the Year 1666,' W G Perrin, editor, *Naval Miscellany*, vol III, London, 1928, contains an extremely spirited and Homeric account of the battle, pp8–13.

52 *NMM*, DAR/2; no copy of a countersignature

53 *NMM*, DAR/13 pp5–10 (letter book copies). A further reissue appears in the National Maritime Museum, DAR/9, where the original twenty-five sailing instructions and the eight additionals are consolidated in thirty-three articles. They are headed as issued by James, but probably belong to the period of the Third Dutch War.

and a general instruction that when the fleet brings to in the open sea, minor war vessels are to anchor to windward of the admiral.[53] At some stage immediately after the war a slightly less advanced issue of sailing instructions was made in which the Commonwealth issue of 1653 was reduced to fourteen articles. It refers to 'Instructions of Articles of Sailing given out by the Generals of the Fleet in '66'.

The Third Dutch War, 1672–3

Superficially the battles of the Third Dutch War, unlike the sea *Iliad* battles of the Second War, have an appearance of greater tactical coherence. This appearance, however, may be an illusion, arising from two quite different causes. First, official reports and admirals' well-kept journals not only exist but have been discovered and printed. Second, the tactical setting of all four of the battles happened to be along or between coasts and shoals, so each occurred within a navigational framework favourable for tactical analysis.

Throughout the war, England was in alliance with France and a substantial French squadron, amounting to practically a fleet, was present at each battle as part of the bargain. Tactical cohesion, as understood in Nelson's time, was precluded by the fact that the flagships outgunned the weakest ships included in the line of battle by more than three to one, and in weight of metal by a still greater proportion. On the Dutch side it was much the same, though the Dutch fleet was

The burning of HMS Royal James *at the Battle of Sole Bay, 28 May 1672, by Willem Van de Velde the Younger.* NMM

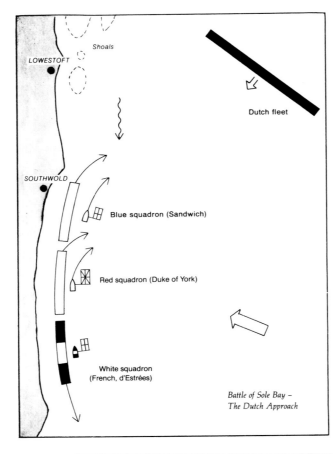

*Battle of Sole Bay –
The Dutch Approach*

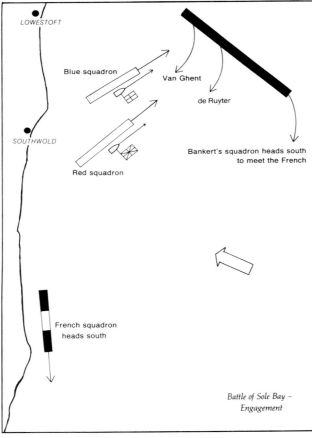

*Battle of Sole Bay –
Engagement*

outnumbered by about eighty-two ships to sixty-two. Control of tactics depended, as usual, largely on divisional leadership. Direct orders could be, and indeed were, sent by boat or by smaller warships, but these again were mainly of local application and often miscarried or otherwise failed to obtain the desired conformity. A commander-in-chief in the midst of a smoke-obscured battle could not exercise any overall tactical initiative once the fleets were fully engaged; hence the importance of the instructions issued before contact with the enemy.

James's 'Instructions for the better ordering of His Majesty's fleet in Fighting' of 1672 contained twenty-six Articles.[54] These were little more than a reissue of his earlier fighting instructions of 10 April 1665, consolidated with the additional instructions of 18 April, though nothing is said about how to deal with disabled enemy ships (Article 10 of 1665). Article 6 of the new instructions enjoins the need to close gaps caused by the falling out of disabled ships from the line, and incorporated Rupert's Article 3 (for a flag officer to shift his flag to another ship).[55] The first thirteen Articles, as in the case of the whole issue of 10 April 1665, are largely a reissue of the Commonwealth instructions of 1653.[56] They are immediately followed by the three 'Further Instructions for Fighting' (signed James and countersigned by Wren) which as noted above, were first issued by Rupert and Albemarle on 18 July 1666.

In the first battle, fought on 28 May 1672 in and off Southwold (or Sole) Bay on the coast of Suffolk, the allied fleet was anchored along the coast in the following order, reading from south to north: the White squadron (30 ships) Admiral Abraham Du Quesne, Admiral Jean d'Estrées (in command), Admiral Treillebois des Rabesnières; the Red (28 ships) Spragge, James (now restored as C-in-C), Sir John Harman; the Blue (24 ships) Sir John Kempthorne, Sandwich (in command), Sir Joseph Jordan. When the Dutch were sighted approaching from the north-east there was a very light breeze from ESE or east by south. The allied fleet was anchored not in a straight line but in a slightly concave formation following the coast, with the flagships somewhat to windward of the rest.

At the start of the battle the allies, though not entirely taken by surprise, were not yet prepared for action. Nevertheless, the Blue and the Red squadrons were soon under sail to the north, that is in reverse order with the rear of the fleet leading. D'Estrées, seeing the English getting under sail, at once sent his chief of staff to ask for the Duke's orders for the White. The French were lying a little too close to sand banks to gain the wake of the Red without falling to leeward of the Dutch. D'Estrées, therefore, felt that he could best keep the wind by heading south, that is in the opposite direction to the two English squadrons. He was not disobeying the Duke in doing so, since no tack had been given him. His ships were in some disorder and, before he was properly lined up, the Dutch van of twenty-one ships under Vice-Admiral Adriaen Bankert was down on him from windward. A stiff action between the two squadrons followed and lasted all day, carrying them well to the south of the main battle area. No blame attaches to d'Estrées for separating himself from the Red and the Blue, nor for any subsequent lack of aggression. The French fought well and des Rabesnières, commanding the rear division, was killed.

54 *NMM*, DAR/13

55 *NMM*, DAR/2; printed in *Fighting Instructions*, p130

56 Another copy in the National Maritime Museum, DAR/9, actually incorporates the three Further Instructions of 18 July 1666 at the end, as Articles 27–29 respectively.

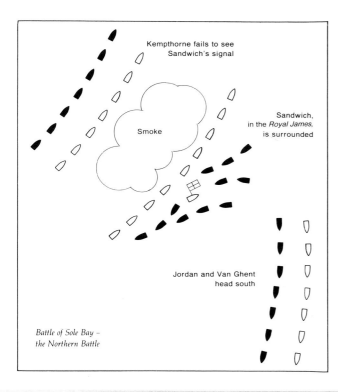

Kempthorne fails to see
Sandwich's signal

Smoke

Sandwich,
in the *Royal James*,
is surrounded

Jordan and Van Ghent
head south

*Battle of Sole Bay –
the Northern Battle*

To the north, things on the whole went ill for the English. De Ruyter brought the Dutch fleet down in line-abreast with a sprinkling of detached ships and fireships ahead. Bankert, seeing the French tacking to the south, turned his squadron into line-ahead on the larboard tack and became detached from his main body. Meanwhile Vice-Admiral Willem Joseph van Ghent with the rear came to the wind on the starboard tack to engage the English Blue squadron. Soon after, de Ruyter with the centre did the same, and engaged the Red. Owing partly to the oblique approach of the Dutch and partly to the slight concavity of the allied line, the English Blue squadron was engaged first. Partly, moreover, because of his more windward position, Lord Sandwich in the *Royal James* met the first severity of the Dutch attack. These facts largely determined the subsequent form of the battle. The Dutch pressed the *Royal James* with all the local strength they could muster. Sandwich sent a message to Jordan with the leading division to tack and weather the Dutch and so return to help him. Jordan had already tacked, his division so far being only lightly engaged; but having successfully weathered the Dutch, he stood on to the south, engaging van Ghent's squadron which was also tacking but keeping away to windward of Sandwich as he passed. Kempthorne came up astern with the Blue rear, passed to leeward of Sandwich and continued to the north, still engaged with the Dutch squadron opposite him at the start of the action. Being unable

The Battle of Sole Bay, 28 May – 7 June 1672, by Willem Van de Velde the Younger. NMM

to see Sandwich's plight because of the smoke, Kempthorne merely sent a boat 'to see what the matter was'.

Meanwhile, James in the *Prince* was hard pressed by the Dutch centre, having been caught somewhat to windward of the rest of the centre division of the Red. Because his flag captain, who was also his chief of staff, was killed, he was forced to shift his flag to the *St Michael*. He was soon heavily engaged once more. Having reached a point only just short of the shoals off Lowestoft, he tacked southward with the ships following him, and soon de Ruyter did the same. As a result both northern squadrons came onto the larboard tack heading south in divisional lines and groups. Jordan still led the English Blue. Van Ghent had been killed and his squadron opposing the Blue was now in some confusion. Astern of them came the English Red, still fighting de Ruyter. The wind meanwhile had veered to south-east, and so carried the fleets straight along the track they had pursued in the morning, the *St Michael* actually passing to leeward of the burning *Royal James*. By the late afternoon, James had been forced to shift his flag again to Spragge's *London,* the fleets continuing to head south in a fierce but tactically confused action. James kept up some kind of communication with the flag officers nearest him, but the general pattern appears still to have been divisional. By now both sides were utterly exhausted and had suffered heavy casualties, so that only a marginal change of fortune was required to bring victory or defeat. Bankert and d'Estrées were now sighted again to the south. De Ruyter judged the moment right to signal a general

withdrawal. Van Ghent was killed, two ships were lost and nearly 2000 killed or wounded. Yet, with an inferior force, he had dealt the allies a very shrewd blow. Their total casualties were also about 2000; the *Royal James* of 100 guns was burnt to the water's edge; and two flag officers, Sandwich and des Rabesnières, were killed.

Sir John Narborough recorded:

His Royal Highness went fore and aft in the ship and cheered up the men to fight, which did encourage them very much. The Duke thought himself never near enough to the enemy, for he was ever calling to the quarter-master which cunded [conned] the ship to luff her nearer, giving me commands to forbear firing till we got up close to them. Between 9 and 10 o'clock Sir John Cox was slain with a great shot, being close by the Duke on the poop. Several gentlemen and others were slain and wounded on the poop and quarterdeck on both sides of the Duke.

Presently when Sir John Cox was slain, I commanded as Captain, observing his Royal Highness's commands in working the ship, striving to get the wind of the enemy. I do absolutely believe no prince upon the whole earth can compare with his Royal Highness in gallant resolution in fighting his enemy, and with so great conduct and knowledge in navigation as never any General understood before him. He is better acquainted in these seas than many Masters which are now in the fleet; he is General, soldier, Pilot, Master, seaman; to say all, he is everything that man can be, and most pleasant when the great shot are thundering about his ears.[57]

Almost exactly the same words were spoken of de Ruyter.

The first Battle of Schooneveldt, 28 May 1673, by Willem Van de Velde the Elder (dated 1684). NMM

57 Narborough's Journal, 28 May 1672; *The Third Dutch War*, pp96–7

The Battles of the Schooneveldt and the Texel

Anglo-French strategy for the year 1673 was simple enough: to defeat the Dutch fleet and land troops behind the Dutch land forces resisting the French. This done, the allies would be able to blockade the southern Dutch ports and capture their returning East India merchant fleet. The allied command had been completely reorganised. Prince Rupert had superseded James as commander-in-chief but was to lead the fleet in the Red squadron; d'Estrées with the White was placed in the centre, apparently because his masters and pilots knew even less about Dutch coastal waters than did the English. Spragge had the Blue in the rear. The Dutch under de Ruyter had Cornelis Tromp in the van and Bankert in the rear.

The Dutch fleet lay to the north of Schooneveldt bank, in an area of open water lying north of Ostend and known as the Schooneveldt anchorage. To describe it as funnel-shaped or of any other particular shape is a little misleading, for whereas it has thin lines of sandbanks running roughly north east and south-west on its north side, its southern edges touch the main system of inshore sandbanks stretching from Flushing to Welcheren. Neither fleet possessed charts based on any kind of trigonometrical surveying. The Dutch charts were the better, but even they were totally unreliable, since they only recorded the traditional knowledge of pilots, shipmasters and fishermen. The French were inexperienced in Channel fighting, and had never fought in Dutch waters at all.

In the first battle, 28 May (OS), each side hoped to profit by the lack of sea room. De Ruyter, in what practically amounted to a defended anchorage, could use his local knowledge to entangle the allies in the shoals and at the same time offset his own disadvantage in numbers. Rupert, though lacking sea room in which to deploy his more numerous fleet to advantage, nevertheless hoped to drive the Dutch onto the Raan shoals by sending forward an advance force drawn from each of his three squadrons and supported by fireships. The wind which had been WSW in the early morning was WNW when the battle began, and was still veering. The allied fleet was in open water in a line abreast ranged roughly north and south, Red, White and Blue and with the Red and Blue somewhat advanced. The fleets were then nine miles apart, the Dutch ranged roughly parallel to the allies, with Tromp, de Ruyter and Bankert in order from north to south. No sooner had firing begun when the allied plan broke down; the advanced squadron proved a failure and the Dutch, instead of retreating towards Flushing, formed a line on the larboard tack. Tromp, to the north, beat off the advanced ships opposed to him and became engaged with Rupert's Red squadron. The French White, now in the centre, engaged a large part of Bankert's Dutch rear, which left the Blue with only the tail end of the Dutch line opposed to it. Meanwhile, de Ruyter, only weakly opposed, tried to push through the allied centre. He was held off by some of the French, who kept to windward. All this time Rupert and Tromp were fighting a detached action to the north. They eventually put about, however, and joined the general action to the south, in which both pairs of opposed squadrons and their respective divisions were becoming mixed. No ships-of-the-line were lost on either side. Fourteen fireships were expended without effect.

The beginning of the second Battle of Schooneveldt, 4 June 1673, from the Instructions for the better Ordering of the Fleet, c1688. MOD

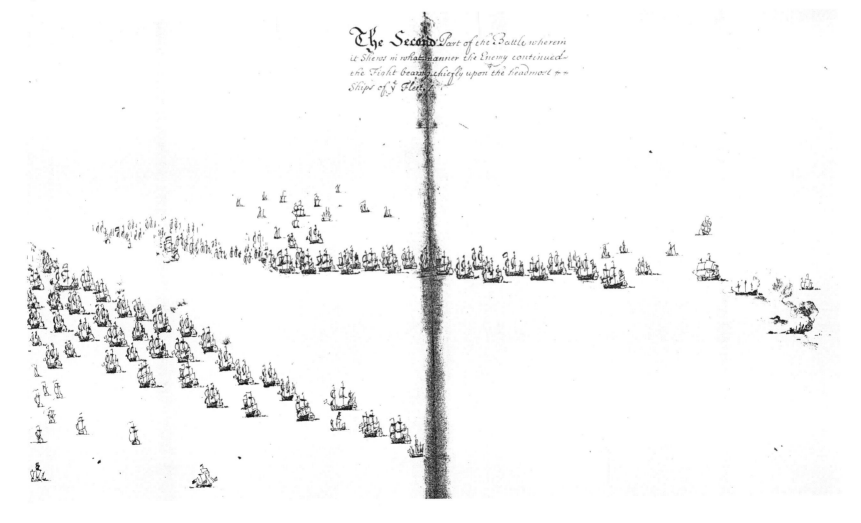

The Second Part of the Battle wherein it Shews in what manner the Enemy continued the Fight bearing chiefly upon the headmost Ships of ye Fleet.

The Second Battle of Schoonveldt was fought on 4 June (OS). The allied fleet was in reversed order, Blue, White and Red from north to south, and the Dutch in the same order as before, Tromp, de Ruyter and Bankert. This time the Dutch had the wind, which was north east. Rupert decided to push ahead but failed to inform d'Estrées, and presently the three Red divisions were ploughing through the middle of the Whites. Meanwhile, Tromp engaged Spragge, though he avoided close action; indeed, it was clearly in the interest of the Dutch not to become so heavily engaged that the allies could use their extra numbers to advantage. At the end of yet another confused and indecisive action in which no ships were lost on either side, the allies retired to the Nore to make repairs and replenish their ammunition. De Ruyter had again scored a tactical success and, in addition, had for the moment broken the allied blockade.

Spragge observed in his journal, on 1st June 1673, that had the Dutch attacked first, the allies would have been caught in disorder for lack of good reconnaissance. 'For which reason and to prevent such surprise you ought always to have your scouts out and your fleet always, if room, to anchor in the line you must sail, and always to sail in the line of battle, if by the wind.' In the case of a large fleet with enough sea room, the squadrons should sail in line abrest one astern of the other, with the divisional vice-admiral and rear-admirals on the starboard and larboard sides respectively.

In the rear or van [division of a particular squadron] I would never have above one ship in the van or the rear of either of the

Flags. My reason is, they being usually the most experienced men will better observe the times of tacking the fleet, and also their ships being commonly the best and strongest ships [of the] force will better endure being in the rear [which] is always the easiest cut off.[58]

This statement is of great interest – first, for its emphasis on reconnaissance; second, because it advocates tactical deployment when not necessarily in sight of the enemy; and third, for its reaffirmation that divisional vice-admirals and rear-admirals should be near the extremities of the squadron and not in the centre of their respective divisions. Only when fleets became sufficiently experienced did it seem wise to station the divisional flag officers in the centre of their respective divisions, as in the case of the squadron commander himself. Spragge also recorded in his journal on 21 July 1673,

The Prince [Rupert] placing himself in the van, the French in the middle, the line-of-battle, being of 89 men-of-war and small frigates, fireships and tenders, is so very long that I cannot see any sign the General Admiral makes, being quite contrary to any custom ever used at sea before, and may prove of ill consequence to us. I know not any reason he has for it except being singular and positive.[59]

This was an extremely apt criticism. Besides, if the fleet had to move in reverse order, the commander-in-chief, being then in the rear, would have less opportunity to gauge the tactical situation than if he were in the centre.

The tactics pursued at the Battle of the Texel, 11 August (OS) 1673, the last battle of the war, are even more difficult to elucidate than those of the first two. Both fleets were larger than before, ninety to sixty, but in respective gunpower much the same. Though at dawn the fleets were close in to

The second part of the second Battle of the Schooneveldt, 4 June 1673, from the Instructions for the better Ordering of the Fleet, c1688. MOD

58 Spragge's Jurnal; *The Third Dutch War*, p321
59 *The Third Dutch War*, p327

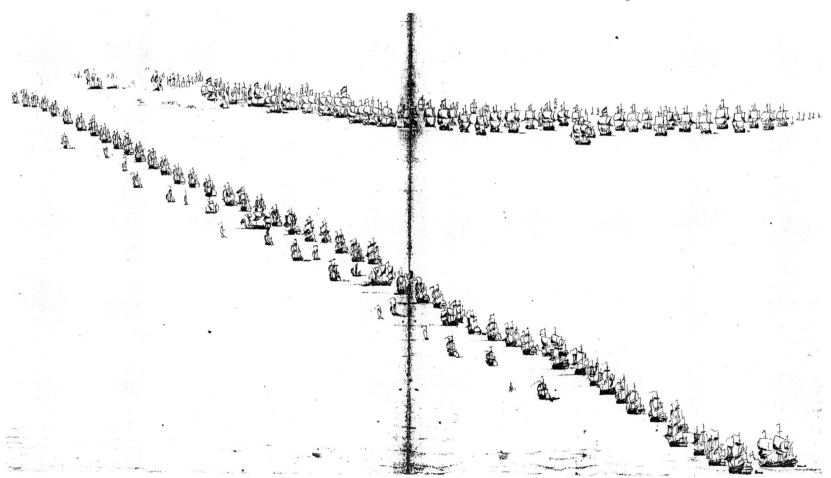

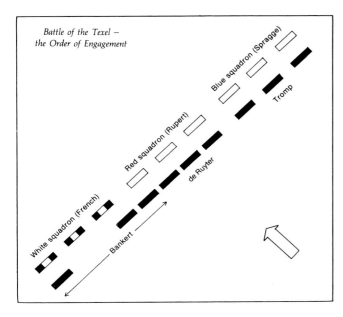

Battle of the Texel – the Order of Engagement

French with Bankert's roughly equal force (the Dutch being in reverse order), and ignoring d'Estrées's in the centre of the White. Thus Bankert's centre and rear, plus the whole of de Ruyter's centre squadron, were able to concentrate on the rear division of the White and the English Red. Roughly speaking, this enabled de Ruyter to bring five weak divisions of Dutch ships against four strong ones. In the rear, Spragge engaged Tromp squadron to squadron. D'Estrées laid great store by getting to windward of Bankert, and was at last successful when the wind gradually veered to south. The French by now were in some confusion, and Bankert was able to drop to leeward to assist de Ruyter against Rupert, these two squadrons having continued fighting together on a south west course regardless of their respective vans. There is no doubt that, with the appearance of a large part of Bankert's squadron, the English Red were hard pressed. Eventually both sides stopped firing and put about, steering north-east towards Spragge and Tromp, de Ruyter being to windward. The French followed well to windward of all the rest. The wind was now south-west, and Spragge and Tromp were still hard at it, disregarding the remainder of their fleets; indeed Spragge hove to quite early on, partly to challenge Tromp's squadron. By the time the remainder of the two fleets reached sufficiently far north to join in the Spragge–Tromp pell-mell, each admiral had had to shift his flag, and Spragge had been killed in the process. Rupert signalled for the whole fleet to form a line on the flagship, but obtained little response. D'Estrées did not obey and possibly misunderstood the signal. Rupert wrote, 'The enemy, when dark came, stood off to their own coast which I had reason to be glad of.' The Dutch had undoubtedly won the battle. They had forced the allies to retire to their base, and so completely broke the blockade of their own ports.

the Dutch coast immediately opposite den Helder and the Texel Channel, the battle was fought in open water. The Dutch squadron commands were as before. In the allied fleet, d'Estrées's White squadron took the van and Rupert with the Red returned to his orthodox station in the centre of the fleet, possibly as a result of Spragge's protest.

When the battle began, the wind was veering from south-east and the fleets were heading south-west by south, with the allied line in good order, the White leading. The Dutch at once showed great skill by engaging the van division of the

The last stage of the second Battle of the Schooneveldt, 4 June 1673, from the Instructions for the better Ordering of the Fleet, c1688. MOD

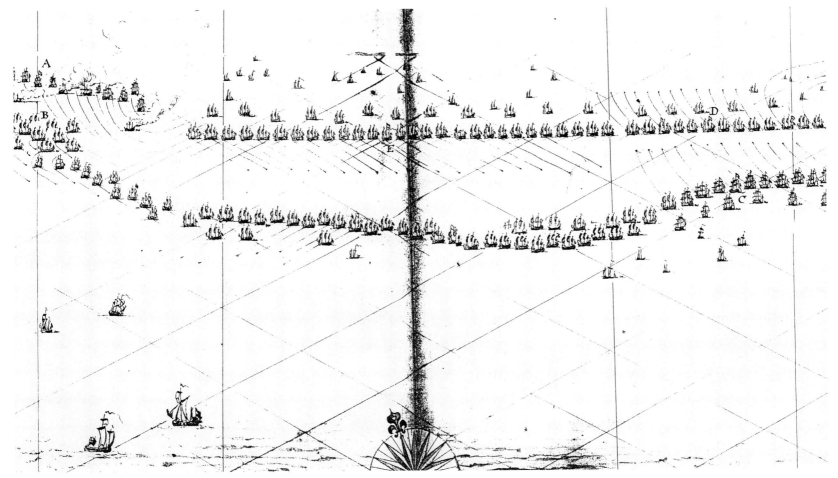

The first book of instructions

The name of James Duke of York, acting in his capacity as Lord High Admiral, appears as the issuing authority of what

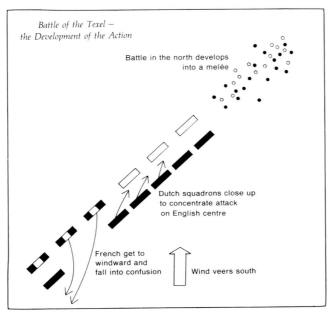

*Battle of the Texel –
the Development of the Action*

Battle in the north develops
into a melée

Dutch squadrons close up
to concentrate attack
on English centre

French get to
windward and
fall into confusion

Wind veers south

*The Battle of the Texel, 11–21
August 1673, by Willem Van de
Velde the Elder. NMM*

must be regarded as the first great tactical book of the Royal Navy.[60] For the first time, the various types of signals and instructions used by the fleet, by day and night, at anchor and at sea, were brought together in consolidated form into a single printed volume. Whose idea this was we do not know, nor do we know its date of publication. James went ashore in September 1672 and never took command of the fleet again. Rupert seems to have been designated his successor even before the Test Act received the Royal Assent on 29 March 1673, though James did not actually resign the office of lord high admiral until 15 June. On 9 July 1673 the office of the admiral was put into the hands of commissioners, with Rupert at their head. Allowing for the time-lag in printing, the book probably appeared somewhere between the early summer of 1672 and the late autumn of 1673.

In detail, it contained nothing very new, and most of the instructions were simple repetitions. It is the form of the book which gives it importance, and the fact that it eventually became the tactical textbook of the Royal Navy in a slightly altered form. It marks the end of instructions written out by hand on odd bits of paper and copied into order books without proper headings or references. When the glorious italic flourishes of the original printing gave way to

60 *NMM*, SIG/A/1 (formerly Sp/78) and *MOD*, NM/104 (formerly *RUSI* 104)

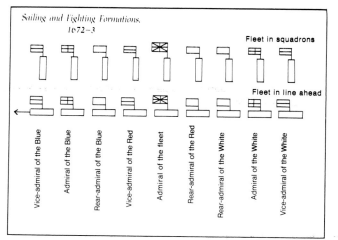

Sailing and Fighting Formations, 1672–3

Fleet in squadrons

Fleet in line ahead

Vice-admiral of the Blue • Admiral of the Blue • Rear-admiral of the Blue • Vice-admiral of the Red • Admiral of the fleet • Rear-admiral of the Red • Rear-admiral of the White • Admiral of the White • Vice-admiral of the White

the more sober roman of the re-issues, the pagination and exact arrangement of words on the pages were carefully preserved. Reprints certainly continued until 1688 and there is some evidence that it was still used officially as late as 1693, that is after Russell's instructions of 1691.

The book begins with twenty-six sailing instructions, which include night and fog signals as well as signals for weighing and anchoring. Then come four instructions for the fireships, signals for the admiral to speak with all of them or with those attached to particular divisions. Fundamentally new is a double-page picture of ships seen from the stern, with spaces for the insertion of names. The format was intended to help scarcely literate seamen recognise where their vessels came in the line of battle. The divisions of the fleet placed the sub-divisions of the vice-admirals of the Blue and White at the respective ends of the line, Blue in the van. The rear-admirals of the Blue and White were placed next to the Red squadron in the centre. This gave the impression that the sailing order intended was always for squadrons abreast, either divisions in single line ahead with the Blue Squadron on the left and White on the right, or with each division in a double column with flagships leading. However, it is unlikely that these rather sophisticated formations, foreshadowing the sailing orders of the late eighteenth century, were used as early as this. Nor, if it was actually in use, is it likely that the order of sailing could have been limited to these formations. The author of the instructions was probably more interested in prescribing the standard order of the ships within each division than in the 'order' in which the fleet was to sail (in the sense of the term in which Tourville and Père Hoste were to use it later). Following four instructions for 'Masters, Pilots, Ketches, Hoys and Smacks attending the Fleet' as auxiliaries, were the twenty-four fighting instructions. The book concludes with four articles of encouragement for the anti-fireship service, which are the same as those of 20 July 1665.

The compiler of these fighting instructions evidently tried to make a selection of the best articles from the Commonwealth onwards. Although the book was issued in the name of the Duke of York, there is no reason to assume that he was personally responsible for the selection, although he must have approved it.[61] The value of the book as a whole was greatly enhanced by separate indices of contents. Each list of contents was by pages, while the signals were indexed under numbers of guns, lights, sails, flags, standards and jacks, with crossheadings: from the admiral, from the chief of any squadron, from other ships. Having brought so many different items together into one volume, the editors evidently felt as much help should be given the users as possible. In fact the whole compilation gives the air of a

standardised codification intended to last for some years. A summary of the contents is given below:

James Duke of York's Instructions of 1672–3

Articles 1 and 2 are those of 1653.

Article 3   is the same as Rupert and Albemarle's no 2 in the instructions of July 1666, 'to divide the enemy's fleet from leeward'.

Article 4   is the same as James's no 3 of 10 April 1665; if to leeward, 'commanders shall endeavour to put themselves in the line close upon a wind'.

Article 5   is no 8 of 1653, to tack or otherwise gain the weather gage.

Article 6   is no 7 of 1653, to provide a signal for use when either to windward or to leeward of the enemy, for the fleet to keep in line with the admiral. An instruction was also included that, if the enemy fled, all the frigates were to try to make captures. The heavy sailers were to be kept in the rear so that they could supply prize crews.

Article 7   is the same as Rupert and Albemarle's first 'Further Instructions' of 18 July 1666, 'to keep the enemy to leeward' when approaching from windward on the opposite tack. The instruction is drafted much more clearly than it was when first issued by Rupert and Albemarle. The only substantive difference is a provision for getting the fleet properly level with the enemy before tacking into battle.

Article 8   is the same as James's no 2 of 10 April 1665. If to windward the leading squadron was 'to steer for the headmast of the enemy's ships'.

Article 9   is the same as James's no 15 of 22 November 1664, for forming line ahead of the starboard or larboard tack.

Article 10   is the same as James's no 9 and 10 of 10 April 1665, signals for the van or the rear to tack first.

Article 11   is the same as Commonwealth no 9, for flagships to come into the admiral's grain or wake.

Article 12   is the same as James's no 2 signal of 27 April 1665, for the other squadrons to make more sail.

Article 13   is the same as James's no 3 of 22 November 1664, to engage in the given order of battle.

Article 14   is the same as James's no 4 of 10 April 1665, ships to fight at half a cable distance.

Article 15   is the same as James's no 16 of 22 November 1664, no firing until near enough to do good execution.

Article 16   is the same as James's no 1 of 10 April 1665, ships to keep in one line and 'preserve the order of battle' as prescribed.

Article 17   is the same as James's no 6, no pursuit of small numbers of the enemy until the main body be disabled or on the run.

61 *BM*, Harleian MS 1247 f. 53 (transcript by Sir Julian Corbett); also in the National Maritime Museum, DAR/15; printed in *Fighting Instructions*, pp139–40

Article 18 is the same as no 11 of 1653, no firing at enemy ships 'laid aboard by any of our own ships'.

Article 19: 'The several commanders in the fleet are to take special care, upon pain of death, that they fire not over any of their own ships.'

Article 20 is the same as no 12 of 1653, for small ships to act against enemy fireships.

Article 21 is the same as no 13, for fireships to keep the wind and signal for small frigates to come under the admiral's stern.

Article 22 is the same as James's no 8 of 10 April 1665, disabled ships to be relieved by the sternmost, the rest to press on. Further paragraphs are the same as nos 4, 5 and 6 of 1653 for relieving ships in real danger, etc, and Rupert's no 3 'additional' of 1666, and also included is a further paragraph summarising James's no 6 of 1672, for closing the line if a flagship has to leave it and for allowing an admiral to shift his flag as convenient.

Article 23 is the same as James's no 7 of 10 April 1665, no chasing beyond sight of the admiral.

Article 24 is the same as no 14 of 1653, anchoring at night and the signal to retreat without anchoring.

This book held the field, so far as tactical doctrine was concerned, for less than twenty years. As regards lay-out, however, the book became the model for all subsequent publications of its type until the innovations of Lord Howe, more than a hundred years later. Clumsy and illogical as its arrangements in sections may have seemed even when it was first issued, it inevitably looked clumsier and still less logical as time went on. Yet such was the sense of conservatism in the sea service that it was not until Howe broke completely with the existing system by giving precedence to signals, and making the various instructions no more than 'explanatory of, and relative to' his new signal book, that the Duke of York's system came to an end. Meanwhile, the original book was twice reprinted but without any of the elaborate section inscriptions such as 'James, Duke of York and Albany,' these spaces being instead left blank. Otherwise, with certain

The Duke of York's printed instructions, with an index of signals and double page of ships allowing individual stations to be written in under each picture.
NMM

When ỹ ADMIRAL would haue all ỹ Ships of ỹ FLEET to fall into ỹ Order of Sayling hereunder prescribed, ỹ RED FLAGG shall bee put on ỹ Missen=Peek of ỹ Admiral's Ship; at sight of which, the Vice Admiral & Rere Admiral his own Squadron, as likewise ỹ Admirals, Vice-Admirals & Rere-Admirals of ỹ other Squadrons are to Answer it, by putting out their own-coloure FLAGGS in ỹ same place, and to Sayl in Order, with their whole Divisions.

The BLEW SQUADRON.   The RED SQUADRON.   The WHITE SQUADRON.

minute but significant exceptions, the reprints were exactly the same, word for word on each line of each page, though by now the glorious renaissance flourishes of the original italics had given way to more sober roman type.[62]

Narborough's instructions

In 1678 Sir John Narborough issued fighting instructions to his relatively small force at Zante (Cephalonia). He had thirty-two weakly-gunned warships, and was expecting to meet a greatly superior French force. His flag was in a ship of only 52 guns, which was nevertheless the strongest in his force, and this no doubt, apart from other reasons, led him to place himself in the centre of his line of battle. Only five other ships mounted more than 40 guns and these he stationed two at each end of his line, with the fifth in the line on the larboard wing when the fleet was in line abreast. The rest, mostly averaging 10 to 30 guns, were stationed in no particular order of gunpower. Article 2 apparently envisaged a line-abreast formation, and would seem to be the precursor of Tourville's 'order of retreat', with the two wings of the fleet thrown forward, the centre ship being at the angular point

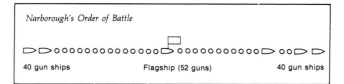

Narborough's Order of Battle

40 gun ships — Flagship (52 guns) — 40 gun ships

and nearest the pursuing enemy. If engaged with a superior force 'and we see it most convenient to fight before the wind, and the enemy follow us, I would have every commander place his ships in this order of sailing prescribed as followeth, and so continue sailing and fighting, doing his utmost to annoy the enemy, so long as shall be required for the defence

---

62 *NMM*, SIG/A/5 and 6 (formerly Sp/1, 3); no 5 is marked in pencil '1689' and this may well be the correct date. The form 'His Majesty's' is used in the main part of the text but in the Encouragement 'Their Majesties' is substituted, confirming that it was printed after James's abdication. No 6 is marked in ink '1693', suggesting that it may still have been in use then. R C Anderson connects no 5 with a Navy Board letter ordering 100 copies of the Fighting Instructions to be sent to Lord Torrington at Spithead, dated 12 June 1689 (*The Mariner's Mirror*, 6 (1920), pp130–5).

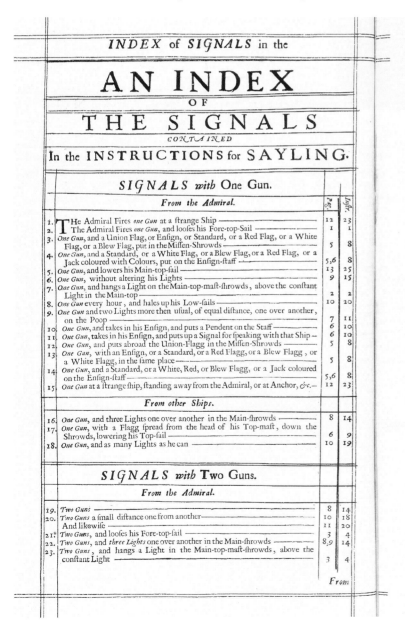

of himself and [the] whole fleet'. In view of the semi-permanency accorded to the Duke of York's instructions, which kept them in use until 1688, Narborough's must be regarded as the earliest known additional fighting instructions. If so, they set the pattern for the future issue of such instructions, valid only for a particular fleet commanded by a particular admiral during the time of his command.[63]

Dartmouth's instructions

In October 1688, James II gave command of a fairly substantial fleet to George Legge, Lord Dartmouth, to prevent a landing by William of Orange. Dartmouth, who might be described as belonging to the third generation of Restoration admirals, had been present at all the battles of the Third Dutch War. On this occasion he issued sailing and fighting instructions which already show a new departure in tactical thought, even before the Revolution had begun.[64]

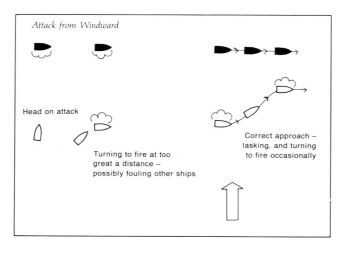

Attack from Windward

Head on attack

*Turning to fire at too
great a distance –
possibly fouling other ships*

*Correct approach –
lasking, and turning
to fire occasionally*

There is little change in the substance of the fighting instructions, but there are important differences in wording. Article 3 seems to suggest that in fighting to leeward, commanders (he does not specify whether this refers to any captain or any divisional flag officer) might divide the enemy's fleet, should they find themselves to windward of a part of it. Article 5, which is supplied with a signal for re-forming the line when to windward of the enemy, indicated that 'each ship or division are not unreasonably to strive for their proper places in the first line of battle given, but they are to form a line, the best that may be with the admiral, and with all the expedition that can be, not regarding what place they fall into or between'. Sir Julian Corbett regarded this as a reaction against formalism, but perhaps it is no more than a repetition of Sandwich's instructions of 29 August 1665, implying that somehow a line of battle must be formed and that difficulty in reaching the precise station must not be an excuse for failing to get into line at all.[65]

Article 6 is an attempt to improve the attack from windward after having approached the enemy on an opposite course and then having tacked when level with them.

> They are not to bear down all at once, but to observe the working of the admiral and to bring to as often as he thinks fit, the better to bring his fleet to fight in good order; and at last only to lask away [sailing with a quartering wind] when they come near within shot towards the enemy as much as may be, and not bringing their heads to bear against the enemy's broadsides.

This comparatively sophisticated direction was intended to deal with what had always been, and would become still more so, the bugbear of all admirals attacking from windward. How was it possible to bring a fleet into action simultaneously and in such a way as to avoid ships being crippled while still at too great a distance, thus holding up the whole advance? How, again, was it possible to prevent ships, when coming under fire as they approached, from turning parallel to the enemy to return fire at too great a distance, again holding up the advance and endangering ships on each side of them? These twin difficulties were never satisfactorily resolved and it is clear that with the unwieldy fleets of the later Stuarts, Dartmouth's most praiseworthy notions could only be properly interpreted by squadron commanders and division flag officers on their own initiative. This, however, was not allowed to occur to the prejudice of the battle as a whole, in the sense of allowing it to degenerate into a disconnected series of fights between separate detached squadrons before any general engagement had taken place, as Article 12 makes clear: when the admiral engages or signals to engage,

> each division shall take the best advantage they can to engage the enemy, according to such order of battle as shall be given them, and no ship or division whatsoever is upon any pretence to lie by to fight or engage the enemy whereby to endanger parting the main body of the fleet till such time as the whole line be brought to fight by this signal.[66]

63 'Instructions for all Commanders to place their ships for their better fighting and securing the whole fleet if a powerful enemy acts upon us,' *Fighting Instructions*, pp155–7

64 *BM*, Sloane MS 3560, printed in *Fighting Instructions*, pp168–72

65 Pencil note in Corbett's file copy of *Fighting Instructions*, p171

66 Corbett connected this instruction with Spragge's action at the Texel, but is it not equally a commentary on nearly all the battles of all three Dutch Wars? Here Dartmouth directs that the fleet as a whole must engage before squadron or division fights begin. Such fights may occur, and indeed should occur, in a real pell-mell, but the fleet as a whole must engage first.

The Naval Library manuscript

The Naval Library possesses a richly bound manuscript volume purporting to be the Duke of York's sailing and fighting instructions.[67] It is not, however, a copy of the printed book of 1672–3, nor, from internal evidence, can it be dated before the Glorious Revolution of 1688. Dating is difficult because the terms 'His Majesty's' and 'Their Majesty's' fleet are used indiscriminately throughout, and the word 'General' is used interchangeably with 'Admiral'. Its chief interest lies in the observations attached to the fighting instructions and in the watercolour sketches of the battles of the Second and Third Dutch Wars. It contains thirty-three sailing instructions made up of the original twenty-five, consolidated with several insertions and rearrangements, which, with the observations, clearly are of a later date. The observations and sketches are important for their rarity, rather than for what they depict or the observations on them. Indeed, apart from the second-hand tittle-tattle quoted by Pepys, we have practically no contemporary evidence about tactical thought other than that included in this official set of instructions.[68]

The twenty night-sailing signals and three fog signals in this work are separated from the day signals, following the method adopted by Russell's book of 1691 and also undated but official sailing and fighting instructions of a slightly later date. Separate contents lists for the day, night and fog signals were provided, as well as an index of the signals themselves, divided into those for guns, lights and flags, and an abstract of the signals used both in the sailing and fighting instructions. This abstract is arranged in three columns, with painted flags on the left, the place at which each was to be flown in the middle, and the signification on the right. In effect, this list amounts to the first known signal book.[69] The fighting instructions, the encouragement and the fireship signals follow the sailing instructions. These are also provided with a separate contents list for each section, and with indexes of the signals.

The commentator's only real contribution to tactical thought was his advocacy of pressing the enemy's headmost ships, later modified to mean windmost ships. The observation made to Article 7, Rupert and Albemarle's first further instruction of 18 July 1666 for engaging the enemy from windward, was that that part of the enemy's line should be attacked which was most to windward, as this would be the most exposed:

The battle fought in 1666, [Four Day's Battle] the headmost or winderly ships were beaten in three hours and put to run before half the rest of the fleet were engaged. We suffered the like on the 4th of June [Second Schooneveldt] for Tromp and de Ruyter never bore down to engage the whole of the fleet, but pressed the leading shops where Spragge and his squadron were like to have been ruined.

Article 8 reads 'If the enemy stay to fight, His Majesty's fleet having the wind, the headmost squadron of His Majesty's fleet shall steer for the headmost of the enemy's ships for this passage'. The manuscript book substitutes ' . . . the windmost ships of the enemy's fleet and endeavour all that can be to force them to leeward', and the commentator observes: 'It may happen that the headmost of their fleet may be the most leewardly, then in such a case you are to follow this instruction'.

The author evidently believed that the enemy's van should be attacked, and from windward. He assumed that, whereas the attackers would always be able to bring down fresh ships to support the leaders, the fleet to leeward would find its headmost and weathermost ships isolated and would not be able to support them with the same ease. This may seem a naive view, later disproven by the tacticians of the American

War, but changes in naval architecture in the intervening century no doubt had a bearing on the capacity of ships to work to windward while fighting their guns. The tactical thinking of the admiralty manuscript does at least represent an advance on the simple alternative of passing, or charging, through gaps between enemy squadrons and back again, or merely engaging the enemy on a parallel course. No doubt, like other tacticians of the time, the commentator was worried by the difficulty of reconciling central and squadron control, as well as by the difficulty of exercising any control at all in the smoke of battle.

Taking the observations and sketches all together, we can see that the author was opposed to any effort to make weathering and doubling the enemy's van, generally from leeward, a frequent and more effective manoeuvre. Article 3, which repeats Rupert and Albemarle's 1666 instruction for dividing the enemy's fleet from leeward, has the observation that this causes disorder and is best omitted unless pressed on a lee shore; ships on either side of the enemy, it was felt, would fire into each other, and furthermore the enemy had only to tack with an equal number and 'then is your fleet divided and not the enemy's'.

Article 5 indicated the signal for gaining the wind of the enemy 'by tacking or otherwise' (1653). The manuscript book added that the ships which gain the wind 'must observe what other signals the general makes', and if they cannot see him must 'press the headmost ships of the enemy all they can, or assist any of ours that are annoyed by them'. In an observation it is noted that,

This signal was wanting in the battle of 11th August 1673 (Texel). The Fourth Squadron followed this instruction and got the wind of the enemy about four in the afternoon, and kept the wind for want of another signal to bear down on the enemy, as Monsieur d'Estrées alleged at the council of war next day. For want of this the enemy left only five or six ships to attend their [the French] motion, and pressed the other squadrons of ours to such a degree they were forced to give way.

Article 6 gave the signal to bear up into the wake or grain of the admiral if he had the wind of the enemy. An additional signal was given for use when any of HM ships had gained wind of the enemy and 'the general or admiral would have them bear down and come to a close fight', thus doubling the enemy. It is noted that 'in the Battle of Sole Bay, Sir Joseph Jordan got the wind and kept it for want of a signal or fireships', the latter presumably to carry a message. Should the admiral be to leeward of the enemy, a signal was provided for ships to leeward of him to come up on a line with him. The instruction adds that if the enemy ran, frigates were to pursue and board them while 'heavy sailers . . . are to keep in a body in the rear of the fleet', and collect and man the prizes.

The 28th May, '73, [it was observed], the [first] Battle fought in the Schooneveld, the rear-admiral of their fleet commanded by Bankart upon a signal from de Ruyter gave way for some time, and being immediately followed by Spragge and his division, it proved only a design to draw us to leeward, and that de Ruyter might have the advantage of weathering us. So that for any small number giving way it is not safe for the like number to go after them, but to press the others which still maintain the fight according to the article following.

67 *MOD*, Ec/48 and *Fighting Instructions*, pp139–40, 152–63. It bears the signature of 'James Luttrell 1784.' This was Captain James Luttrell, Surveyor General of Ordnance.

68 Discussed at length in *Fighting Instructions*, pp120–2

69 See W G Perrin, *British Flags, Their Early History, and their Development at Sea; with an Account of the Origin of the Flag as a National Device*, Cambridge, 1922, p162.

'Instructions for the better Ordering his Majesties Fleet', no date [c1688]. MOD

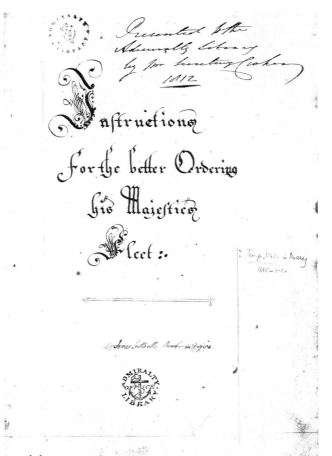

Instruct. 10.

Obser.

The 28 May 73. This Battle fought in this Crowns with the Rere Adm.ll of their Fleet towarded by Banks cast upon a Signall from de Ruyter gave way for some time, & being immediately followed by Spragg & his Division it proved only a designe to draw us to Leeward, y.t De Ruyter might have y.e advantage of Weathering us. So for any small number giving way it is not safe for the like number to go after them, but to preserve y.e others to still maintain the fight according to the Articles following.

If it shall please God that the Enemy shall be put to Run, all the Frigatts are to make all y.e Sail that possibly they can, after them, and to Run directly up their Broad sides, and to take y.e best Opportunity they can of Laying them on board, and some Ships which are the heavy Saylers (with some persons appointed to Command them) are to keep in a body in y.e Rere of the Fleet, that soe they may take care of the Enemyes Ships which shall have yeilded, and Look after y.e manning of the Prizes.

Diversions in case this Enemy be put to Ruin.

Instruct. 11.

Directions when y.e Head-have y.e Wind, and the Enemy stand towards them or they towards y.e Enemy.

In case his Maj.es Fleet hath y.e Wind of the Enemy, & that y.e Enemy stand towards them, & they towards y.e Enemy, then y.e Van of his Maj.es Fleet shall keep the Wind, & when they are come within a convenient distance from y.e Enemys Rere, they shall stay untill their own whole Line is come up within y.e same distance from y.e Enemys Van, & then their whole Line is to Tack (every Ship in his own place, & to bear down upon them soe near as they can (without endangering their Loss of the Wind) and to stand along with them, the same Tacks abroad still keeping y.e Enemy to Leeward, and not suffering them to Tack in their Van, And in case y.e Enemy Tack in y.e Rere first, He who is in the Rere of their Maj.—

Majesties Fleet is to Tack first with as many Ships, Divisions or Squadrons as are those of the Enemys, and if all y.e Enemyes Ships Tack, their whole Line is to follow, standing along with the same Tacks abroad as y.e Enemy hath.

Obser.

In bearing down upon an Enemy when you have the Wind or standing towards them or they towards you, if it is then in your power to fall upon any part of their Ships which to windward will be the most exposed, therefore you must use your utmost Endeavours to ruin that part.
This Battle fought in 66. the headmost or windmost Ships were beaten in 3 hours and put to ruin before halfe the rest of y.e Fleet Engaged. Wee suffered y.e like on y.e 4.th of June for Example y.e De Ruyter never bore down to Engage the body of our Fleet but yea'd the Leading Ships where Spragg and his Squadron had like to have been ruin'd.

Instruct. 12.

If the Enemy stay to Fight his Majesties Fleet having the Wind the headmost Squadron of his Majesties Fleet shall steer for the windmost Ships of the Enemyes Fleet and Endeavour all that can be to force them to Leeward.

Obser.

It may happen that the headmost of their Fleet may be most Leewardly then in such case you are to follow this Instruction, whereas before it was said to stand with the headmost Ships of y.e Enemys.

This is clearly meant as justification for discouraging disorganised pursuit.

The author seems to have been worried by the apparent need for the general to make the determining signal, especially in a case where the commander of the van division could not see his own squadron commander, let alone the commander-in-chief. The first observation, which was by way of an introduction, suggested that the general should have more battle signals to make known his 'pleasure'. It also proposed that 'a person fitly qualified command the reserve who shall by signals make known to the general in what condition or posture the other parts of the fleet are in, he having his station where the whole can best be discovered, and his signals, answering the general's may also be discerned by the rest of the fleet'. Here we have the idea of a repeating frigate, but how could such a ship signal *back* to the general without a signalling code more akin to Admiral Popham's code of the Napoleonic period? And who, moreover, would trust the complete impartiality of such an influential intelligencer? Besides, what is meant by the 'reserve'? That two-way signalling was being recognised as a need is evident from an instruction inserted after the fourth article, which indicated that a close-hauled line of battle should be formed if the enemy had the wind. The additional instruction stated that if any flagship or squadron ahead of the fleet were in a position to weather the enemy, it was to make the signal the general would make for the fleet to tack. Only if the general repeated the signal would it then become an order to the fleet.

Articles 9 and 10 were signals for the 'vice-admiral and the ships of the starboard quarter' to come to the starboard tack, or for the 'rear-admiral and the ships of the larboard quarter', and for the van or rear of the fleet to tack first. The inadequacy of this arrangement had become apparent. The author suggested that there ought to be a separate signal for the different squadrons:

It may happen that by the winds shifting there may be neither van nor rear; then in that case a signal for each squadron would be better understood, so that you may follow the 14th and 15th of the sailing instructions. For in the Battle of August '73 [Texel] the wind shifted and put the whole line out of order.

Accordingly, a new signal was added, to be used 'If the general would have those ships to windward of the enemy to bear down through their line to join the body of the fleet' (Manuscript Article 17).

The note inserted in Article 16 for preserving the order of battle, indicates that the purpose was 'having regard to press the weathermost ships [of the enemy] and relieve such [of ours] as are in distress'.

The observation beside Article 20, describing the duties of commanders of small frigates, ketches and smacks in action against enemy fireships, is that 'The reward of saving a friend [should] be equal to that of destroying an enemy'.[70]

A caution was appended to the provision in Article 22 that a flag officer might shift his flag to any ship in his division if his own were disabled:

In changing ships be as careful as you can not to give the enemy any advantage or knowledge thereof by striking the flag. In case of the death of any flag officer, the flag to be continued aloft till the fight be over, notice to be given to the next commander-in-chief [squadron commander or C-in-C of the fleet], and not to bear out of the line unless in very great danger. It has been observed what very great encouragement the bare shooting of an admiral's flag gives the enemy, but this

may be prevented by taking in all the flags before going to engage. It was the ruin of Spragge in the battle of August '73 [Texel] by taking his flag in his boat, which gave the enemy an opportunity to discover his motive, when at the same [time] he saw three flags flying on board the main topmast-head of three ships which Tromp had quitted.

The first of the accompanying sketches shows two fleets, the one to windward in line-abreast, the other to leeward close-hauled. The author observed that, in 1665 and St James's Day 1666, the English were to windward and pressed the headmost of the Dutch fleet. In 1672 the English lost Sole Bay by being to leeward. On 4 June 1673 the enemy 'came out upon us' and beat the English van 'never minding the rest of the fleet'. 'The same manner the 11th of August we lay to receive the enemy in which we had the same success as ever attended this order of battle.' The second sketch is of the start of the St James's Day battle, with the remark that the English had the wind with which to beat the Dutch. The third is a later stage in the same battle, with the English pressing towards the Dutch in disorder, and Tromp, keeping the Dutch rear well together, weathering some of the English ships and destroying the *Resolution* with a fireship. It is observed: 'This was for not pressing them successively as each ship gave way; for in case of a defeat it is not pursuing the first disabled but to press the next in condition.' The remaining three sketches are of the second battle of the Schooneveldt, 4 June 1673. In the first the Dutch are to windward in line abreast and the allies are close-hauled. The author notes that the French, being in the rear, tacked by order and stood south, and the 'Dutch sent one third of their fleet to attend them'. This appears to be a confusion with the events of Sole Bay. The next depicts a later stage in the battle, with the Dutch pressing the Blue and Red allied squadrons. The observation attached to the last sketch is that Tromp was beaten but nevertheless carried on the fight and was caught to leeward. Spragge, pressing Bankert in the rear, was nearly cut off by de Ruyter. The question is asked that if 'when we have had all the advantage imaginable of [the enemy] and yet do not them the least hurt, whether the same order might not be further improved and kept to'.

A brief reflection on all three of the Dutch wars reveals the huge difficulty of bringing any battle to a decisive conclusion. Though cutting through the enemy's line was held in theory to be dangerous, it happened often and sometimes involuntarily, through the separation of the squadrons from each other. Nor did either side necessarily gain or lose as a result, since the ships which broke through were seldom capable of making an effective concentration against the isolated part of the enemy's line. Cutting the enemy's line could only be really effective if it enabled the cutters to isolate an enemy group weaker then themselves, or at least to place these ships at a tactical disadvantage. A more promising idea was to organise the divisions and squadrons so effectively that they were always ready to cut through and double on an ill-formed line opposed to them, in so far as this action might seem promising. Alternatively, they could present a continuous line of broadside fire to the enemy, and so prevent their own line being dismembered. As long as their own line was preserved intact they were ready for anything and could exploit any situation, compared to a line with gaps in it or with ships not properly aligned with the admiral. Gradually, however, the more defensive aspects of the line theory came to be seen as the most valuable; those who had striven hardest to make the line an effective offensive formation soon came to be regarded as mere formalists, while those who put less stress on warlike order, preferring the old haphazard method of fighting in loose squadrons, came to be regarded as the truly offensive fighters.

---

70 Article 12 in 1653, and no 7 in 1650 to Penn (as numbered in the copy printed in *The First Dutch War*, vol i, p37, and *not* as numbered in *Fighting Instructions*.

HMS Royal Prince *and other vessels at the Four Days' Battle, 1–4 June 1666, by Abraham Storck. NMM*

CHAPTER TWO:

Tactics become a Science

IN THE last years of the seventeenth century, during the wars between England and France under Louis XIV which followed the Glorious Revolution of 1688 in England, naval tactics developed rapidly. Writing in 1780, Kempenfelt observed French progress:

What an extraordinary genius prevailed amongst the French in Louis XIV's reign, to push up everything to the summit of perfection. They had no sooner created a great Navy in that reign (for they never had any before), to the astonishment of all Europe, as at that time they had scarce any commerce at sea . . . than they had the quickness of discernment immediately to see (what we have never been able to see yet) the great advantage that would result in sea fights from a system of naval tactics . . . They judiciously perceived that military tactics might be adapted to naval tactics in the arrangement and evolution of fleets. They set to work about it, and completed it so well as to exhaust the subject, and so were as astonishingly rapid in perfecting the discipline and manoeuvres of a fleet as they were in forming one.[1]

On the English side of the Channel, reform was stimulated by pragmatism and battle experience. During this period

Amiral *Anne-Hilarion de Contentin, Comte de Tourville (1642–1701), by Roger-Viollet.* MM

were published the English 'permanent' Sailing and Fighting Instructions, Tourville's *Signals and Instructions,* and Père Paul Hoste's theoretical study of tactical formations. Tactical practice was developed by the restructuring of the battle line, and the first signal books appeared.

Tourville gave the greatest impetus to the development of tactics and signalling during the wars of the late seventeenth century, mainly because he held the chief command of the French fleet most of the time. Anne-Hilarion de Contentin, Comte de Tourville, was born in 1642 and first distinguished himself against the Turks in 1665. He commanded a ship of the line in d'Estrées' French squadron in the Third Dutch War both in 1672 and 1673, and thereafter commanded a squadron under Du Quesne in the Mediterranean operations against the Dutch and Spaniards.

Tourville did for the French navy, and ultimately for all navies, what *maréchal* Jean Martinet did for the army. His contribution to the science of naval tactics lay chiefly in drilling his very large fleet into a disciplined and controlled force which could deploy from a relatively sophisticated order of sailing to other sailing formations, or into the line of battle. He seems to have been the first naval commander to issue detailed sailing instructions for a large fleet. His order of sailing in six columns was not copied by Howe until ninety years later. He was an adept at fleet organisation, and his aim was to facilitate changes from one order of sailing to another or into line of battle for tactical purposes rather than to create a mere parade-like series of evolutions. His use of printed signals and printed forms for the orders of sailing shows the progress already made by the French in fleet organisation. Similar businesslike methods were not used in the British service for at least another half century. One would imagine also that any fleet commanded by Tourville would be well practised in gunnery as well as in station keeping and seamanship.

As a tactician, however, he does not seem to have striven for anything new, not even to the extent of developing the ideas already initiated by Prince Rupert and the Duke of York. While wisely rejecting the idea that the commander-in-chief should be expected to exercise control over the whole battle, he made no suggestions about squadron initiative. His signal books were in advance of the times in the sense that he exploited the traditional system of 'place were flown' and 'signification' to an unprecedented extent. Apart from all this he was the first of the great French admirals to command a genuinely large fleet with outstanding success in a war against the combined fleets of England and Holland.

Tourville's signed and printed 'Signaux Généraux pour les Vaisseaux de l'Armée du Roy', issued from his flagship the *Conquérant* on 12 May 1689, is a printed book of instructions, in which the signals were somewhat inconveniently

---

1 Kempenfelt to Middleton, 18 January [1780], printed in *The Letters and Papers of Charles, Lord Barham,* Sir John Knox Laughton, editor, London, 1907–19 (hereafter cited: *Barham*), vol i, pp309–13

expressed in words, embedded in the text of the instructions. However, at the end there was also a separate table of signals. There were twenty general signals by day, seven signals for sailing by day and thirteen by night, thirty-four 'pour le combat' (fighting instructions), eight for chasing, and five for fog. There were only three for calling officers, because Tourville preferred to keep these quite separate from his instructions, and no 'encouragements' for the fireships, their activities being included in the fighting instructions. These, though slightly more elaborate than the contemporary English forms, and certainly better arranged, are equally conservative in character. Seven are instructions only, requiring no signal. The main difference seems to be in the eight signals for chase, which, addressed both to private ships and to the fleet as a whole, suggested possible future developments in tactical initiative. An important manuscript addition on a loose sheet of paper, dated 4 August 1689, stated that when weighing, captains were to proceed ahead to take their station in the line, passing any ship which failed to obey the signal, even a flagship. A separately printed book of night signals, signed and dated from the *Conquérant*, 9 June 1689, required use of guns, lights and fuzées (rockets); they also included compass signals.[2]

Tourville's instructions for 1690, the year of his great victory over the Anglo-Dutch fleet at Beachy Head, have survived in the form of a pocket manuscript signal book. This is probably the oldest pocket size, thumb-indexed, naval signal book in existence. It is thumb-indexed for the positions in which the flags should be flown, not for the flags themselves. Under each position are grouped the various flags flown from it, each hand coloured, and each with its signification and a page and article reference to the instructions. What seems surprising is that in addition to signals made with a flag and a pendant, there were eighty-five signals made with two flags, the second flag hoisted in a different position to the first; no doubt this governs the arrangement of the book.

By noting the page and article reference, it is possible to go some way towards reconstructing the instructions, except for those which required no signal. There were about sixty-seven general instructions, of which five were without signals, covering getting under way, boats, sailing, calling officers to speak with the admiral, making the land, danger, mooring, watering and landing the sick. There were, apparently, two series of sailing instructions, one of about thirty-five articles and the other of about twenty-three, the latter being chiefly concerned with sailing in six columns. There were some nine separate chasing signals. References to the night and fog signals are confused, but there were clearly some thirty-six fighting instructions for which none of the signals actually given are new or specially interesting.

The doubling of the number of instructions included in the 1689 book is mainly accounted for by the increase in the general instructions and the sailing instructions. It is in this latter group that the real significance of Tourville's work is found. For whereas the fighting instructions scarcely mention the enemy, except for engaging and breaking off action, together with doubling (number 8) and boarding (number 33), the sailing instructions indicate elaborate evolutions for a large fleet; hence the instructions for the order of sailing in six as well as in three columns. Evidently Tourville was aiming at the maximum degree of tactical control over his force of more than seventy ships of the line. Wisely, perhaps, he eschewed fighting instructions, being content to ensure a variety of sailing formations and methods of changing from one to another. In this way he could dispose his fleet in the manner best suited to meet the enemy, subject to prevailing conditions of wind and weather, and also the tides, currents and coastal

'Signeaux out'on tire du canon', in Admiral Tourville's 'Signaux ordres de combat, etc' of 1690. NMM

restrictions in the Channel and its western approaches. Flexibility of manoeuvre thus implied quickness and precision in forming the line of battle. Once this was formed and the enemy engaged, little further attempt could be made at detailed tactical control. A fleet of seventy ships in line of battle ahead, even at half a cable interval between ships, would cover a distance of five miles; fighting at a cable's interval, the distance might be nearly ten miles, according to the length of the individual ships. Whatever theoretical assumptions lay behind Tourville's instructions and signals for the year 1690, they were amply justified in practice by the victory off Beachy Head.

Tourville also issued printed 'Ordres et Signaux Généraux', specifically described as being for use, in 1690.[3] These consisted of fourteen printed diagrams of orders of sailing for a fleet of eighty-four ships-of-the-line and thirty fireships, organised in three squadrons. Actually, sixty ships-of-the-line only were shown, with their names inserted in ink. The squadron colours were White, White and Blue, and Blue. In

2 *NNM*, HOL/1 and 2 (formerly S(H) 1 and 2). See also HOL/ (formerly S(H)), printed sheets of signals for calling officers, mostly undated.

3 *NMM*, HOL/8 (formerly S(H) 8)

the close-hauled single line ahead, the *Soleil Royal* led the fleet. In three examples of the fleet sailing close-hauled in three parallel columns, each in line ahead, the commanders again led their respective squadrons. Unless otherwise stated, the accustomed arrangement, as in the English fleet, was for the C-in-C's White squadron to be in the centre, the second in command's White and Blue to be to starboard of the C-in-C, and the third-in-command's Blue to be to larboard, thus preserving the traditional order of military precedence. With the fleet in six parallel columns, each squadron in two divisions and the wind astern, the three flagships are shown ahead of and equidistant from the two leading ships of their respective pairs of divisions. Other variants are a wedge formation, close-hauled line of battle ahead with the White in the centre or leading, line of battle abreast with the wind astern, and line of bearing.

Tourville's order of sailing and battle for 1690 contains an unusual item: an 'Ordre de retraitte [sic] sur deux lignes et un front'. Unlike the usual inverted wedge formation, this shows the fleet with the general's squadron in the centre in line abreast, and the other two squadrons each disposed in line ahead directly ahead of the two respective flanking ships of the centre, the formation thus approximating to three sides of a square with the front side open.

The original English 'Sailing & Fighting Instructions'

At some date as yet unknown a new book was officially issued by William and Mary's government with the unadorned title of 'Sailing & Fighting Instrctions' [sic].[4] In general form and layout it looks remarkably like a larger version of the standard folio volume of the eighteenth century, which is indeed what it is. In the main it retains the arrangements of sections first made in the Duke of York's book of 1672–3, but it has a completely new set of fighting instructions. These, moreover, are the instructions which, with only marginal changes, become the official Fighting Instructions of the eighteenth century.

The new book contained nine day, seven night and eight fog signals for weighing and anchoring, and fourteen day and nine night sailing instructions. The day sailing instructions showed a substantial advance on Commonwealth longwindedness, and their separation from the night signals reduced their number. Their form suggests an attempt to exercise more general control over the fleet when at sea, partly by signal and partly by standing orders, thus obviating the ancient, inefficient and time-wasting business of sending ordinary routine orders by boat. Three separate instructions carefully laid down the ceremonial precedence to be accorded to senior captains by juniors, who are not to pass to windward of the latter. There were eighteen separate signals for calling flag officers, captains and officers of various grades on board the flagship, and for the now familiar instructions for the masters and pilots, etc, attending the fleet. The twenty-seven fighting instructions were arranged in an entirely new way, and are the same as those issued, soon after, by Admiral Edward Russell.

Taken as a whole, these instructions tended to concentrate more power in the hands of the admiral, while giving him a wider tactical initiative. Except for Article 17, with its

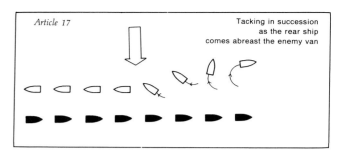

Article 17   Tacking in succession as the rear ship comes abreast the enemy van

notorious later history, the trend was away from the hypothetical cases postulated in earlier instructions and more towards *ad hoc* solutions expressed through a more precise signalling system. The effect of the anonymity of the whole book was to give it greater authority than that of the individual admirals using it. We are no longer dealing with the instructions of the Duke of York for his fleet, or with those of Rupert and Albemarle for theirs, but rather with official sailing and fighting instructions, issued anonymously and not specifically addressed even to His, Her or Their Majesties' Fleet. Even if intended at the start to be no more than a guide book for admirals, it could and did, by entirely natural processes, come to be regarded as an order book for all admirals to obey.

Article 17 of the fighting instructions is a new version of Rupert and Albemarle's further instructions no 1, combined with James's 1672 (OS) no 8, for engaging from windward when approaching on opposite tacks. Here, when the two fleets are level, 'then he that is in the rear of our fleet is to tack first, and every ship one after another, as fast as they can, throughout the line, that they may engage on the same tack with the enemy . . .' Rupert and Albemarle had directed: 'Our whole line is to stand along with them [having first tacked], and same tacks on board [as the enemy]'. James put it more clearly: 'and their [our] whole line is to tack, every ship in his own place and to bear down upon them . . .' The new instructions followed James rather than Rupert and Albemarle in directing an equal number of ships to tack first if the enemy's rear tacks. The importance of this article lies in its uniquely specific direction as to how, under certain given circumstances, the battle was to be fought. Once the instructions had become 'Admiralty', the article became mandatory for any admiral in command of a British fleet in any part of the world. It was in the light of this interpretation of the fighting instructions that Admiral John Byng fought La Galissonière in 1756.

The second printing of the book was different from the first in that it left out the double-page ship plan for the order of sailing. Maybe it was felt that, as these orders were often changed, it was difficult to record them properly on a printed plan. Possibly there is also some connection here with the incorporation of the Dutch in the confederate fleet.

The fact that a copy of the first printing has three further fighting instructions (Articles 28–30) inserted in manuscript, and that these are the same as those added by Admiral Edward Russell to his own printed instructions of 1691, suggests a date somewhere between 1689 and 1691, most probably 1690. We know from Admiral Arthur Herbert, Earl of Torrington's defence of his conduct that at Beachy Head, 30 June 1690, he was still using the Duke of York's book of 1672–3. Russell replaced Torrington in December 1690 and his own instructions of 1691 presumably appeared in time for the summer operations of the following year. It would seem, therefore, that the anonymous Sailing & Fighting Instructions must have been drafted in the light of experience at the battles of Bantry Bay (1689) and Beachy Head (1690), sometime in the late summer or autumn of

---

4 *NMM*, SIG/A/3 (formerly L52/60); the letter U is omitted in this printing. *NMM*, SIG/A/4 (formerly Sp/2) is a later printing with the word 'Instructions' correctly spelt. R C Anderson makes an interesting comparison between SIG/A/4 and 5 (formerly Sp/1), suggesting the SIG/A/4 was drawn up by Torrington in 1690; *The Mariner's Mirror*, 6 (1920) pp130–35.

1690, and that they remained in force even after Russell's personal adoption of them for the 1691-2 campaign.

How long the book remained authoritative we do not know, but the copy of the first printing in the National Maritime Museum has manuscript insertions showing that it was issued by Admiral Sir George Rooke as late as 12 September 1695, when he was appointed to succeed Russell as commander-in-chief in the Mediterranean. These include the three further fighting instructions mentioned above. Quite probably the original book never completely lost its authority and was the actual parent rather than the ancestor of the eighteenth-century work, Russell's and Rooke's books being only personal and temporary adaptations. The 1690 instructions, printed and issued anonymously under the simple title 'Sailing & Fighting Instructions', are summarised below:

Summary of the 1690 Sailing & Fighting Instructions

Article 1 Signal for the line ahead.
Article 2 Signal for the line abreast.
Article 3 Signal for White squadron to tack and gain the wind of the enemy.
Article 4 Blue squadron ditto.
Article 5 Vice-admirals of the Red, White or Blue ditto.
Article 6 Rear-admirals ditto.
Article 7 If admiral to leeward of fleet or port of it, signal for them to come into his grain or wake.
Article 8 Same as 1653 Article 7, second part.
Article 9 Signal for the van to tack first.
Article 10 Rear ditto.
Article 11 Signal for all flagships to come into the admiral's wake or grain.
Article 12 Signal for the White or Blue to make more sail.
Article 13 Signal to engage.
Article 14 Signal for small frigates to come under admiral's stern.
Article 15 Signal to brace head sails if in line of battle by a wind: ships in rear to brace first.
Article 16 Signal to fill and stand on when with head sails to the mast in line of battle: van ships to fill first.
Article 17 A new version of Rupert and Albemarle's Further Instruction no 1, combined with James's 1672 (OS) no VIII for engaging from to windward when approaching on opposite tacks. The new instructions followed James rather than Rupert and Albemarle in directing an equal number of ships to tack first if the enemy's rear tacked.
Article 18 Restated the simpler position, 'If the admiral and his fleet have the wind of the enemy and they have stretched themselves in a line of battle, the van of the admiral's fleet is to steer with the van of the enemy's and there to engage them.' This replaces Article 8 of James's instructions, which reads: 'If the enemy stay to fight (His Majesty's fleet having the wind) the headmost squadron of His Majesty's fleet shall steer for the headmost of the enemy's ships.'
Article 19 No firing until at point-blank range.
Article 20 No pursuit of small numbers, as in James's additional no 6 of 1665.
Article 21 Signal to be made by ship in distress: Next ship to relieve.
Article 22 Signal that the admiral or another flagship is in distress: fleet to cover him.
Article 23 Next ships to close gap left by ship forced out of the line. This and no 21 have the same sense as James's 1672 (OS) no 12 paras v and vi.

Admiral Sir George Rooke (1650–1709), by Michael Dahl. NMM

Article 24 If a flagship was disabled, the flag might go on board any ship in his own squadron.
Article 25 Signal for the whole fleet to pursue beaten enemy.
Article 26 Signal for flagship, squadron or division to chase, the same as for tacking to weather the enemy.
Article 27 Signal to stop chasing.

Thomas Herbert (1656–1733), eighth Earl of Pembroke and Lord High Admiral, by William Wissing. NMM

Vice-Amiral *François Louis de Rousselet, Marquis de Chateau-Renault (1637–1716), contemporary engraving.* MM

The Battle of Bantry Bay, 1 May 1689; French and English ships in action during the War of the English Succession, by Adriaen van Diest. NMM

The Battles of Bantry Bay and Beachy Head

On 1 May 1689 the French and English fleets met in fleet action in Bantry Bay for the first time since 1545, except for the inshore operations of 1627–8, when the English were attempting to relieve the Huguenots at La Rochelle. This was the first battle of the War of the English Succession, now merged with the War of the League of Augsburg. Both fleets were supporting their respective land forces in Ireland. Torrington, with a weakly-gunned fleet of eighteen of the line, a 36 and two fireships, sighted a French fleet of twenty-four of the line, also weakly-gunned but with five frigates and ten fireships, under Vice-Admiral François Louis de Rousselet, Marquis de Château-Renault. The French were at a disadvantage, as they were still landing troops and stores when the English were sighted. Torrington was to leeward and when he saw the French at last coming down in good order to attack him, he left the confined waters of the bay so as to form his own line and to try to gain the weather gage. The ensuing battle was a victory for the French despite the fact that Château-Renault was not properly supported by his rear-admiral.[5]

5 Charles de la Roncière, *Historie de la Marine Française*, 6 vols, Paris, 1899–1932 (hereafter cited: de la Roncière), vol vi, pp47–50, where the chief French and English sources are ably summarised.

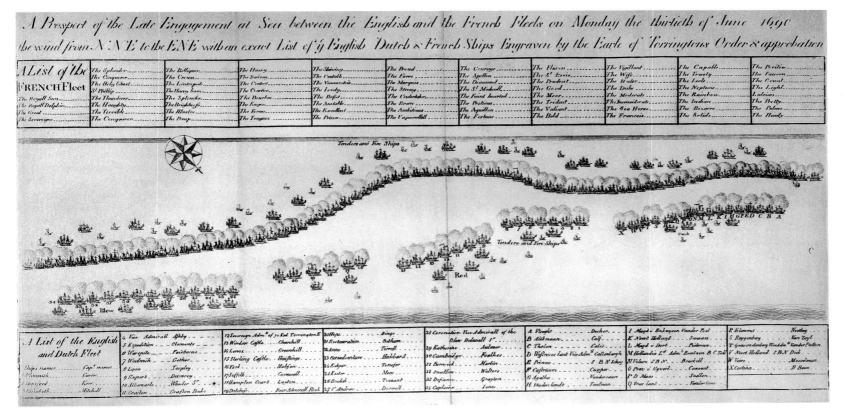

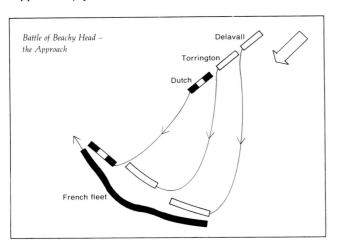

Battle of Beachy Head –
the Approach

The next main engagement was in the English Channel off Beachy Head on the morning of 30 June 1690. The wind was north-east and the confederate fleet bore down on the French, who could be seen in line of battle on the starboard tack, heading NNW towards the English coast with their head sails aback. Whether by design or not, the French line sagged to leeward in the centre, exactly where the French commander, Tourville, was stationed. He had sixty-eight ships in his line of battle but only sixteen of 70 guns or over. The Dutch, numbering twenty-two, formed the confederate van, but with only six of 70 guns or more. Torrington had the Red, centre squadron, and fifteen of his twenty-one ships carried upwards of 70 guns. Sir Ralph Delavall's Blue squadron, in the rear, consisted of only thirteen ships, of which, however, no less than nine were of 70 guns or more. In each fleet, the squadron commanders were in the centre of their respective squadrons and the division flag officers in the centre of their divisions. Each flag officer, except in the Dutch squadron, was in a major ship but not all were supported by powerful seconds.

The Dutch opened the battle at about 9am by turning into line ahead on the starboard tack to engage the French van on a parallel course. In his defence before the House of Commons, Torrington reported that 'About eight I ordered the signal for battle, to prevent the Dutch steering to the southward, as I did; for by the eighth Article of the Fighting Instructions when that signal is made, the headmost ships of our fleet are to steer away with the headmost ships of the enemy.'[6] This shows that he was still using the Duke of York's instructions of 1672 or one of the later reprints and not the new anonymous official book. Nor was he using Lord Dartmouth's instructions, in which this particular article is numbered 7.

Torrington, fearing that his rear might be overlapped and doubled, avoided steering straight for Tourville, according to traditional precepts, and edged to larboard, so as to engage the rear division of Tourville's centre squadron. This allowed the English Blue squadron to measure its length more favourably with the French rear. His second reason, closely connected with the first, was that, since the French lay somewhat off the wind, their rear ships were their weathermost. If left unengaged, they might have weathered the confederate centre as well as the rear. He also gave as a reason that, being inferior in numbers, he judged it better to leave a gap in the centre rather than be overlapped at his extremities. Vice-Admiral Sir John Ashby, commanding Torrington's van division of the centre, seeing the gap becoming too wide, began to steer to close the Dutch. Torrington, with the remainder of the Red squadron, soon followed him. Meanwhile, Delavall was engaging the centre and rear divisions of the French rear squadron on fairly even terms. When Torrington brought the centre squadron into action he found difficulty in getting close enough because of the sag to leeward in the French line.

The Battle of Beachy Head, 10 July 1690. NMM

6 *The Earl of Torrington's Speech to the House of Commons in November, 1690*, (1690) 1710, p32 [NNM, TUN/167]

Meanwhile Tourville had begun to exploit his superior numbers and tactical skill. Finding himself less heavily opposed than he expected, he managed to bring most of his own centre squadron against Ashby. He then pushed ahead to join in the fight against the Dutch, the van of the French fleet having begun to overlap, then weather, and finally double on them. The Dutch were now opposed by the whole of the French van squadron and the van and centre divisions of the French centre. Ashby's division immediately astern of the Dutch were also in a poor way, having several ships towed out of the line. Torrington meanwhile tried to push ahead to come up level with Tourville. By now he could see eleven ships of the French van to windward of the Dutch, and soon likely to weather his own squadron also. Tourville himself, with 'four or five more, [came] pretty near to two disabled ships of the Dutch, and galled them cruelly'.

The crisis of the battle had arrived just as the wind dropped. Damaged ships were towed away by boats, and undamaged ships were towed into better fighting positions. Torrington ordered the nearest damaged ships to anchor and had his own flagship towed between them and the French. The whole confederate fleet then began to anchor with all sail set, thus causing the French to drift away to the south-west on the ebb tide, which had now begun to run strongly. Tourville was not quick enough to appreciate what was happening. When at last he also anchored, the French fleet had drifted out of gunshot to a distance of about three miles.

The battle had lasted eight hours. One small Dutch ship was lost through lack of anchors, but there were no other actual losses on either side, though many Dutch and English ships were severely damaged, and eight or nine of these were subsequently sunk or destroyed.

A French engraving of the Battle of Bévéziers (Beachy Head), 10 July 1690. MM

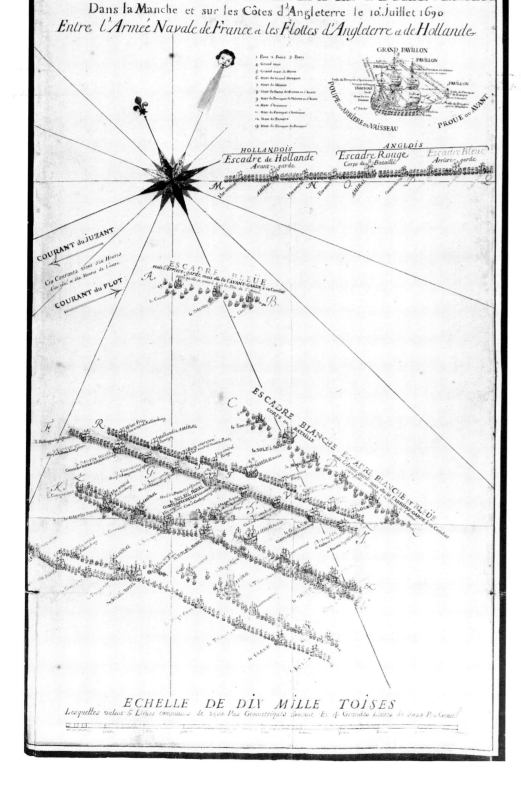

Russell's additions of 1691

On 23 December 1690, Admiral Edward Russell was named commander-in-chief of the confederate fleet. Sometime in the following year he issued a set of instructions which were an exact reprint of the second printing of the original Sailing & Fighting Instructions, except for three new and one extended fighting instructions.[7] His book is the first to bear an admiral's name and the date on the title page since that of the Duke of York, but Russell himself had no hand in compiling the book as a whole, and there is no clue to the names of those responsible for the changes. The issue of the instructions under Russell's name suggests that they were liable to change when he was superseded, but in fact the National Maritime Museum's earliest copy is endorsed in manuscript as issued by Rooke on 10 June 1702, before setting out for the Cadiz and Vigo operations. Russell by then had retired altogether both from the fleet and the Admiralty, Rooke himself having been appointed an Admiralty commissioner on 20 May.

Summary of Russell's 1691 additions

Article 17 Russell added a second paragraph which provided a signal indicating that the admiral wished the ship that led the fleet in line of battle to make some change of sail. The precise nature of the intended sail change was indicated by the admiral making that change on his own ship. The subordinate flagships were to repeat the signal.

Article 28 A disabled ship may be towed by the ship next ahead, with permission of a flag officer.

Article 29 Signal for a flagship to cut or slip anchor by day, the same as that for tacking and weathering the enemy.

7 'Instructions made by the Right Honourable Edward Russell, Admiral, in the year 1691, for the better ordering of the Fleet in sailing by day and by night and in fighting,' *NMM*, SIG/A/2 (formerly L58/120). A second printing, also dated 1691, on folio paper but not quite so large and in smaller type, was made at some unknown date. There is also a third printing, in the same type as the second but with slightly different spacing on the title page. There is variation in the references to 'His' 'Her' and 'Their' majesties, but these probably do not help to establish the date. 'His' was sufficient for William III even though Mary exercised the royal power during William's absences in Ireland and the Netherlands.

Article 30 Signals for the Red, White or Blue suqadrons to
form line of battle ahead or abreast, and ditto
for vice-admirals and rear-admirals of
squadrons.

The Battle of Barfleur

The Battle of Barfleur (19 May 1692) is the first example of a
major fleet action in which one side completely outnumbered
and outgunned the other. The confederate fleet, commanded
by Edward Russell (later Earl of Orford), mustered at least
eighty-two ships-of-the-line, and may have included as many
as ninety, together with a strong detachment of frigates and
fireships. Tourville had only forty-four of the line, including
eight mounting between 90 and 104 guns.

In both fleets the order of battle was similar, with commanders-in-chief and squadron commanders in large ships in
the centre of their respective squadrons, supported by a pair
of powerful seconds. Division flag officers were also in the
centre of their respective divisions, also supported by powerful seconds, though in the French fleet this was not possible
for lack of large ships. The French squadrons, moreover,
were so small, fourteen, fifteen and fourteen respectively, that
the division flag officers were very near the extremities of
their division lines.

Russell says that by 8am he had formed 'an indifferent line,
stretching from the SSW to the NNE the Dutch in the van'.[9]
By 9am the French fleet were seen to be formed, and at 10am
they bore down to attack. Russell sent orders to the Dutch
van to weather the French as soon as possible, and then to
tack 'and get to the westward of them', so as to cut them off
from Brest. He also ordered the Blue squadron in the rear to
close up. A sudden calm, however, stopped this manoeuvre.

Tourville, having the wind, attacked with the best skill he
could muster, instructing his van squadron to avoid close
action so as to prevent his whole fleet being doubled and
completely enveloped. He then brought his centre to close
action with the confederate centre, while spacing out his rear
squadron with wide intervals between ships so as to hold off
the confederate rear. The rearmost French ships kept out of
gunshot, presumably on orders, so as to avoid being doubled.

The two centres were heavily engaged for about three
hours (times throughout being extremely vague). Tourville
in the *Soleil Royal* was towed out of action to windward
'much gauled'. At about 2pm a breeze sprang up roughly
north-west by west and, soon after, five ships from the
French van ranged themselves between Tourville and the
English led by Russell himself and his two seconds. After
heavy fighting for about an hour, a thick fog stopped the
battle at 4pm. Meanwhile, Sir Clowdisley Shovell, commanding the van division of the centre (Red) squadron, had
managed to weather Tourville's centre division of the French
centre and double on him. When next seen, Tourville was
being towed to the north to avoid being completely
enveloped. Russell ordered his own division to head north
under tow, and at about 5.30pm, when a light easterly breeze
gave him the weather gauge, he 'made the signal for the fleet
to chase, sending notice to all the ships about me that the
enemy was running'. Shovell's division could be heard firing
to westward, though still hidden by fog. Tourville, seeing
that the tide was running strongly to the north-east, anchored
together with his nearest supporters. When the wind dropped
again, Shovell and the ships with him also anchored, but not
until they had been carried some distance with the tide, just as
had happened to Tourville at Beachy Head.

9 Russell's official letter to Lord Nottingham, *Harleian Miscellany*, vol XII
p43

Admiral Edward Russell (1653–1727), first Earl of Orford, by Thomas Gibson. NMM

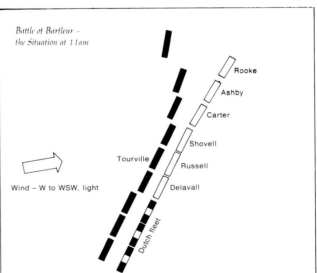

Battle of Barfleur – the Situation at 11am

Rooke
Ashby
Carter
Shovell
Tourville
Russell
Delavall
Dutch fleet

Wind – W to WSW, light

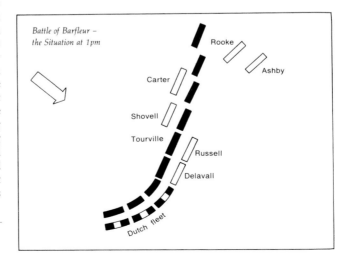

Battle of Barfleur – the Situation at 1pm

Rooke
Ashby
Carter
Shovell
Tourville
Russell
Delavall
Dutch fleet

Fog and smoke made it impossible for either commander-in-chief to direct his fleet. Russell ordered all ships near him to pursue the French to Brest. There was a confused action later in the evening between Rear-Admiral Carter's rear division of the rear (Blue) squadron and some of the retreating French; Carter himself was killed.[10]

Next day the chase continued, but from a tactical point of view the final defeat of the French was no more than a series of widespread piecemeal encounters, with the confederates always attacking and chasing in superior numbers. Nor does the subsequent destruction of part of the French fleet in the Bay of La Hogue offer tactical insight of any significance.

Tourville's signal books, 1691–3

Tourville faced no large-scale tactical encounters in 1691, the year of his famous *campagne du large*, a masterly series of strategic movements at the western entrance to the Channel designed to disrupt Anglo-Dutch trade. His 'Signaux pour le Combat' for 1691, with unnumbered pages, was printed

---

10 The contemporary chart showing 'Les quatre differentes attitudes' of the battle is quite unrealistic as it represents the enveloping of the French fleet in a series of orderly movements; see de la Roncière, vol vi, p112.
11 *NMM*, HOL/3 (formerly S(H)3), dated '1691' on the vellum cover in a contemporary hand
12 *NMM*, HOL/6 (formerly S(H)6)

separately from his general and sailing instructions.[11] This collection of signals contains thirty-six articles and covers much the same ground as those of 1689 and 1690, though the articles are differently numbered and arranged. The only new signals of interest are one for the general in the van wishing to resume his station in the centre (Article 3), and for the van only, or the rear only, to tack in succession (Articles 13 and 14).

The National Maritime Museum has no material for 1692, the year of Tourville's disastrous defeat at Barfleur and later at La Hogue. This may well be due to the destruction of the *Soleil Royal* and also of the *Ambitieux*, to which Tourville shifted his flag. For the year 1693, however, there is plenty of material. Indeed, nothing impresses one more about the resilience of the French navy at this time than the fact that exactly a year after one of the most crushing series of defeats ever inflicted in naval history, Tourville was at sea again with fifty ships-of-the-line.

The year following Barfleur, Tourville issued another signal book containing 229 printed items,[12] by far the most elaborate body of instructions with signals to have appeared up to that time. Unlike all previous and subsequent codes, at least until the end of the Seven Years War, it aimed at covering all the admiral's needs in comprehensive fashion. Even so, administrative signals were omitted, except in the first section of fifty general day signals for use at anchor or under sail. The fourteen signals (of which some were instruc-

The Battle of Barfleur, 19 May 1692, by Richard Paton. NMM

tions only), entitled 'Pour les Ordres de March[e] au plus près du vent sur une ligne,' covered tacking, wearing and change of distance from the wind. Normally, the general was to lead the fleet, his own squadron, the White, occupying its normal position in the centre. The twenty-seven signals for sailing close-hauled, of which again some were instructions only, allowed highly sophisticated manipulation of the fleet by changing the relative position of the squadrons and forming the line of battle in different orders of columns. Great emphasis was laid on correct station-keeping. Four signals covered the order of sailing with the wind large. The twenty-one signals 'Pour Tous les Ordres de Bataille' when close-hauled, provided for manoeuvring the line and making more or less sail when coming closer to or further from the wind. Twelve more signals dealt with these evolutions with the wind aft. Eleven chasing signals then followed, including the lowering and hoisting of flags to indicate the number of ships sighted.

There were only eight fog signals. The night signals were more elaborate; getting under sail, general and special, when in line of battle or otherwise; sailing, tacking and wearing as a fleet or by squadrons and in three columns; movements in line of battle close-hauled or before the wind. There were fifty-one night signals in all.

The thirty-one 'Signaux de Combat' appear to be tactically limited, cautious and uninspired. However, it would be unwise to dismiss them without knowing more precisely what they were intended to achieve. It is probably more useful to ask what kind of fleet Tourville commanded. Was it a fleet sufficiently well trained both to understand and obey the signals, always assuming that they could be seen by the actual flag officers and captains concerned? If the answer is 'yes', these signals can probably be regarded as sufficient. These signals and instructions are summarised below.

Tourville's 1693 'Signaux de Combat'

Article 1 The ships to fight at half a cable distance, weather permitting. No ship to quit the line unless disabled and then to be assisted by the nearest ship to her. (No signal)

Article 2 Fireships to be stationed sufficiently far on the disengaged side of the line to avoid being disabled but near enough to be ready for action. (No signal)

Article 3 Storeships and merchantmen to keep on the disengaged side. (No signal)

Article 4 Small ships to be ready to stop enemy fireships by boarding them and attacking their boats; such conduct to be rewarded; this was an exactly similar provision to the British 'encouragement'. (No signal)

Article 5 Captains not to open fire except at close range. (No signal)

Article 6 If to leeward, signal for the van to crowd sail and double on the enemy.

Article 7 If to windward, signal for part of the van to double.

Article 8 If to windward, signal by rear-admiral for the rear to double.

Article 9 Signal for squadron commanders to their squadrons to tack or wear together.

Article 10 Signal to general asking approval for Article no 9.

Article 11 Signal for the whole fleet to put about if the van or rear has already done so.

Article 12 Signal by rear-admiral for disengaged ships to come to the wind and form a reserve corps to assist disabled ships.

SIGNAUX POUR LE Combat.

I

Pour mettre l'Armée en bataille au plus prés du vent.

Outes les fois que le Général voudra mettre l'Armée du Roy en bataille au plus prés du vent, il mettra un petit Pavillon blanc au bout de la vergue d'artimon ; les Vaisseaux portant Pavillons feront le méme Sgnal. Le Commandant de l'Escadre blanche & bleüe au milieu de son Escadre se mettra a l'avant garde ; & le Commandant de l'Escadre bleüe au milieu de la sienne a l'arriere garde, les Brûlots feront disposez suivant l'ordre de bataille que l'on a donné.

Un petit pavillon blanc a la vergue d'Artimon.

II

Quand le General voudra prendre l'avant garde.

Quand le Général voudra prendre l'avant garde au milieu de son Escadre, l'Armée étant en bataille au plus prés du vent, il mettra un Pavillon bleu a croix blanche au bout de la vergue d'Artimon, à quoy tous les Vaisseaux portant Pavillons répondront par un Pavillon semblable au méme endroit: alors les vaisseaux de l'avant-garde mettront les voiles de l'arriere sur panne, laissant le petit Lunier a porter, & arriveront afin de laisser passer au vent l'Escadre du Général, qui fera force de voile & ira se mettre a l'avant-garde, l'arriere-garde fera aussi force de voile pour gaigner la queüe de l'Escadre blanche & bleüe.

Un pavillon bleu a croix blanche au bout de la vergue d'Artimon.

III

Et lors que l'Armée sera en bataille au plus prés du vent le Général étant a l'avant-garde ou milieu de son Escadre, & qu'il voudra se remettre au corps de bataille au milieu de son Escadre, il en fera le Signal par un Pavillon blanc au bout de la vergue d'Artimon, a quoy les Vaisseaux portant Pavillons répondront par un semblable au méme endroit: alors les Vaisseaux de l'Escadre du Général mettront les voiles de l'arriere sur panne laissant le petit Lunier a porter, & feront arriver pour laisser passer

A

Article 13 Fireships to perform 'plus grands services' by boarding enemy ships trying to cut the line from leeward, for which they will deserve 'une plus grand recompense.'

Article 14 If the enemy are to windward, the fleet to form a close-hauled line according to the order of battle.

Admiral Tourville's 'Signaux pour le combat' of 1691. NMM

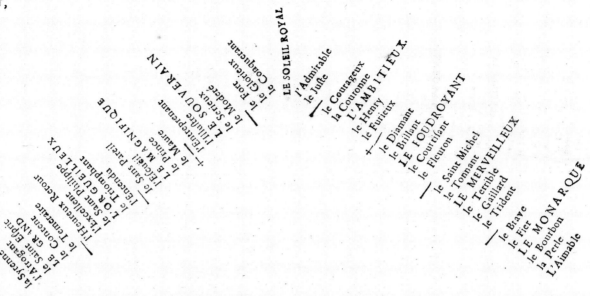

AUTRE ORDRE DE BATAILLE VENT ARRIERE OV LARGVE.

QUAND le Général mettra le Yac à la Vergne d'artimon , à quoy tous les pavillons & les Vaisseaux qui doivent répéter les signaux , répondront par le méme signal , l'Escadre blanche le Général au milieu se mettra au Corps de bataille, & tous les Vaisseaux de son Escadre qui feront à sa droite, comme aussi tous les Vaisseaux de l'Escadre qui se trouvera à sa droite, seront tous les uns à l'égard des autres, dans l'air de vent qu'il faut étre pour se trouver en ligne dans l'Ordre de bataille au plus prés du vent, l'amure à stribord, & les Vaisseaux de l'Escadre du Général qui feront à sa gauche , & tous les Vaisseaux de l'Escadre qui se trouvera à sa gauche , seront les uns à l'égard des autres , dans l'air de vent qu'il faut étre pour se trouver en ligne dans l'Ordre de bataille, la'mure à basbord ,

One of Admiral Tourville's orders of sailing for 1963. NMM

Article 15 Signals for ships out of station when to windward or leeward of the enemy.
Article 16 For the fireships to prime and to attack.
Article 17 For calling a particular fireship.
Article 18 For fireships to return to their station.
Article 19 No special ship will be appointed to escort each fireship but any ship whose bow is crossed by a fireship en route for the enemy is to send armed boats to escort her; the two other nearest ships (of the line) to do the same.
Article 20 Fireships to keep a little to windward of their escorts and to lay the enemy aboard at the fore shrouds, and to avoid masking the fire of the escort.
Article 21 Escort to bring to so as to facilitate fireship attack.
Article 22 Further directions for fireships.
Article 23 Signal to engage.
Article 24 If to windward, for van to bear away.
Article 25 For ships following the general to bear away; to board the enemy.
Article 26 To stop the battle.
Article 27 Signal by flag officer, if disabled, for his division to assist.
Article 28 For a ship in danger of sinking to signal for her division to assist.
Article 29 For anchoring in an action lasting until night.
Article 30 For the fleet to close without anchoring.
Article 31 When in the presence of the enemy, for the fleet to form in the same order of battle as on the previous day; and for ships to follow in the same order.

A further batch of documents is an exact repeat of the fourteen printed diagrams of orders of sailings which

Tourville issued in 1690, except that it includes only fifty ships, all with their names printed. By 17 May, however, Tourville had seventy ships, as given in his anchoring list for Brest harbour issued on that day.[13]

The Smyrna convoy action

The attack on the Smyrna convoy, which Charles de la Roncière describes as *le coup de filet*, is part of *la guerre d'escadre* rather than of *la guerre de course*.[14] The tactical details of his immensely important stroke against the economy of the allied states are impossible to establish. At the beginning of June 1693 a huge convoy of some four hundred English, Dutch, German, Danish and Swedish merchantmen left the English Channel for Spanish, Portuguese and Mediterranean ports, escorted by a force of twenty-two English and Dutch warships and eight bombs, fireships and smaller vessels, commanded by Sir George Rooke. The main Anglo-Dutch fleet covered the movement as far as some distance to the south-west of Ushant and then returned to the Channel. French intelligence was good, and Tourville with the Brest fleet gathered at Lagos Bay with twenty ships from Toulon under the younger d'Estrées. Rooke's reconnaissance seems to have been extremely inefficient. He was badly surprised, and in a confused series of piecemeal actions, 17–18 June, lost two Dutch ships-of-the-line and over ninety merchantmen.

Tourville's secret instructions for the attack on the Smyrna convoy are in printed form, 'Fait à bord du *Soleil Royal*, Est & Oüest de Cap de Finisterre le 29 May, 1693'.[15] (This

---

13 *NMM*, HOL/8 (formerly S(H)8), *Ordres et Signaux Generaux de Monsieur le Comte de Tourville*, dated 1693 in pencil on the cover.

14 de la Roncière, *op cit*

15 *NMM*, HOL/8 (formerly S(H)8). This was the new *Soleil Royal* built to replace the one burnt at La Hogue.

suggests the existence of a printing press in the flagship!) Tourville expected two hundred allied and neutral merchantmen to enter the Mediterranean escorted by thirty to thirty-two ships of war. The admiral's ship was expected to be of 70 guns and to have the Union flag at the main. Tourville exhorted his captains: 'Il seroit plus avantageux de prendre cette Flotte que de gagner une bataille . . . Les Capitaines ne communiqueront cét ordre qu'aux officiers qu'ils jugeront à propos.'

The next day, 30 May, Tourville issued his *ordre de marche,* fully printed, with the ships' names. His fleet then consisted of sixty-five of the line and twenty-five fireships, but a further printed *ordre de march* in three columns, close-hauled, dated 3 June, gives the names of seventy ships-of-the-line and twenty-five fireships, which shows that Tourville's fleet was at full strength again. His printed order of battle issued on 7 July, immediately after his successful attack, lists eighty-seven ships-of-the-line and thirty fireships, as does his printed set of signals for calling officers, issued on the same day. Even as late in the season as 20 September, his order of sailing and battle included seventy ships, some having already returned to Brest.[16]

Père Paul Hoste

The greatest of all the French tactical theorists was Paul Hoste, born at Brest in 1652. He was trained by the Jesuits and, like so many members of his order, became a leading figure in contemporary mathematics and science. He became Tourville's *aumônier* and was eventually appointed Professor of Mathematics at the Royal Seminary at Toulon, where he died in 1700 at the age of forty-eight. He spent twelve years at sea with d'Estrées and Montemart (Duc de Vienne), and also with Tourville, from whom he learned many tactical ideas. It was on Tourville's suggestion, or command, that Hoste produced the first major work on naval tactics. *L'Art des Armées Navales ou Traité des Evolutions Navales* was published at Lyon in 1697 and dedicated to Louis XIV, who rewarded Hoste generously.[10] Its success was immediate and lasting. It was republished in 1727 in its original form, and fifty years later was still a sound text-book.

Hoste's claim to have described all the chief sea battles since galleys were superseded by *gross vaisseaux* or broadside battleships, is fully sustained by accounts, sometimes first-hand, of all the important actions between the French, Dutch and English navies. In his accounts Hoste is extremely fair in commending the skill of Tromp, de Ruyter, Blake, Prince Rupert and the Duke of York. The book is illustrated with 130 magnificently engraved plates, the ships in each case carrying the sail appropriate for the particular evolution.

In his preface (dated 1696), Hoste declared that without evolutions, fleets were like barbarians who waged war

without knowledge and without order, everything depending on caprice and chance. Evolutions provide a framework without which tactical opportunities cannot be seized. *The Art of Evolutions* showed the generals and other officers not only what was necessary but also what was possible.

Hoste's whole system of sailing and battle formations was based on his five *ordres de marche.* These retained their primacy in the French service throughout the age of sail. The basis of these orders of sailing was that each provided a means of forming a close-hauled line of battle, which Hoste strongly favoured. If it was not close-hauled Hoste believed, the windward fleet might lose its position to windward. He also maintained that the leeward fleet should form up close-hauled, because only thus was it able to defend itself effectively and to profit from a sudden change of wind or a mistake by the enemy which enabled it to gain the weather gage.

Hoste's first order of sailing was a close-hauled line of bearing on either tack, one of the most difficult formations both to understand and to execute. The ships were disposed in such a manner that a straight line drawn through their centres in a windward direction would give a close-hauled line. The ships themselves might point in any direction the admiral pleased; it was their bearing from each other which

TRAITE' DES EVOLUTIONS NAVALES.

TROISIE'ME PARTIE,

Rétablir les Ordres quand le vent change.

EXPLICATION DU SUJET.

IL est fort ordinaire à la mer d'avoir des change-mens de vent, & ils sont tous capables de mettre le désordre dans une Armée, qui n'est pas bien disciplinée. Les Ordres étant établis par rapport au vent, ils sont troublez quand le vent change, & l'Armée se trouveroit dans une terrible confusion, si elle n'avoit pas des régles aisées, pour rétablir l'Ordre que le changement de vent lui a fait perdre. Je sçai qu'en ces rencontres on peut rétablir l'Ordre par les mêmes voyes qu'on le forme, lorsque l'Armée n'en a encore point : mais on voit aisément que ce seroit là une source de fâcheux accidens ; les Escadres se sépareroient, les Vaisseaux s'aborderoient, toute l'Armée demeureroit un temps infini à se ranger : au lieu que si on suit les régles que nous allons donner, le changement de vent ne dérange nullement l'Armée; chaque Vaisseau ne laisse pas de se trouver dans son poste, & un petit mouvement qui se fait d'une maniere également exacte & impercepible, remet l'Armée dans l'Ordre qu'elle avoit perdu. Je ne sçai si je me trompe, quand je me persuade que le bon Ordre d'une Armée exige que le poste & la maneuvre de chaque Vaisseau soient si exac-

O   tement

Père Paul Hoste SJ (1652–1700): frontispiece of L'Art des Armées Navales, 1697, *showing Hoste in his library.* MM

16 *NMM,* HOL/8 (formerly S(H)8). There are two copies of the order of battle, one endorsed on the back *Le Magnifique.* This was Coëtlogen's ship and he was possibly the original owner of the Tourville collection in the National Maritime Museum.

17 *L'Art des Armées Navales ou Traité des Evolutions Navales, qui contient des regles utiles aux officiers généraux, et particuliers d'une Armée Navale; avec des examples itez de ce qui s'est passé de considérable sur la mer depuis cinquante ans. NNM,* TUN/82 Hoste was also the author of *Théorie de la Construction des Vaisseaux,* published at Lyons at the same time, the two works being often bound together in single volume with a composite title page. The latter is a highly scientific work. Unlike the tactics, it has the royal imprimatur, dated 7 November 1696, and also the permission of the Reverend Father Provincial of the Jesuits of the Province of Lyons, dated in Rome, 13 November 1696. In the prefeace Hoste states that he was encouraged to write the book by Tourville. The book concludes with the words 'A plus grande gloire de Dieu'. Evidently ship construction was within the scope of superior religious jurisdiction whereas tactics were not!

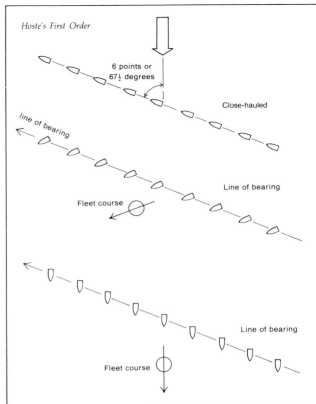

A plate from Père Paul Hoste's L'Art Des Armées Navales ou Traité Des Evolutions Navales, Lyon, 1697, showing his third order of sailing, a formation in which the fleet on any point of sailing formed an oblique angle with the flag at the apex and to leeward. NMM

mattered, not their course. The point of the formation was that when the admiral signalled the fleet to come to the wind on the tack in the direction of the imaginary straight line, the ships would automatically find themselves in a close-hauled line ahead, regardless of their previous course. Other lines of bearing could be set by compass direction, as came to be done in the middle of the eighteenth century, but the principle of the relationship was always the same: as long as the bearing was retained, the ships could always be brought into a line ahead on that bearing.

The second order of sailing was what is known as a 'perpendicular of the wind'. This meant that a straight line drawn through the centre of the ships would be at right angles to the direction of the wind. Again the course was immaterial, though generally the second order was used for a fleet with the wind abeam or abaft the beam.

The third order was with the ships in two divisions, ranged along the two sides of an obtuse angle of 135 degrees, exactly bisected by the direction of the wind. The admiral was in the centre, at the angle and therefore at the most leewardly point. When sailing close-hauled in this order the ships of the leading division would already be in line ahead.

The fourth order was for the van, centre and rear of the fleet to be disposed each in two columns with the division commanders between and slightly ahead of the two leading ships in the respective pairs of columns. The centre pair of columns, the admiral's division, would be ahead of the other two, thus reproducing the obtuse angle arrangement of the previous order. Any course before the wind could be steered, and there were variations of the order enabling all three divisions to be level with each other, that is in line abreast as regards divisions.

The fifth order was for three divisions of the fleet in three separate parallel columns in line ahead, close-hauled, with the admiral's column in the centre, led by him. In the case of a large fleet, each division could be in two parallel columns. The formation then came to resemble one of the variations of the fourth order, except that the fleet would be close-hauled instead of before the wind.

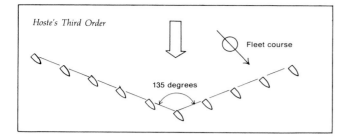

Hoste's Third Order

Fleet course

135 degrees

Hoste's system became highly complex when he attempted to show how squadrons or divisions could be changed about when the fleet was in a particular order. The treatment was more theoretical and mathematical, and he felt compelled to defend himself; these evolutions, he claimed, might have little practical application, but it remained valuable for commanders to study them. At the least they would have training value, and at most they could be used seriously to disconcert the enemy.[18] In subsequent chapters he demonstrated more practical manoeuvres to re-establish the line of battle when the wind changed, and to change the order of battle into one of the various orders of sailing, with or without transposing the squadrons. This, in some cases, involved the fleet moving in three columns, which again necessitated a more complicated treatment.[19]

Hoste did not feel competent to give more than examples

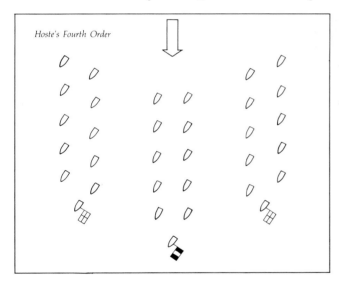

Hoste's Fourth Order

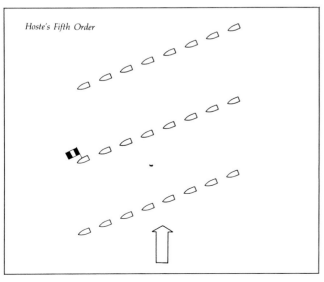

Hoste's Fifth Order

and some general ideas about battle tactics: 'On ne peut presque rien décider dans la theorie: c'est le grand genie d'un Heros avec beaucoup d'expérience, qui en doit faire un grand Général.'[20] He did, however, begin his tactical expositions with a sound practical warning on the dangers of allowing the fleet to be surprised at anchor, with Sole Bay (1672) and Palermo (1676) as examples. He also permitted himself one important generalisation, his first main axiom: if two fleets remained for a long time in sight of each other, either should be able, *régulièrement*, to force an engagement, if it so wished.[21] Later in the book he corrected himself; fleets of equal strength, he admitted, might avoid an engagement (and he described methods of avoiding battle with or without the weather gage), but a fleet of inferior strength caught in the open sea had no recourse but to fight as best it could.[22] Unlike the situation on land, where an inferior force might sometimes occupy a strong defensive position, covered by woods and rivers, an inferior fleet, he argued, was like an inferior army in open country without either defendable ground or the time available for entrenching itself.

The commander of a superior fleet could either detach his best sailers to pin down part of the enemy fleet, and so bring on a general action, or he could afford to divide his fleet into three widely separated squadrons to maximise the chances of catching the enemy fleet. In the summer season, with long days and short nights, forced marches were of little value to a fleet attempting to evade battle, while to try to escape by carrying a press of sail at night only invited dispersion. If the fleet wishing to force a battle were to leeward, and could not win the weather gage, it had only to keep on the same course as the enemy until the wind changed, as Tourville did at the Battle of Beachy Head in 1690. To force a battle from windward was in some respects more difficult, since it was necessary for the attacking fleet to approach level with the enemy fleet and close to them, before actually going down to attack. Otherwise the attacking fleet would lose distance in going down too far before the wind, with the result that the van would fall short of the enemy's van and the enemy would be able to avoid *un combat décicif*.[23] Hoste carefully illustrated navigational methods for ensuring an effective approach.

He summarised the respective advantages of fighting from the windward and leeward positions. The windward position gave a fleet the initiative in approaching, and allowed a superior fleet to double the enemy's rear. It facilitated attack by fireships on disabled enemy ships, attack on enemy ships attempting to escape, and further facilitated attempts to cut off the enemy's van or rear. Finally, the windward position carried smoke and burning matter from the guns towards the enemy, thus avoiding confusion and the danger of fire.

The advantages of the leeward position were that in a strong blow it was safer for the leeward fleet to use its lower-deck guns because the side presented to the enemy would be healed up, clear of the sea. The enemy fleet to windward might not be able to use its lower-deck guns at all without risk of the sea flooding in. The leeward fleet also benefited from the ability of its disabled ships to escape capture by dropping further to leeward, whereas the disabled ships of the windward fleet might not be able to claw to windward away from the enemy. If the admiral of the leeward fleet were obliged to break off action he could signal his entire fleet to

18 Part II, pp104–44
19 Part III, pp145–82; part IV, pp185–330
20 p331
21 p363
22 pp358–63, 375
23 p369

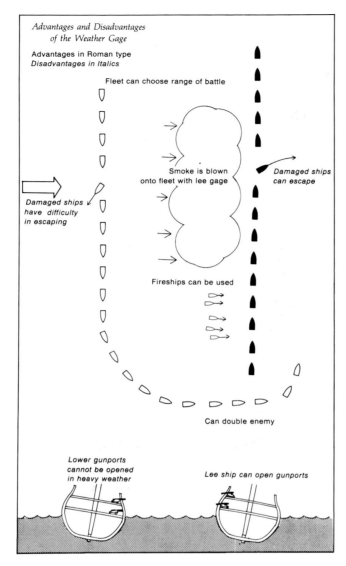

*Advantages and Disadvantages
of the Weather Gage*

Advantages in Roman type
Disadvantages in Italics

Fleet can choose range of battle

Smoke is blown
onto fleet with lee gage

*Damaged ships
can escape*

*Damaged ships
have difficulty
in escaping*

Fireships can be used

Can double enemy

*Lower gunports
cannot be opened
in heavy weather*

Lee ship can open gunports

bear away. The admiral of the windward fleet could only retreat by breaking through the leeward one, 'Ce qui est d'une tres dangereuse conséquence.'[24] Retreat by the leeward fleet might also be dangerous, but less so in the case of night coming on, the wind freshening, the sea getting up, or the enemy to windward being embarrassed by a convoy.

Hoste's major tactical expedient was doubling. A superior fleet, he advised, should seek to take the enemy between two fires by dividing and coming up on both sides of the end of his line.[25] If the superior fleet were to windward, it could double the enemy more easily; if to leeward, it should still overlap the enemy astern so as to take advantage of a change of wind. The leeward fleet in this position could gradually thin its line so as to build up a mass at the rear for use should the enemy haul its wind.[26]

Many experienced commanders, Hoste noted, preferred doubling the enemy's van, because a disordered van would throw the remainder of the enemy fleet into disorder as it came up. He regarded this notion, however, as erroneous: if

---

24 Part 1, p54

25 p376

26 p376

27 pp381–82

28 pp382–88

the leading ship were dismasted, the centre and rear could come to her assistance without causing any disorder, and this was easier still when the doubled fleet was to leeward, as the damaged ship would simply drop still further to leeward. Nor could the doubling ships pursue the enemy vessels they had disabled without placing themselves in danger of being cut off. If, however, a fleet were doubled from the rear, their disabled ships could hardly avoid capture either by the enemy's doubling ships or by the rest of the enemy rear. Ships which had undertaken the doubling manoeuvre and were themselves disabled could easily escape capture if they were in the rear, but not if they had doubled the enemy's van. At Barfleur in 1692 Russell succeeded in doubling Tourville's rear simply because he had more ships and so overlapped the French line.[27] This same observation was made by John Clerk of Eldin over eighty years later.

Doubling the rear could only be avoided by preventing the enemy from overlapping the rear. There were five ways of doing this from an inferior position:[28]

1. A fleet to windward should engage the enemy centre and rear only; the enemy van would then be useless. If it tried to put about and take part in the battle it would lose time and be in danger of being cut off from its fleet by the calm which Hoste, in common with many others, believed was the inevitable result of cannon fire. A space might also be left in the centre of the fleet provided care was taken not to allow the van to be cut off. This was the manoeuvre followed by Torrington at Beachy Head in 1690, when the Anglo-Dutch fleet was inferior to Tourville's.
2. A fleet to leeward should leave a wider gap in the centre and allow less overlap to the enemy van, though there had to be ships and fireships ready to stop the enemy breaking through the centre.
3. An alternative was for flag officers to engage their respective opposite numbers in the enemy line (with the same intervals between ships as the enemy). This would mean that gaps were left between the van and centre and between the centre and rear, and thus a number of enemy ships could not engage. The danger here was that the enemy would overlap both the van and rear. The remedy was to station the largest ships at the head and tail of each division, and to ensure that the rear was not overlapped by the enemy rear.
4. A further possibility was for each squadron to attack its opposite squadron in the enemy line, but in such a manner as to leave no enemy ships astern, rather leaving more ships ahead. This amounts to the leeward version of the first tactic.
5. The remaining tactic was to space out the whole fleet evenly along the enemy line, so as to make a line equal in length. This was the least satisfactory method because it allowed the enemy to employ all their ships against the inferior fleet. It might nevertheless prove useful, Hoste suggested, in certain circumstances, for example when the enemy ships were individually weaker. To enable a fleet to get into order to receive an attack from windward, all ships were advised to bear away a little to gain time, and then each captain to attempt to establish his ship abreast the enemy ship indicated to him by the general.

The strongly defensive tone of Hoste's work is obvious. Much attention was given to avoiding an engagement. Defence against doubling was treated at greater length than the art of doubling itself, and seen exclusively in terms of the inferior fleet. Above all, his defensive cast of mind is revealed by the continuous acceptance of the leeward station as a basis for tactical demonstration.

Finally Hoste dealt with breaking the enemy line, a manoeuvre he felt was much used in the Anglo-Dutch Wars. A fleet to leeward seeking to break the enemy line should

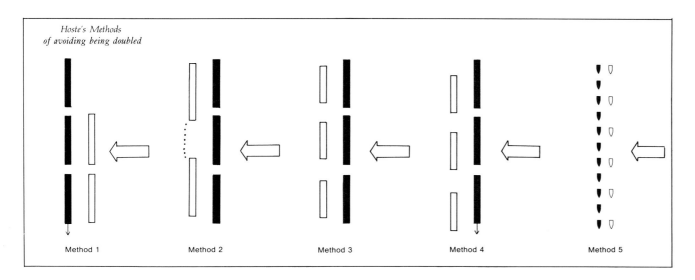

Hoste's Methods of avoiding being doubled

Method 1   Method 2   Method 3   Method 4   Method 5

tack in succession. As the leading ships passed through the enemy line the fleet should tack once more and resume its original course, now to windward of the enemy and roughly parallel. At this point the enemy could be expected to reply in kind, and in this manner the two fleets would pass through each other several times, giving each the chance to cut off, capture and destroy many ships. Hoste argued that this manoeuvre would have to be executed with great technical skill in order to succeed as happily as did d'Estrées in the Battle of the Texel in 1673, in which, according to Hoste, the French squadron passed through the Zealand squadron, weathered it, broke it up, and put the Dutch into such confusion that complete defeat resulted.

Hoste's defence against being cut from leeward was for the windward fleet to tack together as soon as the leeward fleet tacked in succession. If this looked too much like retreat, the windward fleet could wait until the van ships of the leeward fleet were passing through their line, and then tack together, thus putting the leading leeward ships between two fires as they attempted to pass through, while at the same time cutting off those ships which succeeded in doing so.

His view was that there should be little to fear from the enemy's breaking through the line, and it therefore followed that commanders should not attempt the manoeuvre except in special circumstances. Under pressure, an admiral might choose to break the enemy line if, by so doing, he avoided a worse danger. If a wide gap opened in the enemy line, so that a part of the fleet could get into action, it might be in the interest of a commander to steer through the gap. If several of the enemy ships were disabled, the ships opposite them might wish to tack together and pass through the enemy line, followed in succession by the ships astern of them, with the object of cutting off the enemy rear.

It might also be necessary to break through the enemy line to rescue ships which the enemy had cut off. This manoeuvre involved some risk, but several precautions could be taken. The ships cutting through were advised to keep closed up and to carry a press of sail to pass the most dangerous point quickly without troubling to fight the enemy. Once through the enemy line, the ships which had broken through should tack as soon as possible, placing themselves parallel to the enemy, in order to prevent the enemy following on the same tack.

Hoste's 'Projets des Signaux', with an engraved plate of signal flags, is, as he claims, simple, clear and extensive, although it could only be effective in a situation where the flagship had all three masts intact. The system as a whole was, like Tourville's, far in advance of the British equivalent.

It was a true signalling system, and not merely a signalling projection of a series of sailing and fighting instructions. The British were relying on the Royal Standard, Jack, Red Ensign and flags of plain colours including yellow, two striped flags, two part-coloured, three crosses and the distinctive yellow and white stripes (for fireships) for signalling. Hoste preferred to create thirty-six signal flags supplemented by broad pendants, pendants, and wefts. The designs of some of these flags appear to be so much alike as to lead to possible confusion. There are six with crosses, six plain colours with borders, six plain colours pierced with a round blob, and six plain colours with a single fesse-wise stripe. On the other hand, as only three colours were used – red, white and blue – and particular types of signals were restricted to particular

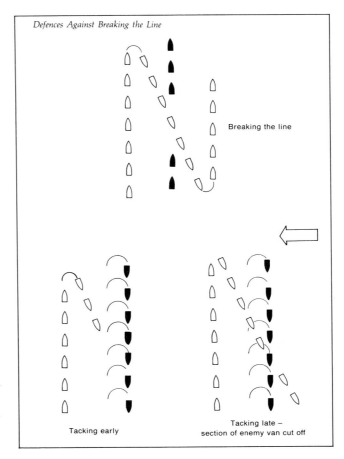

Defences Against Breaking the Line

Breaking the line

Tacking early

Tacking late – section of enemy van cut off

positions recognition would have been straightforward. As Hoste himself claimed, 'On multiple les Pavillons par le mélange des trois couleurs qu'on peut distinguer à la mer: Ce qui fournit un très-grand nombre des Signaux.'

Hoste's first signals were intended to indicate the address to which the message was directed, the whole fleet, a particular squadron, a division, or a single ship. Signals for the various orders of sailing were each indicated by a single flag at the mizzen yard. Those for increasing or decreasing sail and/or for changes of course, etc, were made by a single flag at the main topmast. Battle signals, including to board and to double, were signalled by single flags at the fore topmast. Councils on board the flagship and consultations by signal alone (whether to fight, to put into port, pursue the enemy or retreat), were all indicated by single flags at the ensign staff. Signals for ships to come to the admiral were made by a single pendant at the mizzen yard. Signals to be made to the admiral (whether by squadron or division commanders or by private ships was not specified) were divided into two groups: operational signals were made by a single flag at the jackstaff, and administrative signals by a single pendant at the ensign staff.

Rooke's reissued instructions, 1702

When Rooke reissued Russell's instructions of 1691 in June 1702, William III had only been dead three months. Probably there had not been time to issue new instructions for Queen Anne's navy, and it was the next year before Rooke was able to do so. In effect, his instructions of June 1702 were the final version of the original and anonymous Sailing & Fighting Instructions.[8] There appears to be no further issue of instructions on a personal basis. Eventually Rooke's name was removed and his instructions became the so-called permanent instructions of the eighteenth century.

In so far as Rooke had any hand in it, his intervention was disastrous. Not only does his version follow Russell in leaving out the double-page ship table, which may or may not have been a wise omission, but he also left out all the separate lists of contents and indexes, thus reducing the book to a mere collection of separate chapters. In all the old issues dating from that of 1672–3 to 1691, the indexes enabled the meaning of signals to be read almost at sight. Their removal meant that the signals remained buried in the text of the articles. While it was still possible for the sender, having decided on the article, to discover the signal for it very nearly as quickly as before, the new book was virtually useless for reading signals quickly. The signals therefore either had to be learnt by heart, or each officer had to draw up his own reference book of signals; very soon they began to do just that.

Had Rooke's removal of the indexes stimulated initiative in signalling, his action might not have been so disastrous. On the contrary, however, it had a restricting effect on tactics and tended not only to encourage defensive methods but defensive thinking as well. A book must be judged as a whole: what Rooke handed on was a collection of material which no doubt the Duke of York, Rupert and Albemarle would have recognised as useful but transitory. In detail it

represented the collective experience of the Anglo-Dutch Wars and the War of the English Succession, but as a whole, it turned its back on the slow progress towards efficient fleet control.

'Sailing & Fighting Instructions for Her Majesty's Fleet'

The earliest known copy of the final form of what eventually came to be called the Permanent Sailing & Fighting Instructions was issued in the reign of Queen Anne.[29] The book consists of thirty-one pages divided into ten chapters or sections:

'Signals to be observed at an anchor, in weighing anchor, in sailing & anchoring in the day time', pages 1–7.

'Distinguishing Lights for the Flags', page 8.

'Signals to be observed at an Anchor, in Weighing Anchor, in Sailing and Anchoring in the Night-time', pages 9–12. There are nineteen articles in this section, limited by their form of communication. Up to four lights were used, hoisted in various patterns in various positions in the ship's rigging, or on the poop or ensign staff. Up to eight signal guns were employed for ordinary occasions. Certain emergencies might require 'a great number of lights' and 'gun after gun'.

'Instructions for Sailing in a Fog', pages 13–14. There are eight articles here, mostly listing signals, but containing some instructional material.

'Instructions to be observed by Younger Captains to the Elder', page 15. The four articles in this section mainly deal with punctilios and compliments to be observed when under sail, junior captains giving way to senior.

'Signals for calling the Flag-Officers, and other Officers, on Board the Admiral', pages 16–18.

'Instructions to be observed by all Masters, Pilots, Ketches, Hoys and Smacks, attending the Fleet', pages 19–20. These four articles include the important anti-fireship duties and reward.

'Fighting Instructions', pages 21–27.

'Encouragement for the Captains and Companies of Fire-Ships, Small Frigates, and Ketches', pages 28–9.

'Instructions for the Fire-Ships', pages 30–1.

Once fighting instructions and sailing instructions were bound together and issued as a single book, even though they were paged separately, naval tactics became an expression of all those parts of the book which dealt with the fleet in motion.

However, nearly half the sailing signals were concerned with weighing and anchoring, and there were five signals for various forms of chasing. The admiral had only a very few signals for handling the fleet as a whole when actually cruising; an interesting question is what order of sailing the fleet assumed under ordinary cruising conditions. In most of the major battles of which we have any reasonable account, the rival fleets appear already formed in line of battle ahead or abreast well before the battle began. Where any deployment is mentioned, it is of the simplest kind, from line-abreast, from converging line-ahead or from converging bow and quarter line.

Granted that admirals were skilful at deploying their fleets well in advance, was not this practice, in some ways, disadvantageous, especially if carried out on the previous day? Did not premature deployment make for clumsiness thereafter, quite apart from revealing the fleet's full battle strength and battle area to the enemy? How, moreover, was it supposed that fleets moved between leaving port and encountering the enemy? Did they sail by squadrons, and if so, how? What facilities did the admiral possess, in the way of

8 'Instructions for directing & governing Her Majesties Fleet in sailing & fighting by the Right Honourable Sir George Rooke, Kt, Vice-Admiral of England, etc, and Admiral and Commander-in-Chief of Her Majesty's Fleet, etc in the year 1703', BM, Add MS 28126, issued and signed by Rooke, *Royal Sovereign*, Spithead, 2 May 1703. *MOD* NM/80 (formerly *RUSI* 80) is a mutilated copy.

29 *MOD*, NM/80 (formerly *RUSI* 80), undated and bound up with Rooke's instructions of 1703

instructions with signals attached, for spreading his fleet over a wide area in search of the enemy, or for keeping it closed up and ready to deploy at any moment? What standard formations had he at his command and how could he change from one of these to another or transpose the relative order or position of his squadrons? To what extent were these orders of sailing also orders of battle and to what extent merely drill exercises, 'evolutions', to improve the skill and promptitude of captains and junior flag officers? These are important questions and the answers to them are central to this study.

Of the thirty-two articles which made up the fighting instructions, only nine, with their respective signals, can be said to have had a truly mandatory character for the admiral, and his subordinate flag-officers and captains.

Article 17: If the Admiral see the Enemies Fleet standing towards him, and he has the wind of them, the Van of the Fleet is to make Sail until they come the Length of the Enemies Rear, and our Rear a-breast of the Enemies Van; then he that is in the Rear of our Fleet is to tack first, and every Ship one after another, as fast as they can throughout the Line; and if the Admiral would have the whole Fleet to tack together, the sooner to put them in a Posture of engaging the Enemy, then he will hoist an Union Flag on the Flag-staves at the Fore and Mizen-top-mast-heads and fire a Gun; and all the Flag-ships in the Fleet are to do the same: But in Case the Enemies Fleet should tack in the Rear, our Fleet is to do the same with an equal Number of Ships; and whilst they are in Fight with the Enemy, To keep within half a Cable's Length one of another; or, if the Weather be bad, according to the discretion of the Commanders.

There can be no doubt that this article is mandatory for an admiral to windward of the enemy and about to pass on an opposite tack. His only discretion lay in choosing the exact moment at which to make the signal to tack and whether to tack in succession, beginning from the rear, or all together. It undoubtedly committed him, if to windward, to a co-linear engagement without chance of variation, except in so far as the two fleets might not be equal in numbers or at the same intervals of distance.

Article 19: If the Admiral and his Fleet have the wind of the Enemy, and they have stretched themselves in a Line of Battle, the Van of the Admiral's Fleet is to steer with the Van of the Enemies, and there to engage them.

This presumably meant that when the fleets were on the same tack an admiral to windward was to force an action by bearing down with his van towards the enemy van, so as to ensure a co-linear engagement.

Article 21: None of the Ships in the Fleet shall pursue any small Number of the Enemies Ships, until the main Body be disabled, or run.

This is a repetition of the Duke of York's instruction.

The remaining mandatory articles are less important. Of the nine mandatory articles, therefore, only two, no 17 and no 19, actually instructed the admiral, in advance, on the conduct of the battle, and even then only under particular circumstances. Article 19, moreover, depended to a considerable extent on the conditions envisaged in Article 17 taking effect. The question remains how far the admiral could arrange matters so that Article 17 did not become applicable. If he was to leeward he was free anyhow, but if he was to windward, was he free to make some other tactical approach, given that the enemy fleet did not make some precipitate move? The answer would seem to be that although he was free enough in theory, the number of signalling options allowed him was insufficient to enable him to deliver a different kind of attack. Apart from signals for tacking (van

or rear) with or without endeavouring 'to gain the wind of the Enemy' and various signals for ensuring the correct formation and regularity of the line of battle, there was nothing in the articles to help him at all. When opposing a markedly inferior force, already in retreat, he could presumably make one or more of the chasing signals given in the sailing signals, combined with the signal for battle, Article 13. In this way he might push home his attack without even forming into line, trusting to his captains to act to the best advantage by exercising individual initiative and providing mutual support. In a full-dress action, however, between roughly equal fleets, with the enemy to leeward and in line ahead, there was little he could do but conform to Article 17.

From all this, it would seem that tactical doctrine favoured the line-ahead formation together with a co-linear engagement with the enemy, mainly because of its defensive value. This being the accepted doctrine, authority, in the shape of the leading flag-officers of the day, took care to keep the system starved of new signals, possibly for fear that some admirals might be tempted to indulge in unprofitable experiments.

The devising and promulgation of additional signals which would increase the admiral's options was always within the competence of the Board of Admiralty or of Prince George of Denmark's Lord High Admiral's Council. We can only conclude that, to the fighting admirals of the period, lack of adequate signals was no more than an inconvenience, and possibly not even that. Torrington, Russell and Rooke, men who had helped put the Instructions into their final form, together with Admiral Sir Clowdisley Shovell, Admiral Sir Stafford Fairborne, Admiral Sir John Leake and Admiral Sir George Byng, who had reached high positions operating them, had all had a hand in Admiralty administration between 1689 and 1714. There is no reason to suppose that the civilian members of the Admiralty Board would have interfered had the admirals wished to extend the signalling system. When in later years the inconvenience became irksome, however, tradition served to put a brake on reform.

There seems reason to suppose that these fighting instructions were regarded mainly as a guide for the admiral, albeit perhaps a strict one. They were of course mandatory for captains and 'inferior officers', since once they were signalled by the admiral they became his instructions. In fact, the authority for these fighting instructions was always derived from a particular admiral, not from the Admiralty. No copy of the instructions as a whole had any authority until it was signed by an admiral and issued to some other admiral or captain. When admirals or captains were tried by court-martial for misconduct or neglect of duty, the charges were always related to the Articles of War rather than to the instructions. Commanders-in-chief, at any rate, seem to have considered themselves entitled to amend the articles of any of the instructions, either by making changes to the printed text, or by issuing separate additional sailing or fighting instructions, or additional signals which, unless cancelled, remained valid in an admiral's own fleet for his own period of command. The fact that a newly appointed admiral was often prepared to take over his predecessor's additional instructions tended to give the more useful instructions a more general acceptance in the service. The earliest known British additional instruction, excluding Narborough's of 1678, date from 1710. The earliest additional signals date from 1711, that is, within a very few years of the issue of the consolidated instructions (of 1703 or later) which for so long remained the accepted pattern. The very fact that changes and additions could be made shows that the system was not as rigid as is frequently claimed. After all, these instructions were drawn up by fighting admirals for fighting admirals and

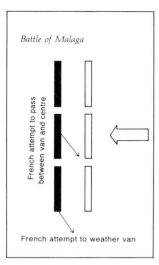

Battle of Malaga

French attempt to pass between van and centre

French attempt to weather van

were kept in force by a succession of fighting admirals appointed to the Board of Admiralty. With Russell, Leake, Sir George Byng, Admiral Sir John Jennings, Admiral Sir John Norris, Anson, Boscawen, Howe, Admiral Sir Charles Saunders, Keppel and Hawke at the Board between 1709 and 1770, it is unrealistic to argue otherwise.

The Battle of Malaga

At the Battle of Malaga (24 August 1704) two powerful fleets pounded each other for a whole day without either side displaying any tactical initiative and without a single ship being actually sunk in the battle. The action was, however, of immense importance. Strategically it settled the future of Gibraltar, which had just been captured by the confederate fleet, ostensibly in the name of Charles III, the Austrian candidate for the territories of the Spanish Empire. Had the French been able to recapture Gibraltar for the future Philip V as a result of the battle, England would have been without a Mediterranean base and in addition might have failed to capture and hold Minorca.

When the two fleets began to engage off Malaga on the morning of 13 August 1704, each in line ahead, heading south, their paper strength of some fifty of the line apiece made them practically equal. The French ships were comparatively clean, being fresh out from Toulon, but many of the confederate ships had been six months at sea. They were, moreover, very short of ammunition, due to the preliminary bombardment of Gibraltar, which had been far heavier than was actually necessary. They were also short of 800 marines, landed to occupy and garrison the fortress. Furthermore the French were between them and Gibraltar, so that their only advantage lay in having the weather gage with a slight easterly breeze.

Descriptions of the battle are vague, contradictory and highly partisan. There seems no doubt that the French van tried to weather the confederate van, and that later in the day they drew off some distance to leeward. There was also some attempt by the French centre to pass between the confederate van and centre. Shovell was much praised for supporting his own van division early on and for supporting the centre squadron later. Conversely, he was also blamed for not

exploiting the retirement of the French van.

No ships were lost on either side during the actual battle, but a number on each side were towed out of the line disabled; on the confederate side this was in some cases due to lack of ammunition. Both fleets spent the next day making repairs, but, with the wind having changed to west, the Franco-Spanish fleet held the weather gage and maintained its position between the confederates and Gibraltar. At a council of war the confederate flag officers decided to share out their remaining ammunition and try to break through. Rooke's bluff was not put to the test. On the following day, 15 August, the wind was round to the east again and the Franco-Spanish fleet departed for Toulon.

Leake's biographer criticised the French 'manner of firing chiefly wounding the masts and rigging (as if to secure a retreat rather than a victory). Whereas our shot, being levelled at the hulls, must consequently kill many more than they did of the Confederates.'[30] This is one of the earliest mentions of what came to be a standard criticism of French gunnery.

With no major fleet actions for forty years, the legend of Malaga was created. When the War of the Austrian Succession (1739–48) began, no senior officer in the British service except Norris had experience of any major fleet action prior to Malaga. Malaga, it was held, proved the efficacy of the line of battle ahead as the best and indeed the only possible method of engaging. Furthermore, it proved that the maintenance of the line intact was the best and only sure method of avoiding defeat. Victory might come if, in the face of a well formed line, the enemy wilted, but the line must be held intact at the start to avoid defeat. Only when a superior force was faced by an enemy unwilling to engage, as was Sir George Byng fourteen years later, was immediate attack without forming a line the way to victory.

On the French side, Malaga seemed to justify the defensive tactical philosophy expounded by Hoste seven years earlier. Mindful of the fearful defeat of Tourville at Barfleur and the

30 Stephen Martin-Leake, (Geoffrey Callender, editor), *Life of Sir John Leake Rear-Admiral of Great Britain,* London, 1920 (hereafter cited *Leake*), vol i, pp364–5

The Battle of Malaga, 13 August, 1704 (detail) by Isaac Sailmaker. NMM

subsequent holocaust at La Hogue, French naval officers of the eighteenth century saw in Malaga the apparent triumph of defensive methods.

The Battle of Marbella

Whatever influence the legend of Malaga may have had on the thinking of officers in Admiral Thomas Mathews's fleet at Toulon forty years later, the flag officers serving under Rooke must have seen the battle for what it was. None of them was therefore likely to try to apply the tactics of Malaga to an entirely different situation, least of all Leake, who was one of the most vigorous and talented sea officers of the age. In command of the confederate fleet destined for the second relief of Gibraltar, he issued instructions on 4 March 1705, as his fleet was about to meet a French force under Rear-Admiral Jean Desjeans, Baron de Pointis, which was supporting the siege.

Leake had twenty-three English ships, four Dutch and eight Portuguese, and his instructions to them were:

If they had intelligence that the enemy were in Gibraltar Bay and the wind westerly, they were to sail in a line of battle abreast of each other, after they were past Tangier until they got near the length of Cape Cabarita [facing the extreme end of Gibraltar on the west side of the Bay]; and if the enemy should be at anchor on the west side of the bay [so they could not weather them, the ship that led with the starboard [larboard?] tacks was, as soon as she got the length of the said cape, to spring her luff and endeavour to get ahead of them, every ship following in a line ahead of each other, till the van got ahead of the innermost [French] ship; and then to make the signal for tacking and to tack accordingly; and as soon as he was about, he was to lay his head sails to the mast, every ship astern doing the same, observing to tack when the ship ahead tacked; and to endeavour all that was possible to annoy the enemy by either laying them athwart the hawse, or otherwise boarding them before they could set their sails; which method was to be observed if they cut or slipped, if any advantage was to be gained before they could form their line of battle.[31]

Admiral of the Fleet Sir Clowdisley Shovell (1650–1707), by Michael Dahl. NMM

Leake's problem was how to enter the bay with the west wind which it was assumed would be necessary to carry him through the Straits, and then work to windward of the French who might well be anchored close to the Spanish shore on the west side of the bay.

Four days later, 'Sir John drew the fleet into a line of battle

31 *Leake*, pp257–8

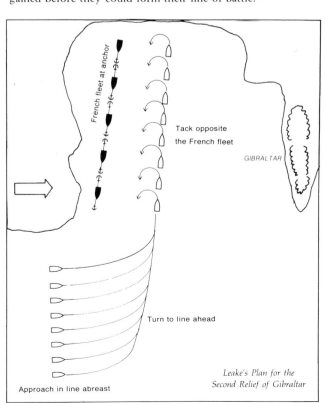

Leake's Plan for the Second Relief of Gibraltar

Vice-Admiral Sir John Leake (1656–1720), by Sir Godfrey Kneller. NMM

to try how the Portuguese would behave, who indeed answered his expectation, he being obliged to bring to for half an hour, to give them time to get into the line; which must have produced fatal consequences had it been to proceed upon action, as he expected very shortly to do.' Luckily for Leake, de Pointis had only five of the line in the Bay of Gibraltar with which to meet him, so that the ensuing Battle of Marbella, fought on 10 March 1705, was a runaway victory with no real tactical features. The interest lies in what Leake was planning to do if faced by a larger force. In neither case could the situation in any way resemble that at Malaga.

Leake's instructions for the projected attack on Cadiz in 1706 include his line of battle for entering the bay with sixteen English and Dutch ships and with his own ship seventh in the line. Each ship's barge was to carry a fire-grapnel to tow off enemy fireships and to be ready to deal with enemy galleys trying to attack confederate fireships. If there was a French squadron in the bay, the fireships were to attack them first. Article no 8 required that 'every ship comes to an anchor the most advantageously he can, for annoying the enemy'.[32] This injunction shows that the doctrine of the line only applied to particular circumstances. As at Marbella, in circumstances different from those at the battle of Malaga, Leake assumed different tactics to be both acceptable and necessary.

The Battle of Cape Passaro

Sir George Byng's defeat of the Spanish Mediterranean fleet in a pell-mell runaway action at Cape Passero on 11 August 1718 is significant less for its tactical features than for its refutation of the idea that the years immediately following the Treaty of Utrecht were years of British tactical decadence. The approach to battle depended on the strategic question of whether or not the Spanish fleet could be treated

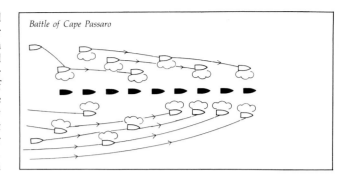

Battle of Cape Passaro

as hostile. Luckily for Byng, the Spaniards opened fire first, largely as a result of the British fleet's threatening demeanour. His instructions ostensibly gave him ample powers, but his position immediately before the battle had nonetheless been highly delicate, and even a slight mistake or misunderstanding on the part of his subordinates might have ruined his career. He called no council of war beforehand, being ready to accept full responsibility for whatever happened, on the grounds that 'a commanding officer should only call a Council of War to screen him from what he has no mind to undertake'.[33] He is said to have given all his captains written orders, presumably before fighting began, that they were 'to attack and destroy the Spanish ships'. If so, this seems to assume that an action could and would be provoked.[34]

From the British and Spanish official accounts it seems clear that Byng's force was superior to that of Vice-Admiral Don Antonio Castañeta to the extent of having twenty-one of the line, including a 90 and two 80s, against eighteen much weaker ships-of-the-line and about ten ships which might be classed as frigates. On the previous evening, Byng had ordered his four fastest ships-of-the-line to keep touch with the Spaniards during the night. In the morning he sent a lieutenant to order the leading ship to engage the nearest Spaniard if she opened fire, which she soon did, and she was eventually taken. When the Spanish rear-admiral, the Marques de Mari, became separated from his fleet and led inshore with some six of the line, nine frigates and various galleys, fireships, bomb-vessels and fleet auxiliaries, Byng ordered Captain George Walton in the *Canterbury* to pursue them with eight ships. Meanwhile Rear-Admiral George Camock, an Irish Jacobite and a rear-admiral in the Spanish fleet, who had strongly advised Castañeta against a fleet action, led away with some other ships so that in the end Castañeta was left with only nine of the line against the remainder of Byng's fleet.

Byng appears to have signalled simultaneously for general chase and to engage. As his fleet gradually caught up with the Spaniards, his ships attacked them piecemeal. While some of Byng's ships stayed by their first opponents, others pressed on towards the head of the Spanish line, after delivering broadsides at the Spanish ships astern as they passed. Altogether, six of the line and a frigate were captured, including the Spanish commander-in-chief and another flagship. Walton was equally successful, capturing two of the line, one of which was a flagship, and three smaller ships, and burning or causing the Spaniards themselves to burn seven more warships and auxiliaries. Byng, however, was far from satisfied, thinking 'there was great reason to blame the conduct of several captains of the Fleet who instead of following and attacking ships of equal force to them, fell, two

Admiral of the Fleet George Byng (1663–1733), first Viscount Torrington, by Sir Godfrey Kneller. NMM

32 *Leake*, i pp322–3

33 J L Cranmer-Byng, editor, *Pattee Byng's Journal*, London, 1950 (hereafter cited: *Pattee Byng's Journal*), p23
34 *Pattee Byng's Journal*, p24

Reordering the line of battle

or three upon one, and so gave to several the opportunity to escape; but as he was crowned with success he would accuse none.'[35] The only captain we can identify for certain as having 'behaved himself very much like an officer and a seaman' was Nicolas Haddock of the *Grafton*. Byng's own second captain in the battle was Richard Lestock.

Byng's dissatisfaction is a tribute to his ardent nature. Aged only fifty-five, he suffered severely from gout, and there had been doubts about his fitness to command the fleet sent to the Baltic the previous summer.[36] However, the Spanish fleet was undoubtedly weaker than the British in every respect, and its commanders seem to have been demoralised from the start, though individual ships fought stoutly. It was this knowledge of superiority, not only in numbers but in fighting organisation and technical skill, and above all in morale, that made Byng look for an even more devastating victory.

This brilliant success belies the idea that during the early years of the eighteenth century the Royal Navy was tactically stagnant. Equally important, what was the source of Byng's tactic of using some ships to attack the enemy rear while other ships passed ahead of them to engage the van? This is exactly the manoeuvre for which Corbett later gave Hawke such high praise. If Byng had been faced by a well-ordered French fleet of equal strength instead of an inferior Spanish one, he would of course have fought the battle differently. But, as Laughton aptly notes, the completeness of his victory, as in the case of Hawke's at Quiberon, masked both his capacity and his determination to exploit a tactical advantage.[37]

During the Second Dutch War the notion prevailed that divisional vice-admirals and rear-admirals should be stationed close to the outer extremities of their divisions. This was partly to strengthen the extremities, since the flag officers were in powerful ships, and partly to ensure authority as well as skill in leading the squadron (and indeed the whole fleet), especially when it came to tacking and wearing. When, under Article 6 of the sailing instructions, the sternmost and leewardmost ships were to tack first, the captain of the sternmost ship of the fleet assumed a highly responsible position in the tacking movement, quite apart from leading the fleet thereafter. According to the older view, the best plan was to discount any possible inefficiency on the part of the headmost and sternmost captains by ensuring that the division flag officers were themselves near the head or tail of the line. Towards the end of the century, however, opinions changed and it became the established practice to station division vice-admirals and rear-admirals in the centre of their respective divisions, supported when possible by powerful seconds. This was in conformity with what by then was an old established custom of stationing squadron commanders in the centre of their own centre divisions and thus in the centre of the squadron as a whole. The view must have been held that, provided the captains of the ships which were at the head and tail of the line were competent men, there was no need to have a flag officer actually present in these positions.

The new organisation depended on having enough ships of force to allow one for each flag, together with his two seconds; that is to say at least nine for each squadron and twenty-seven for the whole fleet. As this last condition could not always be fulfilled, allowances had to be made. With the increased construction of first rates and second rates in the Royal Navy, however, and their counterparts in the French, the formation of solid blocks of defensive fire in the centres of each division of each squadron became more practicable. Hence the defensive aspects of the line tended to be enhanced by developments in naval construction.

The allied line of battle in 1672, soon after Sole Bay,

The Battle of Cape Passaro, 11 August 1718 (detail), by Richard Paton. NMM

35 *Pattee Byng's Journal*, p23

36 Brian Tunstall, *The Byng Papers*, 3 vols, London, 1930-2 (hereafter cited: *Byng*), III, pp241-2

37 Sir John Laughton to Sir Julian Corbett, 19 December 1905, *NMM*, Corbett Papers, Deed Box 'C': 'The entire and easy success of the method has taken away its credit – just as in Quiberon Bay – but I have often thought that if people would only look, they would find that George Byng was one of our tactical masters'.

shows that the heavy ship principle had been carried further in the Royal Navy than in the French navy, though in this case only part of the French fleet was present.[38] In the French White squadron, the flag officers were all in heavy ships and in both the van and centre the admiral had powerful seconds. Each division was led by a 70, which also meant that the sternmost ship would be a 70 with the fleet in reverse order. In the English Red and Blue squadrons the process was more pronounced. Each of the six flag officers, except the rear admiral of the Blue, was in a very powerful ship. In the centre of the Red, the commander-in-chief and his seconds formed an exceptionally strong combination. In no case were division flag officers stationed as near to the extremities of their respective divisions as the English had been in 1665. Nor did there seem any inclination to station ships of force at the actual extremities of divisions except in the case of the van of the French White squadron, which would lead the whole fleet when on the starboard tack.

In the war of the English Succession a new trend appeared which was to have a strong influence on future English tactics. This was to station the two most senior captains, if not themselves flag captains, at the head and tail of the fleet, no doubt as a tactical compensation for the centre stationing of division flag officers. This arrangement gave both seniority and superior gunpower to the head and tail of the line

38 *The Third Dutch War*, pp184–6

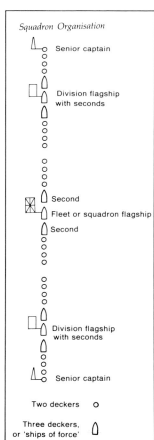

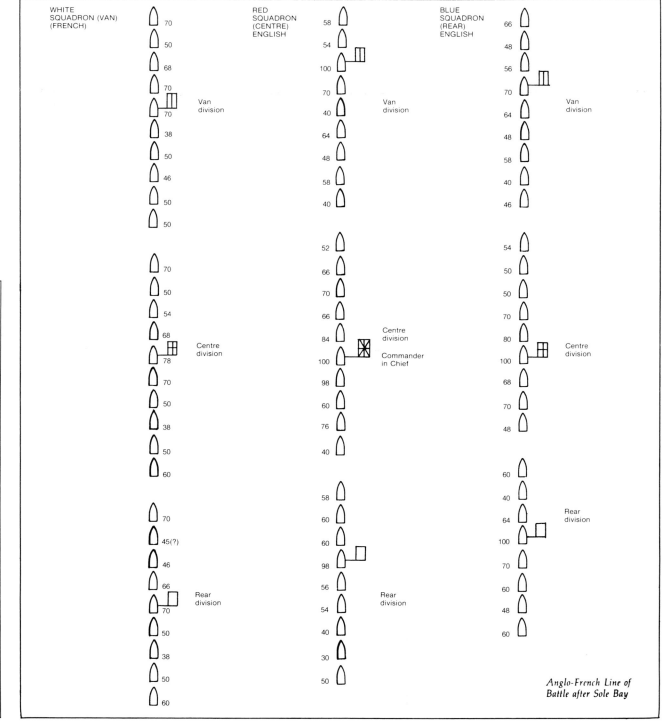

Anglo-French Line of Battle after Sole Bay

because, once the flag ships had been apportioned, the senior captains were more likely to be posted to the heavier ships. The difficulty in tracing this development lies in the lack of reliable lists of orders of battle prior to the war of the Spanish Succession. Even where lists appear in published works, no reliance can be placed on their accuracy.

At the Battle of Bantry Bay in 1689, Admiral Herbert, Lord Torrington, had seven third rates in a weak fleet of nineteen of the line. He stationed these first, third, seventeenth and nineteenth in his line, and positioned the remaining three in the centre with his flag. The six most senior captains were in the first, third, fourteenth, sixteenth (senior captain), seventeenth and nineteenth ships. Hence there was a rough approximation between seniority and strong ships at the head and tail of the line.

At Beachy Head in 1690, Tourville had only eight of his nine flag officers in a ship of at least 70 guns and he did not possess enough first, second and third rates to form a really strong flag officer group anywhere in his line except in the very centre. Here he had a 70, 98 (*Le Soleil Royal*), 80 and another 80. In the English Red and Blue squadrons there were only four flag officers; all, however, in first or second rates (100 or 90 guns) and strongly supported. The Dutch White squadron was comparatively weak in guns: the vice-admiral was in an 82, supported by a 92, but the centre, 60, 70 (flag), 60, was weak. The English Blue with one flag officers had only a 70 at the head and tail of the squadron. Otherwise there is no sign of heavy ships at the extremities in either fleet. In the Red squadron, the leading captain was second in seniority. In the Blue, the two senior captains were at the head and tail respectively, each in a 70-gun ship.

At Barfleur in 1692 Tourville's fleet, though weak in numbers, was better gunned than previously. All his nine flag officers were in first or second rates and each squadron commander had second rates as his seconds. The English were in overwhelming strength and the Dutch had nine second rates or better. In the English Red and Blue squadrons the most senior captains, though in heavy ships, were not stationed at the extremities of their respective squadrons but in most cases as seconds to flagships.

Sir George Rooke is the only English admiral whose treatment of the line of battle can be traced with reasonable accuracy over any length of time. In 1679 he succeeded Lord Berkeley in command of the confederate fleet employed against France in the Channel and Bay of Biscay, and on 21 June he issued a line of battle for the English part of the fleet.[39] The Red squadron in the centre consisted of eighteen ships-of-the-line, the vice-admiral's division having eight and Rooke himself ten, and there was no rear division. The vice-admiral's division had three first and second rates, and Rooke four, including flagships and in each case they were stationed in the respective centres. Each division was headed (on the starboard tack) by a third rate and tailed by a fourth rate. The head and tail captains were comparatively junior, being in ships of less force. The three divisions of the Blue squadron consisted of eight, nine and nine ships respectively. There were seven first and second rates, distributed two, three and two and in each case in the centre of divisions. The rear-admiral's and the vice-admiral's divisions were headed by fourth rates. All the other head and tail ships of divisions were third rates; as before, the head and tail captains were comparatively junior.

Rooke's Baltic squadron in 1700 was so small, consisting of only ten of the line, that no conclusions can be drawn from his line of battle. He and his rear-admiral, Thomas Hopson, had the two strongest ships, and the three most senior captains were their flag captains and Rooke's first captain. When joined by the Dutch, however, Rooke changed his

order of battle so as to have his two captains next in seniority at the head and tail of his own squadron. It is noteworthy here that the Swedish fleet was organised on the old principle, with the division flag officers at the extreme ends of the line, each being next but one to the headmost and sternmost ship.[40]

In the line of battle for Rooke's Anglo-Dutch fleet of 1701 we find the idea of strong centres fully worked out. In the English Red and Blue squadrons there were five flag ships, each supported by strong seconds. In the Dutch White squadron the three strongest ships were the flagships. No clear pattern emerges as regards the captains, several of the most senior being flag captains or flag officers' seconds.[41]

Rooke's Anglo-Dutch line of battle for 1702 shows quite a different pattern. Although the five English flag officers were not over-strongly posted as regards flagships and seconds, the English squadrons when joined together had a 90-gun ship at one end and an 80 at the other. Neither of these captains, however, was the most senior after the first captain and flag captains had been discounted. Three more second rates were stationed in the line at points not apparently having any special significance. The Dutch had only three second rates for their five flagships.[42] For the attack on the treasure ships at Vigo, a composite squadron was formed with four English and four Dutch flag officers.[43]

In Shovell's line of battle in the Mediterranean, 12 May 1704, he stationed his two senior captains at the head and tail of the line, one of them in a 96, though he only had two other second rates, his own flagship and one of his seconds. In his line of battle on 8 June 1704 he again had his senior captains (different from before) at the head and tail of the line in an 80 and 70 respectively, his 96-gun ship (Captain John Jennings) which had previously led now being a flagship.[44]

At Malaga in 1704, the French line was organised in what by now had become the established order. The English van had only two flag officers, but otherwise the line also followed the established order. The Dutch had only two flag officers. Taking the two English squadrons as one, the two senior captains were respectively at the head and tail of the line, though in 70-gun ships.

In Sir George Byng's squadron of sixteen ships for operations against the Jacobites in 1715, three of his captains were senior by ten years and more to all the rest.[45] Of these, one was his flag captain, one his second, and the third had the leading ship on the larboard tack.

In Norris's fleet of twenty-one of the line in 1740, the two senior captains were at the head and tail of the line, each in an 80. Each of the three flag officers was in an 80 and two more 80s were seconds to flagships. A fifth 80 carried the third senior captain whose ship was used to strengthen the centre.

At the Battle of Toulon in 1744, Mathews stationed all his thirteen heavy ships except one in the centres of his three squadrons. The head and tail ships were only 70s and commanded by captains who were by no means amongst the most senior. Apparently, therefore, Mathews did not believe in giving senior captains the leadership of the fleet. Or was it that he wanted their heavier ships to strengthen the squadron

---

39 *NMM*, SIG/A/3, a line of battle signed by Rooke loosely inserted in a copy of the Sailing & Fighting Instructions

40 Oscar Browning, editor, *The Journal of Sir George Rooke, Admiral of the Fleet 1700–1702*, London, 1897 (hereafter cited: *Rooke*) pp18–9, 24–5, 73

41 *Rooke*, pp128–9

42 *Rooke*, pp160–1

43 *Rooke*, pp231–2

44 *Byng*, I, pp44–5

45 *Byng*, III, p138

centres? This seems the most likely explanation, as he does not seem to have had any special reason for choosing either Captain Berkeley of the *Revenge* or Captain Cooper of the *Stirling Castle* to lead the fleet on either tack. Whether he regarded strength in the centres from an offensive or defensive standpoint is uncertain. Writing in 1762, Christopher O'Bryan favoured the earlier view, that where their ships were not too small, the senior captains should be posted either as leaders of the fleet on either tack or as seconds to flag officers.[46]

Fireships

Fireships could be used in two ways, either against shipping in a crowded anchorage, or at sea as part of the general tactics employed in a fleet action. 'Holmes's bonfire', in the Texel in 1666, de Ruyter's destruction of shipping in the Medway in 1667 and Rooke's attack on the French at La Hogue in 1692 are well-known examples of the first. The tactical use of fireships in fleet actions depended on having enough of them ready to hand and enough targets for them to attack, as at Sole Bay, in 1672, which explains why they played very little part in small-scale battles. They were much favoured by the Dutch, no doubt because their shallow draught enabled them to operate in shoal waters.

In battle they were stationed on the disengaged side of the line, generally close to the admiral and other flag officers. The time to set them in motion was when an enemy ship was seen to be disabled and yet still in the line of battle. If the fleet employing the fireships had the leeward position they would probably have to be towed by boats into action. If the fleet was to windward, however, a fireship could slip through the gap between two of its own ships-of-the-line and bear straight down on the disabled enemy.

The fireship was carefully prepared before the attack, filled inside with all kinds of combustible material and with special firing material on the deck. As the fireship approached its target, one or more tenders, generally brigs, ketches, smacks or small 'frigates', stood down with it to take off the crew, who, having set fire to the ship, escaped at the last minute by boat. A few volunteers sometimes remained on board to make sure of steering the fireship right against the enemy, and hooking on with grapnels. They then tried to escape by boat or by swimming.

As the fireship bore down with flames streaming out of it to leeward, the crew of its intended victim often panicked, thus prejudicing last-minute avoiding tactics and attempts to sink it by gunfire. Meanwhile there was even more desperate work for the defenders, whose ship's boats, assisted by their own ketches, smacks and other auxiliaries, attempted to sink the fireship with gunfire, or grapple it and tow it clear, or else board it and divert its course. Even if it struck home there

was still a chance of dragging it clear before its flames really took hold of its victim.

Naturally the attackers were unwilling to allow the action of their fireship to be rendered useless. They, therefore, bore down with their own tenders, ketches, smacks and brigs, not only to cover the retreat of the fireship's crew, but to attack the boats and small vessels of the enemy trying to intercept it. Thus an extensive flotilla battle might take place between the small ships and boats of both sides as a result of an attack launched by only one fireship. The attack might often come as a surprise, the fireship being generally stationed abreast of the flagships or other ships-of-the-line and not opposite the gaps between them. This was to give them protection from the enemy's fire, but it also helped to conceal their presence until they edged through the line. Thus the fireships and their tenders, together with the flotilla opposed to them, had a role analogous to that of the torpedo boats and torpedo boat destroyers under steam. Success depended on the bravery and coolness of the officers and seamen involved, especially those of the fireships themselves, who must not abandon ship too soon, even though the danger of instant death should a lucky shot from the enemy set off all the combustibles and cause the fireship to explode. In the Royal Navy of the Restoration the dangers of the fireship and anti-fireship service were reckoned so great that 'encouragement' was required in the form of the promise of lavish and specified rewards, including a gold medal for successful captains and compensation for relatives of those killed.[47] This was soon afterwards considerably extended, and the rewards increased in value and became incorporated as a separate chapter in the Permanent Sailing & Fighting Instructions.

That the advance of a fireship could be the occasion for a prolonged skirmish between heterogeneous flotillas, with the big ships joining in with gunfire, confirms the slow pace of a naval battle, with ships moving perhaps at only two or three knots. Captain Nathaniel Butler (or Boteler), writing in about 1634, thought fireships were of little value except in a narrow straight with wind and tide or current in their favour, despite their disruptive success against the Armada; he conceded, however, that if disguised as ordinary fighting ships they might score by engaging an enemy at very close quarters and then being fired and abandoned by their crews.[48] Pepys states that the Dutch had proposed to England that both sides should give up using them but that England, on Lord Sandwich's advice, had refused.[49]

Père Hoste, writing in 1697, recommended that the fireships should be stationed to windward of the fleet when cruising. They would then be in less danger from attack, and within easy reach of the admiral. Being better sailers before the wind than on it they would thus have more in hand and would not delay the fleet, while if forced to drop astern through lack of speed they could catch up again whenever the fleet started manoeuvring.[50] Similar views prevailed in the Royal Navy, and a special flag, striped diagonally yellow and white, was appropriated to fireship signals, in contradiction to the general usage of signal flags.

After the Treaty of Ryswick in 1697, fireships fell into disuse. None was used at Malaga, though each side had a number in attendance. Admiral Vernon hoped to make use of

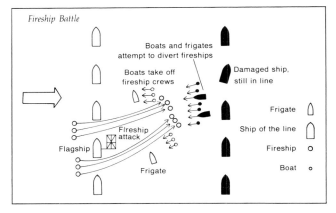

Fireship Battle

Boats and frigates
attempt to divert fireships

Boats take off
fireship crews

Damaged ship,
still in line

Fireship
attack

Frigate

Ship of the line

Flagship

Fireship

Frigate

Boat

46 *qv* p124 below

47 See the Duke of York's 'Encouragement for the captains and companies of fireships, small frigates and ketches,' *Fighting Instructions*, pp149–51

48 *Boteler's Dialogues*, pp311–2

49 J R Tanner, editor, *Samuel Pepys's Naval Minutes*, London, 1926 (hereafter cited: *Pepys Naval Minutes*) p360

50 Hoste, pp403–4

them while commanding in the West Indies in 1739–42, and issued various orders and signals to their captains and the captains of their tenders to ensure full co-operation and the safe recovery of the fireship's crews.[51] Apart from the abortive attempt by the *Anne* galley to burn the *Real Felipe* at the Battle of Toulon, they were not regarded very seriously again until the time of Howe, who strongly favoured their use.[52] Corbett notes Howe's predilection (as evidenced by his fighting instructions) with some surprise, on the grounds 'that the use of fireships in action had gone out of fashion'.[53] This is true mainly because, for various reasons, the tactical circumstances under which most sea battles were fought after 1744 were unfavourable for the use of fireships. Nevertheless, many British officers seem to have agreed with Howe to the extent that they regarded fireships as having great potential value. Kempenfelt favoured using them at the start of a battle as a means of upsetting the enemy, regardless of whether a particular enemy ship was disabled. He also emphasised the need for frigates to support them.[54]

The first signal books

The National Maritime Museum possesses what would appear to be the earliest surviving English signal book.[55] It is a vellum-bound folio manuscript, bearing the date, 1711, at the beginning of the day signals. It has an index of signals by flags, followed by a subject index. Each signal page has on it a painted picture of two ships, one above and one below, each flying a particular signal in addition to the jack, ensign and commander-in-chief's union flag at the main. The ships are painted accurately and with great spirit and in varied forms. One or more guns are painted above the ships as the particular signals may require. The order of signals starts with flags flown from the maintopmast head and descending down the mainmast; followed by signals flown from the foremast and mizzen in the same manner. Each signal has its meaning written below the picture of the ship in the fewest possible words. There are 134 day signals, six fog signals made with guns, and twenty-one night signals, the entries being shown with the lanterns painted in their various positions. The day signals thus include more than the full total of those given in the Permanent Sailing & Fighting Instructions even counting all the possible variations.

Amongst the extras are eight signals 'made by Sir John Norris in the *Ranelagh* in the Straits 1710', when he was commander-in-chief in the Mediterranean. His first four chasing signals were soon incorporated in the Permanent Sailing & Fighting Instructions under the title of 'Additional Signals to be added to the XVIth Article of the Sailing Instructions by Day and observed as the other Signals'.[56] These included signals to board the enemy; for the squadron to chase to the NE/NW/SE/SW; for the squadron to stand on the contrary tack to that the admiral was on; for the fleet or any particular ship to chase to leeward; and, in case of engaging in the night, for each ship to wear a light (at the mizzen peak).

In 1714 Jonathan Greenwood, a member of the Stationers'

Company, and presumably a printer-publisher with naval connections and a good deal of enterprise, produced a book which he sold to naval officers. It was entitled *The Sailing and Fighting Instructions or Signals, as They are Observed in the Royal Navy of Great Britain*. The book was entirely unofficial, and was dedicated to the Board of Admiralty which took office in October 1714. It is, in fact, a signal book, as stated in the title. Greenwood's intention is clearly set out in his dedication:

May it please your Lordships. My publishing this per-

A page from a manuscript signal book, c1710–11. NMM

For yᵉ Fleet to draw into a line of Battle ahead

For yᵉ Fleet to Draw into a line of Battle a Brest

51 Brian McL Ranft, *The Vernon Papers*, London, 1958 (hereafter cited: *Vernon*), pp295–6

52 Lord Sandwich to Keppel, 3 Oct 1778; 'Lord Howe, who when he left England left it with me as a sort of legacy that that useful though horrid instrument might not be laid aside.' *Sandwich Papers*, II, p177

53 *Fighting Instructions*, pp248n and 274n

54 See Kempenfelt to Middleton, 5 September [1779], *Barham*, I, p296

55 *NMM*, SIG/B/1 (formerly Sm/2)

56 p33; the exact date when this happened is uncertain.

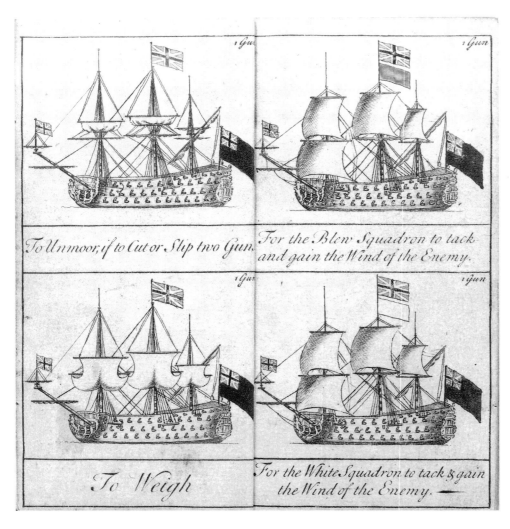

particular connection with each other, apart from the signal for a particular instruction to the White squadron being naturally followed by a similar signal for the Blue.

Jonathan Greenwood has never received the recognition he deserves for his pioneer work, his civilian status and unrecorded life causing him to be treated by many writers in a perfunctory and even supercilious manner. He was, however, a real innovator, anticipating the next printed signal book (by Millan – also a commercial publication) by thirty-four years. Moreover, he either invented new signals himself, or at any rate printed and thus publicised additional signals devised by some unrecorded admiral. These signals were incorporated by Millan in his book published in 1748, with Greenwood's name as the authority and included signals for flagships in line ahead to keep two leagues apart; ditto in line abreast; and for flagships in line ahead to keep one league from each, ditto in line abreast.

Mathews adopted these signals while in command of the Mediterranean Fleet, 1742–4. Greenwood also incorporated the six chasing signals of Sir John Norris, which were recorded in the book of 1711, though without any attribution. The use of false fires is indicated in the night signals, though no mention of these had so far been made in the Sailing & Fighting Instructions. Greenwood's book is clearly derived from the manuscript books, both as regards the order of signals by masts and the method of depicting the signals as flying from the rigging of a particular ship. Possibly he did no more than put into printed form what was already being done in manuscript.

It appears from the number of copies which have survived that the sale of Greenwood's book may not have been restricted to naval officers. His remark about its value to 'the Inferior Officers who cannot have recourse to the printed ones [the *Instructions*]' suggests a security element. This, however, may have been only one factor in the situation; the need for marking the superiority of superiors and the inferiority of inferiors may also have been considered important.

Further evidence of the popularity of this type of personal signal book is provided by a tiny pocket signal book which from internal evidence can be dated to between 1719 and 1721.[57] This is in manuscript and organised on the same lines as the Greenwood book, though mainly with painted flags and guns; the place where the signal was flown and the signification were written in words. There are only a few complete pictures of ships with signals flying from them; these, however, are very vigorously painted. It contains very much the same signals, though with a few omissions and additions. Amongst the additions are Norris's additional signals of 1710 'To board the enemy', and 'To stand on the contrary tack the admiral is on'. It also includes the new signals given by Greenwood and afterwards incorporated by Millan, and includes all six of Norris's chasing signals.

All this represents very slow progress compared with the French. Under the Sailing & Fighting Instructions the total number of signals available to the admiral had nowhere near reached that used by Tourville, nor has any example yet been found of a British thumb-indexed pocket signal book comparable with the French book of 1690. By separating signals from instructions at an early stage, the French had gained an advantage which they did not begin to lose until the American War.

A page from Jonathan Greenwood's The Sailing and Fighting Instructions or Signals As they are Observed in the Royal Navy of Great Britain, *London, 1715(?). NMM*

formance is not with any design of derogating from the value and usefulness of the *Printed Instructions;* this being an exact *copy* of them. But what I have endeavour'd at, is the putting of them in such a method, as may make them more easily and more readily found out and likewise to supply the Inferior Officers who cannot have recourse to the printed ones. And as I have disposed matters in such a manner that any *Instruction* may be found out in half a minute; so I have made it a pocket volume that it may be at hand on all occasions. As to the dangerous consequences of mistaking any signall especially by night, I need not mention them to your *Lordships*.

The plan is extremely simple.

For the more immediate finding any signall that shall be made, note that in the first pages are all the signals about the mainmast taking it from the masthead downwards. In the next pages are those about the foremast and after them the signals about the mizenmast and ensign staff. In the latter part are the signals in a fog and by night. The number of guns fired at the making any signals [*sic*] is inserted at the top of each page.

A standardised picture of a warship is printed at the top and bottom of each page, so that when the book is open, four ships can be seen simultaneously. From these ships are seen flying the various signals in addition to the jack, captain's pendant and ensign. The flags are left blank to be coloured by the buyer.

The book makes no pretension to be more than a quick means of reading signals. At the time, it was an obvious boon to all officers, 'inferior' or otherwise. It could not be used with the same facility for sending signals since the arrangement by masts involved lumping together signals having no

57 'R T's' Book: *NMM*, Tunstall Collection, TUN/31 (formerly S/MS/SS/1). The attribution of dates is easily fixed by paintings of the flagships of 'Sir George Byng', (peer in 1721) and of 'Admiral Hopson' as Rear-Admiral of the Blue, (promoted 1719). The book, therefore, must have been made after the battle of Cape Passaro. A possible clue to attribution is supplied by a page entitled 'A View of the *Launceston*' with the opposite page left blank for a picture of the ship.

Convoy tactics

Big battle snobbery and big battle romanticism have long denied convoy actions the full status of naval battles. Our ancestors knew better than this. In fact, they sometimes tended to go to the other extreme, measuring tactical results solely in terms of their broader, strategic influence on seaborne trade. For the merchants, seaborne trade was to a major extent what the war was about, so that in a sense every battle was a convoy battle, whether a convoy was directly involved or not.

Reading the recommendations in *The Seaman's Vade-Mecum and Defensive War at Sea*, published in 1700 by William Mountaine, who was mathematical examiner of the Honourable Corporation of Trinity House, one gathers the impression that by a combination of individual dexterity, mutual support, and sheer weight of numbers, no well organised convoy need fear penetration even by a number of privateers. The section of Mountaine's book dealing with defensive warfare was apparently written by Captain Robert Park of Ipswich in 1704, and was not reprinted; it was deemed relevant seventy-four years later.

If well handled, a merchant ship could, in theory at least, make things very difficult for a would-be captor. Detailed specifications existed for fortifying and barricading the ship against boarders. Equally detailed specifications existed for how to avoid capture either from the windward or leeward position, taking into acount tides, currents and shoals. In addition to sailing techniques there was the extra need for keeping out of range of the chasing warships' guns. Even at close quarters a great variety of strategems were open to the skilful captain to avoid being boarded. Against a single privateer they might well prevail, and indeed the ship might be successfully defended from barricaded positions even if driven ashore.

In theory, a convoy could put up a good fight even if were penetrated by privateers. Convoys generally sailed in three columns, and in more if very large. By keeping well closed up and using such guns and small arms as they possessed the convoy ships could relieve each other by taking a position on the bow and quarter of any privateer attempting to board their next ahead or astern. Warships were a different matter.

In practice, however, convoys were not generally well-organised or well-drilled enough to offer serious resistance to attack, except in the case of the well-armed British, French or Dutch Indiamen. Escort commanders knew this, so that when attacked they generally signalled the convoy to head for safety or scatter. There is no well authenticated case of a convoy or part of a convoy keeping together when attacked and offering organised resistance. Examples of signals for

---

146   *The Seaman's* Vade-mecum.

IX. *Discharge, but get not in the Cannon open; both must be done in your Close-quarters, and on the contrary Side to the Enemy.*

BUT if the Enemy sails better than the Ship he attacks, as is evident most Privateers out-go the generality of our Merchant-men, then will they Board him maugre all Opposition, in what Place they please, except thwart the Hawse; which is not to be done but by Accident, or Want of Conduct in the Ship so boarded.

But before he is on Board, the Commander must order all his Guns in the Waste and upon his Quarter-deck to be discharged without letting them run in; for if they should be loaded when the Enemy enters, and they should traverse any of them Fore and Aft, they would soon level the Bulk-heads with your own Cannon; and if they are run in they are soon loaded; whereas if they are out, and the Tackle-falls moused, or a running and standing Part seized together, their Men will be the more exposed before they can accomplish any such Design; and left under the Covert of your Smoke, or any other favourable Accident, they should get in a Gun, the designed Advantage may prove their Ruin, by leaving in every Gun when you retire to your Close-quarters, a Piece of lighted Match.

And those on the Side from the Enemy when engaged, in the Close-quarters, must not only be discharged, but got in, that the Enemy do not toss in Hand-Granadoes or Stink-pots, to destroy or suffocate the Men in those Quarters; and that they should be discharged is necessary, because otherwise an Hand-Granade, Fire-pot, or some such thing may discharge it in your Quarters, and do more Damage to your

Ship

The Seaman's Vade-mecum.   147

Ship than the Enemy; or by carrying away a Port make a Vacancy, where the industrious Enemy may toss in Showers of Hand-Granadoes; besides, in discharging your Cannon, they run in of themselves, whereas at such a Time the Hands cannot be spared to get them in.

X. *How to act when a Ship comes up your Wake, and lays you aboard upon the Quarter.*

THE Enemy, in his Approach to Board you, comes either up your Wake, upon your Quarter, upon your Broad-side, or lastly upon your Bow.

If the Enemy come up your Wake, ply him briskly with your Chace-Guns loaded with round and Cross-bar; and as soon as he is within Pistol-shot, give him your Fore-Chace-Guns loaded with a Double-headed-shot and a Bag of Case-shot; the former may spoil his Masts and Rigging, and the latter destroy his Men: Next let your Guns upon the Quarter be ready loaded with double and Case-shot, that as the Enemy range up your Quarter with his Men ready to enter, they may be discharged among them; let likewise your Powder-Tubs be ready, and just as the Enemy is going to share aboard, set fire to the Fuze, hoist it up to the Yard-Arm, and then let it run amain among his Men upon the Decks: If he still persist in his Resolution and board you, let all your Ports be lashed in, lest the Enemy wedge them, which is of ill Consequence, as has been before observed: Keep firing your Blunderbusses out of the Look-holes in the Quarter among his Men, as they stand thick and ready to enter; as soon as he is aboard, spring your Powder Chests upon the Quarter, for then his Men will, mounting your Quarter, be numerous. Let your Men in the Round-house be ready with their Small-Arms to give the Enemy a Volley as soon as they come upon your Quarter-Deck,

O 2

A page from William Mountain's The Seaman's Vade-Mecum and Defensive War by Sea, London, 1747. NMM

convoys from 1700 onwards exist in large numbers but they seldom have any tactical significance, either as regards the escort or the merchantmen. *The Seaman's Vade-Mecum* gathers together from various sources what must be the most extensive collection of convoy signals prior to 1790, comprising twenty for day, sixteen for night and eight for fog. However, these are solely concerned with governing the convoy's movements under ordinary conditions. No 17 (by day), which is really an instruction, pertinently remarks:

'Signals made by the Admiral in Chief, the other Flags, the several convoys of private men of war [ie, escorts], and the Captains under those convoys; the care the Trade ought to have in following their respective Commodores and the methods they ought to take when the Convoys are attacked, are peculiar to the Men-of-War, and therefore need not be inserted here'.[58]

The convoy signals issued by Admiral Lord Keith for his Cape of Good Hope expedition, 1795–6, appear rather more sophisticated.[59] Single flags flown from particular positions in the traditional style, provided fifty-eight signals, of four different types (though unclassified as such): signals by the Admiral to ship or ships of the escort, to ships of the convoy, and from warships and merchantmen to the admiral. Although convoy movements are still on a very simple basis, there are hints both of stricter control and of greater responsibility given to individual merchantmen.

Convoys of the ships of the British East India Company were notoriously difficult to handle; their captains often knew the eastern seas far better than did the escort commander, and they treated arrogant naval officers with derision and contempt. If bullied with signals and written reprimands, they risked capture by leaving the convoy altogether. Nor could they be dismissed by the Company as a result of naval complaints to the Admiralty, since they were appointed by the owners of the ships, who treated them with respect and deference. Under such conditions effective signalling depended on tact, politeness and discreet conviviality.[60]

From the tactical standpoint, there were two kinds of convoy battles. The first, such as the Saints, the First of June, St Vincent, and even in a sense the Nile, were those in which convoys are directly or indirectly the cause of the fleets coming together, but no more than that. The second, such as the two battles of Finisterre in 1747, were those in which a fleet of merchantmen protected by an escort force were the subject of direct attack by a superior force of warships or privateers. In the first category the battle might take place a long way from the convoy; in the second, it was fought with the convoy actually in sight and with its preservation or destruction the immediate, as well as the overriding, tactical consideration.

Certain general principles governed all French and British convoy tactics under sail. The commander of the escort force was obliged to maintain good reconnaissance directly ahead of the convoy's course and in any quarter from which an attack might be expected. Ships comprising what today would be called the close escort were generally stationed a little ahead and to windward of the weather column of the convoy. The escorts' minor warships were stationed between the columns of merchantmen to regulate their station-keeping and observation of signals. At night the look-out ships ahead closed in so that they did not lose sight of the lights carried by the escort commander. These were counsels of perfection, however, as very often the escort commander had only a tiny handful of warships at his disposal. Further complications arose if he was meeting a home-bound convoy unescorted as far as the pre-arranged rendezvous, or if he was detached with all or part of a convoy hitherto covered by a larger force.

The first duty of the escort commander was to his convoy. If attacked by a force which he was able to repulse, he should never pursue further than the strict protection of his convoy justified. No capture, however attractive, should be attempted if it in any way prejudiced the convoy's safety. In case of a threatened attack by a superior force, the commander of the escort should be prepared, if necessary, to sacrifice his whole force in defence of the convoy. This meant using his own judgment as to how to engage the enemy, so that having once provided for the convoy's safety by completely engaging the enemy's attention, he was free to fight in any manner he chose. If the convoy had no time in which to escape, he might have to form in line-ahead between his convoy and the attackers, or he might choose to retreat slowly in line-abreast so as to prolong his resistance as much as possible. Much depended on whether the convoy had plenty of sea room ahead or whether it was approaching some restricted area or some terminal port.

French requirements were even more exacting, as might be expected. Escort commanders were subject to very severe penalties for deserting their convoys, and merchant ships could be heavily fined for deserting their convoy or for failing to sail in one.

In every convoy action the commander of the attacking force held two advantages, whether in superior force or not. First, the ships of the convoy were forced to sail at the speed of the slowest, while in convoy, and since the speed of even the average merchantman was slower than that of a warship, the attacking commander should have no difficulty in keeping them in sight under normal weather conditions. Second, the commander of the escort, being encumbered by the convoy, was bound to fight under tactical limitations. If the convoy was not near a home or allied port, the attacker could perhaps afford to wait, striving meanwhile for the weather gage and a chance to attack the convoy direct. If, being in superior force, he decided to fight the escort at once, he might detach one or more ships to harry the convoy while the escort was fully engaged. Even when the escort force was beaten or eluded, much skill was necessary to ensure the maximum capture of merchantmen; the available stock of ships' boats and prize crews could be soon exhausted, while to damage a merchantman too much by the usual disabling operation of cutting bowsprits and rigging might mean never

---

58 p267

59 *NMM*, Tunstall Collection, TUN/17 (formerly S/MS/R/2)

60 See C Northcote Parkinson, *Trade in the Eastern Seas, 1793–1815*, Cambridge, 1937, chapter 10

Convoy Organisation

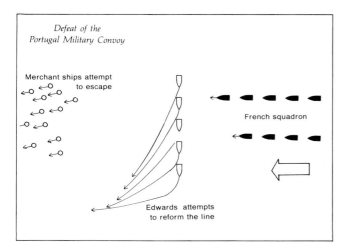

Defeat of the
Portugal Military Convoy

Merchant ships attempt
to escape

French squadron

Edwards attempts
to reform the line

René Duguay-Trouin (1673–
1736), by Largillière. MM

getting her home to the attacker's base.

In the First Dutch War the differentiation between battles brought about by the movements of convoys and those fought in their direct support is scarcely apparent, all the battles being directly or indirectly concerned with convoys sailing or about to sail. With the long series of Anglo-French wars which begin at the end of the seventeenth century, operations were no longer confined to a small area close to the terminal ports of the belligerents. Tourville's success in intercepting the Smyrna convoy in 1693 was gained off Cape St Vincent. Rooke's successful attack on the Spanish treasure ships in Vigo Bay with part of the Anglo-Dutch fleet in 1702 was in a sense a convoy action.

The most important convoy battle of the period, and one quoted as an example more than fifty years later, was the overwhelming success of René Duguay-Trouin and Claude, Chevalier de Forbin against the Portuguese military convoy of some 130 sail on 10 October 1707. The escort force consisted of the *Cumberland* (80) (Captain Richard Edwards), the *Devonshire* (80), the *Royal Oak* (74), the *Chester* (54) and the *Ruby* (80). They were attacked by Duguay-Trouin with four of the line and two frigates, followed by Forbin with five weak ships-of-the-line and a frigate. The setting was simple enough: the French attacked from the east with the wind, the convoy was scattered to the west with the escort interposed between them, forming in line ahead to the south on the larboard tack. Captain Edwards, tried to get his squadron together by turning west into line-abreast and sent a message by boat that they were then to fight in line-ahead, a little from the wind and as close together as possible so as to give mutual support. Heavy seas and the quick approach of the French prevented the *Royal Oak* getting the message. The *Devonshire* sank; the *Cumberland, Chester* and *Ruby* surrendered. Only the *Royal Oak* escaped. Twelve valuable transports were captured.

It was a great triumph for French arms. The *Cumberland, Devonshire* and *Royal Oak,* all three-deckers, were difficult to board. On the other hand, the *Lys* and the *Achille* were up to British 70-gun standard, while the *Amazone* was as big as a British 50, and the state of the sea prevented the British three-deckers using their lower-deck guns when firing to leeward. The French also had the double advantage of having twelve ships to five and a greater superiority in men, both for boarding and gun crews. More than half a century later Christopher O'Bryan was to make this battle the subject of a searching tactical analysis, urging that Captain Edwards should have made a slow retreat in line-abreast. The real test of the battle, however, seems to lie in its purpose and results. The duty of the British squadron was to save the convoy at all costs and in this it succeeded to a very considerable

measure. The French purpose was to get at the merchantmen, of which in the end they took less than 10 per cent. True, they sank a British three-decker and captured three out of the remaining four of the squadron and this remained a great feat of arms, but in a full tactical sense their success was limited.

When, in February 1708, Leake was preparing to take a large Anglo-Dutch fleet of merchantmen to sea *en route* for Lisbon, he arranged to have an escort ship two leagues ahead of him by day and another two leagues ahead of her:

. . . and in sight of the Admiral's lights by night. Four sail to

Claude, Chevalier de Forbin
(1656–1733), by Sergent. MM

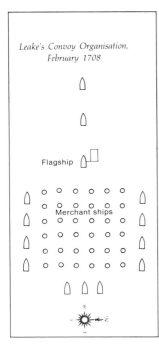

Leake's Convoy Organisation,
February 1708

Flagship

Merchant ships

keep to the southward of the fleet, at a convenient distance from each other; four more sail to the northward of the fleet in a like manner, and three sail to keep astern to secure the rear. No ship was to chase without a signal, nor to lose company as they would answer to the contrary.[61]

So important did the whole business become that a Cruisers and Convoys Act was passed very quickly through the British Parliament in 1708. No doubt arising out of this, Queen Anne's husband, Prince George of Denmark, now Lord High Admiral of Great Britain, issued special instructions for captains of warships and masters of merchant ships 'for the better security of the latter when they shall from time to time proceed under the convoy of the former'.[62] The twenty-three day signals were in fact instructions of a fairly simple kind, including Article XVII which required ships in danger of capture to cut 'jeers' [gears, ie, halyards], or otherwise disable their masts and sails to prevent their being carried off before being relieved by their escort. There are also twenty-four night signals.

Captain Thomas Fox's success against the French convoy from San Domingo in 1747 provides a good example of tactical anti-climax. Very early on the morning of 20 June, Fox, cruising off the west coast of France with a squadron of two 70s, two 60s, a 50, a 40, a frigate and a fireship, sighted a great body of ships to windward. Soon, more than 120 were counted, together with three ships-of-the-line, commanded by Commodore Du Bois de La Motte-Picquet. The wind was

NNE and all the ships on both sides were steering roughly east. After an interval of dense fog, the French escort could be seen drawing into a line-abreast and lying to, but on observing Fox's strength they filled and stood on, keeping about a point off the wind and to leeward of the convoy. Fox, by crowding sail and keeping close-hauled, managed to draw up within five miles of them. During the afternoon the French also crowded sail and steered closer to the wind, and at the same time signalled to the convoy to disperse, at which point the rear ships of the convoy tacked and headed away to the NW while the leading section continued to windward of their escort on an easterly course. So ended the first day.

At daylight on 21 June Fox could see a large part of the convoy about twelve miles away to the north-east but there was no sign of de La Motte's warships. Fox detached one of his 60s to the north-west to look for them. He gradually gained on the convoy, but only managed to capture one ship during the day. On 22 June thirteen more were taken. On 24 June a strong south-westerly gale helped the convoy to make for home ports, but altogether from start to finish Fox's squadron took forty-four of them. An additional four were intercepted by Admiral Warren before they could reach a French port. Thus, through unexplained mismanagement by de La Motte, in a purely tactical issue the French lost a third of their whole convoy without a single shot being fired in its defence.

A typical example of ill-informed criticism of an escort commander is provided by Commodore de Ternay's refusal to engage Admiral Sir William Cornwallis. On 20 June 1780 de Ternay with an 80, two 74s and four 64s was escorting a convoy of 6000 troops bound for Rhode Island. Off Monte Christe he sighted Cornwallis's squadron, which comprised two 74s, two 64s and a 50. De Ternay, seeing his convoy well on its way to leeward, hauled to the wind in line-ahead to block Cornwallis, who was still further to windward. Not only was Cornwallis greatly inferior in strength, but one of his 64s, the *Ruby,* was so far detached and to leeward that the French had a good chance by holding their course to cut her off. Alternatively, they could force Cornwallis to fight at a numerical disadvantage in order to save her. When de Ternay saw that this was what Cornwallis was prepared to do, and that in fact the *Ruby* had managed to rejoin him, he withdrew after a desultory and long-distance cannonade, and rejoined his convoy. He was aware of the presence in the vicinity of another British squadron equal or superior to his own force in strength, and to have been tempted to fight a weaker force, even with the hope of capturing the *Ruby,* would have been contrary to his instructions.

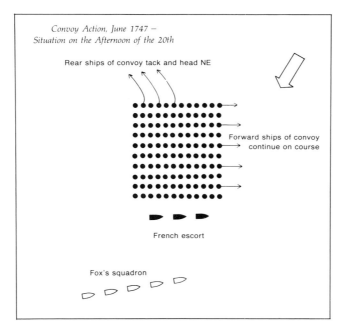

Convoy Action, June 1747 –
Situation on the Afternoon of the 20th

Rear ships of convoy tack and head NE

Forward ships of convoy
continue on course

French escort

Fox's squadron

---

61 *Leake,* II, p180

62 *NMM,* uncatalogued

CHAPTER THREE:

English Additional Signals

THE period 1714 to 1739 has often been described as one of naval stagnation, but this can fairly only be said of the years after 1728. Until then, at any rate, there was considerable naval activity, including the almost yearly appearance of a British fleet in the Baltic in connection with the Northern War. There was the war against Spain in the Mediterranean, 1718–20, including the Battle of Cape Passaro, the short war with Spain in 1728, involving the British blockade of Spanish ports in the West Indies, and the abortive siege of Gibraltar.

None of this led to a major revision of the Sailing & Fighting Instructions, presumably because no operations lasted long enough to require further development. Thus there was no tactical experience of sufficient importance to modify the system which, having passed through a long formative period, was at last consolidated. However, there was tactical development of a more limited nature. It was open to commanders-in-chief to issue additional sailing or fighting instructions.[1] Alternatively, and more easily, they issued additional signals, devoid of instructions. Such signals and instructions, though official, were only valid for the particular fleet in the particular area concerned, and ceased to operate in the case of a change of command.

How extensive the practice was at this time it is hard to say. *A Narrative of the Proceedings of His Majesty's Fleet in the Mediterranean . . . By a Sea-officer,* published in 1744 in support of Admiral Richard Lestock during the controversy following the battle of Toulon, comments:

Our Signal Book at present, has been found by constant practice in many necessary points to be defective and insufficient. . . . Men in the highest stations at sea, will not deny but that our Sailing and Fighting Instructions might be amended and many added to them, which by every day's experience are found to be absolutely necessary, tho' this truth is universally acknowledged, and the necessity of the Royal Navy very urgent, yet since the institution of these signals, nothing has been added to them, except the chasing signals excellent in their kind by the Right Honourable Sir J[ohn] N[orris].[2]

Norris's additional chasing instructions of 1710 are the earliest known; Vernon's additional signals, issued while he was commander-in-chief in the West Indies, are the most important. In this way, by piecemeal experiment and innovation, the now established British tactical and signalling system gradually threw out fresh shoots. Eventually a selection from these was published by the Admiralty, but individual admirals still continued to give out official instructions and signals of their own with temporary and local force.

Vernon's additional fighting instructions

Unlike some naval officers at that time, Edward Vernon was a man of education and property, and still full of life at the age of fifty-five, when he accepted the West Indies command. He not only issued the first substantial body of additions to the corpus of sailing and fighting instructions, but saw them generally accepted by the navy, if not incorporated in an Admiralty publication. Possibly his additions were no more than a compilation of various individual additions, examples of which are now lost, emanating from Rooke and Shovell, as well as Norris, in all of whose flagships Vernon had at one time served.[3] Their piecemeal issue, however, suggests personal ideas worked out gradually, rather than the re-issue of an existing body of lore, and the language in which they are phrased is highly personal. The 'sea-officer' was willing to acknowledge Vernon's exceptional contribution to tactical development:

Mr V[ernon], that provident great admiral, who never suffered any useful precaution to escape him, concerted some signals for so good a purpose, wisely foreseeing their use and necessity, giving them to the captains of the squadron under his command.[4]

Vernon was critical of the refusal of British seamen to study tactics:

Admiral Edward Vernon (1684–1757), by Charles Philips.
NMM

---

1 In 1735 Sir Charles Wager issued Captain Swale of the *Rippon* with the Sailing & Fighting Instructions which include Norris's additions, but without any additions or changes in ms. The printing is slightly different from that of earlier and later forms. *MOD*, Ec/39

2 pp74–5

3 *Vernon*, pp285–86. Brian McL Ranft's conclusions as editor on this point seem very reasonable.

4 Quoted in *Vernon*, pp74–5

Our sea officers despise theory so much, and by trusting only to their genius at the instant they are to act, have neither time, nor foundation whereby to proceed on; their consultations are all in a hurry; and by want of either theory or experience, which should furnish them with a competent number of ideas distinct and clear, their thoughts are puzzled, perplexed and confounded.

However, that fault was a lesser of evils. Tactics were no substitute for resolution:

Where officers are determined to fight in great fleets, 'tis much of the least of the matter what order they fight in, more especially when they are determined to make that to depend on accident at last, which if thoroughly well pursued, is unavoidably connected with science. All formality therefore as matters are circumstanced only tends to keep the main point out of the question, and to give knaves and fools an opportunity to justify themselves on the credit of jargon and nonsense.[5]

With such attitudes it is not surprising that Vernon, although he did not fight a fleet action during his period of command in the West Indies, made a substantial contribution to the development of British naval tactics.

His first additional instruction was issued on 26 July 1739, during the voyage out from England. It states that, if the enemy's force proved to be smaller than the British, and the admiral wished any of his own ships to quit the line, he should signal them to do so by flying a striped yellow and white flag at the flagstaff at the main topmast head. The ships astern were to close the line. As signals might be difficult to see, pendants indicating particular ships would be flown from the topgallant shrouds. Since the ships which were ordered to leave the line could not be expected to see the admiral's signals once the battle had begun, they are automatically

to demean themselves as a corps de reserve to the main squadron, and to place themselves in the best situation for giving relief to any ship of the squadron that may be disabled or hardest pressed by the enemy, having in the first place a regard to the ship I shall have my flag on board as where the honour of His Majesty's flag is principally concerned, and so it is morally impossible to fix any general rules to occurrences that must be regulated from the weather and the enemy's disposition, this is left to the respective captain's judgment that shall be ordered out of the line to govern himself by, as becomes an officer of prudence and resolution, and as he will answer the contrary to his peril.[6]

The second, issued 8 August 1739, established the union flag at the main topmast head as a signal to indicate 'come to a close engagement', ships 'taking their distance from the centre'. In the case of the signal being made to individual ships, their own identification pendant would be flown in addition. If to leeward of the enemy, the admiral would also 'hoist the yellow flag at the fore topmast head for filling and making sail to windward'.[7]

Next day he issued a table of the signals most likely to be used in battle with the respective flags indicated in a column opposite, under the heading of main topmast head, fore

topmast head, mizzen topmast head, and mizzen peak. The new signals for particular ships to quit the line and for coming to a closer engagement were included.[8] On the same day he issued signals for the line of battle with ships at half a mile distance from each other, and another for ships to close the admiral 'for our not losing company in the night'. Vernon made amendments to the official night signals, reducing dramatically the number of gun signals employed. The official signals, meant for bad weather, called for a number of guns 'which in moderate weather and our present cruising station might be inconvenient'.

On leaving Jamaica for his famous capture of Porto Bello, Vernon issued 'Several Additional Instructions and Signals to those of the General Instructions'.[9] To these were added a new group of 'Signals in case seeing ships in the night', which covered a strange ship seen in the north-east, north-west, south-east and south-west quarters, with the admiral's own acknowledging signal. The reporting ship was to fire a gun if her signal was not acknowledged. If more than one ship were seen, false fires up to the number seen were to be made after the original signal had been acknowledged. A signal was established for the admiral to order chase of the strange sail, and to leave off chasing. A recognition signal was established 'in case of engaging an enemy in the night', and a signal was provided for forming a line of battle at night with individual ships carrying the recognition signal. A special recognition signal was also established for individual ships carrying the recognition signal. A special recognition signal was also established for individual ships which parted company and then rejoined.

Whereas the night signal in the general instructions is by hailing, which in m[an]y case[s] might prove inconvenient, the signal for knowing each other in the night shall be for the weathermost ship to hoist the distinguishing lights where they can best be seen, which are, two lights of equal height, and the leewardmost in like manner, to show three lights of equal height where they can be best seen.[10]

It would appear that Vernon was making provision for night operations, which was something never yet envisaged in home waters, mainly because of the different climatic and weather conditions, and no doubt because summer nights in the English Channel could be as little as four hours long.

For the actual attack on Porto Bello, Vernon issued orders embodying an entirely new principle:

And you are to take notice, that in what side soever the wind be for favouring our attempt to lead into the harbour, I shall make no alteration in my line of battle now strictly enjoined you to follow, which is for Commodore Brown to lead and for all the ships to follow in the order of battle as is directed in my present line of battle, for those to lead who are to lead on the starboard tack.

The signal to form the line was to be a red flag at the main topmast head.[11]

Vernon had only six ships in his line, but this does not invalidate the radical nature of his order, which cancelled the dependence of the first formation of the order of battle on the wind direction, and instead gave permanent preference to the starboard, right-wing order. This of course would not have prevented Vernon from reversing his line by signalling it to tack or wear together. What he was trying to ensure was, first, that there should be no muddle in forming line at the start, and second, that in any event Commodore Charles Brown should lead in his 70-gun ship.

This arrangement may have appeared rather revolutionary to seamen at the time because of the confusion which existed between the formal and the practical aspects of fleet organisation. In formal terms, the second-in-command was always

5 *An Enquiry into the Conduct of Captain Savage Mostyn,* 1754; the first part quoted in *Vernon,* pp287–88, and by Sir Herbert W Richmond, *The Navy in the War of 1739–48,* Cambridge, 1920 (hereafter cited: Richmond, *The Navy in the War of 1739–48*), III, pp254–5.

6 *Vernon,* pp290–1.

7 *Vernon,* p291

8 *Vernon,* p292

9 *Vernon,* p293

10 *Vernon,* pp294–5

11 *Vernon,* p33

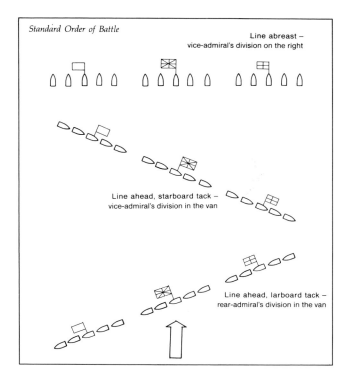

Standard Order of Battle

Line abreast –
vice-admiral's division on the right

Line ahead, starboard tack –
vice-admiral's division in the van

Line ahead, larboard tack –
rear-admiral's division in the van

deemed to be in command of the starboard division, or right wing, of the fleet when in line abreast, just as he was of an army drawn up in line on land. Just as the army, when it moved off in column, usually turned to the right, thus placing the second-in-command in the van, so at sea the second-in-command's starboard division took the lead when the fleet was on the starboard tack. The formal assumption clearly was that with the wind on the starboard side of the fleet being drawn up in line-abreast, the starboard division would lead because it was to windward. If, however, the wind happened to be on the larboard side of the fleet, then obviously it was for the third-in-command's left wing, or larboard division, otherwise the rear, to lead the fleet on the larboard tack.

This was the formal position, but it could also be applied in practice when the fleet was signalled to form in line ahead while cruising in some loose extended order or when huddled close together and lying by, even if this meant delay in the event that the ships due to take the lead happened to be to leeward of the rest. This formal organisation was also a practicable one when the fleet was signalled to form line-ahead from some other formation or from no formation at all. Adherence to the formal routine, however, could be very impractical if an admiral lacked enough heavy ships and experienced subordinates to make both the van and the rear of his battle-line strong. It could also be confusing if a shift in the wind, or the need to work to windward, obliged the fleet to tack. Adherence to the formal routine would then require that responsibility for leading the attack be passed back and forth between the van and rear commanders.

At later dates Vernon issued two orders which developed the precedent of making Commodore Brown the leader in the attack on Porto Bello into a general instruction for the van division to lead in battle, regardless of the tack the fleet was on. Vernon's Sailing Instructions contained signals for the 'Weathermost and headmost ships to tack first', union flag at the fore topmast head, for the 'Sternmost and leewardmost

ships to tack first', union flag at the mizzen topmast head, and for 'The whole fleet to tack together', in which case the union flag would be flown from both the fore and mizzen topmast heads. In his Fighting Instructions, the first two signals were repeated, as 'Van of the fleet to tack first' and 'Rear-Admiral of the fleet to tack first'. Tacking together is a manoeuvre which is easily understood; it reverses the order of sailing, making the ship which previously was sternmost, lead. What is less clear but makes sense in the light of Vernon's orders at Porto Bello, are the orders for the weathermost, or headmost ships, or the van to tack first; and, conversely, for the sternmost; leedwardmost, or rear-admiral to do the same. If each ship tacked in the wake of the ship ahead, that is in succession, the fleet would then be in its original order, but on the opposite tack. This would conflict with the usual order of battle, under which the leading ship of the second-in-command's division, being normally the van, led the fleet on the starboard tack; and the sternmost ship of the third-in-command's, normally the rear, led on the larboard tack.

When I make the signal in line of battle [ahead] for the van of the fleet to tack first in order to gain the wind of the enemy, then each ship is to tack in the headmost ship's wake for losing no ground. But when I shall tack in order to come to engagement with the enemy on the other tack, that the proper ship may lead as appointed for that tack, I will make the signal for the rear of the fleet to tack first, when each ship is to go about as fast as he can after the ship astern of him to form into a line the sooner and with the least loss of ground.[12]

Here is a very clear distinction. The van tacking first means tacking in succession, so that the original van continues to lead though the fleet is now on the opposite tack. But when Vernon wished to engage the enemy in reverse order he indicated that his intention was to signal for the rear to tack first.

The Princess Amelia [leading ship of his second-in-command, Sir Chaloner Ogle] to lead with the Starboard and the Suffolk [sternmost ship of his third-in-command, Commodore Lestock] with the Larboard tacks aboard, and if I shall find it necessary from the different motions of the enemy to change our Order of Battle, and to have those who are now appointed to lead on the Starboard tack to continue to lead on the Larboard tack on our going about: or those now to lead the fleet on the Starboard tack on the contrary [tack] to do the same, as the exigency of the service may require, I will for my signals for tacking hoist a Dutch Flag on the Flag Staff under the Union Flag, the usual signal for tacking; when they are to lead the fleet on the different tacks accordingly.[13]

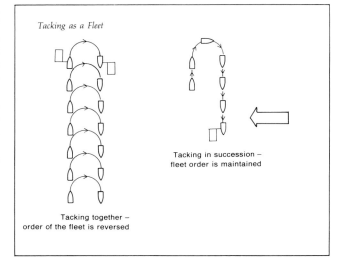

Tacking as a Fleet

Tacking in succession –
fleet order is maintained

Tacking together –
order of the fleet is reversed

---

12 *Vernon*, p298

13 *Vernon*, p298, 13 Jan 1740. Vernon's fleet had been considerably augmented.

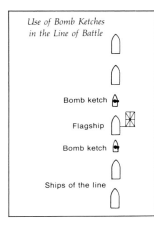

Use of Bomb Ketches in the Line of Battle

Bomb ketch

Flagship

Bomb ketch

Ships of the line

Vernon's battle line placed four 80-gun ships in the van. It may have been this consideration which led to his development of the tactical idea of having the van squadron lead in any battle, even if on the larboard tack. It may also have reflected a greater confidence in the vice-admiral, Ogle.

Vernon was a great stickler for attention to signals, proper station-keeping, and for 'daily, as the weather will permit it', practising the forming of the fleet in line of battle so as to train young officers and prevent confusion in battle.[14] He was strongly opposed to 'hasty firing' in battle, 'which only serves to embolden the enemy instead of discouraging them'. Guns were to be carefully aimed and strict fire discipline maintained.[15] He was a great believer in fireships, and issued detailed orders and signals both to their captains and to those of their tenders. Vernon also favoured the tactical use of bomb ketches in fleet actions (as distinct from shore bombardments), ordering their captains to lower their 'howitzers' and 'large mortars' to 'any level you shall judge proper for disabling the enemy's ships in their masts and rigging [which] may greatly contribute to our success in any such action'. In

battle a bomb vessel was to station itself in the interval between his flagship and the ship next ahead and another bomb vessel in the interval immediately astern of the flagship. They were to use their 'best endeavours there for disabling in her rigging the admiral or commander-in-chief of the enemy's squadron or one of his seconds either with grapes or shells'.[16]

Apart from being a tactical innovator, Vernon had a largeness of vision and pungency of expression that gives his sayings more than a contemporary interest. Though a strict disciplinarian, he trusted his officers. He was neither haughty nor secretive. Writing to the Duke of Newcastle in 1739, just after leaving Portsmouth for the West Indies, he mentioned his tactical scheme for ships in excess of the enemy's to form a *corps de réserve*, 'and which I explained more fully to them in the verbal orders I gave them upon it yesterday, having called all my captains together to deliver them their orders, and advise with them on the execution of them.'[17] 'It is,' he also wrote, 'from my knowledge of the experience of my captains and my confidence in their resolution, that I have my chief reliance successfully to execute His Majesty's orders'. For this reason he was prepared to give them a wide discretion, as seen in the concluding words of his additional instruction about the so-called *corps de réserve*:

But as I am apprehensive that it can't well be expected, you should be able to discern such signals through the cloud of smoke we may then be in, I principally rely upon their

14 *Vernon*, pp29, 165, 301–3, 413

15 *Vernon*, pp182–3, 197

16 *Vernon*, p297

17 Quoted by Richmond, *The Navy in the War of 1739–48*, I, pp41–2

18 *Vernon*, p295, also pp296 and 301

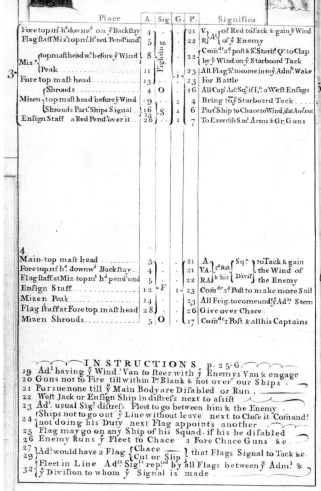

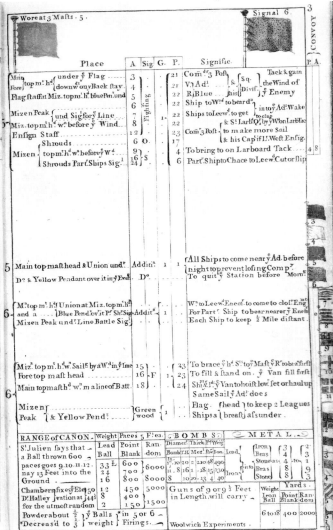

A page from John Millan's Signals for the Royal Navy and Ships under Convoy Sailing & Fighting Instructions. Articles of Warrr [sic] Given by ye Lords of ye Admiralty to Flag & other Officers . . . , London, 1749. NMM

prudence and resolution in observing where such service lie open for your execution, or require your relief.[18]

His orders to the captains of fireships and bomb ketches similarly expressed the duty of captains to act independently.

Millan's signal book

In 1748, thirty-four years after Greenwood, John Millan published 'Signals for the Royal Navy and Ships under Convoy – Sailing and Fighting Instructions etc'.[19] Millan was a commercial publisher and bookseller with a flourishing business near the Admiralty, 'next Scotland Yard', who produced not only a great deal of naval and military material, but also books on heraldry, coins, card playing, mathematics, architecture and other technical and semi-technical subjects. Although the signals form the first part of the book, they only occupy eight out of its twenty-two pages. It measures 6¾in by 4, only a trifle larger than Greenwood's, and is much thinner, at ⅝in including calf binding. It was an excellent signal book and officer's *vade-mecum* in pocket book form, but cost five shillings by itself, and eight if bound up with the 'Coins and Weights and Measures of all Nations'.

The signals were arranged in an entirely new way. Two, three or four flags were shown at the top of each page, and the flags were numbered serially throughout, beginning with the Royal Standard (then used as a signal flag) as number 1. On the middle and lower part of each page the number of the flag was printed on the extreme left; opposite it, the various places in which the flag was flown for signalling purposes. Beside each specification of place flown are four columns for (1) the number of the article, (2) the class of signal, (3) the number of guns to be fired with it, if any, and (4) the page on which the signal occurs in the *Sailing and Fighting Instructions*. Finally, on the right-hand side, is a column showing the signification of the signal. The very fact that Millan numbered his flags as well as depicting them in the usual heraldic hatching with a key, shows that he was pointing the way towards a numerical system not officially adopted for another half century.

One of Millan's most useful innovations was his classification of signals under four main headings, Fighting, Fireships, Officers and Sailing, as shown in the *Sailing and Fighting Instructions*. He also introduced a new class, Additional; these signals had of course no reference number to article or page, and prove to be those issued by Vernon. Vernon's night signals were included, and two additional fog signals which are not specifically known to have been issued by Vernon, but were certainly adopted by Admiral Sir William Rowley in the Mediterranean when he succeeded Mathews. Millan also included four signals marked 'Greenwood', taken from his book, and Norris's chasing signals.

Like Greenwood's signal book, Millan's was intended as a quick means of reading signals, not of sending them. Also like Greenwood's, it makes nonsense of the idea that there was anything secret about either the signals or the instructions to which they referred, since the nature of these latter was sufficiently indicated. Nor was the nature of those other instructions which did not require a signal of sufficient importance to prevent the French, at least, from knowing or guessing every word of them.

The Battle of Toulon

The Battle of Toulon (11 February 1744) is the greatest example of tactical disorder in British naval history. Never had a Board of Admiralty deliberately yoked together two such mutually antipathetic officers as Admiral Thomas Mathews and his second-in-command Vice-Admiral Richard

Admiral Thomas Mathews (1676–1751), by Claude Arnulphi. NMM

Lestock, the haughty secretiveness of the one provoking a devastating retaliation of tactical bloody-mindedness on the part of the other. Never had so many captains, on the whole brave and experienced men, worked according to rule with such exasperating results. Never had any battle resulted in so many courts-martial. Never were so many naval careers ruined, either temporarily or permanently, several with no

19 Although 1746 is the generally accepted date of publication and appears at the foot of page 1, on page 20 appears the date 10 February 1747/8, opposite 'Military Honours'.

Admiral Richard Lestock (1679?–1746), by John Wollaston. NMM

*Admiral Sir William Rowley
(1690?–1768), artist unknown
(British school, eighteenth
century).* NMM

20 Home Office Records, Admiralty 102, quoted in: Sir John Knox
Laughton, editor, *The Naval Miscellany*, vol II, London, 1912 (hereafter
cited: *Naval Miscellany*, vol II, p214.

justification whatever. Never was such a clear case of
delinquincy absolved by such a combination of forgery,
perjury and general crookedness. Never had professional
rancour burnt so strongly with so little political ardour to
blow the flame.

Yet Admiral Mathews was an able commander-in-chief
who strove hard and conscientiously, and in the main
successfully, to carry out a multitude of duties assigned to
him in the Mediterranean with forces inadequate in number,
manning and equipment. At the time of the battle he was
aged sixty-seven, and suffering from gravel. Worn out by
service and exasperated by his grievances against the Admiralty, he was only hanging on to his command 'until this
Hurly-Burly shall be over'.

Lestock, his designated successor, was a highly experienced officer, aged sixty-four or more. Mathews wrote that
Lestock was

a cripple at best, subject to very severe fits of the gout . . . and
is, at this very instant, so very weak that it was with great
difficulty he could get out of his ship to come on board me to
be sworn for his commission. He returned much fatigued,
though supported by his Captain up and down the ladder.
Should any accident happen to me when we come to action, I
humbly conceive he is by no means able to go through the
fatigue which such a command and in such a conjunction will
require. It's true he can sit in a chair, but that is all he can
possibly do, except God should work a miracle.[20]

This was probably true; it was never contradicted. Today
it is likely that neither admiral would be regarded as employ-
able on active service.

Rear-Admiral William Rowley, third-in-command, was
aged about fifty-three. He was able enough, and when he
succeeded to the command he introduced the signals for lack
of which Mathews had failed.

A chart of Hyères Bay from A
Narrative of the Proceedings
of His Majesty's Fleet in the
Mediterranean Including an
accurated Account of the late
Fight near Toulon, and the
Causes of our Miscarriage, *by
'a Sea-Officer' (sometimes
attributed to Admiral Lestock but
certainly inspired by him).*
NMM

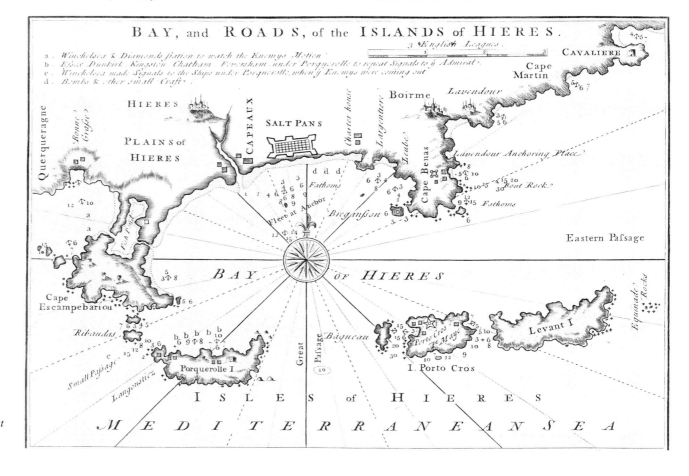

England had been at war with Spain since 1739 and Mathews's Mediterranean fleet was blockading a Spanish squadron which had been lying for some considerable time in the neutral port of Toulon. French intervention in support of Spain was imminent, and Mathews had instructions to attack the French warships if they put to sea in company with the Spaniards. He was particularly instructed to prevent a Spanish invasion of the territories of Britain's allies in Italy, to co-operate with the allied land forces, to protect British merchant ships and business interests, especially at Villefranche, and to seize the merchant ships of Spain. His station, at this time for watching Toulon was Hyères Bay, within sight of the French coast.

Inside Toulon, Admiral La Bruyère de Court, a veteran of nearly eighty, had a more clearly defined task. He was instructed to put to sea with a French squadron, taking the Spanish ships under his own command. He was to break the British blockade by battle, being careful, however, not to open fire first. Supposing the British withheld their fire against the French squadron, he was to allow the Spaniards to begin the battle.

De Court planned to attack the British in Hyères Bay. The combined Franco-Spanish fleet began to leave Toulon on 8 February 1744 with a light northerly breeze. This, however, was unfavourable for working east towards Hyères Bay. Next day the Spanish squadron was still not clear of land and, though the lightish wind had gone a little to the west of north, de Court had lost any chance he might have had of bottling up the British. The wind soon veered to south-west making it difficult for Mathews to get his fleet clear of Hyères Bay, as there was a heavy swell and a strong current setting east along the coast. By 10 February, de Court was clear of land with his whole fleet, formed in line ahead standing to the south with a light breeze at WNW. The British were in view to the east and to leeward. This was important, as de Court had apparently given the Spanish Commodore, Don Juan José Navarro, some kind of undertaking that he would not allow the Spaniards to become engaged in a battle to leeward.[21]

The French led with sixteen of the line, de Court, in the *Terrible* (74) being fourteenth, counting from the van. His seconds were another 74 and a 50. Indeed the French possessed nothing stronger than their five 74s. De Court's station set him in the centre of the combined fleet. The Spanish squadron consisted of twelve of the line, mostly weak ships, except for the rearmost ship of 80 guns and Navarro's flagship, the colossal *Real Felipe* of 114 guns, stationed sixth in the Spanish part of the line and twenty-second in the line as a whole.

Passing east of Porquerolles Island on the morning of 10 February, with the more favourable breeze at WNW, the British were still in considerable confusion when at noon de Court led the combined fleet towards them in line abreast. The breeze, however, died away and then changed to the east, on which de Court turned his fleet once more into line ahead and stood to the south on the larboard tack. He was under easy sail with his line well formed. Though making no attempt to force an action, he was certainly not running away.

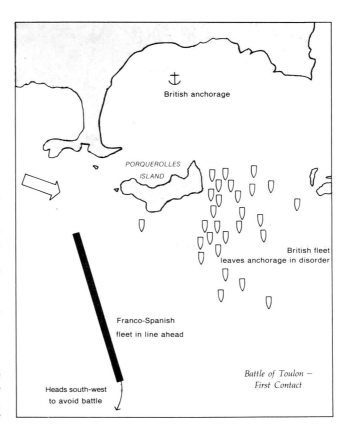

British anchorage

PORQUEROLLES ISLAND

British fleet leaves anchorage in disorder

Franco-Spanish fleet in line ahead

Heads south-west to avoid battle

Battle of Toulon — First Contact

Mathews's fleet was still in some confusion. Despite the easterly breeze which at last enabled him to clear the restricted waters of Hyères Bay, there was still a heavy westerly swell which made it difficult to form the line of battle ahead.

In terms of traditional naval precedence, Lestock's squadron was the van, Lestock being senior to Rowley. By Mathews's orders, following common practice (and in contrast to Vernon's innovations), Lestock's squadron was to lead the fleet when on the starboard tack and Rowley's when on the larboard. When the fleet first emerged from Hyères Bay on the starboard tack, with the wind westerly, Lestock had been in the lead, with his leading ship, the *Revenge*, at the head of the line. Now, with the easterly breeze, the larboard tacks were on board. This meant that not only had Rowley's squadron to push ahead and take the lead, but that the order of ships in each squadron had to be completely reversed. Thus the *Stirling Castle*, hitherto endeavouring to get into station as the rear ship of Rowley's squadron, and hence of the whole fleet, had now to push ahead and lead the fleet. Conversely Lestock's *Revenge* had to drop back and take her place in the sternmost station of all. No wonder that, with a light wind and a heavy swell, the fleet suffered further confusion and delay.

It was probably this unfortunate experience which led Rowley, when he succeeded Mathews in command of the fleet, to promulgate Vernon's additional signal for 'same ships to continue to lead on different tracks'.[22] Vice-Admiral Henry Medley, who succeeded Rowley, had an additional signal with much the same significance: 'Ships leading in a line of battle, to continue to lead on the different tracks'.[23]

By contrast, de Court had a well formed line. Whatever his intentions he now held the initiative and with clean ships he could accept or refuse battle as he pleased. At about 3pm Mathews, fearing that the combined fleet would slip right away from him, signalled for the line of battle abreast. The result was disappointing. Though Mathews managed to

21 *Naval Miscellany*, vol II, p272–3. See also *A Letter to a Friend in the Country occasioned by the late Naval Engagement in the Mediterranean*, 1744, p13.

22 The signal was a Union over a Dutch jack at the main topmast head. See Trelawney's Mediterranean Signal Book, *NMM*, Tunstall Collection, TUN/12 (formerly S/MS/AS/1).

23 The signal was a Union at the fore topmast head. See Trelawney's Mediterranean Signal Book, *NMM*, TUN/12.

bring the centre squadron to within four miles of the enemy's centre, with Rowley's van some distance to windward of him and about five miles from the enemy, Lestock's squadron trailed a long way astern and to windward. His leading ship was about three miles to the north and east of the rear ship in Mathews's squadron. On Mathews's approach, de Court turned his fleet away towards the south-west, no doubt to avoid bringing the French squadron into direct contact with the British. By 6.30pm it was getting dark and Mathews made the night signal for the whole fleet to bring to on the port tack, that is with the ships pointing roughly south.[24] He assumed that both Rowley and Lestock would bring their respective squadrons into line with the centre before bringing to, thus completing the execution of the original day signal for the line abreast. Neither did so; on the contrary, they brought to at once, with the result that next morning they were both out of alignment, Lestock being far astern. At the subsequent courts-martial it was argued that Rowley and Lestock should have stood on with their respective divisions until properly in line. Rodney commented later that Lestock should certainly have come to 'in the wake' of his commander-in-chief 'agreeable to the known practice'. Nothing, apparently, was officially laid down and the point was never settled at the time. Lestock's action, however, was to have a disastrous effect on the battle the following day.

During the night the ships of the combined fleet, also lying with their heads to the south, drifted in a south-westerly direction and more rapidly than Mathews's and Rowley's squadrons which, moreover, drifted rather more to the west. Lestock's squadron, being nearest the Gulf of Giens, was caught by the easterly current running near the coast; by dawn on 11 February was seven or eight miles away from the centre as well as to windward. Rowley was also out of alignment with Mathews and also to windward of him, as he had been on the previous night, but relatively no worse than he had been before. The combined fleet could be seen to the south-west under easy sail, still heading south on the larboard tack.

Lestock made sail as soon as it was light, followed by Mathews and then by Rowley. At 6am Mathews signalled for the line of battle ahead, and at 7am for the line of battle abreast, and at 7.30am he signalled for both the vice-admiral's and rear-admiral's squadrons to make more sail. Rowley obeyed but Lestock did not, even when Mathews later sent a lieutenant to him in a boat with an order. Mathews sent another lieutenant to order the rear ships of the centre to close up and at the same time to order Lestock to make more sail. In response to this Lestock not only showed no hurry but actually shortened sail on two occasions. Rowley, though he made more sail, was slow in getting into his station, being still too much to the east and windward of the centre. At about 7.30am Mathews made the signal for the line of battle ahead and kept it flying until after the battle had virtually ended.

Meanwhile de Court was standing on slowly to the south, having easily the legs of Mathews's fleet. When Mathews himself had to shorten sail to allow the rear ships of the centre to come up, the French also shortened sail, though they were still gaining on the British. When Mathews once more made sail, the French did the same. When Mathews seemed to be approaching them they bore up and drew away, keeping their original distance of about three miles, and then resumed their southerly course. Having clean ships, well handled, de Court could easily prevent Mathews from bringing his own line close enough for a direct van to van and centre to centre attack.

Mathews realised all this well enough, and at about 11am he made the signal to engage, but a quarter of an hour later he cancelled it, realising that it was impossible to engage in the accepted manner by bringing his van level with the enemy's van at close range. Rowley was still two and a half miles east, and to windward, of the enemy's van and his leading ships were still further to windward; nor were they even level with the leading ships of the enemy. It seemed impossible for them by making more sail to stretch far enough ahead to turn to starboard and attack the enemy's van ship for ship, unless the French were prepared to wait for them. Nor did an approach in bow and quarter line offer better prospects: in any sailing match the French must win. Mathews's own centre squadron, though in reasonably good order, was two miles to windward of the enemy line and in the rear of the enemy centre. Lestock's squadron was a long way astern, his leading ship being about three miles from the sternmost and nearest ship of the enemy.

Mathews was now extremely worried. The further he was drawn south, away from the French coast, the more he risked letting the Spaniards ferry their troops across from Barcelona to Spezia. In addition he had intelligence that the French Brest fleet had put to sea on 26 January, the assumption being that it would enter the Mediterranean and make a junction with de Court. Was it not highly probable that at this very moment de Court was luring him to meet the Brest fleet, against which he would be heavily outnumbered? Mathews surely felt that he must fight at once as best he could. His only chance of forcing a general action seemed to lie in attacking such ships of the combined fleet as he could reach within the next hour, 'judging it absolutely necessary for His Majesty's service to come to an engagement with any of the fleet I could come at, tho' in never so irregular a manner', before the junction of the Brest squadron took place, which was by all my intelligence hourly expected'. This reads like the honest opinion of an intelligent and forthright naval commander.

Having made up his mind, Mathews once more made the signal to engage, at about noon. This time he turned his flagship direct to starboard, carrying his whole squadron with him in what was more like a line abreast than a line ahead, and thus making what was practically a right-angled approach towards the enemy's line. His avowed object was to attack the Spanish squadron at the rear of the combined fleet and if possible to force the French to shorten sail to help them and so 'bring on a general engagement'.

If the French continued to draw ahead, Mathews argued,

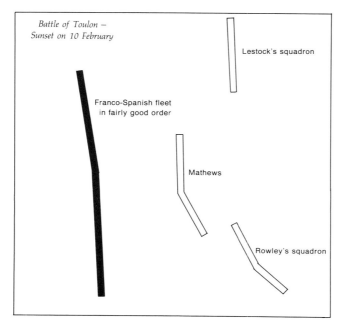

Battle of Toulon –
Sunset on 10 February

Lestock's squadron

Franco-Spanish fleet
in fairly good order

Mathews

Rowley's squadron

---

24 The signal was four lights in the fore shrouds and eight guns.

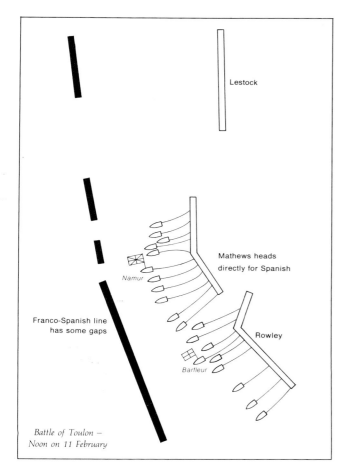

*Battle of Toulon –
Noon on 11 February*

All this must have seemed as obvious to Mathews as it is to us today. Yet, determined as he was to come to an irregular engagement, he seems unwittingly to have encouraged pedantry by keeping the signal flying for the line of battle ahead. His fleet commanders thus beheld the spectacle of their admiral bearing straight down to engage the enemy and carrying most of his division with him in what was roughly a line of battle abreast, though his two signals taken together seemed to demand that they should seek to engage by keeping roughly the line-ahead and gradually edging to starboard so as to reach the enemy. By the approach apparently signalled, the British fleet would certainly have had a better chance of engaging the enemy van to van, centre to centre and rear to rear. The fact that from the point of view of relative sailing speeds this was quite impossible, unless the French co-operated, did not seem to trouble the British commanders. All they worried about was that the admiral was going down to engage with his line of battle still unformed and his van not yet level with the enemy's nor his centre and rear with the enemy centre and rear.

Tactical disarray took over; Rowley, uncertain of Mathews's intention for the van, at once bore down in the *Barfleur* with his seconds to engage the *Terrible*, carrying with him all the ships of his squadron astern of the *Barfleur*. What should the four leading ships of the van have done? According to Article 19 of the Fighting Instructions they should have stretched ahead to engage the twelve leading ships of the enemy, now spread out over a distance of about four miles. This seemed difficult as well as dangerous. According to an additional fighting instruction promulgated by Mathews himself, they should have kept the same distance from each other as 'those ships do which are next the admiral', his seconds. What happened? The *Chichester* (80), next but one to the *Barfleur*, half obeyed this additional instruction and engaged the weak 50-gun *Diamant*, but only at some distance. The captains of the three leading ships bore down to within a mile of the enemy and then stretched ahead and kept their wind, deeming it their duty under the circumstances to deter the French van ships from tacking and so doubling the rest of the British van and centre. For this sensible act of individual initiative, they were all tried by court-martial and cashiered, as also was the captain of the *Chichester*.

the Spaniards were doomed since Lestock's squadron would eventually close up to complete their defeat. Indeed it looked as if he would succeed in making what came later to be regarded as one of the most effective and most dreaded forms of tactical approach, a concentration on the enemy's rear.

He was helped, moreover, by disorder in the Spanish line. De Court had been too slick for his allies, who were unable to make sail quickly enough after each shortening of sail by which the French ships had confused and 'amused' the British. The Spaniards were now straggling astern with a gap of about a half mile between the second and third ships and another of one and three quarter miles between the seventh and eighth. It is doubtful whether this was a deciding factor in Mathews's calculations, though it certainly helped his plan because the Spanish rear ships would be sooner at Lestock's mercy. At the same time they were too far astern to be rescued by the French, supposing de Court decided to tack and double the British van.

Having made his decision, Mathews signalled to engage while still keeping the signal flying for the line-ahead. He shouted to the captain of his second, the *Marlborough* (which was astern), that he himself would engage the *Real Felipe*. For this he had three good reasons. First, she was well on his starboard quarter at a distance of about two and a half miles so that he was bound to close her if he made his quickest possible approach to the enemy line by bearing down before the wind on a course a little south of west. Second, according to tradition it was for flagships to engage flagships, and if he could not get up with de Court in the *Terrible* then the Spanish flagship was his next natural opponent. Third, for purely practical reasons it was best that a ship of the *Real Felipe's* unusual gunpower should be engaged by the strongest British ships available, the admiral and his seconds, the *Norfolk* (80) and the *Marlborough* (90).

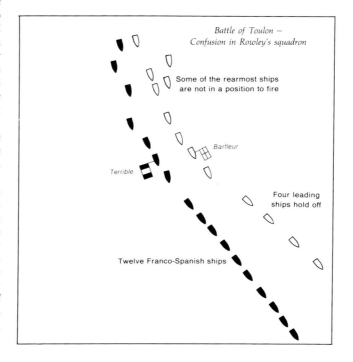

*Battle of Toulon –
Confusion in Rowley's squadron*

Some of the rearmost ships
are not in a position to fire

Barfleur

Terrible

Four leading
ships hold off

Twelve Franco-Spanish ships

Meanwhile the French van ships stood on unengaged and were soon followed by the *Terrible*, her seconds and the sternmost French ship, after their preliminary fight with Rowley's ships. These sternmost Frenchmen edged away a little to leeward and easily cleared their British opponents. In rear of the *Barfleur* and her seconds, Rowley's three remaining ships and two of his 50s at first found only two French ships to oppose them. They engaged the Spanish van ships as they began to come up, but nothing decisive happened for some time.

In the centre, Mathews and his seconds, together with the two ships ahead of them, engaged their five opposite numbers very closely, damaging the *Poder* (60) severely and driving Navarro's two seconds out of the line. The fifth was forced to make sail to catch up with the rest of the fleet. Even the *Real Felipe* was badly damaged, though she fought back magnificently and inflicted severe damage on both the *Namur* and the *Marlborough*. With the enemy dropping to leeward in disorder, there was a splendid chance for the British ships to press them harder and make at least one capture. No captain, however, was prepared to pass to leeward of the admiral with the signal for the line still flying. The remaining four ships of the British centre could equally have joined in and smashed the *Real Felipe* (only one had so far opened fire and they were now practically unopposed, as the five remaining Spaniards were still straggling into action). Similarly, the British centre ships, so far not seriously engaged, had the chance to concentrate on the remaining Spaniards as they came up, being able to see that Lestock's rear squadron was at last beginning to close up to support them. Yet, apart from the *Dorsetshire*, they were ineffective, holding off to windward and making little attempt even to engage the Spaniards at long range.

It was now nearly 4pm and, as Lestock came up with his

squadron, his two leading ships opened distant fire on the two sternmost Spanish ships. His flagship, the *Neptune*, stationed third in his squadron, fired a few shots at the retreating Spaniards ahead. He made no attempt whatever to bear down and cut them off; on the contrary he kept his whole squadron well to windward and failed completely to join the battle. Meanwhile, the leading French ships were still in good order ahead of the British van and about a mile to leeward of it.

Captain Hawke of the *Berwick*, stationed two astern of the *Barfleur*, had been very active all through the fight. As the *Poder*, much damaged, emerged from the pell-mell round the *Real Felipe*, making sail to catch up with the French, he bore down and, after a battle of two hours, forced her to strike. This was the only capture of the day. There was a gap of a mile and a half between the *Poder* and the *Real Felipe*, which was lying to leeward with another Spanish ship and no doubt waiting for the remaining Spaniards to come up. Mathews tried to organise a fireship attack on the Spanish flagship, but the fireship was not properly covered by the ships astern of the *Namur*, though it was blown up by the Spaniards only a few yards from its goal.

De Court had already seen that only three Spaniards were following him and easily divined that the rest were in danger of capture by the British centre and rear. Sometime between 3 and 4pm he signalled his van to tack in succession. The three leading ships at once obeyed and began to steer north, with the wind still at about ENE. The three leading ships in Rowley's division did the same, so as to protect their main body from being doubled. The rest of the French ships failed to see the signal and made sail as before, on which the three leaders returned to their stations, though keeping a little to windward. De Court then signalled the whole French division of the fleet to tack together.[25] This time they saw the signal, and at once went about, hauling fairly close to the wind on the starboard tack. Rowley quickly signalled his own squadron to tack. Although his three leading ships actually anticipated his signal they were nearly caught by the French owing to their inferior speed. De Court, however, did not waste time on them, nor powder and shot, and, passing under their sterns within musket shot, bore up and steered large to cover the Spaniards.

If Mathews's intention had been to 'bring on a general engagement', his wish seemed about to be fulfilled, though not in the manner he had originally intended. So quick were the French that they retook the *Poder* before Hawke could even withdraw his prize crew. Closing the gap between her and the *Real Felipe*, they re-established close contact with the Spanish squadron. Mathews made no attempt to force a close action for possession of the *Poder*; and nor did he try to draw de Court into a pell-mell round the *Real Felipe*. Nor, with damaged ships and his general tactical disarray, could he be expected to accept the risks of what must become a night action. Instead, he signalled his own squadron to tack. Owing to the heavy swell, he and most of the centre had to wear. They then stood away to the north under easy sail with Lestock's squadron also in retreat, immediately ahead of them, and Rowley's astern. Some of the British centre and van engaged for a time the rear Spanish ships pursuing their original southerly course, with the two forces passing each other on opposite tacks, but it was now dark and both fleets brought to for the night and repaired damage.

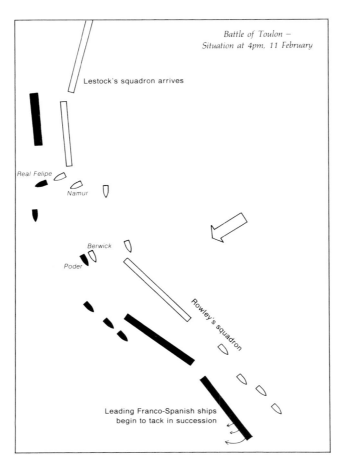

*Battle of Toulon –
Situation at 4pm, 11 February*

Lestock's squadron arrives

Real Felipe

Namur

Berwick

Poder

Rowley's squadron

Leading Franco-Spanish ships
begin to tack in succession

25 *A Narrative of the Proceedings of His Majesty's Fleet in the Mediterranean, 1744* (hereafter cited: *A Narrative of the Proceedings . . .*), p102. (*NMM, TUN/179*).

The courts-martial

The court-martial which sentenced Mathews to be cashiered listened to fifteen separate charges against him, the most important of which were that he made the signal to engage before the fleet was ready and formed in line of battle; that he failed to steer a proper course so as to bring his van and centre against those of the enemy, which it was possible for him to have done; that he failed to support the *Marlborough* sufficiently when engaged with the *Real Felipe;* and that he failed to direct the battle effectively after it had begun, especially as regards those captains directly astern of him who neglected their duty. His conduct on the days following the battle was also heavily censured.

It would seem that the attack Mathews devised was the best possible under the circumstances, but, being irregular, it had left some of the captains in his own squadron puzzled as to what they should do, with the result that they did very little. Ironically, it was the straggling of the rear Spanish ships (all five of which should have been taken) which left the puzzled British with nothing to fire at.

Although the court-martial found no fault with Mathews for keeping the signal for the line flying during the battle, this really does seem to have been a major mistake, since it invoked the taboo on breaking the line enshrined in Article 21 of the Fighting Instructions. If he had hauled down the signal for the line and kept the signal to engage flying as soon as the *Namur* came to close quarters with the Spaniards, his captains would have been free to fight pell-mell. There was no signal ordering the fleet to come to a closer engagement when to windward, though Rowley later added one for the 'Leewardmost ships to come to a closer Engagement', the precise meaning of which is itself unclear.[26]

When Lestock was tried he was charged with all the shortcomings that might have been expected from the way in which he hung back on 10 February and during the battle, and the evidence against him was voluminous, well-attested and absolutely damning. He was, however, acquitted on all charges. His defence statement was extremely well-argued, and based on a literal and pedantic adherence to established principles. The evidence he called was, in Sir Herbert Richmond's view, not only inferior in authority and character, but was supported by seemingly perjured statements, wholesale falsification of journals and the admission of arguments and statements which should have been rejected by the court as contradictory and inadmissible. It would seem that the nature of the men who composed the court inclined them more towards condoning superior backwardness than superior hastiness. For inferior backwardness, that is in the case of captains, they had no mercy, nor for the display of individual initiative by the three leading captains of the van.

Today it is easy to see where Mathews went wrong. Neither he nor Lestock, while in temporary command, seem to have done enough, or indeed any, exercising of the fleet in forming the line of battle ahead and abreast, at varying distances between the ships, and in bearing down together on a supposed enemy.[27] His relations with Lestock were hopeless from the start. According to Beatson,[28] he had made Lestock's early recall a condition of his taking the command and had publicly rebuked him at their very first meeting in the Mediterranean. On the evening of 9 February Lestock had gone on board the *Namur* to ask if Mathews had any commands for him in the coming battle, to which Mathews is said to have replied that it was a cold night and wished him good evening. If this story is true it shows an unpardonable attitude on Mathews's part. Here was someone he disliked, whom he had himself described as a cripple, making the tiring ship-to-ship visit in an apparent spirit of humility. No wonder Lestock felt bloody-minded next day and the day after, though it was equally unpardonable on his part that His Majesty's Service should suffer thereby.

The second charge against Mathews at his trial was that, although it was his duty 'to appoint necessary and proper signals, for the better conducting the said fleet under his command, as well by night as by day, according to the various exigencies of His Majesty's Service; yet the said Thomas Mathews did not direct and appoint necessary and proper signals'. In particular, he had made no provisions for a night signal ordering the fleet to bring to in line of battle. Instead,

> On the tenth of the said month of February His Majesty's fleet bearing down on the said combined fleets of France and Spain, then laying-to in a regular and well-formed line of battle, in full sight, and within four, five or six miles distance of His Majesty's fleet, the said Thomas Mathews, in the night, did make the night-signal for the fleet to bring to; by which signal the windermost ships of the fleet [Lestock's squadron] were to bring to first; and did not appoint and make a night-signal to form the line of battle, and to bring to, and keep in the line.

To all this Mathews answered that he had given the necessary and usual signals in the same manner and form as he himself had received them from Leake, Admiral James Berkeley and Torrington (that is, George Byng, Viscount Torrington); that if Lestock had not shortened sail during the afternoon of 10 February, he would have been well up with the centre and in line abreast by nightfall; and that while it was true that the night signal to bring to directed the weathermost ships to bring to first, (and rightly so, to prevent the weathermost ships colliding with ships to leeward), nevertheless if Lestock had been in his proper station in line abreast, there would have been no ships to leeward of him except the enemy's.

The exact meaning of the charge, and Mathews's answers, are not at all clear. The charge seems to imply that Mathews, as commander-in-chief, should have issued his own set of additional signals and instructions to be used by the ships under his command. In particular he should have been ready with additional night signals. Mathews seems to have contended that this was unnecessary, and that if Lestock had behaved properly everything would have gone satisfactorily.

The court decided in Mathews's favour on the grounds that he was not compelled to issue any additional signals. Nevertheless he should have contrived things better by making a night signal for the line of battle. As, however, no such signal officially existed, he should have sent officers in boats through the fleet, the night being fair and calm.

Asked about this by the court, Mathews replied with remarkable candour that he had had

> very little rest for the two nights and days before, and was thereby much fatigued; therefore as soon as I had brought the ship [*Namur*] to, I left the deck in order to rest myself, that I might the better be enabled to go through the next day's expected fatigue: not imagining that an officer of Vice-Admiral Lestock's rank, and very great experience, should require a message sent to him to do his duty.[29]

26 A yellow flag at the main topmast head combined with a Union at the mizzen topmast head. Trelawney's Mediterranean Signal Book, *NMM*, TUN/12.

27 *A Narrative of the Proceedings . . .*, p83

28 R Beatson, *Naval and Military Memoirs of Great Britain, 1727–83*, London, 1790, 3 vols, I, p153

29 *Admiral Mathews's Remarks on the Evidence given and the Proceedings Had, on his Trial and Relative thereto,* 1746 (hereafter cited: *Admiral Mathews's Remarks*) pp50 1. (*NMM*, TUN/191).

We know that Mathews had in fact made at least one addition to the Fighting Instructions. This was to Article 1, concerning station-keeping 'save only in time of fight.' It reads: 'Every ship is to observe and keep the same distance those ships do which are next to the Admiral'; meaning that the distance between the admiral and his seconds was to be the determining distance for all. Mathews also issued signals for flagships in a line of battle or abreast to be a league or two leagues apart.[30] He may well have issued more, and the whole situation thus remains obscure. What the court was really saying was that there was a need for more signals to cope with particular needs; this was a fact apparent to all.

The charges, answers, counter-charges and exculpations made in the controversy reveal exactly the attitude of mind satirised by Vernon when he wrote, 'Our enquiry since [the Dutch Wars] has not been whether a Commander be bold, daring and intrepid, but whether his ship lay in such and such a disposition; and what he said and she said; whether a line of battle was properly formed, and if such a signal did not represent this or that.'[31]

Mathews's case was that Lestock failed to support him. Lestock's case was that Mathews had rashly committed the British fleet to battle before the line was properly formed and that it was impossible to support him as long as he kept the signal for the line flying.

The main interest of the controversy lies in the issues raised about the line of battle and the methods of engagement. On the line itself the pro-Lestock *History of the Mediterranean Fleet*[32] is superbly authoritative:

A line of Battle is the basis and foundation of all discipline in sea fights, and is universally practised by all nations that are masters of any power at sea; it has had the test of a long experience, and stood before the stroke of time, pure, and unaltered, handed down by our predecessors as the most prudent, and best-concerted disposition that can possibly be made at sea.

The *History* continues with the observation that 'no temptation, no allurement of any sudden advantage to destroy the enemy, no exploit, let it be ever so considerable, and well executed, can extenuate the guilt of disobedience' involved in leaving the line without permission (Article 24) or to 'pursue any small number of the enemies ships until the main body be disabled or run' (Article 21).

Was there, however, anything to prevent the admiral from signalling 'general chase' together with the signal to engage, as Anson and Hawke did later in their respective battles of Finisterre? This question was not fully considered at the time, but the answer would seem to have been that, where the

enemy were of equal force and in a well-formed line, it would have been imprudent and even culpable to try the experiment merely because the attempt to form the line of battle had broken down. Against a weaker enemy, already in retreat, a pell-mell irregular attack might be necessary as well as justifiable.

Mathews also accused Lestock of failing to repeat his signal to engage; in his answer, Lestock contrived to raise wider issues than the comparatively simple one of accepted practice. He argued, first, that the signal, being immediate and executive, required no repetition by other flag-officers, nor was any provision made for repetition in the article itself (no 13), 'for where repeating is necessary, every article expresses it shall be repeated.'[33] Lestock had been in two general battles and the signal to engage was not repeated in either. At Malaga 'he was Lieutenant to the Admiral of the *Namur* [Shovell] who did not repeat it, nor did any other flagship, either English or Dutch'. If, continued Lestock, Mathews had really intended the signal to be repeated, he would have taken the opportunity to say so at the time of his addition to the article itself, which read '. . . and strictly charged not to fire before the signal be given by the Admiral'.[34] At Malaga Rooke did not make this signal before he was within gunshot of the French admiral; in the Battle of Beachy Head it was not made until just before the action began, long after the fleet had been drawn into a proper order of battle. At La Hogue (Barfleur) Russell did not signal to engage until 'within three quarters musket shot'.[35] To make the signal to engage as Mathews did, with his fleet at a distance and his line unformed was clearly reprehensible. If, however, Mathews had hauled down the signal for the line, and instead signalled to Rowley 'to keep his wind' then the provisions of Article 19 would have become operational, that is, 'the van of the admiral's fleet is to steer with the van of the enemies, and there to engage them'.[36]

Mathews was questioned at his trial as to why he did not change the signal for the line ahead to the line abreast when he bore down to engage the *Real Felipe*, since the manoeuvre required what was practically a line-abreast approach. To this he answered that at the time of attacking he consulted Captain Cornwall of the *Marlborough*, who was killed in the battle, and they agreed that the signal for the line abreast 'would not be so proper as that which was then flying', and that had he hoisted it 'confusion and disorder' would have been 'unavoidable'.[37]

The trials and the resulting pamphlets also raise interesting questions about station-keeping in battle. Mathews had made an addition to Article 1 of the Fighting Instructions which is in fact no more than the signal for forming the line of battle ahead. It reads: 'And every ship is to observe the same distance those ships do which are next the Admiral, always taking it from the centre'. Though well-intentioned, this addition must have been difficult to follow with a large fleet. How could ships at the very head of the line judge if they were under or over the distance between the admiral and his seconds? Signals for particular distances would have been better. Armed with this extra fact, Lestock denied Mathews's accusation that he had failed to obey the signal for Article 12 'where the Admiral would have the Admiral of the White and his squadron make more sail, though himself shorten sail [waiting for Lestock to come up], he will hoist a white flag at the ensign staff'.[38]

As to Lestock's failure to cut off the rear ships of the Spanish division, why did not Mathews make the signal for Article 27, 'If the Admiral would have any particular flag ship and his squadron or division give chase to the enemy . . .'[39] As the signal for line ahead was flying he could not possibly have got properly into action by running more to

30 Trelawney's Mediterranean Signal Book, Tunstall Collection, *NMM*, TUN/12 (formerly S/MS/AS/1)

31 Quoted by Richmond in *The Navy in the War of 1739–48*, III, p254

32 *The History of the Mediterranean Fleet from 1741 to 1744*, 1745, p.48 (*NMM*, TUN/180). See also *Admiral Mathews's Charge Against Vice Admiral Lestock*, 1745 (hereafter cited: *Admiral Mathews's Charge*), pp11–3 (*NMM*, TUN/181).

33 *Admiral Mathews's Charge*, p9

34 *Admiral Mathews's Charge*, p10

35 *Admiral Mathews's Charge*, p12. See also *Vice-Admiral Lestock's Defence to the Court Martial, Giving a short View of the Nature of his Evidence*, 1746 (hereafter cited: *Vice-Admiral Lestock's Defence*), pp38–9 (*NMM*, TUN/182).

36 *Admiral Mathews's Charge*, p35

37 *Admiral Mathews's Remarks*, pp54–5

38 *Admiral Mathews's Charge*, p37

39 *Admiral Mathews's Charge*, p21. See also *Vice-Admiral Lestock's Defence*, pp45–6.

leeward, nor could he make more sail than he did. Yet, as Sir Herbert Richmond so cogently pointed out, Lestock only two years before had reprimanded various captains for not making sail quickly enough to join the line of battle irrespective of their divisional commanders. When one of them, Captain Barnett, asserted that he ought to take station on his divisional commander, Lestock had strongly denied it, stating that it was the duty of captains to join the battle as quickly as possible. 'Is it your duty to see two-thirds of the squadron sacrificed to the enemy when you could and did not join the battle?' Here speaks a more vigorous Lestock, even if the general tone of his letter is 'passionate and ridiculous'.[40] Besides, as Richmond again points out, all this was in conformity with the traditional doctrine of the duty of all captains 'to keep up with the admiral of the fleet'.[41]

An interesting suggestion appears quite casually in the *History of the Mediterranean Fleet*, that in battle the admiral should hoist his flag in a frigate, with a boat from every ship ready alongside 'to wait his immediate commands'.[42] He would thus avoid being exposed to the enemy's fire and avoid having his view obscured by smoke.

Tactical development following the Battle of Toulon

The battle of Toulon began a period of tactical development in the English fleet, carried out by successive admirals and commanders of squadrons, trying to find a means of obtaining decisive results from the fleet actions. It seems at least arguable that the battle of Toulon was a providential happening, in that it revived that earlier spirit of continuous innovation which had long been in decline in the Royal Navy, largely due to the static condition of naval architecture and gunnery and the static relationship of the chief maritime states. Scandal provoked reform, and the Battle of Toulon can now be regarded as a great sea-mark in the history of British tactics. From it stemmed new ideas about bringing a fleet into action as quickly and effectively as possible while at the same time avoiding confusion and misunderstanding. Vernon's ideas were taken up and exploited by Anson and Hawke and the course was laid for the triumph of Lagos and Quiberon.

In August 1744 Mathews handed over the command of the British Mediterranean Fleet to Rowley, Lestock having already gone home. Morale was low. A number of captains and other officers had already been ordered home or were expecting to go home, either to face trial by court-martial or to give evidence as witnesses. Rowley rose to the occasion. On 17 September, within a month of taking command, he issued 'Observations and Instructions in Relation to Engaging the Enemy'. These were accompanied (presumably at the same moment) by some fifty new signals (new that is to the Mediterranean fleet).[43]

Rowley apparently decided that decisive action could be obtained only if the traditional concept of station-keeping were modified. After quoting Mathews's additional instruction on station-keeping, he gave it as his opinion that when forming line of battle each ship should get into station in her own division, and that it should be the responsibility of division and squadron commanders to get their own ships into station and align them with the centre of the fleet. He revised Article 29, which required ships to keep their station in action, laying it down that where 'any ship shall be overmatched or pressed by the enemy, that the next ship to her breaks the line and goes to her assistance'.[44] If the next astern failed to act, then the next but one was to do so, after first sending an officer to ask the captain of the ship ahead of him why he had not supported the ship in distress. 'All ships are to keep their barges alongside, manned and armed, for this

or any other service during the time of an engagement.' Mutual support, in effect, was to take precedence over line-discipline. Attacks by enemy fireships were to be prevented by the barges of the ships directly ahead and astern of the ship endangered. If the fireship was already too close to her intended victim, the next ahead and astern were 'to close the line between her and the enemy so as to cover her from the fire [from the enemy ships of the line, presumably, not the fireship] while her own people are endeavouring to save her'. Frigates and small tenders were to assist.

As well as these injunctions to break the line of battle, under particular circumstances, Rowley sought to improve the degree of tactical control. He ordered that, if a flag officer were obliged to shift his flag because of a fireship attack, his second was 'to have his [that is, the admiral's] flag ready to hoist the moment the Admiral comes alongside'. Repeating frigates were ordered to fire a gun at ten minute intervals to call attention to any signal not observed by the private ships. Every ship was to have a midshipman on the poop 'to watch and write down all signals, etc, as pass in the engagement'. The master was to be responsible for insuring this was done; he was to report every quarter of an hour, and to replace the midshipman if he were killed or wounded.

Rowley's new day, night and fog signals included all the Additional Signals issued by Vernon in the West Indies operations of 1739–42. Their introduction into the Mediterranean, however, reflects well on Rowley's capacity as commander-in-chief. These signals include those for closer engagement, for closing up scattered squadrons, for small ships to leave the line and form a reserve corps, for the slowest ships to drop astern, and for the same ship to lead even if the fleet tacked, to prevent confusion such as was suffered by Mathews at Hyères Bay. His night signals include those for forming a line of battle, and for identification of friends. They foreshadow the possibility of night action.

The gap between Millan's summary of Vernon's and Rowley's innovations and the later stages of the war in the Mediterranean is covered by a manuscript signal book signed W Trelawney.[45] The middle double page is entitled, 'The Sailing & Fighting Instructions, As they are observed in the Royal Navy of Great Britain, with the Additional Signals of Admiral Mathews, Rowley, Medley and Byng'. The additional signals of the admirals are of great interest, especially as none of the admirals left much reputation as a tactical innovator. Rowley is credited with twenty-two signals, together with variations to cover different squadrons or points of the compass; Byng has nine, Medley five and Mathews four.

---

40 The full correspondence is printed in Charnock's *Biographia Navalis*, LV, pp213–17. Barnett refused to be put down and answered Lestock in a good-tempered and carefully argued letter to which Lestock did not reply.

41 Richmond, *The Navy in the War of 1739–48*, II, pp18–19. An example of Barnett's interest in tactics is provided by a comment in Christopher O'Bryen's translation of Hoste, *Naval Evolutions . . .*, p17, following Hoste's method for chasing a ship from leeward (p35, Remargue no 3) 'We now practise this method in the navy, when chasing an enemy, as it is found to be mathematically true. It was first put in practice by Admiral Barnett in the Mediterranean, upon reading the treatise of L'Hoste.'

42 pp21–2

43 Richmond, *The Navy in the War of 1739–48*, II, pp261–3

44 Rowley wrote article '29' in error.

45 *NMM*, Tunstall Collection, TUN/12 (formerly S/MS/AS/1). This is a vellum-bound pocket book which may well have belonged to Captain William Trelawney, later a baronet and Governor of Jamaica (1767–72) and not to be confused with Edward Trelawney, also Governor of Jamaica (1736–52). Since the additional day signals of all four admirals are worked in amongst the rest, the book could not have been compiled until Byng issued his signals on taking over from Medley.

Rowley's are by far the most important; they include all those already issued by Vernon and later correctly credited to him in the pro-Lestock booklet *A Narrative of the Proceedings of His Majesty's Fleet in the Mediterrean.* Those of Rowley's signals not derived from Vernon are either variations or extensions of existing signals, or signals of a less important nature, such as for taking tenders in tow, and for victuallers going into port. There is a separate list of Rowley's night signals, fourteen in all, including Vernon's. Rowley's fog signals are listed separately.

Peyton and La Bourdonnais

On 25 June 1746 an action took place off Negapatam between a British squadron under Commodore Edward Peyton and a French squadron under that redoubtable and colourful figure, Mahé de La Bourdonnais, already something of a St Malo corsair cum East India nabob, and now Governor of Mauritius and Bourbon, as well as commander of the French navy in those waters. He was, moreover, as far as we can tell, the first sea officer to devise and use an absolutely unqualified numerical system of signals. Peyton had had no real fighting experience, and he had been only two months in command as successor to Commodore Barnett.

At daylight, with the wind south-west and Peyton's ships in loose order, the French squadron was sighted to the SE. Peyton at once steered a little south of east to cut across them to windward, signalling for the line ahead as he went. The French meanwhile were steering north, and on sighting the British, La Bourdonnais also began to form his ships in line ahead. The interesting feature of this battle was the unusual balance of the two squadrons. Peyton had his broad pendant in the *Medway* (60) with three 50s and a 40. As the *Medway* was leaking at the rate of 30in per half hour, Peyton was in serious danger if heavily attacked. La Bourdonnais had his flag in a 70, and commanded a squadron of eight Indiamen of from 36 to 28 guns. In both gunpower and manpower, he

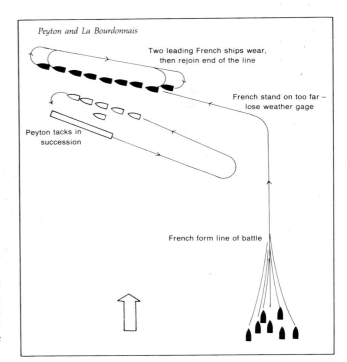

outnumbered Peyton very considerably.

Neither commander showed much inspiration. La Bourdonnais stood so far north before hauling his wind that by the time he brought to and formed a line, tightly closed up pointing NW, Peyton had been able to pass to the SW of him, and so gain the weather gage. La Bourdonnais' *Achille* (70) was the centre ship. Peyton now tacked in succession and began to edge down so as to attack the French van. At about 4.30pm the British closed in, starting to engage more or less ship to ship with the French centre and rear, and passing on towards the French van. The two leading French ships were quickly driven to leeward, on which they wore round the lee side of their own line, then tacked and formed up again in the rear. At 5.30 Peyton signalled his two leaders to tack, and then the remainder, all in succession. By 6.30 the five British ships had tacked, and were passing the French line on the opposite tack. By 7pm Peyton had cleared the French line and the battle had ended.

Next morning the French were still in sight to the NE. Peyton made sail and formed in line of battle. However, the wind dropped by evening without the action being resumed. A council of war aboard the British ships decided in favour of steering south to Trincomalee for repairs. Although La Bourdonnais claimed a victory, he made no attempt to renew the fight, having only provisions for twenty-four hours, and fearing that he might be carried to leeward of Pondicherry by the strong northerly current.

The first convoy battle of Finisterre

The two battles fought off Cape Finisterre in 1747 have markedly similar features. In both battles a stronger British force completely defeated a weaker French force seeking to cover the retreat of a threatened convoy. In both battles the French were overwhelmed by a pell-mell attack from the rear.

In the first battle, that of 3 May, the British admiral, George Anson, profited from having worked up his squadron very efficiently after leaving Portsmouth barely a month before. He had been continually exercising his ships in various tactical and scouting formations, and had them constantly ready for battle. As in the previous year when in

Amiral Bertrand François Mahé de La Bourdonnais (1699–1753), by Lepaulle. MM

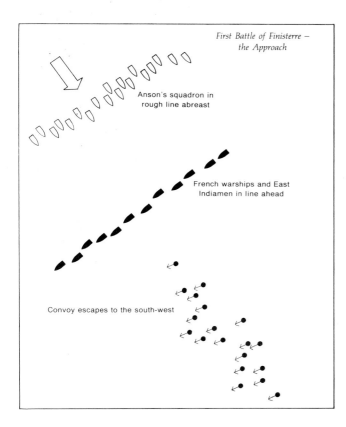

First Battle of Finisterre – the Approach

Anson's squadron in rough line abreast

French warships and East Indiamen in line ahead

Convoy escapes to the south-west

command of a squadron, he was continually signalling for the line ahead and abreast, and for exercising the guns. All this meant additional signals, and at least some reasonable degree of mutual confidence.

Early on 3 May his squadron was spread out in a loose line abreast covering an area of about nine miles, steering large before a NNW wind, when he sighted straight ahead of him the French convoy of which he already had had intelligence. He at once signalled 'general chase', and at 11am called in his cruisers. By noon he was near enough to distinguish details. The French convoy was in flight to the south-west. Astern of it was the escort, about twelve French warships commanded by the Marquis de La Jonquière, now forming in line abreast. By 1pm the British squadron was only three miles from the French warships, which were now brought to with their heads to the north-west in line ahead, directly barring the British approach to the convoy. Anson signalled for the line of battle abreast, and called his second-in-command,

Rear-Admiral Sir Peter Warren, to come on board the flagship. Warren is said to have urged an immediate pell-mell attack, the weakness of the French both in size and numbers being now clearly observable. Anson apparently rejected this proposal and signalled for the line ahead on the starboard tack, so as to bring his ships roughly parallel to the French. This manoeuvre took two hours to complete. When, at about 3pm, Anson was at last able to signal his ships to fill and

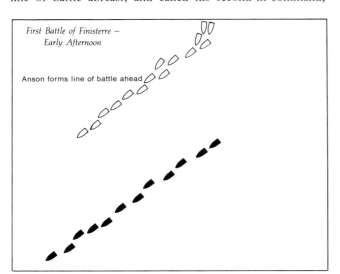

First Battle of Finisterre – Early Afternoon

Anson forms line of battle ahead

The Battle of Toulon, 1744
(see pp 83–90).
NMM

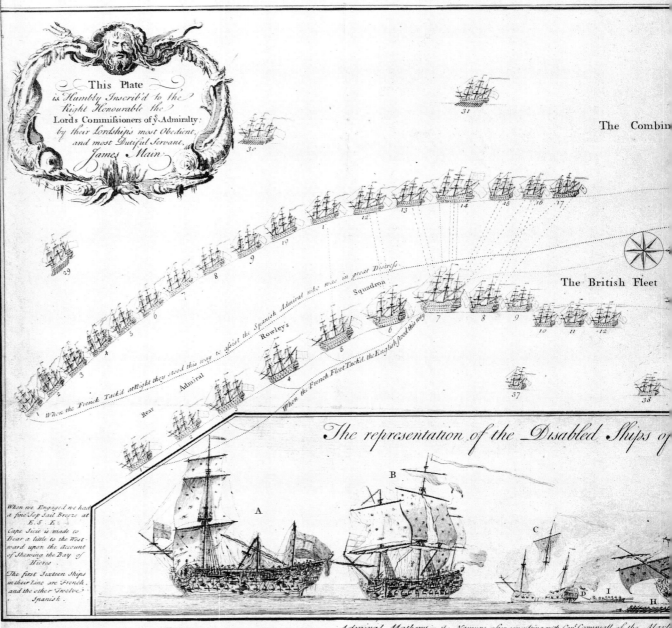

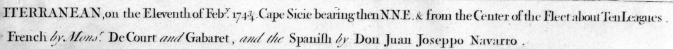

ITERRANEAN, on the Eleventh of Feb.ʸ 1743/4. Cape Sicie bearing then N.N.E. & from the Center of the Fleet about Ten Leagues.

French by Mons.ʳ De Court and Gabaret, and the Spanish by Don Juan Joseppo Navarro.

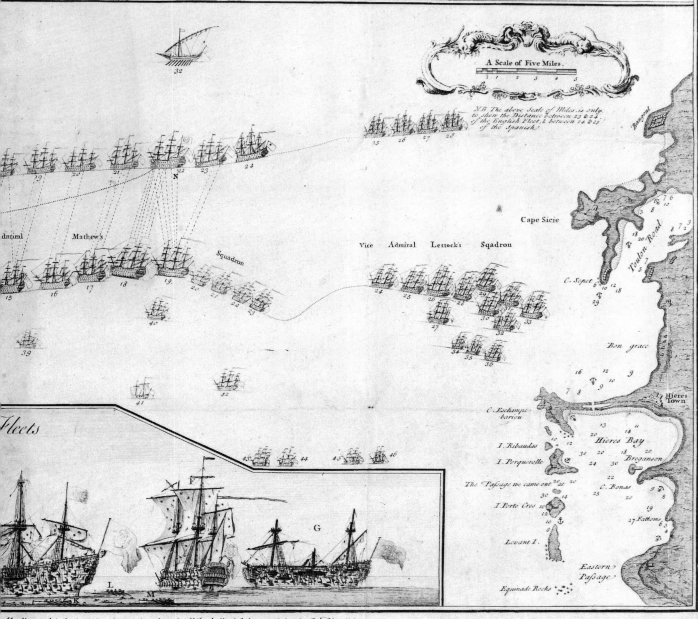

A Scale of Five Miles.

N.B. The above Scale of Miles is only to shew the Distance between 23 & 24 of the English Fleet, & between 24 & 25 of the Spanish.

Cape Sicie

admiral   Mathew's   Squadron

Vice Admiral Lestock's Sqadron

C. Sepet

Toulon Road

Bon grace

Hieres Town

Fleets

C. Eschampe bariou

I. Ribaudes

I. Porquerolle

Hieres Bay

Bregançon

The Passage we came out at

C. Bonas

I. Porto Cros

Levant I.

Eastern Passage

Equinade Rocks

Marlborough in the Condition she was in when the St Isabella left her to Sink, the Ann Fire Ship
... Mizen Masts gone with her Ensign Nail'd to the Stump of the Mizen Mast, she had 43 Men
& 128 Wounded; amongst the first was Cap.ᵗ Cornwall of the Marlborough & Cap.ᵗ Godfrey of the
... amongst the Latter was Lieutenant Cornwall who lost his right Arm: Night coming on
... comitted to the care of the Salisbury & Feversham, next Morning was sent to Mahone ...
... protection of the Oxford.

Princefs in the Condition she was in at the Second or third Broadside from y Poder, she was oblig'd to go
... the Line having her fore Top Gallant Mast gone her Main Stay Cut & her Main Top Sail split
head to foot.

... prefentation of the Ann Galley Fire Ship when she blew up, being within 30 Yards of the Spanish
... als Weather Quarter, & within Musket Shot of the Namure, and Marlborough
Boatswain throwing himself off the Jafford, & was taken up by the Long Boat: the Cap.ᵗ Lieutenant,
... unner, with four Men Perish'd in the Flames.

Royal Philip in the Condition she was in when Day light parted us, her Rigging in a Shatter'd
... on, not a Top Mast on End, her Main Yard on the Deck, & her Top Sail Yards in their Slings, not a Sail
... ld Set but a Sprit-Sail which she bore away with all Right in Tow of another Ship
... d 238 Men Kill'd and 262 Wounded.

Constant, when she was beat out of the Line by the Norfolk, her fore Top Gallant Mast gone,
... r Rigging in a Shatter'd Condition; She had 25 Men Kill'd and 43 Wounded.

Poder when she Struck to the Berwick, having neither a Mast on End or a Yard a Cross.
... Launch that the Spanish Admiral sent to
... the Ann Galley Fire Ship.

... Ann Galleys Boat.

M. The Royal Philip's Boats taking up
... ten that had Leap'd into the Sea to avoid
... blown up with her.

... place where the Ann Galley blew up

The Somerset Engaged Number 18 and 19
but they run from her and joind the French
20.21.22 & 23. Engaged these four Sail astern of
the Spanish Admiral when they came up.
N.B. The Scale of Miles is only to shew the
distance between Number 23 & 24 of the English Fleet:
and between 24 and 25 of the Spanish.
The Dotted Lines from Ship to Ship, is to represent
those that Fought together.

... 1743/4 according to Act of Parliament, & Sold by W.H. Toms Engraver in Union Court near Hatton Garden Holborn, & by the Printsellers of London & Westminster. Price Plain 2ˢ & 4ˢ Colour'd

French & Spanish Line of Battle.

Hercules to Lead with the Starboard & the Borée with y Larboard Tack.

Ships	Comanders	Guns	Men		Ships	Comanders	Guns	Men
1 Le Borée	Damaquart	64	650	23	St Isabella	Don Ignatio Dutablt	80	900
2 Toulouse	Drasture	60	600	24	El Schiere	Don Juan Baley Castro	60	600
3 Duc d'Orleans	Douvez	74	800	25	St Ferdinando	Vel Conde de Vega Florida	64	650
4 L'Esperance	Gabaret			26	Brilliante	Blass de la Barrida	60	600
5 Le Trident	Cailcux	64	650	27	Alcion	Joseph de la Rentiria	58	600
6 L'Alcion	Lanacasca	54	500	28	Hercules	Coom Alcares	64	650
7 L'Aquilon	Vandevici	48	500			Mons De Gladeres		
8 L'Erle	De Albert	64	650	29	Le Saphire	to attend Mon Gabaret	32	505
9 Le Furieux	Gravier	60	600			in the 3ᵈ Post		
10 Le Serieux	Chelues	64	650			Mons De la Clochet		
11 Le Ferme	Desorquart	74	800	30	L'Atalante	to attend the French	32	503
12 Le Tyre	Saurin	50	550			Admiral		
13 Le Diamond	Marritart	64	650			Chev De Beaumont		
14 Le Terrible	De Court	74	850	31	La Bolante	to attend D.ᵒ	24	502
15 Le St Esprit	Puisson	74	800			to attend the Spanish		
16 Le Solibe	Chatainneuf	64	650	32	A Tattan	Admiral	30	500
17 El Orient	Paccom Man St Filens	60	600					
18 El America	Anibal Petruche	60	600					
19 El Neptune	Henrico Oliveres	60	600					
20 El Poder	Roderigo Urrutia	60	600					
21 El Constant	Augustine Eturiaga	70	750					
22 Royal Philip	Juan Joseppo Navarro	114	1350					
							1954	21232

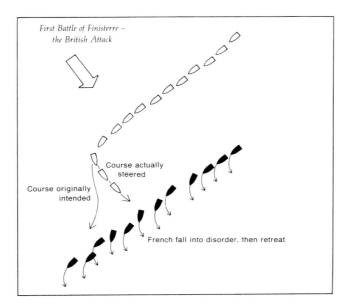

First Battle of Finisterre –
the British Attack

Course actually
steered

Course originally
intended

French fall into disorder, then retreat

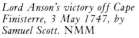

Lord Anson's victory off Cape
Finisterre, 3 May 1747, by
Samuel Scott. NMM

stand on, he also signalled for the van, Warren's division, to lead large (in succession), that is to steer more to port and so more towards the French van, as otherwise he might have run too far ahead of them. Warren and several of the leading ships responded by steering individually towards the French, thus breaking the line and chasing, instead of bringing the line round in a steady curve. Anson, though no pedant, liked precision. He at once signalled Warren to re-form the line and for the leading ships to lead large, thus allowing an almost perpendicular approach towards the centre of the enemy line.

All this time La Jonquière had been lying to awaiting the British attack. Having forced Anson to deploy he had gained valuable time for his convoy's escape and as the British bore down, apparently to cross his bows, he signalled his squadron to fill and stand on. The moment, however, that he saw the leading British ship bearing up to lead more directly towards his centre, he bore up himself and took his squadron away as fast as they could go to the south-west, following the convoy. Being inferior in force, he had no wish to fight an ordinary line of battle engagement in which he was certain to be doubled, as well as to have his weaker ships immediately overwhelmed. Very soon his ships became strung out in an irregular formation more like a triple line ahead than a single line ahead or abreast.

Anson hauled down the signal for the line of battle ahead, and hoisted 'general chase' and 'engage the enemy,' both together, thus leaving his ships to get up with the enemy in the best order their respective sailing qualities permitted. Whether La Jonquière deliberately shortened sail to allow Anson to come up is uncertain. On balance it seems that he did not; in any case, by about 4pm, he was overhauled and brought to action, which showed the immense tactical value of superior speed in pursuit. As the leading British ships closed the French rear, they came under heavy fire, often from two ships at once. After an hour's fighting only six British ships were actually engaged, but by the end of another hour the sternmost French ships were beginning to surrender.

The leading British ships had aimed at crippling the nearest enemy they could find and then making sail to catch up with the next unengaged enemy ship ahead, leaving the damaged enemy astern to be dealt with by the heavier sailing ships as they gradually worked their way into the battle. So great was Anson's confidence that, soon after the battle began, he

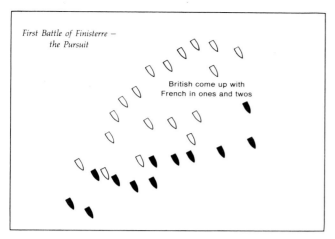

First Battle of Finisterre — the Pursuit

British come up with French in ones and twos

detached two of his weaker ships and a sloop to pursue the convoy and shadow it during the night. By sunset he had captured every one of La Jonquière's squadron and, though his own ships were badly damaged aloft, he was able to detach three of them to chase the convoy, of which eighteen out of a total of thirty-eight were subsequently taken.

The most remarkable features of the battle were Anson's refusal to be hurried at the start, followed by his quick decision to chase as soon as the French retreated. La Jonquière's ships fought well individually, several with great gallantry, but having lost their formation and taken to flight they lacked tactical cohesion and acted only out of a sense of self-preservation. In the end a 74, a 66, a 56, two 52s, a 46 and three East Indiamen of 30 guns each, surrendered to a 90, a 74, a 66, four 64s, five 60s, a 54 and a 50. The odds seemed heavy but the difference in comparative strength between British and French ships, mounting the same number of guns, must be kept in mind. It was because of the strength of French ships that some of them were able to hold out so long.

'Anson's' additional fighting instructions, 1747

When, on 1 August 1747, Sir Peter Warren arrived back at Plymouth, too ill with scurvy to continue at sea any longer, the newly appointed Rear-Admiral, Edward Hawke, was named as his replacement. On 12 August Hawke issued a set of additional fighting instructions which are of tremendous importance in the history of naval tactics. Probably they originated with the commander-in-chief, Anson, and apparently there once was a set of additional sailing and fighting instructions in the Naval Library, attributed to Anson, and probably dated 1747 or even 1746.[46] This may have been the first appearance of the new instructions. There is also a manuscript signal book in the National Maritime Museum, dated 19 April 1747, which credits Anson with the idea of forming an emergency line when in pursuit.[47] Hawke himself issued various individual additional instructions, but never again a complete set comparable with that of 1747. Anson's authorship of the latter is not firmly established, and Hawke continued to issue them, but these additional instructions can be attributed to Anson with some reasonable certainty.

There is no doubt that Articles 8, 9 and 10 of Anson's instructions are of the highest importance. They envisaged the maximum degree of offensive effort, and they codified

Byng's tactics at Cape Passaro. Article 8 ordered that, if engaged with an inferior enemy, the overlapping ships either ahead or astern were to leave the line without waiting for a signal, and rake the enemy's van or rear ships. Article 9 ordered that, when chasing an enemy, the five or seven ships nearest the enemy were to draw into a line ahead of the main force 'and to engage the ships in their rear endeavouring at the same time to get up to their van till the rest of my squadron can come up with them'. Article 10 completes the picture by instructing the leading ship in a chase to form an *ad hoc* line of battle, with other ships joining the end of it as they came up. These dynamic instructions, it must be remembered, were only for use against an inferior, beaten or flying enemy. The difficulties of lines tactics where fleets were equally or nearly equally matches were still unresolved.

A summary of the additional fighting instructions of 1747 is given below.

Summary of Anson's additional instructions, 1747

Article 1 Signal for smaller ships to quit the line, to hold themselves in readiness to assist disabled ships, the strongest of those withdrawn to be ready to fill a gap in the line caused by a ship having to withdraw. (This is a repeat of Vernon's instruction in a slightly different form.)

Article 2 Signal for frigates to attack a convoy and for ships to leave the line and do so. (Richmond thought this to be original. Both 1 and 2 are referred to in May 1748 as the first and second articles of Warren's additional fighting instructions.)[48]

Admiral Lord Edward Hawke (1709–81), by Francis Cotes. NMM

46 Letter from Richmond to Corbett, 21 January [1914]; Corbett, *Fighting Instructions*, Corbett's file copy (*NMM*, MS 81/143, Deed Box A)

47 *NMM*, CLE/2/19

48 D Bonner-Smith, editor, *The Barrington Papers*, 2 vols, London, 1937-

Article 3   Closer engagement (Vernon's instruction but communicated by a different signal).

Article 4   Signal for a line of battle ahead and abreast at half a cable, a cable and two cables. (Much the same as Mathews's, Rowley's and Byng's instruction.)

Article 5   Signal for the leading ship to alter course to starboard, with gun signals to indicate the number of points.

Article 6   Ditto larboard. (Richmond attributes both these to Anson.)

Article 7   If the fleet was in two divisions, to form two separate lines ahead at a cable's distance between ships and a cable and a half between divisions. The signal was a checkered blue and yellow flag at the mizzen peak. (This was a new development, though a similar instruction was, of course, given in Hoste.)

Article 8   If engaged with an inferior enemy, the overlapping ships either ahead or astern were to leave the line without waiting for a signal, and to rake the enemy's van or rear ships as the case may be, 'notwithstanding the first part of the 24th Article of the Fighting Instructions to the contrary', that no ship was to quit the line without leave. (This was extremely bold and interesting instruction.)

Article 9   When chasing an enemy, the five or seven ships nearest the enemy were to draw into a line ahead of the main force 'and to engage the ships in their rear endeavouring at the same time to get up to their van until the rest of my squadron can come up with them.'

Article 10 These ships were to form the line, without regard to seniority, in their order of distance from the enemy, 'that no time may be lost in the pursuit'; and all the rest of the ships were 'to form and strengthen that line as soon as they can come up with them without regard to my general form of the order of battle'. (9 and 10 taken together introduce a phase in the authorisation of *ad hoc* line formation when in pursuit.)

Article 11 Signal that ships ahead of the admiral in line of battle were at too great a distance: this was to be repeated by other ships ahead if they found the ship ahead of them at too great a distance.

Article 12 Ditto for ships astern.

Article 13 Ditto for any captain who thought the ship beyond him from the centre was at too great a distance. (These three are interesting as they allow individual captains to correct each other's intervals.)

Article 14 Signal for fireships to prime. When in the chase, the fireships were to prime as fast as possible, whether signalled to do so or not. (Richmond attributes this to Anson.)

Article 15 Boats were to be kept armed and manned on the off side in action to assist friendly fireships or intercept the enemy's. (Attributed by Richmond to Admiral Martin in 1746 and to Anson 20 May 1747, but in fact Rowley's 'Observations' of 1744 amount to much the same thing.)[49]

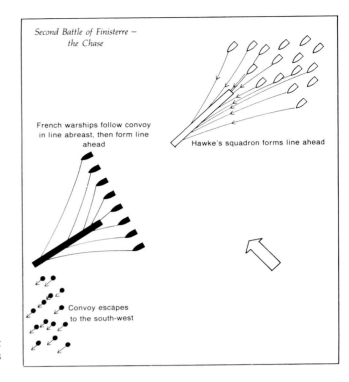

Second Battle of Finisterre – the Chase

French warships follow convoy in line abreast, then form line ahead

Hawke's squadron forms line ahead

Convoy escapes to the south-west

The second convoy battle of Finisterre

In the second Battle of Finisterre, fought on the morning of 14 October, a British fleet under Hawke sighted a large French convoy to the south-west, covered by a squadron of eight powerful ships-of-the-line and a 60-gun East Indiaman, commanded by Admiral des Herbières de l'Etrenduère. Hawke had fifteen of the line, of which three were weak in gunpower. The situation was different from the first battle in that the wind was about south-east with Hawke slightly to leeward. He at once signalled 'general chase' and after only about three hours was up to within four miles of the French warships. At this point he stopped chasing and signalled for the line ahead on the port track, so that his ships could proceed straight ahead on the shortest course towards the enemy. This manoeuvre, however, took nearly an hour to execute, the heavier sailers having dropped astern during the chase and at least one of the leading ships having to shorten sail and tack so as to fall into station at the rear of the line.

While this was going on, l'Etenduère sent his convoy ahead on a course a little south of west, accompanied by his frigates and the Indiaman. Having followed them so far with his eight of the line in line abreast, he now formed his squadron in line ahead rather closer to the wind. His object seems to have been to draw Hawke to attack him and so save the convoy, now crowding sail to leeward. Alternatively, he would still be to windward and capable of giving trouble if Hawke decided to ignore him and steer directly for the convoy. Unlike La Jonquière, he did not attempt a battle in retreat. At about 11am Hawke signalled 'general chase' again in place of the line, judging that the enemy were trying to escape him. As a result his fleet came up with the French rear in a loose irregular formation, three ships managing to scrape round onto the weather side of the French and the remainder attacking from leeward. The French rear, which included three of the weaker ships, was thus taken between two fires,

41 (hereafter cited: *Barrington*), I, p36

49 Richmond, *The Navy in the War of 1739–48*, III, p266 and n

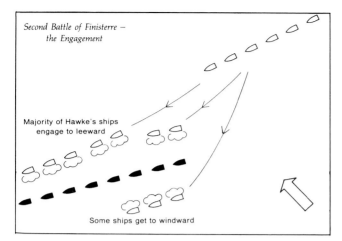

Second Battle of Finisterre – the Engagement

Majority of Hawke's ships engage to leeward

Some ships get to windward

but fought magnificently. By 1.30pm two of the sternmost French ships had surrendered, but two more, including the *Neptune* (58) and the *Monarque* (74), held out until 3.30pm. Meanwhile the British ships, following the same procedure as in the first battle, were gradually working up the French line. A French account states that the British began the action with case shot so as to disable the French masts and rigging, and that the French ships were issued with only four charges of case shot per gun.[50]

A pell-mell ensued. The leading French ship, the *Intrépide* (74), having been less heavily engaged then the rest, tacked and worked her way back to support the French admiral in the *Tonnant* (80), which, though badly damaged in masts and rigging, was holding her own. These two ships eventually withdrew to leeward and, though pursued, made good their escape to Brest; all the rest surrendered. In the closing stages of the battle several of the leading British ships had tacked or worn, like the *Intrépide*, and had thus been able to help in forcing the surrender of those French ships still holding out. The convoy, however, was saved, since Hawke's fleet was too damaged to pursue.

Hawke had done well. His two 70s, one 66, two 64s, seven 60s, two 50s and a 44 had been splendidly handled against the French 80, three 74s, two 64s, one 58, and a 50, each of which was much stronger than any British ship of equal number of guns. As in the first battle, and just as in the Battle of Toulon, clean ships were important. Under different circumstances both Anson and Hawke might have been badly caught owing to the slowness of their ships in forming the line. Apparently the idea of forming the line on the leading ship, with other ships taking station in appropriate order as they came up, did not find favour on either of these occasions. Whether Anson or Hawke had so far issued an instruction with a signal for this is uncertain.

Hawke seems to have made more effort than Anson to exercise some control over the middle and later stages of the battle by signals to individual ships. In this he was only partially successful. The fact that he made such attempts, however, disposes of the notion that he was a mere fighting admiral who cared for nothing but the pell-mell.

Additional signals and instructions of 1748

In May 1748 Captain George Edgecumbe, under Warren's command, issued additional signals of which four are attributed to Mathews, though with different flags.[51] The Blue Peter now appeared, possibly for the first time, at the mizzen topmast head to indicate an order 'to form the line of battle according to seniority'. Presumably this instruction applied whenever an order of battle had not been issued to

ships sailing together. New flags were employed, checkered blue and white, and quartered red and white. Apparently there was a shortage of flags: white, Dutch and French jacks, French, Danish and Swedish ensigns were mentioned as possible substitutes.[52]

On 3 August 1748, at Fort St David (Cuddalore), Rear-Admiral the Hon Edward Boscawen, commander-in-chief East Indies, issued a set of additional fighting instructions immediately before sailing to attack Mauritius.[53] They were unnumbered but corresponded exactly with numbers 1 to 3 and 7 to 10 of Anson's set of the previous year. However, in the equivalent of Anson's Article 9, Boscawen omitted the highly important words 'endeavouring at the same time to get up with their van'. This would seem an early confirmation of the later hypothesis that, although he was one of the leading and most successful admirals of the period, Boscawen was a little behind the Anson-Hawke-Martin-Warren group in tactical innovation.

The line of bearing

At some time during the War of 1739–48, Anson issued an additional fighting instruction, together with signals, for forming the line of battle on a particular compass bearing, as

50 Charles Ekins, *Naval Battles from 1744 to the Peace in 1814, critically reviewed and illustrated by Charles Ekins*, 1824 (hereafter cited: Ekins, *Naval Battles*), p17 (*NMM*, TUN/94)

51 *Barrington*, I, pp32–3

52 See W G Perrin, *British Flags*, p164; *Barrington Papers*, p36

53 *NMM*, SP/73. The instructions are in manuscript inside an ordinary copy of the printed Sailing & Fighting Instructions.

Admiral Edward Boscawen (1711–61), by Sir Joshua Reynolds. NMM

well as signals for ships to 'bear from each other on any of the points'. This appears to be the earliest British provision for 'line of bearing', or 'bow and quarter line' as it was also called because the signal indicated the quarter of the compass on

which the line was to bear. This instruction was reissued by Hawke in the Seven Years War with attribution to Anson.[54]

The line of bearing formation attained great importance during the American War, but also became the subject of some confusion, requiring many clarifying additional instructions. It embodied two quite separate requirements. First, that all the ships in the fleet should simultaneously steer the same course. Second, that each ship should bear the same number of compass points from each other, in the sense that if a line were drawn through the centre point of each ship, the line would not only be straight but it would be north and south or north-west and south-west, or north-east and south-west, exactly as directed. The term starboard or larboard line of bearing, without further qualification except the course to be steered, meant that the ships were to bear from each other in such a manner that if they altered course so as to form a line ahead they would be close-hauled on one or other tack. This was simple enough, provided no change was made, since it required no more than good station-keeping for the ships to

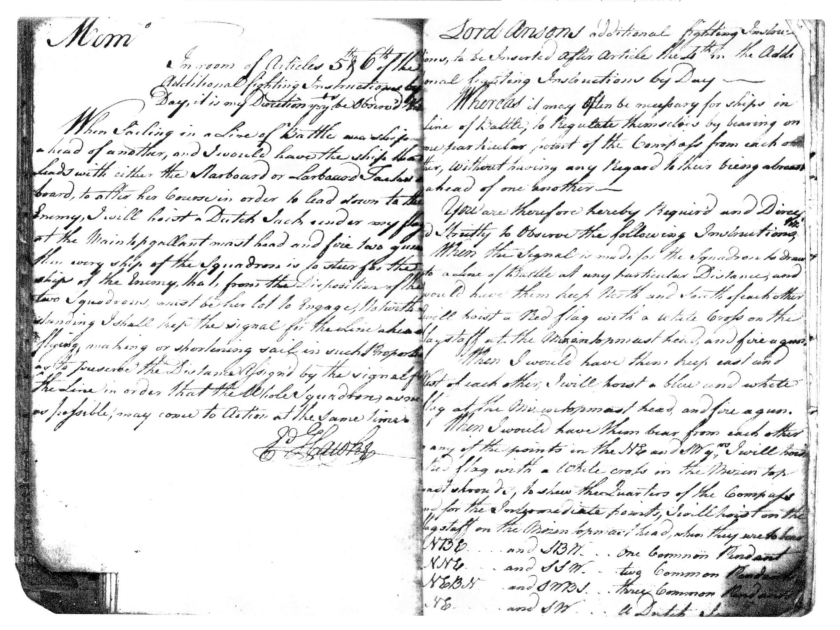

Example of a Line of Bearing

Line of bearing – north-east

Fleet course – north

Admiral Hawke's reissue of Anson's instructions for forming a bow and quarter line, in which all ships steered a prescribed course and simultaneously maintained a bearing on the next ship ahead. MOD

54 *MOD, NM/29 (formerly RUSI 29)*

retain both their course and their relative positions. To change the course of the fleet while retaining the line of bearing was not difficult; the ships merely altered course together. Other and more elaborate variations could be made, provided the line of bearing was not changed. What could not be done, except with great difficulty, until the days of steam, was to keep the course but change the bearing. This could only be carried out successfully after a long series of manoeuvres, based on specification of which ship's course was to determine the new line of bearing. In fact, this was seldom attempted, and changes in the line of bearing generally only took place as a result of some intermediate movement, such as altering course in succession.

Knowles's new signals

In 1777 Admiral Sir Charles Henry Knowles published 'A Set of Signals for a Fleet on a Plan Entirely New', in which he included observations on fleet tactics which he had gathered from his father, also Admiral Sir Charles. The elder Knowles was a man of all-round ability and an enquiring mind, although he was certainly a difficult man to serve under, and a real reformer. In writing to Anson about ship construction, masts, sails, rigging and the treatment of timber in the dockyards he anticipated much of what Kempenfelt was to say nearly forty years later.

The need, which Vernon had perceived, to subordinate the formal, honorific station of officers to the tactical requirements of the battle line evidently took a very long time to be recognised by the service in general. According to Knowles, when a fleet left port in three squadrons, with the commander-in-chief in the centre, his second-in-command would be to starboard, and his third-in-command to larboard. The captains in each squadron would be disposed alternately to starboard and larboard of their respective flagships in order of seniority 'without the least regard had to the station of each ship prescribed in the line of battle given out by the Commander-in-Chief'. 'This,' he said, 'however absurd, has been the constant uniform practice for almost all ages.'[55] The elder Knowles was quoted as saying that the advantage of gaining the wind was often lost through time wasted and 'the great distance many ships have to sail to get into their stations, the disorder and confusion it always will and does create by different ships steering different courses, and crossing each other, some crowding all the sail they can, whilst others are lying to, until the line is formed'.[56] This observation, if it was realistic, demonstrates that unless the admiral was prepared to allow seniority alone to dictate his order of battle, regardless of the gunpower and sailing capacity of particular ships, he could not translate the order of sailing into the order of battle without much confusion and waste of time. In his semi-autobiographical *Observations on Naval Tactics*, written in 1830, the younger Knowles wrote that in his earlier days it had still been the practice for the senior captain in the commander-in-chief's centre squadron to be stationed immediately to starboard of him, but, in forming line of battle ahead, to be stationed at the head of the fleet on the starboard tack. Similarly the next senior officer would take sailing position to larboard of the admiral, but would be posted to the head of the line on the larboard tack. Thus, whenever a line of battle was formed, the two ships immediately next to the commander-in-chief would be compelled to detach themselves from the centre squadron and cut right through the fleet so as to take their respective stations at the head and tail of the line.[57] In the absence of documentation giving contemporary orders of sailing, Knowles's assertion cannot be proved. Perhaps, however, it should be

Admiral Sir Charles Henry Knowles (1754–1831), by Beerseraapen. NMM

assumed that the non-survival of orders of sailing for this period confirms the statement, since captains knew their respective seniority and thus did not require a written list.

This revelation of pedantry and obscurantism in the ordinary cruising methods of fleets emphasises the revolutionary implications of an order of sailing in three or more columns, especially if the flag officers led their respective columns. Not only was this a device to facilitate tactical deployment, but it was also a means of introducing to the Royal Navy the whole range of sailing and tactical evolutions as already practised by the French. With this new sense of manoeuvrability, based on squadron and division positions determined solely on the wind, the whole outmoded notion of the fighting power and fighting position of ships being subordinated to the relative rank of their captains passed away. Senior captains still took charge in the absence of flag officers, of course, but in fleet organisation ships were considered in terms of their gunpower and sailing capacity, as fighting units.

Who it was who set the fleet free from this traditional punctiliousness we may never know for certain. The younger Knowles, using some of his father's arguments, claimed the credit. In tactical matters, however, it is difficult to distinguish clearly between father and son. More important than determining authorship, however, is an attempt to establish the date at which Royal Navy fleets began to employ the columnar sailing order.

Knowles and Reggio

Early on the morning of 1 October 1748, Admiral Don Andres Reggio, with six of the line and a 36, steering north

55 p111
56 *A Set of Signals*, ppiii–iv
57 Admiral Sir Charles Knowles, *Obvservations on Naval Tactics and on the claims of Mr Clerk*, 1830, pp20–2 (*NMM*, TUN/75)

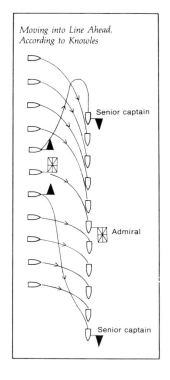

Moving into Line Ahead, According to Knowles

Senior captain

Admiral

Senior captain

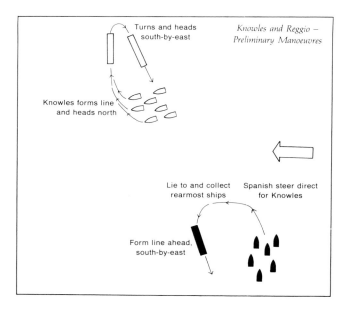

Knowles's own line was thinning out, although at the time he had tacked it had been well formed. His five leading ships were still fairly well together, but the sternmost two were nearly a mile in rear owing to the slowness of the penultimate vessel, the *Warwick*. Knowles was all for beginning the battle at once, as he was now well to windward of the enemy, who were waiting his attack. He therefore made the signal for 'ships that lead on the larboard tack to lead large'. This simply meant that his three leading ships (he himself being fourth in the line) were to stop steering on a parallel course with the enemy and edge down together to engage them. What the leading ships actually did is not very clear but, to Knowles's extreme annoyance, the three sternmost ships showed signs of obeying the signal instead. Their captains mistakenly assumed that, having been originally designated to lead the squadron on the larboard tack, they should now do so, despite Knowles's previous signal for the squadron to tack in succession and for the same ships which led on the starboard tack to lead on the larboard. By the time this was corrected more distance had been lost, and Knowles had lost his opportunity to signal the *Canterbury*, the sternmost ship, to pass the *Warwick*, as he had particularly wished her to engage the last in the Spanish line. Instead he shortened sail himself, with the result that his three leading ships shot ahead before they also brought to. At this point Reggio, who up until then had also been fourth in his own line, shifted his position to third; Knowles began to do the same. Reggio now dropped his 36 out of the line altogether, and Knowles dropped the *Oxford* (60) so it could act as a reserve according to Vernon's practice.

After all this, it might seem surprising that Knowles ever got into action at all, but Reggio was still waiting for him and it was only noon. The wind dropped and Knowles, instead of gaining on the Spaniards as he had done earlier, began to average less than two knots. In the early afternoon the wind freshened again, blowing from slightly north of east, and the two leading British ships drew near enough to the Spanish centre for Reggio to open fire at long range. The leading British ship replied, without any signal from Knowles to engage and contrary to his standing orders.

Seeing what was happening, Knowles now signalled the two leading ships to lead large and engage. The captain of the *Tilbury*, the leader, failed to see the signal and continued on a gradually converging course as if to draw level with the enemy's leading ship before going down. When Knowles in desperation signalled for him to come to a closer engagement, he failed to make out what the signal was, and continued on his converging course, firing at long range. After waiting in vain for the *Tilbury* to obey the two signals, Captain David Brodie of the *Strafford*, second in the line, steered straight for the enemy. Knowles himself and his next astern, the *Lenox*, did not get into action until 4pm. At 4.30 the *Conquistador* struck to the *Strafford*, but when no British boats appeared to take possession of her, Reggio forced her crew to rehoist her colours by firing at her from his own flagship.

Although four British ships had been sustaining the fire of six Spanish ships, Knowles was now doing well enough to drive Reggio out of the line. The other Spanish ships followed him, running before the wind to the west. Knowles at once hauled down the signal for the line and hoisted the signal for the general chase, sending his own captain in a boat to order each British ship to chase and come to close action, and threatening instant dismissal to any captain who 'lay firing at so great a distance as they did'. Knowles's flagship was hamstrung by the loss of her fore topsail and main topmast, and the chase was led by the *Strafford*.

Knowles was much dissatisfied with the conduct of the other three ships which had been up with him since he began

on an east wind, sighted Knowles's squadron of seven of the line to the north-west. Thinking at first they were a convoy, Reggio chased without waiting to collect his force together. He had the weather gage and success looked easy. Knowles, with his ships somewhat scattered, brought to and then wore to the north, making the signal for the line ahead. He was neither running away from the Spaniards nor trying to lure them on in disorder. His purpose was to reach far enough ahead of them to gain the weather gage before attempting to close. Realising his mistake, Reggio bore up and steered with his leading ships to leeward to pick up three more of his ships which were struggling astern and to the south-west of him; by this he sacrificed any chance he had of retaining the weather gage.

At 6.30am Knowles tacked in succession, signalling for the same ships which led on the starboard tack to lead on the larboard. As his ships led round, one astern of the other, steering about south by east on the larboard tack he was almost in a position to weather the Spaniards, who could now be seen forming a line and retiring on very much the same course. By 8.30am Knowles's leading ship was only about three miles north-east of the sternmost Spaniard, Reggio standing on under easy sail in good order. By now, however,

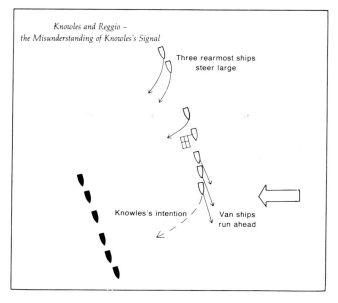

to engage. As for the *Warwick*, she had once more interpreted his more recent signal for the van ships to lead large as applying to herself and only arrived late in the battle. Her captain infuriated Knowles by claiming to take the *Conquistador*, which had at last struck to him after previously striking to the *Strafford*. Meanwhile the rear ship of all, the *Canterbury*, had finally got into action. Knowles had signalled her to push ahead and pass the *Warwick* only as late as 3pm; once in action, however, she fought extremely well.[58] The pursuit went on into the night, a rare event in those days, especially in view of the dangerous Cuban lee shores, which finally forced the British to withdraw at about 10pm.

Four Spanish ships reached Havana. The *Invencible* (74), flagship of the rear-admiral, was a wreck and had very narrowly avoided capture. Reggio's flagship, the *Africa* (74) ran into a small bay and was burnt next day by her crew.

Knowles wrote so stingingly of all his captains, other than Brodie of the *Strafford* and Captain Edward Clark of the *Canterbury*, that they pressed successfully for him to be court-martialled. He had undoubtedly made some serious mistakes, and this laid him open both to genuine, as well as to merely bloody-minded, accusations. On the other hand, he had forced and won a battle. The Spanish authorities were greatly incensed with Reggio, and he was court-martialled on thirty separate charges covering every possible aspect of the battle. He was also separately tried for the burning of the *Africa*.[59]

From a tactical point of view the battle showed yet again the weaknesses on the attackers' part of opening fire too soon and failing to come quickly enough to a close engagement. If Knowles's opponents had been French rather than Spanish, his success might have been much less assured.

58 An anonymous account from the *Lenox* suggests that this ship played a more active part in the battle than Knowles suggested; see W H Long, editor, *Naval Yarns*, pp30-3.

59 *Defense Mylytar por el Capitan de Fragata Don Juan Antonio . . . e Justyfycacyon de la Conducta . . . de Don Andres Reggyo . . . 12 de Octobre de 1748* and ditto *15 de Octobre de 1748* (NMM, TUN/229)

The beginning of Knowles's action off Havana, 1 October 1748, by Samuel Scott. NMM

CHAPTER FOUR

The Seven Years War

THE Seven Years War produced surprisingly little tactical development, despite its great tactical triumphs and the emergence of younger commanders such as Rodney, Admiral Augustus Keppel and Lord Howe. The reason lies in the nature of the war itself. Apart from Admiral John Byng's battle off Minorca in 1756, and Pocock's defensive operations in the East Indies, the only real battles were Lagos and Quiberon, in which the fighting was too disorganised for any true display of tactical virtuosity. The real triumphs of the war were those arising from the general weakness of the French and Spanish at sea; the capture of so many *places d'armes* and valuable colonial settlements throughout the world were made without a single full-dress naval action. Byng's tactical mistakes and sheer bad luck were quickly understood, and provided for. His strategic actions and attitude of mind were tactically irrelevant, so that his personal tragedy had no tactical significance.

The greatest tactical development which took place during the war, and which obviously was successful, was in the technique of combined operations. Admiral Sir Charles Saunders's instructions for carrying his great fleet of warships and transports up the St Lawrence to Quebec in 1759 are extant.[1] So are Rodney's for the capture of Fort Royal, Martinique, in 1761–2, and Keppel's for the capture of Belle Isle in 1761.[2] These operations involved elaborate arrangements for disembarking troops, for the protection of troop transports *en route* and for the protection of military convoys generally. There was much progress in the development of cruising formations used by blockading and observation squadrons, ordered to intercept French warships and merchantmen trying to leave or enter home ports, particularly after the Battle of Quiberon.

Just as it seems no longer necessary to ask how soon after Rooke's consolidation of the Sailing & Fighting Instructions in 1703 the additions begin to appear – since they began almost at once – so it is no longer necessary to ask the same question about signals. The early manuscript private signal books of the period 1744 to 1762 invariably quoted the page and article for every signal recorded, with the names of the admirals promulgating additional signals. Many of these additional signals were retained by succeeding admirals, who re-issued them as their own, and the private signal books sometimes quoted as many as three authorities for a single addition. When the Admiralty additions appeared in 1756 the spate of additions by individual admirals was not checked, since the Admiralty set was both conservative and limited in its scope.

By the end of the Seven Years War there was already a considerable body of signals in existence, at any rate in the private manuscript signal books, for which no specific Admiralty authority could be found. These signals, however, were often extensions or variations of instructions or direct signals embodied in one or other of the two Admiralty publications. Such signals could be operational, (or quasi-operational), or administrative. The first group included extensions of the various chasing and recognition signals, tacking and wearing, making more or less sail and bringing to, distance variations of the line ahead and abreast, the handling of prizes and ships in convoy, taking ships in tow, sighting land, taking soundings, and the movement of hospital ships. The administrative group included calling various ranks of officers on board the admiral, calling the carpenters, blacksmiths, sailmakers or caulkers from various ships with their gear to board the admiral or other ship needing repair, and for weekly returns to be made of various stores. The trend was thus towards giving the admiral more control and enabling him to explain his intentions more clearly. Not much appeared in the way of new fighting instructions, apart from the innovations of Howe, while the Admiralty evidently thought enough had been done when, in 1756, they officially adopted Anson's additionals of 1747.

Almost unperceived, a revolution was occurring in the great increase of signals to be made by private ships, either in answer to the admiral or in reporting strange ships, sighting of land, discovering shoal water, and so on. Even Saunders's conservative signal book records twenty-seven signals for private ships by day, requiring the use of five flags and a pendant. These signals, moreover, had often to be repeated by one ship to another until they reached the admiral, who might be out of sight of the originating ship. Hence, through the new discretion allowed them, as for instance in judging their own capacity to overtake and deal with a chase and in the repetition of signals, private ships were gradually acquiring a signalling system of their own, parallel with though strictly subordinate to that used by the admiral. The further development of this system was encouraged by Howe through his signals for frigates appointed to watch the movement of the enemy, until it blossomed into the great reconnaissance system operated by Nelson immediately before the Battle of Trafalgar.

British additional signals and instructions of the Seven Years War

On 19 July 1755, with war in the offing, Hawke issued additional signals two days before sailing from Spithead in command of the western squadron.[3] On 12 August 1755, Vice-Admiral Boscawen, now in command of the British fleet at Halifax, issued a set of additional signals and additional fighting instructions.[4] As in 1748, these were not numbered. In fact, Boscawen's additionals were Anson's additionals of September 1747, omitting Articles 5, 6, 14 and 15, and adding six chasing signals and a number of cruising

1 Julian S Corbett, editor, *Signals & Instructions*, London, 1908 (hereafter cited: *Signals & Instructions*), pp351–2; and see William Wood, editor, *Logs of the Conquest of Canada*, Toronto, 1909 pp97–106.

2 *Signals & Instructions*, pp355–63 and *Barrington*, I, pp295–323

3 *NMM*, (formerly Sp/130 and Sm/168 [ms folio bound in with printed instructions]; *NMM*, SIG/B/7, (formerly Sm/5), 'the General printed signals with additionals,' Hawke's additional signals, dated *c*1760

4 *Barrington*, I, pp128–40

signals. As he did when he employed Anson's additionals in 1748, Boscawen omitted from Article 9, which ordered the formation of an *ad hoc* line of battle in chase, the important words 'endeavouring at the same time to get up to their van'.

As soon as the war began, the Admiralty also intervened by issuing a printed folio set of instructions.[5] Although the copy in the National Maritime Museum is dated 1757, the folio was clearly issued as early as March 1756, just before Byng sailed for Minorca. His secretary, George Lawrence, quotes them in his evidence at Byng's court-martial. The signals reflect the ideas behind the additional signals issued by Rowley and others in the Mediterranean in 1744 to 1748. The accompanying fighting instructions, with one tiny exception, are word for word the same as Anson's issue of 1747, omitting, however, in Article 10 the words 'rest of the' in the sentence 'and all the rest of the ships are to form and strengthen that line . . .'

The Admiralty compilation was strictly conservative, and it is not surprising that admirals, as well as captains in command of squadrons, continued to issue their own additional signals and instructions just as before. Hawke's and Boscawen's sets of additional signals and instructions are the precursors of the great spate of 'additionals' issued during the next seven years. On 3 May 1756 Boscawen, then commanding a squadron blockading Brest, issued a long unnumbered list of additional signals, starting with the first three of Anson's 1747 issue, followed by a second list embodying Anson's numbers 7 to 13.[6] Neither Boscawen nor Hawke himself, however, were prepared to go beyond the fighting instructions issued over his signature eight years previously.[7]

On 23 June 1756 Captain the Hon Richard Howe, then aged thirty-two, issued additional signals for a squadron of eight minor warships cruising off Granville.[8] These signals appear to be the earliest issued by the man destined to reform the whole signalling system, and have a lasting influence on British naval tactics. Nevertheless, they do not stray far from the Admiralty instructions, except that in Howe's no 16, ordering the formation of an *ad hoc* line when chasing, the order to attempt to reach the head of the enemy line is omitted. This may reflect on no more than the fact that Howe's ships were few, and of moderate fighting strength.

A manuscript signal book used by Hawke during the Seven Years War includes a memorandum, possibly issued in July 1756 when he superseded Byng in the Mediterranean.[9] It reads as if designed to avoid Byng's particular difficulty by providing a signal ordering every ship

to steer for the ship of the enemy that from the disposition of the two squadrons it must be her lot to engage, notwithstanding I shall keep the signal for the line ahead flying, making or shortening sail in such proportion as to preserve the distance assigned by the signal for the line, in order that the whole squadron as soon as possible may come to action as the same time.

This instruction in signal form with a slightly different flag arrangement appears in a manuscript signal book inscribed, 'C[harles?] Hanby, Sept. 15th 1756.'[10] The signal is there described as additional signal number 18, issued by Admiral Henry Osborn (who may have taken it over from Hawke) on succeeding to the Mediterranean command. The most interesting signal in the book is Osborn's twenty-seventh additional, 'for a line of retreat'. This revival, or adoption, of Hoste's order of sailing antedates its hitherto assumed adoption in the British service by twenty years. Another undated manuscript signal book contains a page of thirteen coloured flags, without signals attached to them, entitled 'Admiral Osborn's Additionall Signall colours'. Six are quartered, one has a cross and four a horizontal stripe. It may well be that

Osborn, about whom we know very little, was something of a tactical and signalling innovator. It is tempting to identify the inspiration for this work as Captain Augustus John Hervey, Osborn's chief of staff in the Mediterranean from May 1757 to August 1758, although the additionals noted by Hanby are before Hervey's time.[11]

The book also includes signals attributed to Admiral Temple West, together with an additional instruction to the effect that, in the case of an enemy ship surrendering in action, no ship should stay by her longer than it takes to send a boat to take possession, and for the commanding officer to disable the prize if this had not been done already. The capturing ship should then go immediately to the support of the nearest British ship in need of assistance. This interesting instruction was probably issued by West when preparing to sail in command of a secret expedition in January 1757, an appointment he immediately resigned on hearing of Byng's condemnation.

One way of giving emphasis to an additional fighting instruction was to issue it in the form of an amendment to an article in the original Sailing & Fighting Instructions, provided a suitable article existed. Thus Hawke, in issuing a copy of the original work on 20 October 1757, altered in manuscript Article 13, the signal for battle, so that it reads: 'Every ship in the fleet is to use their utmost endeavour to engage the Enemy as close as possible, and therefore on no account to fire until they shall be within pistol shot.' This replaced the usual form, '. . . in the order the Admiral has prescribed unto them.'[12] The point here seems to be that the duty to fight at the closest possible range is made to supersede any other notion of tactical approach.

Captain the Hon Augustus Keppel, when aged thirty-one and commanding a small cruiser squadron off Finisterre in October 1756, issued additional signals which in reality were instructions. Although he became a post-captain at an early age, through influence, he was evidently at this time a man of initiative. Twenty years later he presented a very different figure, energetic still but much sapped by ill-health, and lacking in tactical initiative. His number 1 reads:

Every ship that chases from the squadron is not to wait for the signal from the *Torbay* [(74), Keppel's ship] to engage, but is to begin when she is at a proper distance, except the enemy is equal or superior to the ships with me. In that case you are not to venture engaging until my force is collected. Then, if no signal is made for the line of battle, every ship is to take the nearest her own size and engage her as near as possible.[13]

Amongst further instructions issued 18 December 1757 is number 18, 'If at any time you are in pursuit of many ships of the enemy and you bring to any of them, without boarding

5 *NMM*, (formerly Sp/130 and Sm/168)

6 *NMM, ibid*

7 A copy of the Sailing & Fighting Instructions in the Ministry of Defence Library, Whitehall (Naval Library), EC41, with manuscript additions provisionally dated by Corbett as about 1756 (*Fighting Instructions* p208). In comparing these with Boscawen's set of 1759 he did not realise that both sets stem from Hawke's 1747 issue.

8 *Signals & Instructions*, pp347-9

9 *MOD*, NM/29 (formerly *RUSI* 29). See also Corbett, *Fighting Instructions*, pp217-8. The book is in several different hands; Hawke's memorandum, signed, is in his own hand, as also is his drafting of the line of bearing instruction with attribution to Anson.

10 *NMM*, Tunstall Collection, TUN032, (formerly S/MS/SY/1)

11 *NMM*, SIG/B/6 (formerly Sm/105). See David Erskine, editor, *Augustus Hervey's Journal*, London, 1954, pp246-92

12 *NMM*, (formerly Sp/130)

13 *Signals & Instructions*, p349

and securing them, in order to your better pursuing the rest you are to acquaint the squadron.'[14]

On 11 March 1758 Hawke issued the printed (that is, Admiralty) Sailing & Fighting Instructions,[15] and on 27 May Anson, in command of the Channel Fleet, issued a slightly different version. The difference, apart from layout, lay in a slight re-arrangement of the additional signals by day. A new number 3 was inserted in Anson's issue, directing ships, when not otherwise ordered, to keep station in such a manner as to enable them to form a line as quickly as possible.[16]

In the summer of 1758 Howe was appointed commodore of a force co-operating with the army in attacks on the French coast, and on 24 July he issued a printed book entitled 'Additional Signals & Instructions to be Observed by the Ships of War.'[17] The issue of a printed book, especially as Howe was not even a flag officer, was an innovation. The phrase 'ships of war' appeared for the first time, to make a clear distinction between the warships and transports, and it became an important feature of all of Howe's future signal books. It was adopted for use in the first 'Admiralty Signal Book' of 1799 and remained in use until 1815.

During August and September 1758 Anson was in temporary command of the Channel Fleet, and exercised it in tactics 'in a great part new'.[18] He issued additional fighting instructions which, at last, provided a remedy for Mathews's dilemma at the battle of Toulon:

If upon coming to action with the Enemy, I should think proper to haul down the Signal for the Line of Battle, every Ship in the Fleet is then to use his utmost endeavours to take or destroy such Ships of the Enemy, as they may be opposed to, by engaging them as close as possible, and pursuing them if they are driven out of their Line, without having any regard to the situation which was prescribed to themselves by the Line of Battle, before the Signal was hauled down.[19]

This is a stronger, or at any rate a more comprehensive, response to the need for decisive results than was Hawke's amendment to Article 13, requiring engagement at pistol shot.

On 13 May 1759 Hawke, in command of the Channel Fleet, issued the new version of the Admiralty's Sailing & Fighting Instructions with manuscript additions. The growing needs of fleets were evidently leading to continuous expansion. Another copy of the same instructions, dated 2 June, has manuscript additions which include Anson's for

ships to regulate themselves by bearing on a compass point from each other, and the signal, dated 28 May, for ships to steer for their immediately opposite opponents 'notwithstanding I shall keep the signal for the line ahead flying . . .'[20] On 16 July he issued an even more emphatic version of Anson's instruction of 30 August 1758:

Whereas many and great inconvenience may arise from every particular ship in the Squadron strictly preserving her situation in the Line, either immediately at the beginning of, or during an Action, in cases where the whole of the Enemy's ships shall not be in a direct, or strict Line, or their Van, Centre or Rear shall alter the position they were first in, You are to observe, that as soon as I shall have led on the Squadron, so as to be within the distance I shall think proper to engage at, the moment I hoist the Signal for Engaging, I will haul down the Signal for the Line; When you are hereby required to continue engaging the Ship of the Enemy that shall be immediately opposed to you, in such close manner, according to her position, as will best enable you to take, sink or destroy her; in either of which if you succeed, you are to go to the assistance of the next of the King's Ships engaged ahead or astern of you, as you shall judge most necessary; on the whole having a particular regard to the 21st Article of the General Printed Fighting Instructions.[21]

Article 21 required that none of the ships in the fleet should pursue any small number of the enemy's ships until the main body was disabled, or had run.

The Naval Library possesses a set of twenty additional fighting instructions, including nineteen signals, issued by Boscawen at Gibraltar on 27 April 1759, when in command of the Mediterranean Fleet.[22] They are in printed form and are largely a re-issue of his Halifax instructions of August 1755. These presumably are also those under which the battle of Lagos was fought. He omitted Anson's number 7, concerning a fleet in two divisions forming separate lines of battle, though he had included it in his 1755 issue. In number 9 he again, as in 1755, omitted the key words used by Hawke, 'endeavouring at the same time to get to their van', an omission that caused him much trouble in the battle. Although Boscawen's instructions are mainly a rehash of signals at least twelve years old, there is one new instruction of great interest: number 18 instructed a captain to break the line if necessary to destroy a damaged enemy, provided the two fleets were about equal in strength. Corbett comments that this article was generally regarded as being too risky, and was subsequently suppressed. In fact, however, it was only a slightly watered down form of the instruction which Anson issued in the previous year.

Apart from that instruction, the set shows no tactical advance. This is proved not only by comparison with Anson's 1747 set, but by the fact that Charles Hanby's signal book of 1756 contained thirteen out of Boscawen's nineteen actual signals of 1759. These thirteen were shown in Hanby's book not as additions but as ordinary signals, indicating that they were already sufficiently well known to merit inclusion. Boscawen, however, did create a new signal for private ships, to inform the admiral how many battle ships were protecting an enemy convoy under observation. This was included in the signal book of Captain Henry Marten, of the *Northumberland*, with additions issued by Sir Charles Hardy, 6 April 1758, and by Boscawen.[23]

Hawke's signals for the Quiberon Bay campaign included eighteen additional night signals. One was 'For knowing each other coming up with or engaging the enemy', an interesting indirect acknowledgement of the possibility of a night action.[24]

14 *NMM*, CLE/2/11

15 *NMM*, CLE/2/5

16 Another version omitting the new number 3 is in the National Maritime Museum, Duff/32 (formerly Sp/127). It is slightly smaller folio, and is bound in bright pink marbled covers instead of the usual red-cherry marbling. In article 9 it omits the words 'in order to engage the ships in their rear, endeavouring at the same time to get up with their van,' yet it includes the words 'rest of the' in article 10.

17 *Signals & Instructions*, pp350-1

18 Manuscript note on p216 of Corbett's file copy of *Fighting Instructions*, *NMM*, MS 81/143, Deed Box A. Quoted from *BM*, Add Ms 35376.

19 *Barrington*, I, pp231-2, (30 August 1758)

20 *NMM*, (formerly Sp/130), and CLE/2/7, 2 June 1759

21 *Barrington*, I, pp259-60

22 *MOD*, Ec/40; see also *Fighting Instructions*, pp219-25

23 *BM*, Add Ms 41362

24 *NMM*, Tunstall Collection, TUN/36, (formerly S/MS/SY/5). This manuscript book, entitled 'Sir Edward Hawke's Signals for 1759,' has at the end 'Lord How's [sic] Night Signals of the River Villaine.' This suggests that it belonged to someone in the *Coventry* or the *Sapphire*, the two frigates detailed under Howe for the abortive attempt to burn those French ships which had escaped up the river.

French signals and instructions

Just as it did in England, the experience of the Battle of Malaga continued to influence French tactical thinking. Malaga had justified the teaching of Père Hoste because, on the French side at any rate, the battle was fought in the manner he taught. In 1727 a new edition was published of *L'Art des Armées Navales* in exactly the same format, and with exactly the same plates. It was probably copies from this edition which found their way into the libraries of British officers of the middle and later eighteenth century; Admiral John Byng possessed a copy. Meanwhile the French mathematicians and scientists were as influential as before. In 1731 H Pitot of the Académie Royal des Sciences, published *La Théorie de la Manoeuvre des Vaisseaux réduite en practique*, another treatise on stability, buoyancy and wind pressure. Similar works, including the great series by Pierre Bouguer, followed throughout the century.

Although Hoste put forward proposals (*projets*) for signals, it would seem that, during the war of the Austrian Succession, the French continued to follow Tourville. This is not surprising since Hoste did not publish fully worked out plans, his *projets* being no more than he claimed them to be. The signals used in the Duc d'Anville's squadron for the disastrous attempt to recapture Louisburg in 1746 seem to be an adaptation of Tourville's, mostly using the same flags but with the impractical addition of a red flag pierced with blue.[25]

The signalling systems used by the admirals of the Seven Years War show marked differences and individualisms. La Galissonière's are strictly conservative. Those of Admiral Anne-Antoine, Comte d'Aché are more sophisticated. La Clue's book is well organised, apart from the clumsy arrangement of folding signal tables. That of Vice-Admiral the Marquis de Conflans is quite different; its division into separate booklets, each with a good deal of explanatory and instructional matter, is a portent of the general consolidation of signals and instructions which took place following the American War.

In 1754 La Galissonière took a squadron of seven ships to sea from Toulon for a training cruise. On this occasion he used a set of signals drawn up by M de Marquisan, promoted to *capitaine de vaisseau* in that year.[26] His twenty-three flags included twenty-two out of Tourville's twenty-five, plus a white one with a red cross which was new. The book was thumb-indexed and set out in the traditional fashion; that is, with flags first, then their position and signification. There were only four signals made with two flags, and in each of these the flags were flown in different positions and not as a single hoist. The references given to page and article make it easy to reconstruct the sailing and fighting instructions for which the signals were made, even though there are some puzzling gaps. The book contains twenty-four beautifully executed colour view plans of various orders of sailing and battle, in some cases with the actual signal flying from the flagship.

The Battle of Minorca

On the morning of 20 May 1756 the fleets were approaching each other in line of battle close-hauled on opposite tacks; as far as could be seen, neither had an advantage. When they were less than a league apart the wind changed two points in Byng's favour so that the British van was able to pass to

Lieutenant-général *Roland-Michel Barrin, Marquis de La Galissonière (1693–1756), medal.* MM

windward of the French, who were forced to give away to larboard. The French admiral, La Galissonière, in his sixty-third year, was not the kind of man to be worried by this. The British fleet made the most of their advantage by holding close to the wind and thus diverging from the French at a fairly sharp angle. Byng, however, soon turned his fleet a little to leeward, so as to come on a course roughly parallel with the French, and the fleets began to pass each other on opposite tacks.

Byng held the initiative, but he was now in the particular situation provided for by Article 17 of the *Fighting Instructions;* one, that is, of only two situations in which the instructions specifically told the admiral in advance how to fight the battle. Under this article, Byng, being to windward, had no choice but to tack when his fleet had 'come the length

Admiral John Byng (1704–57), by Thomas Hudson. NMM

---

25 *NMM*, SIG/C/1 (formerly Sm4). *La Rose Rubrique pour les Signaux de l'Escadre commandée par Mons le Comte de la Gallissonière Chef d'Escadre des Armées Navales à l'usage de Mons de Marquisan, Captaine de Vaisseau de Roy. Par Ardisson pilote sur la dite Frigate en 1754.*

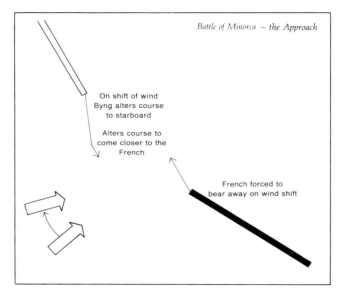

Battle of Minorca – the Approach

On shift of wind
Byng alters course
to starboard

Alters course to
come closer to the
French

French forced to
bear away on wind shift

ship to lead slanting down on the one she was to engage, and they would not be so liable to be raked by the enemy's shot'.

Here Byng reveals himself as a thoughtful and intelligent tactician, anxious to avoid the typical disadvantage of a perpendicular attack from windward. When, a few minutes later, he signalled the fleet to tack in succession, beginning with the rear, La Galissonnière seems to have interpreted Byng's delay quite differently, fearing a concentration against the French rear. He, therefore, signalled the French fleet to throw all aback, so checking their forward movement. It seems that La Galissonnière had some justification for his fears. He had twelve of the line and two junior flag officers. Byng also had twelve of the line and a 50, which he ordered to withdraw during the battle so as to equalise with the French. Instead of taking station in the centre, he had his only three-decker, his flagship the *Ramillies,* fourth in the line when on the starboard tack and in the exact centre of his own division of seven ships. The remaining five were formed in a rear division under his only other flag officer, Rear-Admiral Temple West. Therefore, when Byng began to tack, Galissonnière saw that the movement would place Byng near the rear of the British line, where he would be well disposed to lead an attack on the French rear.

Byng, seeing that the French had thrown all aback, decided it would now take too long to reach beyond their rear. Wishing to hurry on the battle, he signalled for the British fleet to tack together, thus bringing them on to the same tack as the French.

The two fleets were now on slightly converging courses on the same tack, the French at first with only steerage way on, but soon beginning to pick up speed. Byng's action was now governed by Article 19 of the Fighting Instructions, 'If the admiral and his fleet have the wind of the enemy, and they have stretched themselves in a line of battle, the van of the admiral's fleet is to steer with the van of the enemy's, and there engage them.' Left to themselves, the British van would

of the enemy's rear', thus his signal to tack could be no more than a time signal for an evolution which in theory he was bound to execute. He did, however, have two choices as to the method of tacking. He could either signal the fleet to tack in succession, the rear to begin, or he could signal to tack together – as circumstances seemed to require. Of course he could have ignored the article altogether and tried something different, which, provided it had proved successful, would probably have received approbation. The difficulty was that in doing so he would almost certainly have confused his captains; in any case, he did not have a sufficient number of signals to express his orders. Yet Byng was no pedant and, when his secretary, standing beside him on the quarter-deck of the *Ramillies,* 'took the liberty of observing that agreeable to that Article [17] the fleet should tack', the admiral answered 'that he should stand rather beyond their rear before he tacked, as it would give an opportunity to every

*Battle of Minorca – British
Fleet Tacks in Succession*

Temple
West

Byng

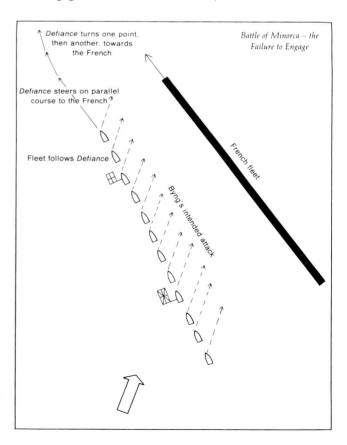

Defiance turns one point,
then another, towards
the French

*Battle of Minorca – the
Failure to Engage*

Defiance steers on parallel
course to the French

Fleet follows *Defiance*

French fleet

Byng's intended attack

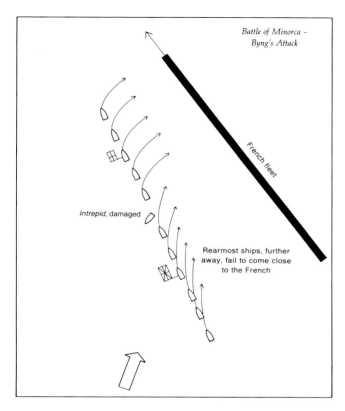

*Battle of Minorca –
Byng's Attack*

French fleet

Intrepid, damaged

Rearmost ships, further
away, fail to come close
to the French

eventually come within range of the French and the two fleets gradually close in battle.

This accepted plan of attack was entirely upset by Byng's leading captain on the larboard tack, Thomas Andrews of the *Defiance,* who kept on a strictly parallel course with the enemy. Seeing the leading French ship a little ahead of him on his starboard bow, he seems to have thought it necessary to make up distance before leading down to attack. Thomas Andrews was an experienced captain of eleven years standing, though by no means the most senior in the fleet. Byng's idea of a slanting attack was quite lost on him, and Byng had no signal with which to express his full intentions. Following Andrews's lead, Temple West's division, now the van in the reversed order of sailing, also kept on a parallel course to the French. As there was no appropriate signal for leading down to the enemy, Byng 'recollected that the fifth article of the Additional Fighting Instructions directed the leading ship to lead to starboard, which he said would answer the purpose he intended.'[27]

The signal was made and a gun was fired to indicate a turn of one point, it being impossible from the *Ramillies* to see exactly how the French van bore from the British. Byng intended the signal to be a general indication and no more, but Andrews interpreted it literally and bore away exactly one point. Seeing that Andrews had failed to understand that this was really a signal to bear down slantwise and engage, Byng made the same signal again, with one gun, and Andrews turned one more point to starboard and towards the enemy. This was still not enough, as he remained a mile to windward of the leading French ship. Instead of repeating the signal yet again and firing, say, three guns, Byng gave up trying and hoisted the signal to engage, on which the fleet as a whole turned to starboard and steered what each captain deemed a direct course for the enemy.

So ended Byng's attempt at a subtle tactical approach. By his original plan of approaching from slightly astern, he had hoped to edge down towards the French, as advocated by Lord Dartmouth in 1688,

not to bear down all at once, but to observe the working of the admiral and to bring to as often as he thinks fit, the better to bring his fleet to fight in good order; and at last only to lask away when they come near within shot towards the enemy as may be, and not bringing their heads to bear against the enemy's broadsides.[28]

His fleet was now exposed to full French broadsides in its approach, each vessel effectively having its T crossed by the French line.

Whether Byng could really have brought off his plan is doubtful, mainly because of his own distance from the van with the line of battle in reversed order. Had he stationed the *Ramillies* in the centre of the fleet and appointed one of his captains a commodore, to balance Temple West, he could both have seen better what was going on and exercised more control. Had he engaged on the opposite tack and with the *Deptford* (50) out of the line, the *Ramillies* would have been the third ship and he could have controlled the angle of attack himself. As it was, his tactical approach was a fiasco.

When he signalled to engage, his leading ship was just under a mile from the leading French ship, with the *Ramillies* about two and a half miles from her opposite number and the sternmost British ship over three miles from the sternmost enemy. The British attack thus became a straggling piecemeal affair, of which the French took full advantage, especially as their ships mounted a far higher proportion of heavy guns than British ships of the same rating. The French moreover were clean ships, fresh out from Toulon, and could easily outsail the British.[29]

Having opened a heavy and accurate fire on the five leading British ships as they approached, the ships in the French van skilfully filled their topsails and stood on a short distance so as to upset the aim of the British gunners as their ships turned into line ahead one after another to engage. The French then began to bear away to leeward one at a time, but not in their order in the line, thus giving the impression that they were beaten. The British van, however, made no attempt at pursuit, having received heavy damage in masts, yards and rigging.

Further astern things went badly for Byng. In terms of the battle as a whole there is much of interest here; in terms of tactical analysis, however, this part of the battle is no more than an illustration of the lack of wisdom of approaching an efficient enemy fleet from windward in a straggling bow and quarter line, which is what Byng's attack had now become. In order to get into action, the centre and rear ships of his fleet had had to keep away a little to larboard, otherwise in attempting a right-angled approach they would have fallen too far astern of the enemy. When, therefore, the sixth ship in his line was damaged aloft and became unmanageable, a gap opened ahead of it and the ships astern were held up. A similar hold-up took place when the eighth ship was damaged. In the confusion caused by smoke obscuring the scene, collisions were only just avoided, as was the danger of ships firing at each other. Such was the delay that the *Ramillies*, tenth in the line, was never properly in action at all.

There was much subsequent debate as to whether the captain of the seventh ship should have passed the damaged

---

27 This refers to the Admiralty Additionals, not his own, and is evidence of their issue as early as 1756. Article V is the same as Hawke's Article V of 1747.

28 Article VI. 'Lasking', ie sailing with a quartering wind, was not quite the same as 'bow and quarter' line, which implied an angular relationship to other ships as well as a special relationship to the wind.

29 Of Byng's thirteen ships only six had been cleaned during the year 1756, one not since May 1755.

Intrepid according to Article 24 of the *Fighting & Instructions,* 'ships astern to close a gap in the line'. All this, however, is beside the point, which is that the British made a poor approach, and the French exacted the appropriate penalty.

When advised by his flag captain to order more sail to be set so as to clear the confusion around him and get into action quicker, Byng made a very significant answer:

You see, Captain Gardiner, that the signal for the line is out [pointing to the signal] and that I am ahead of the *Princess*

Louisa [eighth in the line] and Durrell [Captain of the *Trident,* ninth in the line], and you would not have me as Admiral of the Fleet, run down as if I was going to engage a single ship. It was Mr Mathews's misfortune to be prejudiced by not carrying his force down together which I shall endeavour to avoid.[30]

Byng's quasi-pedantry, and the very clear pedantry of his captains, shows the inhibiting effect of the signal for the line of battle. As long as this signal was flying, no-one seemed capable of any obvious or useful initiative, even to the extent of closing gaps in the line as required by Article 19 of the Fighting Instructions. Had Byng hauled down the signal for line and hoisted the signal for closer action, he might have

---

30 Evidence of Captain Arthur Gardiner at Byng's trial.

The Battle of Quiberon Bay, 20 November 1759, by Dominic Serres the Elder (see pp 115–6). NMM

succeeded at least in carrying the *Ramillies* into action against the French rear, though distance was against him and the French ships could always outsail him.

La Clue's signals and orders, 1757

Admiral de La Clue Sabran commanded the French rear under La Galissonière in the Battle of Minorca, and succeeded him as commander-in-chief of the Toulon fleet. In 1757 he was involved in a series of ignominious operations which led to the blockade of his forces in the neutral port of Cartagena by Admiral Osborn, and eventually returning to Toulon. His 'Signaux et Ordres' for that year are the

equivalent of the British Sailing & Fighting Instructions, with separate tables of signals.[31] Though officially printed, it appears to represent La Clue's own ideas.

31 *Signaux et Ordres qui seront observés par l'Escadre du Roi, commandée par M De La Clue, Chef d'Escadre des Armées Navales. De Sa Majesté; Armée à Toulon en mil Sept cens cinquante Sept A Toulon Chez Jean Louis Mallard, Imprimeur ordinaire du Roi, & de la Marine M.DCC.LVII; NMM*, Tunstall Collection, TUN/46, (formerly S/P/F/1). This particular copy is inscribed 'M Le Chev De Fabri cap & état major de l'escadre de M De la Clue.' [sic] The Chevalier de Fabri had been a lieutenant in Galissonière's *Foudroyant* at the Battle of Minorca, and thereafter commanded the frigate *Gracieuse* under La Clue until after the Battle of Lagos Bay. After the war he was associated with Morogues in the reorganisation of the Navy.

The lay-out of the book is clear and simple. Down the centre of the page are the articles, headed with Roman numerals, in the form of 'Lorsque le Géneral . . .' or 'Si le Géneral . . .', giving the order followed by the signal and then by any instructional words necessary. In the left-hand margin is the short title, and in the right margin the signal, written in words. There are 240 day signals and seventy night signals; folding tables of signals are provided for each. These are ruled in squares, each for a signal, in which the words and figures are clearly indicated. Down the left-hand column are the seven positions from which signals could be flown, and along the top the various flags, ten on each page. Reference, therefore, to the square indicated by the flag and its position is easy. The signals were numbered serially from page to page, taking the top line of each page first, then passing to all the second lines, and so on. Each square has the number of the signal at its top, followed by its short title, and the page reference and number of the article of the instructions to which it relates. Eighty-six out of the 280 squares are left blank for extra signals, since only 194 of the 240 articles of instruction required signals. Although the book, a thin folio, could be easily handled on deck, the same cannot be said of the folding tables of signals, on which all reading of signals would depend. Hence the book can really only have been suitable for cabin use.

Thirty-four flags were used and six pendants, compared with the thirty-six flags, three broad pendants, three pen-

32 *NMM*, SIG/C/2 (formerly Sm/170); contemporary English translation and copy
33 Sir Julian S Corbett, *England in the Seven Years War*, two vols, London, 1918 (hereafter cited: Corbett, *Seven Years War*), I, p347

Admiral Sir George Pocock (1706–1792); engraving for Hervey's Naval History. BM

dants and a weft employed by Hoste in 1697. The number of flags of the same pattern (though of different colours) had been reduced, and new patterns had been introduced. La Clue used only four crosses, four bordered flags and four *facié* (single stripe or heraldic fesse across a plain colour) in place of six of each as before. All six of Hoste's *percées* (plain colour with a circular blob in the centre) were gone, no doubt because of their obvious confusion with the *bordées*. In place of these were four plain yellow *sautoir* flags (plain colour with a saltire), four *tranchées* (parti-coloured fessewise), and two chequered flags. The night signals appear to be much too simple, consisting of lights in various positions, with or without guns, or of gun-signals alone. Apart from the difficulty of differentiating between the various positions in the ship from which the lights were to be shown, some of the signals by gun were duplicated. It would seem very difficult to operate this system without previous knowledge of the course of action contemplated by the general.

D'Aché's signals 1758–9

D'Aché's signal book, which he used in his three East Indies battles against Sir George Pocock, shows a higher degree of tactical sophistication than anything displayed by Galissonière, Conflans or La Clue.[32] His twenty general orders included an instruction, number 10, that the amount of sail carried by the admiral was not to be made a rule for others. If the admiral shortened sail this was a chance for others to recover lost distance. When the signal was made for ships to correct station, it was not sufficient to set a little more sail; ships were instructed to crowd on sail until they had gained position. Any ship noticing that another ship, especially when chasing, had not seen the admiral's signal, was instructed (Article 16) to repeat it (presumably the entire fleet would have been provided with the necessary flags). Article 18 instructed that in battle no ship was to fire until the admiral had made the signal. No ship was to run a greater risk than the rest, but all should go down to engage together. If the enemy were defeated, the best sailers should go in pursuit, but should take note of the admiral's movements and signals. Several signals were devoted to the possibility that the admiral might wish to avoid close action, either by bearing away (number 7), or by making a running fight (number 18). On the other hand, provision was also made for surrounding a fleeing enemy (number 11), for attacking the enemy's rear (number 24), for the squadron to engage all together (number 26), to close the enemy to within musket shot (number 27), to board (number 28), and even to continue an action into the night (number 29). Flag officers were directed to repeat the admiral's signals, and to take the initiative themselves in battle when necessary by making signals to their own divisions.

Pocock and d'Aché

The three battles in the East Indies fought between Sir George Pocock and the Comte d'Aché in the years 1758–9 show the limitations placed on tactical skill both by internal and external circumstances. In the first battle, off Cuddalore on 29 April 1758, Pocock had only five of the line and two 50s against d'Aché's own 74 and seven Indiamen, of which the strongest was a 60. The battle began with Pocock chasing d'Aché, who eventually formed a line and lay to, waiting. 'Under these circumstances,' wrote Corbett, 'seeing that he was in numerical inferiority, Pocock had practically no choice either by the Fighting Instructions or the tactical science of the time, but to form line himself and attack van to van.'[33] The qualification implied here is important: had he been superior in numbers he might have tried a concentration on the French van. Pocock, therefore, signalled for the line of battle ahead at half a cable distance. Even so, his well-understood form of attack was a failure. For various reasons the three near ships were never fully engaged, and their captains were severely dealt with after the battle. D'Aché also was not properly supported, and after an explosion in his flagship and the failure of several of his ships to stay in action when signalled to do so, he bore away after them. Pocock now signalled general chase but his four leading ships, which had done practically all the fighting, were so damaged aloft that he had to desist as night closed in.

In the Battle of Negapatam, 3 August 1758, Pocock was able to force action because of a sudden change of wind in his favour. He had the same fleet, less one 50, while d'Aché was minus the wrecked Indiaman but plus a 74. Pocock won this battle by hard fighting. The leading French ship lost her mizzen; D'Aché's flagship lost her wheel, left the line for repairs and later fell on board another ship. Two other French ships left the line and d'Aché himself now bore away. Pocock signalled for closer action and then for general chase as the French crowded sail. By nightfall they were out of range, having suffered 850 casualties, against only 200 in Pocock's ships.

In commenting on Pocock's failure to obtain a decisive victory in either action, Corbett points to the general success of his main strategy, adding that it is sounder to dwell on this 'than to disparage him, as the custom is, because he knew no other way of bringing an unwilling enemy to a decisive action than by attacking in general chase'.[34]

In the Battle of Ponchicherry, 10 September 1759, Pocock had seven of the line and two 50s. D'Aché had four king's ships plus a 68 and six 54s of the Company's – over 100 guns more than Pocock. Despite his inferiority, Pocock closed from windward, with the French waiting to receive him in line ahead. Bearing down in line abreast, he seems to have had little choice as to how he should attack except by forming line ahead on a parallel course. As it was, he made an excellent approach. His next ahead and second-in-command, Rear-Admiral Charles Stevens, ran up the French line and drove d'Aché's next ahead out of the line. In this way four leading British ships more than dominated the five leading French vessels, though their respective total armament was 242 to 320 guns (121 to 160 in terms of broadsides), exclusive of the higher proportion of heavier guns mounted by the French ships. Pocock and his two next astern now came into action against the French centre, but, through poor sailing capacity, the two rearmost British ships came very late into the action. After five hours' fighting only six French ships were still left firing in the line, and soon afterwards two of these quitted it. D'Aché was badly wounded and his flag-captain killed, and the officer in command wore the French flagship. The three remaining ships still engaged soon followed, crowding sail. Of Pocock's fleet, only the two rearmost ships were capable of pursuit, the rest having received heavy damage to their masts, yards and rigging, so that he could do nothing more than lie to. The French casualties were 1500 and the British 570.

D'Aché had preserved his fleet and kept it mobile, but only at the cost of heavy casualties and hull damage. Had Pocock attempted a more ambitious form of attack in any of his three battles, it is unlikely that he would have succeeded as well as he did, the French being not only superior in numbers but far more adroit in specialised manoeuvres.

The Battle of Lagos Bay

The Battle of Lagos Bay, 18 August 1759, presents few tactical features but raises some interesting tactical and

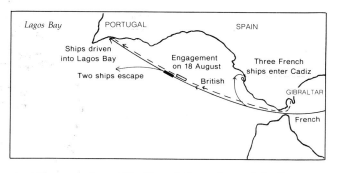

signalling questions. The French fleet of ten of the line, two 50s and three frigates, commanded by La Clue, was attempting to reach Brest from Toulon. They had already succeeded in passing Gibraltar on the night of 17 August 1759, pursued by Admiral Boscawen in command of a more powerful fleet issuing from Gibraltar in a great hurry and in no special order. La Clue had given out Cadiz as his first rendezvous, apparently thinking his ships would not be able to keep together over the whole course, which would involve at least 1700 miles of sailing. At midnight, being near Cadiz but having his whole fleet with him and a favourable wind at a little south of east, he decided to lead on round Cape St Vincent.

French authors are in doubt as to what signal he made, but Corbett guessed it was 'sail large on the starboard tack'.[35] His printed signal book issued in 1757 suggests that he made night signal number 27, 'Prendre chasse les amures à tribord, au plus-près au largue'. The signal for this was for the admiral to extinguish his lights and fire three guns or three rockets. If he actually made this signal, it is likely that he fired the guns as these were less likely than rockets to give away his exact position to the pursuing British. The signal continues 'Les vaisseaux seront la même route que le Général, et auront grand soin qu'il ne paroisse aucun feu'.[36]

At 2am he could still count his fleet; soon after, however, Boscawen's ships began to creep up and fire on his rear. According to some French accounts, he then put out his own poop lanterns and so continued till daybreak. Meanwhile three of his line ships, the two 50s and all three frigates (including Captain de Fabri's *Gracieuse*) had lost contact and worked their way into Cadiz, following their original orders; no doubt they had missed his signal.

There seems some confusion here, as the putting out of the poop lights should have been part of the earlier signal. Moreover, the first article of the 'Advertisement' to La Clue's night signals says 'Lorsque le Général éteindra ses feux de poupe, on redoublera d'attention dans tous les vaisseaux pour ne laisser voir aucun feu'. If he made signal number 27 he must have put out his lights anyhow, which would have indicated that a change of course was being signalled. It seems almost certain from this that the ships which broke away to Cadiz did so because they had already lost contact, not merely with La Clue, but with the ships immediately ahead of them who dowsed their own lights in conformity with the admiral's signal.

The running fight on 18 August occurred, first, because La Clue mistook the British for his own ships which had run into Cadiz and so allowed them time to come up, and second, because the British were able to come up on a strong easterly breeze while the French lay almost becalmed. Beatson says

34 Corbett, *Seven Years War*, I, p350

35 Corbett, *Seven Years War*, II, p36n

36 Copy belonging to the Chevalier de Fabri; *NMM*, Tunstall Collection, TUN/46

that when La Clue saw the British going for his van and centre he

made his fleet luff up as much as they possibly could, so as to form a sort of crescent, by which position, the whole of his ships in their van and centre were enabled by their fire not only to assist the rear, but each other, in their endeavours to repel the attack which they looked for every moment from the British Admiral. By this manoeuvre of M de la Clue's, such of our ships as first got up with the enemy's rear, and to leeward of their line, were thrown out of action; while, for want of a sufficient breeze of wind [which had dropped again], they could not get into it again[37]

Boscawen's flagship, after engaging La Clue's flagship for half an hour, lost her mizzen and both topsail yards and fell astern, so that Boscawen had to transfer his flag. This success, however, was of no avail. The French were being clearly overpowered and knew it. La Clue signalled a retreat on a freshening breeze and left his rear ship, which had already been heavily hit, to be partially dismasted and taken. La Clue's difficulty, just like that of Conflans and later Villeneuve, was that he had been ordered to go from one place to another. Had he been instructed to break Boscawen's aggressive blockade of Toulon by boldly attacking him as he lay refitting at Gibraltar, or as he tried to come out, things might have been different.

Boscawen later complained that his leading ships had concentrated on the French rear instead of 'attacking in inverted order, by which each ship as she came up passed on under cover of the one already engaged and got alongside the next ahead, until all the enemy from rear to van were tackled in succession'.[38] As he had not issued any instruction for this, he could only signal his leading ships individually to make more sail, the real point of which they failed to interpret. He had, indeed, just issued additional instructions (numbers 9 and 10) to the effect that when chasing a fleeing enemy he might signal for five or seven of his leading ships to start engaging the enemy by forming a line without regard to seniority or the order of battle but according to their distance from the enemy.[39] This, however, as Corbett points out, was not what Boscawen now required. It was merely a provision

for an *ad hoc* line of battle when in pursuit, instead of a means for trying to get at the enemy as far ahead as possible. In any case, these instructions were merely a repeat of those issued by Hawke in 1747. The anonymous sea-officer who translated Morogues, suggested in a footnote that Boscawen would have done better to have divided his fleet 'into three small divisions, the better to surround the enemy'.[40] He noted that Mathews and Haddock ran right through the retreating Spanish fleet at Cape Passaro (1718) until they were able to engage the two leading Spanish ships, thus holding up the Spanish retreat until Sir George Byng came up with the main body of the British.

The obvious conclusion to be drawn from all this is that Boscawen had doubtless made adequate plans for fighting La Clue off Toulon, but not for a long running fight up the Atlantic coast of Spain and Portugal. When the moment came, he was caught unawares, being at dinner several miles away with the Spanish Governor of San Roque, together with some of his senior officers. He had failed to make known to his captains in advance what he wished done in a running fight, and in the event it was too late. Doubtless some of his captains behaved badly, or at least unimaginatively, but no blame attaches to the official Sailing & Fighting Instructions. Boscawen could either have added pursuit signals to his repeat of Hawke's additional instructions of 1747, or he could, like Nelson, have worked out plans with his captains to cover pursuit contingencies. He apparently had done neither and his captains merely acted according to the more generally accepted 'custom of the service'. Boscawen was no innovator, and in relying on Hawke's twelve-year-old instructions he suffered the inevitable consequences.

---

37 Robert Beatson, *Naval and Military Memoirs of Great Britain from 1727 to 1783*, six vols, Boston, 1972 (hereafter cited Beatson, *Naval and Military Memoirs*), II, p318.

38 Corbett, *Seven Years War*, II, pp37-8

39 Corbett, *Fighting Instructions*, p221

40 p16

The Battle of Lagos, 18 August 1759, by Thomas Luny. NMM

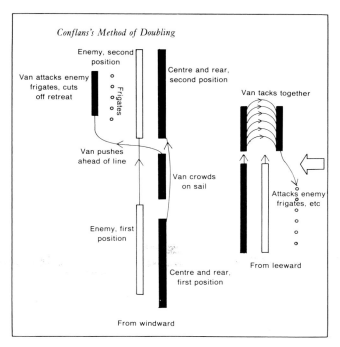

Conflans's Method of Doubling

Enemy, second position

Centre and rear, second position

Van attacks enemy frigates, cuts off retreat

Frigates

Van tacks together

Van pushes ahead of line

Van crowds on sail

Attacks enemy frigates, etc

Enemy, first position

From leeward

Centre and rear, first position

From windward

During the night, two French 74s slipped away, one eventually making Rochefort and the other the Canaries. Next morning's destruction of La Clue's remnant of four on the neutral shore of Lagos Bay and within range of Portuguese forts merely demonstrates the final demoralisation of the French after their navigational failure to weather Cape St Vincent.

The Marquis de Conflans's signals, 1759

Conflans's various sets of signals, bound together in a single volume, mark yet another stage in the attempt to provide instructions and signals for as many eventualities as it seemed appropriate to envisage.[41] The volume might be likened to a series of textbooks for a course of instruction rather than a handy guide to action. Throughout the volume, a standard form is followed, with the order on the left of the page, the signal in the middle and the observation on the right. Each group of signals is preceded by some *avertissements,* a *plan général* or an *order général.* At the end of the book are twenty-three pages of tactical diagrams.

In the day signals, under sail, Conflans reserved the foretopmast head for chasing signals, the mizzen topmast head for simple navigation, and the ensign staff for all non-navigational signals. Forty-four flags and three *flammes* were used. Six bore saltires, six were plain colours with a border, six were striped, eight had crosses and eight included yellow. Reserved for the *général* were the red, white and blue and all flags of these three colours. Yellow mixed with any of the three colours was reserved for signals by private ships.

Apart from the means of tactical control, a great deal can be learned from this volume about Conflans's tactical system. Half a cable was his standard distance between ships in line of battle, weather permitting. No ship was permitted to quit or disorder the line unless disabled, in which case she was to be assisted by ships not in the line. Ships were not to open fire until *à la petite partée du canon.* If the enemy tried to break the line from leeward, ships at the threatened point were to close up so that the enemy was obliged to run aboard and foul one

41 *NMM,* HOL/16 (formerly S(H)16): Brest, Chez à Malassis, Imprimeur du Roi et de la Marine, 1759

or more. If the enemy was outnumbered, his van might be doubled from windward or leeward. If the fleet was to windward, and presumably if the attack was launched from windward, the overlapping ships of the French van were to pass to leeward of the enemy van, and then put about in succession so as to regain their original course, now parallel with the enemy van. The leading ships were to shorten sail until the whole detachment was properly formed in line. Meanwhile, these detached ships should not approach too close to the enemy because of the danger of being hit by gunfire from ships on the enemy's windward side, or of hitting them. The mere presence of the detachment was considered sufficient to worry the enemy and make it possible to cut off the retreat of any of his ships driven to leeward. In the meantime the detached ships were expected to attack the enemy frigates, fireships and storeships, stationed on the unengaged side.

If the enemy were to windward, Conflans wanted the French van to crowd sail so as to be sure of weathering theirs, and then put about together, thus coming on a parallel course with the enemy's van, and to attack their frigates, fireships and storeships. To avoid the danger of the enemy doubling the French van from windward, the French van was to attack the enemy's as they came round, ship for ship. To double the enemy's rear from windward, the overlapping ships were required to pass as close as possible across the stern of the enemy's rearmost ship and crowd sail so as to draw up level with them on their leeward side. To double from leeward, the ships had to tack and pass the enemy's rear in bow and quarter line. Once to windward, they could tack again and crowd sail to draw up level.

The mere recital of these recommendations raises the image of an accommodating and docile enemy. Conflans's *signeax de combat,* however, show a great deal of tactical sophistication. Number 6, for instance, is a signal requesting advice about the nature of the enemy as seen by scouts. The night signals include number 40, to prepare for night action, and number 41, to begin a night action.

Conflans retained the traditional system under which the order of sailing in line ahead was the same as the order of battle, except that the *général* was to lead the whole fleet. Orders of sailing in two and three columns followed, the latter described in great detail, as also were the usual single line, line of bearing and wedge formations. Great emphasis was laid on precision in executing all evolutions, and on correct station-keeping. A strong mathematical bias was evident in the evolutions, apparently requiring great skill in steering and sail-handling.

The Battle of Quiberon Bay

A tourist's view of Quiberon Bay from Carnac Plage, St Gildas or le Croisic suggests a huge expanse of sea, bounded only very distantly by the Quiberon Peninsula, Belle-Ile, and La Pointe de Croisic, with some tiny islands of no particular account dotted here and there, and scarcely visible from the mainland. Yet this was the scene of one of the greatest battles in sea history, the special feature of which was heavy fighting in heavy and chaotic seas with the wind against the tide, amidst dangerous shoals and rocks.

When, on the morning of 20 November 1759, a frigate of Hawke's fleet sighted the French fleet off Quiberon Bay, the whole British fleet of twenty-three of the line knew that they were at the end of their long chase from Torbay and also at the end of their long blockade of Brest. What the French knew or thought is uncertain, but their admiral, Conflans, appreciated the situation clearly enough. He had twenty-one of the line. His first duty, having broken out of Brest while

Hawke was driven to raise his blockade by extremely bad weather, was to make contact with the large assembly of transports and victuallers gathered in the great land-locked gulf of Morbihan within Quiberon Bay. These forces were for the invasion of England, and Conflans's fleet was to provide the escort. Like La Clue only three months earlier, he found himself committed to getting from one port to another, with fighting the enemy only a secondary consideration.

At dawn on 20 November Conflans had sighted a small British force, under Commodore Robert Duff, keeping watch on the invasion forces in the Morbihan. The wind being NW, Conflans decided to attack them, signalling 'close on the general' (the usual dawn signal), 'clear for action', 'pay attention to battle signals', 'prepare for action', and 'general chase'. When he saw Duff's force splitting up as it retreated inshore, he sent his van after one part, took his centre after the other and left his rear close-handed to observe some distant ships seen to be approaching, and which turned out to be Hawke's fleet. When this was realised, Conflans at once signalled his whole fleet to 'abandon chase', 'close', 'order of sailing in a single line', 'pay attention to battle signals', and 'prepare for action'. He then took station at the head of his line and steered for the entrance to the bay, which lay roughly ENE of him. The wind was now WNW, so he could make it quite comfortably, though surprising as it may seem when viewed from the shore or a pleasure boat, the safe entrance for sailing warships was by then only seven miles wide. To starboard, between Conflans and Le Croisic lay the Four shoal, and to port between him and the Île Hoëdic an irregular scattering of sharp-toothed rocky islets, known as Les Cardinaux. Conflans calculated that once inside the bay he could haul his wind and use the local knowledge of his pilots and masters to assume a strong defensive position protected by rocks and shoals. Should Hawke dare to follow, he had only to shift his own position to the centre of the centre squadron to put his fleet in 'the natural order of battle'.

Hawke meanwhile was in loose order south of Belle Île. On sighting the French, he signalled for line abreast so as to close up his fleet, having earlier sent the *Magnanime* (74), commanded by Lord Howe, ahead to make the land. Seeing the French withdraw towards the bay, Hawke made his 1747 additional signal (Article 9) for the seven ships nearest the enemy to draw into a line ahead to engage them until the rest of the fleet could come up. He could see Conflans 'under such sail as all his squadron could carry, and at the time keep

together; while we crowded after him with every sail our ships could bear'. This is really the key sentence in Hawke's despatch, denoting an absolute determination to engage the enemy despite a rising gale, sudden squalls, and heavy seas at the entrance to the bay. Even so, it was not until 2.30pm that a few shots were fired at the French rear, and Hawke made the signal to engage. The shots stopped and the chase continued. Presently Conflans was seen to have led round Les Cardinaux and into the bay. In theory he had had plenty of time to get inside with his whole fleet; his rear, however, not only failed to do so but equally failed a proper line. He had never imagined that the British would carry such a press of sail under such dangerous conditions, with the light of the winter afternoon dimming in the squalls. A very severe rain squall brought the three leading British ships into collision, but they managed to clear each other after some damage and continue the chase.

What happened next no-one really knows, except that the *Magnanime, Warspite, Montague* and other ships which forced themselves into the lead attacked the French rear, the *Magnanime* going as far ahead as possible to get up with the French centre.

The supposition that Howe and others pushed ahead without a special signal is proved by the eye-witness account of Patrick Renny, MD, surgeon of the frigate *Coventry*, stationed ahead of the fleet:

Lord Howe, in the *Magnanime*, had led the chase the greatest part of the day, but was now passed by Sir Peter Dennis in the *Dorsetshire* and Captain Patrick Baird in the old *Defiance*, who ran along the French line to windward, receiving the fire of every ship that passed without making any return, intending to stop and engage the van of the enemy. Lord Howe followed in the same glorious career, but coming abreast of the Rear Admiral [St André du Verger] in the *Formidable* was disabled from going further, his foreyard being carried away in the slings. He immediately bore down upon the rear admiral.[42]

In this kind of fighting, devoid of tactical order or control, British superiority in fighting power was decisive. The wind shifted suddenly to the NW, and the French line of battle fell into confusion. At about 4pm the French rear-admiral struck; two French ships were sunk by gunfire, and at about 5pm another ship struck but could not be boarded. As Hawke rounded Les Cardinaux in the *Royal George*, he met Conflans in the *Soleil Royal* trying to lead his fleet out to sea again. They exchanged broadsides, and amidst a general pell-mell and a double collision Conflans fell to leeward of the Four shoal and anchored off Le Croisic.

It was now almost dark and Hawke had no pilot. With the wind blowing hard he was on a lee shore with the Pointe du Castelli and Rade du Croisic just under his lee, the Four shoal to his south, and the dangerous Île Dumet just ahead. Some of the French had retreated towards the Vilaine Estuary in the extreme north-eastern corner of the bay, beyond Le Croisic. More British ships were entering the bay and the wind was kicking up terrifying seas against the tremendous ebb tide fed by the Morbihan and the Rivers Crach, Auray and Vilaine. Hawke made the signal to anchor, with his own position two to three miles west of Ile Dumet.

The chaos of the night and the further destruction inflicted on the French next day and subsequently, are without tactical significance. For the price of two British ships wrecked on the Four shoal, Conflans's fleet was completely destroyed as a fighting organisation. The French invasion scheme was

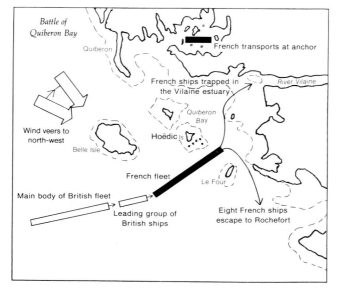

Battle of Quiberon Bay

Quiberon

French transports at anchor

French ships trapped in the Vilaine estuary

River Vilaine

Quiberon Bay

Wind veers to north-west

Hoëdic Is

Belle Isle

French fleet

Le Four

Main body of British fleet

Leading group of British ships

Eight French ships escape to Rochefort

---

42 From the transcription by General Henry Renny. An incomplete and inaccurate version of Renny's memoirs was printed in W H Long's *Naval Yarns* in 1899.

ruined and, with the defeat of Lagos and the loss of Quebec, the Seven Years War began to change in character from a war fought for the command of the sea to a war in which British naval pressure began to be applied directly to the centres of French military power.

When I consider the season of the year [wrote Hawke in his despatch], the hard gales on the day of action, a flying enemy, the shortness of the day, and the coast they were on, I can boldly affirm that all that could possibly be done was done. As to the loss we have sustained, let it be placed to the account of the necessity I was under of running all risks to break this strong force of the enemy.

Hawke was no longer a tactical innovator. In signalling for his seven leading ships to draw into a line ahead to engage the enemy's rear, he was merely following his own additional instruction of twelve years before. Nevertheless it was enough, partly because the French could retreat no further into the bay, so that their whole fleet was involved, and partly because Hawke's captains had developed such a highly aggressive spirit through the long blockade that they did not require special signals to urge them on.

Additional signals, 1760–2

In the spring of 1760, after his return from the Quebec campaign, Vice-Admiral Sir Charles Saunders was appointed to command the Mediterranean Fleet. On 12 May 1760 he issued the new version of the Admiralty's additional sailing and fighting instructions, and again on 12 July.[43] Saunders's signals, 'extracted from the General Printed Sailing & Fighting Instructions', were formed into a special signal book 'so as to dispose all signals that a Commander-in-Chief may see what Signal is to be made'. The book is in two parts of folio size, separately bound, beautifully written and painted. The first volume is labelled 'Admiral Saunders's Signals for Commanding Officers', possibly indicating that it was specially prepared for his divisional flag officers or commanders of detached squadrons, though the phrase might also apply to captains.[44] In effect, however, the first volume serves as the index to the second volume, which contains most of the signals. References to the general printed Sailing & Fighting Instructions were printed in black, and those to additional signals in red. The second volume was the signal book itself, arranged in the usual pattern, with references also in black and red. The two volumes provide the best example of signal book layout certainly up to the end of the American War, and possibly even until the appearance of the first official Admiralty Signal Book in 1799.

On 8 March 1762 Commodore Howe, having succeeded to the title Lord Howe, issued a printed book of 'Additional Signals & Instructions to be Observed by the Ships of War' to his cruiser squadron in the Basque Roads watching Rochefort.[45] This book provides the link between his earlier essay of 1756 and his epoch-making Signal Book & Instructions of exactly twenty years later. The additionals of 1762 contain sixty-one signals for day, fifteen for night and seven for fog. No attempt was made to separate sailing from fighting signals, with all lumped together under the day signals. A special and distinguishing feature of the book is that it is laid out in signal-book form, except for the purely instructional articles. Each page is ruled into three columns,

the first two headed Flag and Place, the signification being on the right with its respective number.

There were no startling innovations; Howe's main purpose was to increase the efficiency with which his squadron could carry out its immediate task, stopping enemy ships leaving and entering Rochefort. For this he needed flexibility and simplicity of manoeuvres, and a more efficient method of reporting a strange sail by private ships. Combined with this energetic direction of effort were characteristic signs of that obscurity of statement which were to make his later tactical reforms so difficult to understand (particularly to modern readers). His signal number 59, for instance, seems dangerously ambiguous, considering that it was intended for use in a sudden emergency:

When the Commander-in-Chief upon making the Signal for Line-ahead or abreast, though meaning to form on the Starboard Tack, would nevertheless have the Ship appointed to lead on the Larboard Tack, and that Division then to lead: Or on the contrary the Ships of the Starboard Division to lead though forming on the Larboard Tack. The same change of Divisions is meant to be made when this Signal is shewn with the Signal for forming in two separate Lines of Battle.

To set against these ambiguities of drafting, there were some striking advances, clearly expressed. Number 12 was the long overdue proviso: 'For the Squadron when in chase to engage the Enemy in Line of Battle according to the Order in which they may successively arrive up with them', which amplified Anson's additional number 9 establishing an *ad hoc* line of battle. Number 50 was an instruction, not a signal: Howe instructed his squadron that, when tacking to windward, the division which happened to be to windward when the signal was made was to stay there until a further change of wind or course enabled it to regain its designated station in the order of sailing, 'The intent of this distribution of the Ships in Sailing Order being only to prevent accidents by their running through the Squadron and crossing one another in the night time.'

Article 1 of the night signals and instructions, which are

Admiral of the Fleet Richard Howe (1726–99), first Earl Howe, by John Singleton Copley. NMM

Admiral Sir Charles Saunders (1713?–79), detail by Sir Joshua Reynolds. NMM

43 *NMM*, CLE/2/8

44 *NMM*, Tunstall Collection, TUN/34 and 35 (formerly S/MS/SY/3 & 4). *NMM*, HOL/17 (formerly S(H)17) is a pocket version of this volume.

45 *Barrington*, I, pp326–56

otherwise of no special interest, reads: 'As it is necessary in cruising that all signals should be made [at night] with as few guns as possible, they will therefore be made in fair weather with lights only, as directed in the Printed Instructions.' Howe specifically indicated which signals should be made using guns. The reduction of gun signals at night as well as by day remained one of his constant preoccupations; Hawke's experience at the Battle of Quiberon Bay, when the gun signal for anchoring was confused by many ships for the distress guns fired by enemy, may have been the origin of Howe's determination to depend primarily upon light signals alone.

Five weeks later, Howe issued three orders of battle. In the first he stationed himself in his 74 in the centre of the first division of five ships. He gave the central position to two 90s by placing them at the rear of the first division and the head of the second. His second order of battle provided for 'being myself opposed to the Commanding Officer of the Enemy placed in the Centre of their Line as is their usual practice'. He took station fifth in the line, and placed his 90s third and eighth on the assumption that numbers were equal and the enemy were to be engaged ship for ship.

His final order of battle provided for the possibility of 'finding them composed, as is possible, of a number of ships greater than that of which the British Squadron consists'. In this case, the order of battle was to be the same as number 2, but, assuming the enemy to number fourteen ships, their fourth, sixth, ninth and eleventh ships were to be left unopposed. Howe's two 90s were to fill the respective gaps.

The idea of trying to contain the whole of the enemy's line instead of concentrating on part of it may seem retrogressive. Nevertheless the mere notion of making any kind of tactical distribution for a specific purpose implies a higher level of tactical thinking than anything so far sanctioned by the 'custom of the service'. His tactical creativity is further revealed by his observation that, in the event of the enemy outnumbering his squadron, 'I may perhaps attempt by studied delays to conceal from them my purpose of bringing them to action, until later in the day, and that opportunity offers of doing it more to advantage.'

What appears to have been Hawke's last signal book of the Seven Years War is entitled 'General Printed & Additional Signals delivered out by Sir Edward Hawke 1762'.[46] Twenty-four flags were used, and two pendants. It incorporates the traditional signals from the Sailing & Fighting Instructions with those in the 1758 version of the Admiralty additionals by means of two pairs of reference columns. An interesting additional day sailing instruction, for ships when not otherwise appointed, is 'to keep the stations in respect of each other (the several Admiral's seconds ahead and astern upon their different quarters respectively) as prescribed by the order of battle'. This suggests the beginnings of a more formalised cruising organisation than had yet been adopted.

Undated Admiralty signals and instructions

The Naval Library possesses a printed set of 'Signals and Instructions for His Majesty's Fleet' of exceptional interest, but to which no certain date can be assigned.[47] There are forty-five fighting instructions in this set, in which are continued the efforts begun after the Battle of Toulon to produce decisive results when the British fleet was superior. Article 38 was for particular ships to quit the line, and to distress the enemy by raking their van or rear, or to assist disabled or hard pressed ships by taking their place in the line; this was a fuller and combined version of existing instructions. Article 34 ordered that, if the admiral hauled down the signal for the line during battle, every ship was to engage as close as possible, and drive the enemy out of their line 'without having regard to the situation which was prescribed to themselves, by the line of battle, before the signal was hauled down'. The intention of this was apparently to allow captains a free hand to act to the best advantage should the course of the battle necessitate such action. Article 43 made the instruction automatic, if British numbers exceeded those of the enemy. Article 40 was a signal for engaging as close as possible, with the British ships at a distance of half a cable or less, should the weather permit, together with the usual warning against firing over the top of British ships.

Sir Julian Corbett conjectured that these instructions were 'possibly never issued to the fleet'.[48] They certainly seem too progressive, as well as to individualistic both in content and arrangement, to have been issued by the Admiralty. The decorative printed designs suggest affinities with the undated and hitherto unknown instructions issued by Howe.[49] Nevertheless the title, text and general arrangement are unlike anything produced by Howe either then or later. They would seem to have emanated from a commander-in-chief, and from someone other than Anson, Hawke, Boscawen, Saunders or Keppel. Apart from Sir George Pocock's, it is difficult to suggest a name. They are quite unlike Keppel's of 1756, and when Rodney sailed for the Capture of Martinique in 1761 he issued the new version of the Admiralty additionals.[50] The date and authorship of this printed set, therefore, remains a mystery.

Howe's Three Orders of Sailing

Flagship

90-gun ship

90-gun ship

Enemy squadron

Second order

First order

Third order

46 *MOD*, Ec/35

47 *MOD*, Ec/42

48 *Signals & Instructions*, p383

49 *NMM*, Tunstall Collection, TUN/111 (formerly FI/5)

50 *NMM*, (formerly Sp/130 and Sm/168)

CHAPTER FIVE:

Signals Development following the Seven Years War

ONE of the main features of the revival of the French navy under Étienne François, Duc de Choiseul, Louis XV's minister of marine, was the advance made in tactics and signalling. The leading figures were Sébastien François de Bigot, Vicomte de Morogues, and Jean François de Cheyron, Chevalier du Pavillon. Jacques Bourdé de Villehuet was a lesser figure, but he publicised the work of Mahé de La Bourdonnais, who anticipated the important development of the numerical signalling system. Although Morogues and Pavillon succeeded in rescuing tactics from the civilian mathematicians and scientists, they followed Hoste in tending to substitute evolutions for fighting, and in failing to emphasise the need for offensive action. This is not really surprising, for just as Hoste wrote in the shadow of Barfleur and La Hogue, so they wrote in the shadow of Lagos and Quiberon. They were also influenced by the *Ordonnance du Roi,* published on 25 March 1765, which was an unrealistic and theoretical exercise allowing captains no freedom of action. They were required to fight 'yardarm to yardarm', and to 'give more attention to defending the flags of the Generals than to preserving their own ships, letting themselves be sunk rather than deserting the flag'. However, when the French intervened in the American War in 1778, the commander of the French main fleet, the Comte d'Orvilliers, was able to exercise tactical control through a signal system devised by Pavillon. Though baroquely complex, it greatly surpassed anything previously available.

Christopher O'Bryan, and possibly his brother Lucius, helped to spread French tactical ideas across the Channel, and they and their father found a market in England for essays on naval discipline and tactics. The printing by Lord Howe of a new set of additional instructions, possibly in 1770, is a milestone in the development of English tactics, and indicates that the victories of the Seven Years War did not have their usual stultifying effect. The Admiralty itself published a companion volume for the Sailing & Fighting Instructions, and, in the opening phase of the American Revolution, Howe had the opportunity further to develop his tactical system.

Morogues's tactical system

As in Hoste's time, naval matters such as ship construction and hydrodynamics, navigation and the determination of longitude, hydrography, chronometers and voyages of discovery, were considered in France as worthy of free discussion between mathematicians, scientists and naval officers. Naval science was something in which civilians could, and indeed were encouraged to, play an active and fruitful part. The Vicomte de Morogues was already eminent in this highly civilised naval-scientific society; he had been chosen as the first director of the Académie Marine when it was founded in 1752. Much of his work *Tactique Navale ou Traité des Evolutions et des Signaux* had been circulating for some years amongst naval officers at Brest before he submitted it in book form to the judgment of commissioners appointed by the Académie Royale des Sciences. Only a month after the signing of the humiliating Treaty of Paris, three members of the Académie signed a laudatory report on his book, published the same year in Paris. It was dedicated to Choiseul, the minister being pointedly invited to give reforming attention to the navy.

Although in some respects modelled on Hoste, whose work Morogues declared in his introduction to be now 'un peu rare' and to contain 'beaucoup de manoeuvres inutiles' though 'excellent dans plusiers parties', his own book was quite different in structure. Whereas Hoste wrote as a mathematical expert with a strong historical bias, Morogues wrote as a naval officer anxious to give the service a complete system both of tactics and signals together with all the theoretical, explanatory and administrative guidance necessary to make the system effective.

During the Seven Years War, Morogues had been active in proposing various strategic measures, and in the disaster of Quiberon Bay he commanded the *Magnifique* (74). This experience must have reinforced whatever prejudice he already had in favour of the over-riding importance of defensive tactics. In his introduction, Morogues wrote that tactical order and discipline were decisive factors in battle. A small well-disciplined force, whose captains kept their line well closed up and gave each other mutual support, could overcome a more numerous but less disciplined force, quite apart from the fact that the smaller the fleet the easier it was to handle. In comparison with fighting on land, the weaker

Amiral *Sébastien François de Bigot, Vicomte de Morogues (1706–81).* MM

force at sea had more opportunities to escape. Tourville's *campagne au large* showed how a force could make up for its inferiority by *vivacité de l'action*. Morogues's thoughts on tactics were doubtless influenced by his belief that sea fighting was not decisive, at least as far as the war as a whole was concerned; for eighteenth-century France this was indeed the case.

Morogues stressed the importance of large ships, as compared with ordinary two-deckers. They mounted a higher proportion of big guns, they sailed better in stiff winds and heavy seas, and they were less easily boarded. In bad weather they were more likely to be able to use their lower-tier guns than the two-deckers, while if conditions would not allow this, they had the advantage of being able to use two upper tiers of guns against the two-decker's one. In battle, the fleet with most large ships should win even if inferior in numbers. A small fleet was more easily handled than a large one, especially if it were not too closed up in formation, though in fine weather one-third of a cable between ships was considered sufficient. A fleet too greatly extended risked having some of its ships fired on by two enemy ships at the same time.

Morogues's approach to the problem of defeating an inferior enemy by doubling his battle line was much more practical than anything in Hoste's academic exercise. To double the enemy's van from windward, the leading ships should crowd sail so as to draw ahead of the enemy van.

Having doubled, they should then drop to leeward and resume their course as soon as the enemy van drew level; this would place the enemy's leading ships between two fires. For the superior fleet to double the enemy van from leeward, the leading ships should tack right across the enemy's line of advance, each giving the enemy's leading ship a broadside as it passed. They should then resume their course, now to windward and abreast of the enemy van. To double the enemy rear from windward, the rear ships should turn to leeward and crowd sail so as to sweep round the enemy rear and bring them between two fires. To double the enemy rear from leeward, the rear ships were required to tack so as to pass astern of the enemy and so gain the wind. They were then required to resume their course and crowd sail so as to draw up level with the enemy rear. On balance Morogues, like Hoste, favoured doubling the enemy rear rather than their van, and for the same reasons. It was easier in this way to cut off the enemy's doubled ships from support, and it involved less risk that one's own ships, if damaged, might be caught by the enemy as they came up.

All this was highly theoretical, in that it assumed a complacency on the part of the enemy which might not be forthcoming even if they were weaker in numbers and hence more easily overlapped at each end of their line. Nevertheless, these were positive ideas recommended for actual use in battle, rather than manoeuvres debated in theoretical terms on the basis of historical examples.

Below: *A page from le Vicomte de Morogues's* Tactique Navale ou Traité des Évolutions et des Signaux, *Paris, 1763, showing ships signalling with skyrockets.* NMM

Below right: *A plate from Morogues's* Tactique Navale, *showing the points of sailing.* NMM

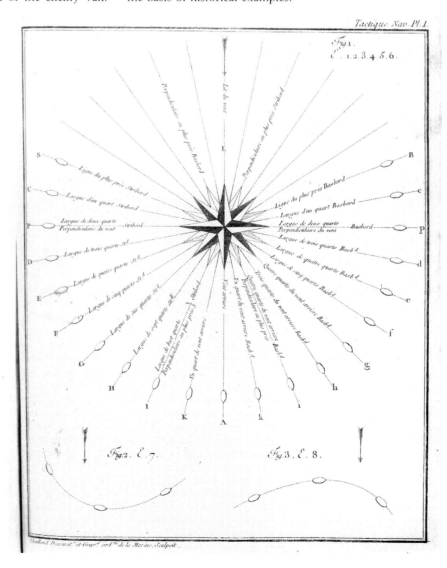

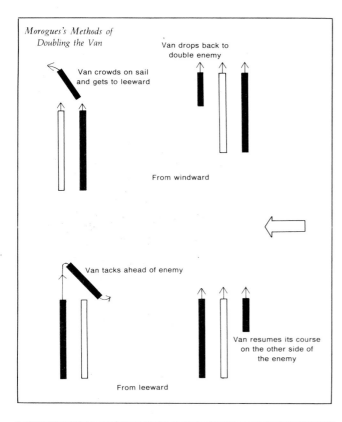

Morogues's Methods of Doubling the Van

Van crowds on sail and gets to leeward

Van drops back to double enemy

From windward

Van tacks ahead of enemy

Van resumes its course on the other side of the enemy

From leeward

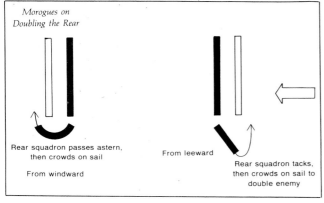

Morogues on Doubling the Rear

Rear squadron passes astern, then crowds on sail

From windward

From leeward

Rear squadron tacks, then crowds on sail to double enemy

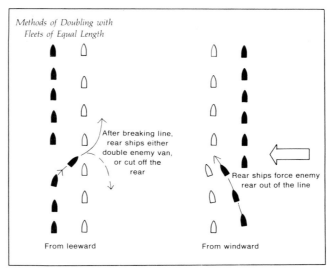

Methods of Doubling with Fleets of Equal Length

After breaking line, rear ships either double enemy van, or cut off the rear

Rear ships force enemy rear out of the line

From leeward

From windward

Morogues suggested another, novel, approach to the tactic of doubling. If the superior fleet were well closed up so that it only equalled the enemy's line in length, it might double the enemy rear from leeward by cutting right through them with its own rear ships. These ships could then fan out, so as either to resume their course to windward of the enemy or to surround the enemy rear completely by passing round them in succession from windward on the opposite tack. Similarly, the superior fleet, if well closed up, might double the enemy rear from windward by bearing down hard with its own rear ships and driving the enemy rear to leeward and right out of their line.

To avoid being doubled when inferior in numbers, Morogues adopted the traditional expedients of spacing the fleet either by individual ships or by squadrons in such a manner as to prevent the enemy from overlapping either end of the line. He also introduced the idea of bringing up the fireships to fill gaps which the enemy might otherwise be tempted to exploit. Large ships should be stationed at the centre and rear of their respective squadrons. He admitted, however, that it would be difficult for an inferior fleet to leeward to avoid being doubled. Much depended upon its bravery and experience.

The second half of the book, entitled 'Des Signaux et Ordres Généraux', is the more important. It consists of over 200 pages of orders with signals annexed, written in the form of sailing and fighting instructions, and containing over 300 administrative and operational articles in addition to explanatory matter. Embedded in these instructions is the real essence of Morogues's new tactical thinking, and both in number and in detail they go far beyond the scope of the standard French system of the Seven Years War. Although the appropriate signals are annexed to the instructions, there is an eighty-page table of signals, for day, night and fog. This forms the draft lay-out of what a signal book should be. The engraved plates at the end illustrate the work as a whole, including both parts and showing the ships in diagrammatic form. In the middle is a forty-page index with detailed references to both parts of the book and to the table of signals and the diagrams. Each part of the book is heavily cross-referenced, so that the instruction, signal and diagram dealing with a particular evolution can be found at once and vice-versa. The diagrams all bear references back to the text.

Morogues proposed employing thirty-three flags with a choice of six positions, and nine pendants with twelve positions, as against what seems to have been the official system of thirty-four flags and six pendants, all flown from seven positions. The flags and pendants were numbered by Morogues in two separate series. Signals for the whole fleet were to be flown from the mizzen yard. Signals for the van, centre and rear squadrons were to be flown from the fore, main and mizzen topmasts respectively, except for the blue and white, white, and blue flags or pendants which indicated the van, centre and rear squadrons respectively and were supernumerary to the numbered flags. Morogues left the colours blank for all his flags or pendants; except for the three flags mentioned, they all were given numbers only. This was done on the grounds that the colours of the flags should be constantly changed for security reasons.

As in the British system, a flag only had a meaning when flown in a particular position, but Morogues gave each meaning a number which corresponded with the number of the instruction to which it applied. This number was quite distinct from the number given to the flag, which indicated its colour. A true numerical system was superimposed on the main signal system, for use when it was wished to draw attention to a particular article in the instructions, or in cases where figures had to be given, as for soundings and latitude and longitude. The flags numbered 1 to 9 were to be flown from the fore topmast head for the hundreds, at the main for tens, and at the mizzen for the digits. Nought was signified by having no flag flying at the appropriate masthead.

In some cases a signal might have more than one meaning according to particular circumstances. In such cases, and indeed in many other cases, the signal was accompanied by signal guns. Morogues devised an ingenious system to increase the possible vocabulary of gun signals, by varying their rate of fire: *lentement,* at twelve to fifteen second intervals, or *coup sur coup* at four to five second intervals. If the two methods were combined in the same signal, an interval of thirty seconds was to be made between each group. Not more than six guns were used for any one signal. Morogues provided a separate table of gun signals, with cross-references to the flag signals. Each grouping of gun signals was given a number which covered all the flag signals, night and fog signals, to which it applied.

The night signals were expressed in exactly the same form, the numbering of the articles being continuous, but the system employed for transmitting signals was extraordinarily complicated. Each night signal had a number printed in the margin which differed from the number of the article itself. In the next column was printed one, two or three letters of the alphabet, ranging from A to Q. These letters were purely notional: they were simply symbols indicating the use of from one to five lanterns at the fore, main or mizzen shrouds. The lights themselves thus indicated the original number of the signal. Several of the signals had more than one meaning, and, as in the day signals, these were differentiated by signal guns. Numbered references to the table of gun signals were in a separate column opposite those covering the lights. After the title of the signal, a final column provided references back to the numbered articles. Under conditions where the respective masts of the flagship could not be distinguished, rockets *en pluie,* to the right number, might be fired to indicate the foremast, rockets *en étoiles* for the main mast, and rockets *en serpentaux* for the mizzen. Thus for signal number 135 two star rockets, four rain rockets and five serpentine rockets might stand for the same number of lanterns at the main, fore and mizzen respectively. This would indicate the purely notional letters B I Q. With the help of gun signal number 32, six guns *coup sur coup,* captains would then know that the signal (number 135, referring to Article number 292) was to anchor. Lanterns could be used in the same manner as numeral flags by day to indicate numbers.

The fog signals were much more restricted. Even so, nearly two hundred separate signals could be made by an ingenious use of hailing by speaking trumpet (for recognition), various bugle calls, drum, bell, trumpet, fife, musket and swivel gun. Some of these were combined with ordinary signal guns and a large number were made by signal guns alone.

From all this, it is apparent that the book was comprehensive, both in its lay-out and in the system it served, but by no means simple to handle. Whereas the official French signal book was a modest folio of just over 100 pages of large print, Morogues's guide book had 480 pages and 49 engraved plates depicting 133 tactical diagrams and weighed 5lb. There is, however, no real evidence that it ever had any immediate practical usage, even though its influence was very considerable. It was reprinted in Paris with the original date of issue, 1763, and a Dutch edition was published in Amsterdam in 1779.

Morogues's reputation must certainly rest more on the ingenuity of his signal system than on the originality of his tactical thought or his desire to discuss battle tactics at length. Despite his own suggestions to the contrary, he strengthened the notion already advanced by Hoste that tactics are mainly evolutions, and that skill in evolutions is almost a substitute for fighting efficiency. He encouraged the idea that greater tactical control, exercised in terms of evolutions, would be the future answer to the British, with their improvised tactical methods, and their tied signal system.

Bourdé de Villehuet and La Bourdonnais

In 1756 a book was published entitled *De la Manoeuvre des Vaisseaux ou Traité de Méchanique et de Dynamique.* It is not surprising to find that the author, Pierre Bouguer, was a member of the Académie Royal des Sciences and sometime hydrographer at the ports of Croisic and le Havre, and had already written books on observing the height of the stars (1729), compass declination (1731), ship construction and movement (1746) and the shape of the earth (1749). Nor is it surprising that Bourdé de Villehuet, writing on tactics in 1765, remarks of Bouguer, 'mais, pour l'entendre, il faut beaucoup de Géometrie et d'expérience: d'ailleurs, combien d'opérations ne peuvent s'apprendre qu'à la mer!' Bourdé acknowledged the value of the theoretical ideas obtained from earlier writers, but his own work was based on his personal sea experience.

Bourdé described himself as an officer in the French East India Company's service, and he entitled his book *Le Manoeuvrier, ou Essai sur la Théorie et la Pratique des Mouvements du Navire et des Évolutions Navales.* It was dedicated to Choiseul and contained a warm eulogy from the Académie Royal. Compared with Morogues's work it is rather disappointing to the historian, being to some extent a restatement of Hoste in simpler form, lacking the lavish historical examples. Nevertheless it is a distinctly more useful and more practical book, produced in handy form, measuring only 7⅞ x 4⅞in. After a section of theory, Bourdé explained simply and clearly how to handle a single ship, and he presented new ideas about chasing which were later adopted by Howe and Kempenfelt.[1]

In Part III Bourdé wrote about the organisation of the crew for fighting purposes and about battle discipline. He strongly enjoined silence so that orders could be heard. He very clearly explained gunnery and training in gunnery, and stressed the importance of getting in the first broadside, 'quand on est de bien près'. 'Si le vaisseau que vous attaquez tiraille de loin, laissez-le faire; il ne vous sera jamais beaucoup de mal.' The enemy, he argued, must be brought to close quarters as quickly as possible and overwhelmed by rapid fire.[2] He provided a good description of the organisation and conduct of boarding.

In Part IV Bourdé discussed evolutions. On the relative advantages of the weather and lee gages he followed Hoste, adding by way of summary that the weather gage offered fewer disadvantages than the lee gage, though it was correspondingly the more dangerous. On balance, the weather gage was best, especially for the stronger fleet which could detach its surplus ships to double on the enemy rear. In fact, a fleet ought not to accept battle from leeward 'à moins que de s'y trouver forcé par quelque fâcheuse circonstances'.[3]

Copying Hoste's fourth order of sailing, and Morogues's fifth order, Bourdé favoured a sailing order in three columns abreast, which he called the order of convoy. He gave examples of changing from this to the order of battle and order of retreat and back again.[4] A longish chapter dealt with how to restore order in shifts of wind. Only in the last tactical

---

1 Part II, Chapter 8; see *Signals & Instructions,* p143

2 Part III, Chapter 11, Art 7

3 Part IV, Chapter 1, Art 5

4 Part IV, Chapters 2 and 3

chapter in the book (Part IV, chapter 5) does the enemy come into the picture, and here the treatment followed Hoste. In Article 9 of this chapter he stated that a fleet engaging from leeward should stand on as soon as the action began. This would force the weather line to set topsails, and if as a result a ship was disabled in the advance, confusion would ensue, as happened to Byng at Minorca.

Bourdé's last chapter describes the signalling system invented and used by Mahé de La Bourdonnais, a very distinguished officer in the French East India Company's service who, during the War of 1739–48, was governor of Mauritius and Bourbon and commanded the French naval forces in the Indian Ocean. It appears that La Bourdonnais had a description of his system printed in 1746. It was, at last, a true numerical system, with ten pendants, numbered 0 to 9, of plain red, white, blue and yellow and parti-coloured. By this simple arrangement thousands of signals could be made. Each signal needed one hoist only, and could be made from any part of the ship most convenient for sending and reception. The bottom pendant signified the digits, the next above the tens, and so on.

In addition, La Bourdonnais propounded a system of long-distance signals involving a combination of the old and new systems. There were to be eight symbols only, four plain flags and four plain pendants of red, white, blue and yellow. First, a pair of pendants or flags would be hoisted anywhere convenient to indicate whether it was a hundred, ten or digit figure which was to be signalled. Then would come the indicative pendant or flag, which would be valued entirely according to its position. In each case the same position was reserved for each number (1 and 6, ensign staff; 2 and 7, mizzen yard; 3 and 8, mizzen topgallant; 4 and 9, main topgallant; 5 and 0, fore topgallant). The shape and colour, however, were different in each case, according to whether the pendant represented a hundred, ten or digit figure. There were five positions for each pendant or flag, so a particular pair could be used to cover 0 to 9 in each category. Thus, although the main topgallant position was reserved for 4 and 9, the 4 of a four-hundred signal was a white pendant, the 4 of a forty signal was a red pendant and a plain 4 was a yellow flag. Provided the flag or pendant could be properly made out at a distance, the signal could not be misread. In addition to the flag's single possible meaning in the position occupied, there was the additional check provided by the preliminary signal indicating whether the substantive signal was a three, two or single figure number.

La Bourdonnais also advocated numerical night signals by combination of lights and guns to be agreed upon beforehand (he did not lay down a scale of his own). He wisely added that lights should be used entirely in numerical terms, and hoisted in one place. The practice of hoisting them in different positions made it difficult to see them all from any particular bearing.

Publication of French tactical studies in England

The first book on tactics in the English language, *Naval Evolutions: or a System of Sea Discipline*, appeared in 1762.[5] It was published by Christopher O'Bryan, at one time a lieutenant in the Royal Navy, who is said to have entered the Russian Navy in 1739, apparently as a midshipman. He was a son of the Christopher O'Bryan who was made post in 1713, commanded the *Rippon* (60) at Cape Passaro, entered the Russian Navy in 1739 and died in 1742. From this there

5 *Naval Evolutions: or a System of Sea Discipline: Extracted from the Celebrated Treatise of P L'Hoste, Professor of Mathematics, in the Royal Seminary of Toulon: Confirmed by experience; illustrated by Examples from the Most Remarkable Sea-Engagements between England and Holland; Embellished with Eighteen Copper-Plates; and Adapted to the use of the British Navy. To which are added, An Abstract of the Theory of Shipbuilding; An Essay on Naval Discipline, by a Late Experienced Sea-Commander; A General Idea of the Armament of the French Navy; with some Practical Observations; By Christopher O'Bryan, Esq; Lieutenant in His Majesty's Navy, 1762 (NMM, Tunstall Collection, TUN/105)*

A plate from Christopher O'Bryan's Naval Evolution or a System of Sea-Discipline, *London, 1762, following the work of Paul Hoste, showing the ease with which a superior fleet, when to windward, could detach ships to double the enemy rear.* NMM

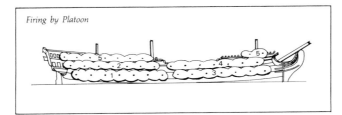

Firing by Platoon

is reasonable inference that Christopher O'Bryan the elder was the author of the 'Essay on Naval Discipline' which his son included in his text.

Translated extracts from Hoste's book amounted to about a quarter of the younger O'Bryan's work. The experience of the Seven Years War may have tended to discredit Hoste's approach, or at any rate the kind of battle tactics derived from it, but there were, no doubt, British officers who regretted the absence of a tactical manual in English. The extracts from Hoste, covering part of the elementary sections at the start and most of the battle sections at the end, were probably those considered most useful for British officers. Quite apart from the indifferent translation, the presentation was too disjointed to enable the reader to ·appreciate either the strength or the weakness of Hoste's system. The plates were redrawn from the originals, and to the same scale. Hoste's *Théorie de la Construction des Vaisseaux* was also abstracted by O'Bryan.

The 'Essay on Naval Discipline', though dull and prolix, is of considerable value to the historian because it discusses current thought and practice. The author believed that it was more effective for ships to fire their guns 'in platoons' than in broadsides. Half the guns on the lower deck, aft, should be fired first, followed by the aft guns on the upper deck, the forward guns on the lower deck, the forward guns on the upper deck, and finally the guns of the quarterdeck and forecastle. The author also believed that at sea, unlike on land, the line should be formed with the main strength in the centre. Ships in excess of the enemy's line should be in a reserve corps, and the admiral should have a repeating frigate stationed opposite to him. The suggestion was dismissed that the admiral should himself be in a ship of the reserve, though keeping his flag flying in the line. Sea battles were not like land battles, it was argued, and the situation of the general and the admiral were not strictly comparable.

In forming a line of battle, O'Bryan wrote that, except in

the rare instance when a more senior captain commanded a less powerful ship, 'seniority should be regarded, either by giving orders to the oldest captains to lead or be seconds'. Where the fleet was in three divisions with three flags, each flag officer should have the senior captains as seconds 'and each division having a leader upon either tack, according to the seniority of the next oldest captain'. The two junior flag officers could not be seconds to the admiral as that would preclude three divisions. In a fleet of at least fifteen ships there should be three divisions, 'because sometimes a division might be ordered to tack to gain the wind of the enemy, as well as to give the two junior flags command and honour, by their share of division and rank'. Where there were only two flags in a fleet, the senior captain should be given a division, 'that the line may thereby be regular and complete. This [he wrote] is what I have seen and learned in the course of my experience in the navy.' This last seems relevant to Byng's action at Minorca, though admittedly he had had only thirteen of the line, including one weak 50-gun ship.

O'Bryan stressed the importance of reconnaissance. Action, once joined, should be at point-blank range, that is about 300yds, but not closer as otherwise the line would fall into a curve. 'Besides, you cannot as well observe your own line: the enemy's ships may block your view of your own ships, especially if the line be long, which may cause a mistake in the signals.' The notion apparently is wrong that British admirals invariably thirsted for the closest possible action, and that the pell-mell was deliberately sought when fleets were of equal strength.

One of the more interesting sections in the book is that on convoy action, for which O'Bryan employed an example taken from an action over half a century previously. In 1707 Commodore Richard Edwards, escorting a large convoy with three 80s and two 50s, was attacked and heavily defeated by René Duguay-Trouin 'with ten or twelve under his command from 30 to 60 guns'. Edwards, according to O'Bryan, instead of lying-to as he actually did, should have retired in a line-abreast under easy sail; a number of ingenious tactical reasons were given to prove that he would thus have made a better defence. Answering the objection that the enemy might come up and penetrate the line, putting some of the escort between two fires, he argued that ships engaging an enemy ship stationed between them would be in danger of hitting each other's masts and yards.

When Morogues's *Tactique-Navale* first appeared in Paris in 1763 it must have aroused immediate interest in England. With the signature of the definitive Treaty of Paris in that year, Englishmen resumed their visits to France in large numbers. There was an obvious need for a good translation. Instead, in 1767 a partial and wholly unsatisfactory one appeared.[6] Only a small part of the original work was in fact used, the selection being highly arbitrary, with the omission of important sentences. Morogues's entire idea of a system of tactics and signalling, expressed through the agency of a book, with the material organised, explained, numbered and cross-referenced in a particular way, was completely ignored. Instead, a selection of his ideas were jumbled together with other material.

The whole treatment of this translation is reminiscent of Christopher O'Bryan. The publisher, W Johnston of 'Ludgate-Street', was also the publisher of the Hoste translation five years earlier. The translator adds footnotes either

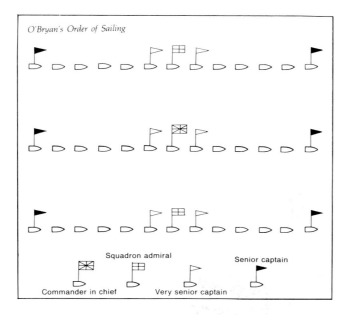

O'Bryan's Order of Sailing

Squadron admiral

Senior captain

Commander in chief   Very senior captain

---

6 *Naval Tactics: or, a Treatise of Evolutions and Signals, with Cuts, Lately Published in France, for the Use of the Cadets, or Guard-Marines of the Academy at Brest; And now Established as a Complete System of the Marine Discipline of that Nation. By Mons De Morogues . . . Translated by a Sea-Officer.* Paris, 1767 NMM, Tunstall Collection, TUN/103

laudatory of Hawke, to whom the book is dedicated, or critical of Mathews, who was the standard whipping boy in contemporary tactical discussions. These suggest another influence, that of Lucius O'Bryan, who was brother of Christopher O'Bryan the younger and commanded the *Essex* at Quiberon. Lucius is also said to have served in the Russian Navy in 1739. He died a rear-admiral in 1770 or 1771. He is the most probable author, since the book includes the claim that its author served as a midshipman in the *Orford* under Sir John Norris when still a captain, and was present at the Battle of Malaga in 1704. A comment that the victory at Lagos would have been more complete had the leading British ships reached ahead and cut off as many of the French as possible, by 'bringing to the headmost of the enemy' while leaving the remaining French ships to be dealt with by the rest of the British fleet as they came up, links the two O'Bryan experiences at Passaro and Quiberon. A section called 'Some Familiar Thoughts and Observations on Sea Fights, [and] Naval Discipline' echoes the earlier Essay on Naval Discipline; presumably both were written by the elder O'Bryan. The author stated that he commanded a ship of the line under Byng in the Mediterranean in 1718, which is certainly true of the elder O'Bryan.

Although the translated excerpts from Hoste and from Morogues, respectively, are poor work, the 'Essay' and 'Some Familiar Thoughts . . .' provide valuable information about tactical thought. If we assume that the selection of incidents over half a century old in the Essay make it rather an anachronistic effort, it becomes difficult to account for its reappearance in 1767. On the contrary, it would seem that the 1762 book was well received, and that this was not entirely due to its wretched presentation of Hoste, with whom many British officers must by then have been familiar, especially after the second edition of 1727. Past events, even half a century old, were relevant to the study of tactics because the *matériel* of navies had changed so little. In both works, and especially in the later, there is frequent insistence, particularly in the footnotes, on the need for better tactical discipline and better signalling.

The British 1768 manoeuvres

In the summer of 1768, the Duke of Cumberland was introduced into the sea service under the tutelage of Captain (later Admiral) Samuel Barrington, hoisting his flag in command of five frigates and two sloops. Barrington seized on this unusual peacetime opportunity to carry out extensive exercises in the Channel, issuing tactical memoranda and additional signals to the squadron over the prince's signature.[7] On one occasion, when the squadron was 'forming lines of battle and different manoeuvres, the better to observe them' the prince boarded another frigate which 'kept away to windward to observe signals'.

In memorandum number 3 Barrington provided for attacking the enemy's rear by inverting the order of the line of battle from an original bow and quarter line formation. An unnumbered memorandum has a signal for when the squadron was 'turning to windward on the enemy in a line of battle and the sternmost ship can't weather them on the other tack'; also for the sternmost captain when the squadron was 'to windward of the enemy and he thinks himself far enough astern of the enemy's rear to tack and lead down to it out of the sternmost ship's line of fire'. A similarly worded memorandum, number 6, gives a signal for the headmost ship to

make to indicate that it should be able to weather the enemy in line of battle when turning to windward.[8]

This isolated instance suggests that Barrington must have shared the general concern of men like Howe, Keppel, Rodney and Bryon for tactical and signalling efficiency, although not all of these were specifically concerned either with tactical innovations or a general improvement in signals. The conservative outlook of such commanders, and their commitment to the line of battle, does not imply that they were necessarily unconcerned with ensuring peacetime tactical efficiency.

The obvious need was for a complete revision of the Admiralty Sailing & Fighting Instructions to include the best of what had so far been produced by individual enterprise; this, however, was too revolutionary a project. The original cause of so many errors and limitations was retained simply because it had acquired the authority of tradition and long usage. Besides, re-writing it might have involved the Board in unpleasant controversies. Instead a new edition was printed slightly smaller in size and with fewer pages. A surviving copy of this edition, signed by James Gambier as commodore on the North American Station, 22 August 1771, aboard the *Salisbury* in Boston Harbour, was issued to Captain Benjamin Cadwell of the *Rose*. It included

A plate from the translation by 'A Sea-Officer' of part of Morogues's Naval Tactics, *London, 1767, illustrating (fig 11) a chasing fleet edging van and rear squadrons around the chase, and (fig 12) a manoeuvre to double the enemy van. NMM*

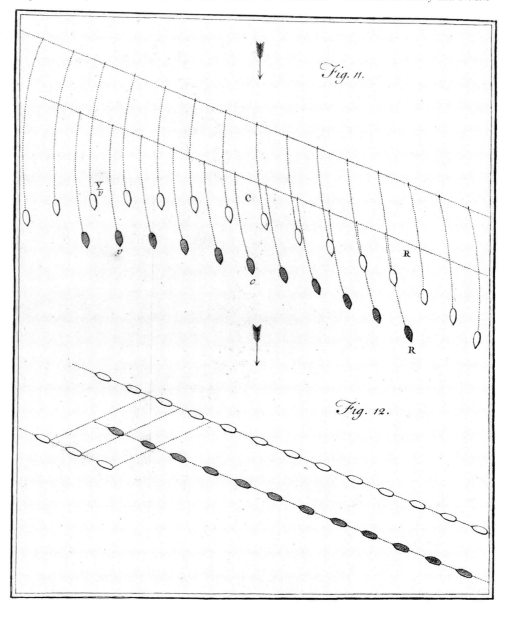

handwritten additional day, night and fog signals on folio pages, signed by Gambier and sewn in. Sir Charles Knowles states that there was a re-issue of this edition in 1772, but it seems more likely that he was referring to that of 1771.[9]

Lord Howe's undated additions

At some unknown date between March 1762 and June 1776 Lord Howe produced a new printed version of his 1762 additionals, entitled 'Signals & Instructions in Addition to the General Printed Sailing & Fighting Instructions by Day and Night'. In November 1770 he was appointed to command a squadron to be employed in the Mediterranean in consequence of the Falkland Islands dispute with Spain. *Barfleur* was ordered fitted out for him, but he never hoisted his flag. It may well be that the book in question was printed with this command in view, and that it incorporated his most recent thinking. Though almost identical in title with the subsequent Admiralty production of the same period, this is undoubtedly a Howe book, employing Howe's special phrase 'to be observed by the Ships of War'.[10] Taken *en bloc*, the new additional fighting instructions reveal a formidable development of tactical thinking, going far beyond anything reached in the Seven Years War, including Howe's own issue of 1762 of which these are extensions.

The book, which is well indexed by flags, by lights and by guns, contains fifty-three additional day sailing signals and instructions, with one more in manuscript; twenty-one night signals, with five more in manuscript; eight fog signals and forty-nine 'Additional Fighting Instructions Respecting the Duty of Frigates in Action'. In the index the flags and pendants are listed with the places flown, and with blank columns opposite, headed 'Art[icle]' and 'Page' in which the relevant numbers were inserted in manuscript. These page numbers refer only to the instructions themselves and not to an accompanying signal book. Similarly the night sections give the number of lights, how and where disposed, the guns, and the article and page inserted in manuscript. In the case of the fog signals, various details were printed, and the number of guns; the article and page numbers once again were inserted in manuscript. It would appear that Howe, possibly as a new idea, was making a serious attempt at secrecy.

The signals were in traditional form, with twenty-five flags and pendants flown in particular positions, their details written into the text of the respective articles. There is no reference of any kind to an accompanying signal book. The unusually large number of sailing articles is accounted for by the inclusion of a number of orders, derived from different sources, in the form of additional signals issued by particular admirals. Much attention was given to cruising distances and changes of direction, but there is no hint of new cruising formations. The additional fighting instructions seem to represent an early version of the signals and instructions issued by Howe when taking command on the North American station in 1775.

9 *NMM*, CAL/130 (formerly Sp/77)

10 *NMM*, Tunstall Collection, TUN/111 (formerly FI/5). This particular copy, unsigned and presumably, therefore, unissued, was part of Lord Sligo's collection of Howe documents. It is a printed folio with one of the usual but undated water marks, and is bound in the usual thick marbled paper covers. The manuscript insertions are in a clerk's hand.

11 *MOD*, Ec/43

12 *MOD*, (formerly *RUSI* 92 and 83[G]. 'not found' at Naval Library, but believed to be NM/92 and 83[G].

13 In the copy issued by Byron, 3 March 1779, *MOD*, NM/92 [?] (formerly *RUSI* 92)

Admiralty signals and instructions

The Falkland Islands crisis may also have been the occasion for the issue of a new official book by the Admiralty, not to supersede the Sailing & Fighting Instructions but to supplement it. The earliest known copy is one issued by Captain the Hon Robert Digby, 7 November 1777, when in command of a squadron cruising off Cape Ortegal.[11] 'Signals & Instructions in Addition to the General Printed Sailing and Fighting Instructions' is arranged similarly to the parent work, but the fact that the word 'Signals' comes first is indicative of a slightly changing view.[12] In the same way, the word 'General' suggests a slightly qualified view of the parent work. Signals and Instructions has forty-four sailing instructions and signals by day, ten by night and eight in fog; and twenty-eight fighting instructions and signals by day, and two by night. There is also the same kind of subject index of signals, with cross references to the instructions, as was in the original Sailing & Fighting Instructions of the late seventeenth century. Nothing else is included. The book was evidently intended to provide a synthesis of all the best and most acceptable developments in tactics since the time of Vernon. Although the parent work was thus no longer the only official book, its ultimate authority was unimpaired; indeed there were specific references to it in the new one. Possible confusions which might arise from this strange dualism does not seem to have worried the Board of Admiralty, since all the signals were different. In fact, by retaining the purely artificial division between sailing and fighting, the Board committed itself to a quadruple series of instructions for the fleet at sea.

The new sailing instructions are not impressive. There were some new chasing signals, but many of the signals were quasi-administrative, the inheritance of the last quarter of a century. No new cruising formations were included.

The new fighting instructions, which were also drafted in the traditional form, included seven articles dealing with the line of bearing, now described as 'bow and quarter line'. The advantage of this as a battle formation was, as Thomas Graves noted on his own copy, that 'the ship ahead will always enfilade the ship that engages her second astern.'[13] Several of the articles merely repeated long-accepted ideas, some as early as Vernon. Article 18, however, provided the instruction and signal which Boscawen had failed to issue at Lagos. It established, and improved upon, Hawke's famous Article 9 of 1747:

If the Commander-in-Chief should chase with the whole squadron, and would have those ships that are nearest attack the enemy, the headmost opposing their sternmost, the next passing on under cover of her fire, and engaging the second from the enemy's rear, and so on in succession as they may happen to get up, without respect to seniority or the prescribed order of battle, he will hoist . . .

This was an improvement on Hawke's additional because it specifically ordered ships coming into action to pass ahead under cover of their own ships already engaged. It actually prescribed what Sir George Byng had managed to do without a signal as long before as 1718.

In other respects the general effect is disappointing to the historian. Hawke's very important instruction of 1759, for pressing home the attack regardless of the line, was omitted. No inspirations seem to have been drawn from Howe's additionals of March 1762. Nothing was added in the way of new cruising formations as a means of increasing the ease and speed of fleet control, apart from the strong emphasis on the bow and quarter line formation. Though reflective of Hawke, the new Admiralty *Signals and Instructions* is on the whole a conservative production. With the sole exception of

Article 18, it did no more than give official sanction to instructions with which the navy had been familiar for twenty years and more, while actually excluding some of the most useful.

The worst structural feature of the compilation is the retention of the original method of indicating signals by embedding them in the text of the instruction. This may have seemed logical because it was a companion volume in the parent work. Nevertheless it was retrograde, not merely on grounds of logic and common sense, but in view of the fact that so many captains, as well as admirals, had become accustomed to issuing their own sets of additional signals, and that, in consequence, a large number of private signal books, constructed as such, were already in use.[14] More serious, though more excusable, was the failure to re-organise the whole signalling system on simpler and more efficient lines. To cope with the new instructions, more flags had

become necessary; this was the only way to make the old system continue to work. To have changed the system would have meant scrapping or completely re-writing the parent work.

Private signal books took note of the new work in various ways. Some had a separate column headed Book, with the reference below, and opposite the signal: 1, for the parent work or 2 for the new one. Others adopted the abbreviation GSA for General and Additional Sailing, GF for General Fighting and AF for Additional Fighting. There is also evidence that occasionally the two books were bound up together, though not necessarily by the Admiralty, and the pages given continuous numbering. This would explain the high figures of some of the page references in the *Terrible's* signal book of 1779.[15]

Pavillon's signalling system

La Bourdonnais's numerical signalling system, if it had been adopted by the French navy, would have been a major tactical development. The Chevalier du Pavillon, however, had the influence necessary to ensure that it was his system which was adopted instead, as it was by a naval council of

14 *MOD*, NM/92[?] (formerly *RUSI* 92) includes a set of the original Sailing & Fighting Instructions, issued by Byron on 26 October 1778, and embellished by the recipient with the flag signals and gun and false fire signals painted in the margins.

15 *NMM*, Tunstall Collection, TUN/1 (formerly S/MS/Am/1)

Chasing signals issued by Admiral Hawke in the Seven Years War. MOD

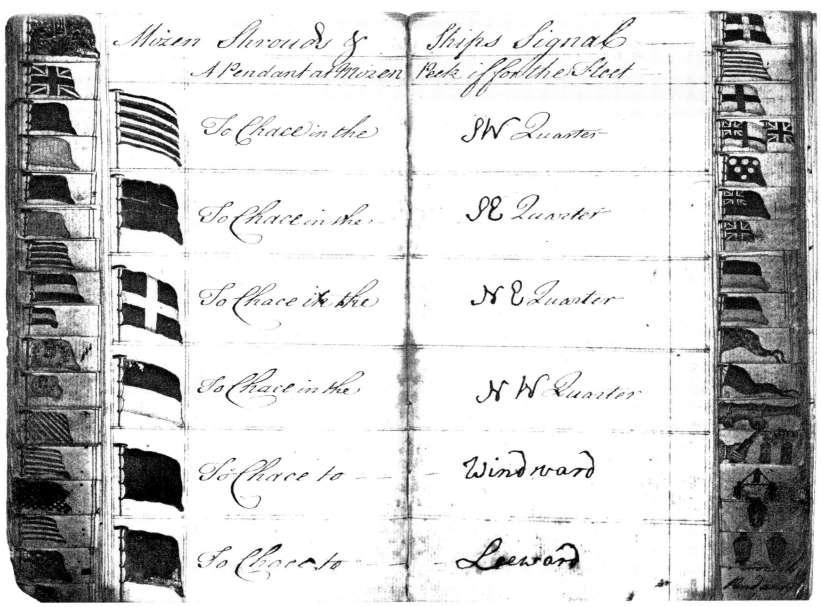

Mizen Shrouds & ... Ships Signal
A Pendant at Mizen ... Peek if for the Fleet

To Chace in the ... SW Quarter
To Chace in the ... SE Quarter
To Chace in the ... NE Quarter
To Chace in the ... NW Quarter
To Chace to ... Windward
To Chace to ... Leeward

fourteen flag officers assembled at Rochefort on 4 June 1773, and again by a naval council held at Versailles on 17 March 1775 which included d'Estaing and d'Orvilliers.[16] During that summer an 'evolutionary squadron' was formed, in which Pavillon commanded a corvette. During the exercises his signals were given a trial. Next year another evolutionary squadron was formed under Admiral Louis Charles, Comte du Chaffault, for whom Pavillon prepared a complete set of signals. He published his system in 1776, and shortly thereafter was appointed chief of staff to d'Orvilliers, who encouraged him to elaborate his system still further, and to impose it on the Brest fleet for the operations for 1778-9. Admiral Luc-Urbain du Bouëxic, Comte de Guichen, later adopted Pavillon's system, but not de Grasse.

table, filled with numbers which could be read off by reference to a series of flags painted along the top and down the left side. The numbers read downward from left to right. The particular number required was signalled by a hoist of two flags, the top indicating the column in which the number was to be found and the bottom indicating the line. The meaning of the number could then be looked up in the signal book. Pavillon's table contained a hundred numbered squares, and as usual in such a system, the same ten flags were used both for the superiors along the top and the inferiors down the side. These were:

1	Union Jack	2	White and red horizontal
3	Red	4	White
5	White and red cross	6	Saltire white red and blue
7	White bordered red	8	Dutch Jack in oblique stripes
9	Yellow and white	10	Chequered red and white

As in all such tables, the result was confusing. Except along the bottom line, where flag 5 over flag 10 read off at 50, and flag 10 over flag 10 at 100, etc, the table was always ten behind the actual flag numbers. Flag 2 over flag 1 indicated the square numbered 11 (not 21) while flag 9 over flag 5 read off at 85 (not 95).

The hundred signals which could be made using the table provided for the entire handling of the fleet when disposed in three columns, changing the relative order and position of the columns, changing into the various orders of sailing and back again, and forming the line of battle from the various orders of sailing and back again. Also included were signals notifying the general's intention to revert to the traditional signalling system, and to return again to the new, to change to *le grand tableau numéraire*, and to signal numbers for expressing soundings, numbers of ships, etc.

The same ten-squared table, with exactly the same flags, was employed to supply a separate series of a hundred signals for exercises and battle. The sixteen squares occupying the top left-hand corner of the table, which would otherwise bear the numbers 1-4, 11-14, 21-24 and 31-34, were occupied instead by sixteen letters of the alphabet: A, B, C, D, E, G, H, I, K, L, M, O, R, S, T, U. The remainder of the table, that is to say the eighty-four numbered squares, contained signals of a general kind, as well as the usual battle signals, including that for doubling and also concentration on the enemy's centre or van, but not rear. There was no signal for cutting the enemy's line. Numbers 82-98 were entirely for individual ships, designated by their number in the line of battle, to correct their position.

In each of the first sixteen alphabetical squares, one of the standard orders of sailing was also printed, ending as usual

with the order of retreat. To signal these, reference had to be made to a separate *grand tableau numéraire*. Instead of flags, the same letters of the alphabet as were used in the main table were printed along the top and down the side, the numbers in the squares reading from 101 to 356. By means of this table, the fleet could be signalled to change from any one of the standard sixteen orders of sailing into another, the top letter indicating the order the fleet was in, and the bottom one the order into which it was required to move.

On yet another sheet Pavillon produced his original ten-flag table, but with the squares numberd 101 to 200. Here each signal was in two halves, the top, in italics, giving the formation of the fleet at the time, the bottom, in roman print, giving the formation required. These changes of formation were more detailed than those envisaged under *le grand tableau numéraire*, which only dealt with the sixteen standard orders of sailing. The new table of 101-200 signals included changes of formation by the fleet as a whole as well as by squadrons, changes of the relative positions of the squadrons while on different tacks, changes of tack and wearing, and reaching the same positions either from line ahead, reversed line ahead, or from three columns. The signals numbered 101 to 164 were specifically for the fleet in three columns. Signals 165 to 200, although apparently overlapping those given in *le grand tableau numéraire*, in fact covered different movements, always with three columns in mind.

Although it must have been easy enough to change from one table to another by means of a preliminary signal made by a single flag, flamme or cornette, the possibilities of confusion would seem to have been immense, especially as regards reading off the numbers in the tables. What seems extraordinary is that, in spite of so much juggling with numbers and letters, the flags, though each bearing a number, had no significance except in relation to one or other of the tables. How easy it would have been to have expressed 300, 500 or even 900 signals by a truly numerical system.

Although it was an advance on the existing official system, Pavillon's tabular system represents a *cul-de-sac* in the development of signalling. He lifted French signalling out of its Seven Years War defeatist traditionalism, and put it on a new and complicated mathematical basis. Given quickness and accuracy both of reading and sending signals, the system was capable of being used to ring the changes on every possible fleet formation. In this sense it gave scope for tactical sophistications far beyond anything as yet regarded as standard in the British service. It was quite possible to transcribe the contents of Pavillon's complex signal sheets into a manuscript signal book; the problem, however, was that once the French navy turned in the direction of this semi- or even pseudo-reform, it became more difficult to cut through the new mathematical tangles set in train by Morogues, and proceed to the ultimate goal of a true numerical system. In the British service, where neither order, logic, simplicity nor common sense had so far prevailed, the breakthrough was achieved partly by Howe's skill and persistence, and partly by the threatened collapse of all tactical development under the old signalling system.

Howe on the North American Station, 1776-8

When Lord Howe arrived at New York on 12 July 1776 to assume command of the North American Station, the War of American Independence had scarcely begun, the Declaration

---

16 *Tactique Navale à l'usage de l'armée du Roi commandée par M le Cte d'Orvilliers . . . 1779*, p40 (copy in the library of the Royal Navy Staff College, Greenwich)

of Independence having been signed only twelve days previously. France had not yet intervened and the Royal Navy was solely concerned with operations against the colonists. Nevertheless, Howe at once set about organising his fleet as if for a full-dress naval war with a major power. His instructions and standing orders issued on the day of his arrival included an important battle order (number 17) directing captains to fire as many guns as possible at the first discharge. Apart from the general effect of doing so, this addressed the problems that firing so many together might subsequently be impossible owing to ships drifting out of range or bearing.[17]

On the same day, 12 July, Howe issued his printed Signal Book for the Ships of War.[18] Despite its unassuming appearance it is one of the most important documents in the whole history of naval tactics. It is a real signal book at last, though not an Admiralty one. In form, lay-out and typography, moreover, it has a special character which persisted through all Howe's later work and can instantly be recognised. Its title gradually permeated the navy and was eventually adopted by the Admiralty for their first official printed signal book, issued in 1799.

Howe's signals were only to be used by ships within his own fleet. They were made in the traditional manner, using twenty-three flags and seven triangular flags of simple design flown from particular positions. However, they had several new features. Signals for engaging, forming line, etc, were to remain in force, even though the signal might be taken in after acknowledgment, until annulled or countermanded.

The signals will generally be made without guns, when it may be done with the same effect; unless when it happens that two guns or more are necessary to constitute a part of the signal, or that it does not appear to have been timely observed. And all signal guns will be fired on the same side, to windward in a fog.

Unlike the earlier private signal books, in which all kinds of signals were thrown into juxtaposition on the same page, (largely as a result of successive admirals filling up unappropriated places for flying particular flags), Howe's signals were roughly grouped together. This had already been done by Saunders in his index volume, but not in the signal book itself which he issued.[19]

Such references as were given in the signal book are not to the Admiralty Sailing & Fighting Instructions, but to the page and article of an accompanying printed book of instructions.[20] The title of this book, 'Instructions for the Conduct of the Ships of War, Explanatory of, and Relative to the Signals contained in the Signal-Book Herewith Delivered', its content and wording demonstrate the revolutionary step which Howe had now taken. On the one hand it marked the separation and independence of the new signalling system from all instructional matter and its supremacy over such instructions as remained. On the other, it provided a new,

though subordinate, system of instruction. Unlike the official book, Howe's book consisted partly of standing orders, partly of orders designed to make the admiral's general tactical ideas more easily understood, and partly of explanations of how certain signals were to be executed so as to standardise evolutions. The actual signals were very sparingly described.

These instructions, far from representing a revolt against the line-ahead formation, reveal Howe as a stickler for its effective application, especially when in Article 14 of the fighting instructions he revived the supposedly retrogressive ban on pursuing beaten enemy ships. Nor was anything said about concentration on part of the enemy's line as directed in the earlier undated book, though because of its appearance in the Signal Book this might now have been considered self-explanatory. In Articles 10 and 22 of the 'instructions respecting the Order of Battle', the insistence on flag officers opposing the flag officers in the enemy's line is extremely conservative. So was Article 16, which forbade any captain to quit his station in battle without permission 'though much pressed'. Only in extreme necessity could they do so, but 'Captains deficient in time of action are to be removed from

A page from Admiral Lord Howe's 1776 Signal Book for the Ships of War. NMM

17 *MOD*, NM/88 (formerly *RUSI* 88). Corbett only prints the first six orders in pp87–92

18 *NMM*, HOL/21 (formerly S(H) 21) signed copy issued to Captain Sir George Collier, the *Rainbow*. HOL/22 (formerly S(H)) 22 is an unsigned copy in the original marbled covers. A further copy issued to Thomas Graves is in the *MOD*, NM/56 (formerly *RUSI* 56), and another issued on 13 June to Captain Cornthwaite Ommaney, the *Tartar*, is in the *MOD*, Ec/153. Manuscript copies are in the NMM, SIG/B/13 (formerly Sm/111) (without name of owner) and *MOD*, NM/59 (formerly *RUSI* 59), copy made by Thomas Graves.

19 *NMM*, Tunstall Collection, TUN/34 and 35 (formerly S/MS/SY3 & 4)

20 *NMM*, SIG/A/8 (formerly Sp/III), issued and signed 13 June 1776, *Eagle*, at sea, to Captain Cornthwaite Ommaney, the *Tartar*. See also *MOD*, NM/76 (formerly *RUSI* 68), and *MOD* Ec/153.

[28]

Signal	Pendants	Place.	Inſtruc. Page	Art.
	Red PENDANT,			
1	For all Barges and Pinnaces manned and armed, to repair on Board the Admiral's Ship—	Fore *top-maſt Head*	.	
	N.B. If meant to Chace on any particular Bearing, the proper Bearing Flag will be ſhewn at the ſame Time, at a little Diſtance beneath the Pendant.			
2	For the Ships of the Fleet to open to a greater Diſtance from each other, for forming any Evolution when in Line of Battle	Main-*top-maſt Head*	.	.
3	For all Launches, or Longboats, manned and armed, to repair to the Admiral's Ship.—If meant to Chace, the Direction will be ſignified as by Sig. 1 preceding	Mizen-*top-maſt Head*		
4				
	Yellow PENDANT.			
5	To cloſe in Line of Battle, back to the Diſtance before in Appointment, when the preceeding Signal 2 was made	Fore-*top maſt Head*	.	.
	Blue *Pendant.*			
6	For all Flat-Boats to aſſemble onboard the Admiral, or the ſhip in which the ſignal is made; With, or without the Troops embarked, according to the Circumſtances at the time of making this ſignal	Fore *Top-Maſt-Head*	.	.

their ships'. The arrangements Howe ordered for the execution of the line battle, however, were innovative because of their practical seamanship, and because they recognised the need for fleets to operate in a different manner from that of armies.

In his general instruction number 2, Howe abandoned the practice by which the junior ship was required to keep clear of the senior, and ordered the captains of all ships to take whatever action was necessary to avoid collision. In Article 1 of the 'Instructions Respecting the Order of Battle', he ordered ships in line of battle to open out when tacking to give each other sufficient room, and to close up afterwards. In Article 4, he ordered slow ships to fall astern and not disorder the fleet by trying to keep their stations. 'The Captains of such ships will not be thereby left in a situation less at liberty to distinguish themselves; as they will have an opportunity to render essential service, by placing their ships to advantage when arrived up with the enemy, already engaged with the other part of the fleet.' In Article 9 captains were instructed to 'studiously endeavour to obtain by repeated observation and experiment, a perfect knowledge of the proportion of sail required for suiting their sets of sailing respectively to that of the Admiral's ship'. The repeating frigate stationed opposite the Admiral was instructed to copy exactly the sail set by the Admiral so that ships at either end of the line would be able to make their own adjustments.

A year later, on 1 July 1777, Howe issued 'Additional Instructions Respecting the Conduct of the Fleet Preparative to and in Action with Enemy'.[21] In these Additionals Howe revealed more fully the advanced thinking which lay behind the executive form of his Signal Book. Though these were clearly designed to give an even higher degree of precision to the forms of attack included in the main instructions, still greater emphasis was laid on the need for tactical sophistication by individual captains. No fleet at that time could have interpreted these additional instructions intelligently without several weeks, and possibly even several months, of exercising together. Article 1 stated that, whereas 'the chief purpose of a regular disposition' in attack is to avoid exposure to the enemy's fire and the danger of collision with friendly ships, and whereas this may hamper captains 'from taking advantage of the favourable incidents which may occur in the progress of a general action, it is the object of these instructions to facilitate the means of improving such opportunity by an authorised deviation from these restrictive appointments'.

Article 2 ordered that the enemy line was to be engaged throughout its full length, so as to prevent them doubling the British van or rear. If the enemy were superior, captains 'commanding ships of greater force [were] to disregard or pass the weaker and worse sailing ships of the enemy and confine their first endeavours solely to disable the stronger and more active, as their accidental situations in the line may afford opportunity'. Article 7 instructed that, if the enemy were inferior in numbers, the overlapping ships in the British van or rear were to quit the line without awaiting a signal, and 'to distress and annoy any of the nearest ships of the enemy in conjunction with the ships of the fleet particularly opposed to them; or otherwise to assist or relieve any disabled ships thereof, as they can be employed with most advantage'.

21 *MOD*, NM/88 (formerly *RUSI* 88), manuscript copy by Thomas Graves. A comparison of the Instructions of 1776 and Additional Instructions of 1777 with the undated but obviously earlier Signals & Instructions gives the following results, the Instructions of 1776 shown in [] and the Additional Instructions of 1777 shown in (): undated book: Article 7 [7]; 8 [8]; 21 (3); 22–3 (5); 24 (6); 25 [18]; 42 [17]; 46 [19]

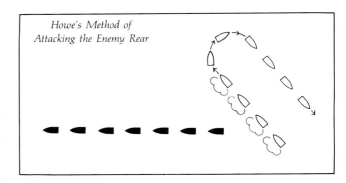

Howe's Method of Attacking the Enemy Rear

Articles 3 to 6 dealt with the situation of the fleet weathering that of the enemy preparatory to a general action. Captains were instructed to expect the admiral to make the signal for the rear to tack first, or for the fleet to tack together, as appeared to be 'most convenient for forming upon the line of bearing best suited to the position of the enemy'. If a number of the enemy's rear ships tacked first, an equal number of British rear ships were to do the same. If the enemy chose to await the British attack, and were on the same tack, 'the Admiral will then most probably choose to continue on upon the same course' until his van was level with theirs before attacking, provided that his ships were at a sufficient distance not to be 'materially injured' by the enemy's fire. However, he might chose to attack the enemy rear. In that case he would signal for the leading ship

to open to a convenient distance, and upon coming up with [the] enemy to give their sternmost ships her fire upon the quarter, then tack or veer and fall into the rear of the [British] line. . . . this method of attack is to be continued in succession by the ships of the fleet, or any particular division, so appointed until further signals, care being taken to leave room as requisite from each other in the execution of this service.

When it was decided to break off this hit-and-run form of attack and bring on a general action, the signal would be made to invert, in succession, the line which had been established primarily 'for the circumstance of coming up at a small distance to windward of the enemy upon the same tack'. The leading ship would then engage the sternmost enemy ship while her next astern would pass her to windward and engage the sternmost but one of the enemy, the whole fleet following the same procedure in turn until the original sternmost British ship engaged the leading ship of the enemy. The object of this manoeuvre was to avoid exposing the leading ships to the fire of the whole fleet of the enemy as they attempted to draw level with the enemy's van, as would be required by the Article 19 of the general printed Fighting Instructions, which stipulated that 'If the admiral and his fleet have the wind of the enemy, and have stretched themselves in a line of battle, the van of the admiral's fleet is to steer with the van of the enemy's and there to engage them.' Much the same tactical idea had earlier been expressed by Barrington in his third memorandum of 1768.

Article 8 provided for the eventuality that 'it may be necessary on some occasions to set the ships of the fleet at liberty to steer for those opposed to them respectively in the enemy's line independent of that necessary regard to the

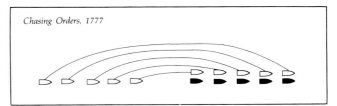

Chasing Orders, 1777

uniformity in distance, course, and movements to be at other times observed'. When the signal was made, captains were 'to keep those ships to which they are then separately directed upon the same constant bearings, if possible, as they advance towards them', care being taken to avoid collisions 'by a proportionable increase or decrease of sail carried'. On reaching the enemy they were free to engage either to windward or to leeward, taking care not to be drawn away too far; 'being in such circumstances to consider wherein they may render the most effectual service against the enemy and govern themselves accordingly'. Taken together with the signals for attacking, respectively, the enemy van, centre or rear, this instruction seems to mean that captains were to keep steering for the enemy ships they were making for at the moment when the signal was made. If there is some doubt about this interpretation, it arises from Howe's own wording which, though somewhat clearer than his worse utterances, is nevertheless ambiguous.

Article 9 was concerned with preparing ships for action. It ends with the recommendation to captains 'whilst advancing . . . to keep their unoccupied men laid close down upon their decks upon the off side from the enemy ships or', in the case of bombardment of land fortresses, 'works'.

The most striking aspect of these additional instructions is the high confidence reposed in the skill, initiative, good sense and personal restraint of the individual captains. Nothing but a 'band of brothers' could have given full effect to these counsels of perfection. The instructions clearly show Howe's tremendous candour in fully revealing himself to his captains; there is no trace of that arrogance and secrecy so common in admirals.

On 21 August 1778, before leaving the North American station for England, Howe issued 'Signals for the Frigates or other Ships of War appointed to observe the motions of a strange fleet discovered or enemy's fleet, during the night'.[22] This rigmarole, in fact, stands for the germ of a system of night reconnaissance; the end product was to be the system by which Nelson was able to keep in touch with his inshore squadron watching Cadiz during the nights immediately before Trafalgar. The ships, detached on night reconnaissance service, were to carry distinguishing lights, shown so as to be visible to their own fleet but not to the enemy. They were provided with eight signals for indicating the enemy's movements. Howe's Observations stressed the need for showing the lights effectively, especially when the enemy were seen to be drawing away from, or bearing down on, the British fleet. Of the importance of these night reconnaissance instructions there can be no doubt. They were re-issued in 1782 on Howe's assumption of the command of the Channel Fleet, and were a permanent feature of all his future additional instructions.

In the same manuscript there were also fog signals and a hitherto neglected numerical signalling table. The fog signals were made with guns, drums and bells, and there were some experimental gun signals by which one or more guns indicated the page in a signal book, followed after a pause by more guns indicating the number on the page. The layout, however, does not conform with Howe's signal book, and does not seem to have been adopted by him.

The numerical table employed ten flags and a substitute flag to produce 100 signals. The signals could be flown anywhere in the ship. A separate table of flags was also provided.[23] As is inevitable with tabular systems, the flag numbers hoisted bore only an indirect relation to the number they indicated on the table, and this limitation severely restricted the utility of numbered flags. If numbers were to be used, it was obviously easier, quicker and safer to use the ten flags for direct numerical signalling, numbering them 0 to 9, and making hoists of two flags, the top representing the tens and the bottom the digits. Pendants could be used for hundreds, as in the tabular system. This was, of course, the system which had to be adopted in the end. What seems strange is that such a progressive admiral as Howe should have allowed himself to be side-tracked into this complicated cul-de-sac.

The accompanying statement, 'by this means the number of any article in the Day Signals may be expressed', implied that the day signals were numbered right through, and that they could now be made by numerical means as well as by the old method of flags flown in particular positions. There is no reason, however, to assume that this numerical table was actually put into use by Howe. The document may have been no more than a project. Sir Charles Knowles claimed that, when he was a lieutenant on the North American Station, he gave Howe a new scheme for signals in 1777, and that he published his own tabular scheme in 1778. It may have been this scheme with which Howe experimented.

By the end of his service on the North American station in 1778, Howe had practically created the tactical and signalling system on which he was to run the Channel Fleet in 1782. So far as we can tell, he had done this without any help from Admiral Kempenfelt, who a few years later was to be deeply involved in the search for tactical reforms.

The National Maritime Museum possesses an undated manuscript volume containing a copy of Howe's signals on the North American Station in 1776, followed by a complete set of numerical signals compiled by the anonymous writer.[24] In his introductory and explanatory remarks this writer states that

The general signals used by the Navy are very deficient in all the requisitions, particularly the letter ones in the confusion of colours in the flags and in the display of the signals; indeed the nature of our signals makes the defect irremediable as the number of the flags used makes the confusion of the colours unavoidable, and the number of orders signified by each flag renders it impossible to show each signal where it would be most visible.

His own proposed scheme was based, he asserts, on French practice and on Howe's signals in America 'with some additions'. These are of interest as an early example of an experiment with a numerical system; the signals themselves exhibit no interesting tactical features.

The younger Knowles's set of signals

In 1777 the younger Sir Charles Knowles published *A Set of Signals for a Fleet on a plan entirely new*.[25] He had entered the navy in 1768 under Barrington in the *Venus*, and served in the West Indies from 1773–6 under Rodney. He was later appointed by Howe to third lieutenant of the *Chatham* on the North American station, and saw much service in the preliminary stages of the war against the American colonists. Knowles's later semi-autobiographical

---

22 *MOD*, NM/34 (formerly *RUSI* 34), manuscript copy presumably by Thomas Graves

23 The flags were, Union, Red, White, Blue, Half Red and Half White, Half Blue and Half Yellow, Dutch Jack, Red Pendant, Yellow Pendant, and Dutch Pendant. The substitute flag was the St George's Cross or Ensign.

24 *NMM*, SIG/A/8 (formerly Sm/111). The date might well be about 1778–80

25 *A Set of Signals for a Fleet on a plan entirely new by Charles Henry Knowles, Bart, Lieutenant in the Royal Navy. London: Printed for J Robinson, New Bond Street. MDCCLXXVII. NMM*, HOL/25 (formerly S(H) 25)

Observations on Naval Tactics, published in 1830 when he was seventy-six, states that he was 'like a son' to Rodney. He also stated that in November 1777, while on a voyage to North America, he was encouraged by Lord Hastings 'to strike out something new [in signalling], and to lose sight of the old plan, and Lord Howe's plan altogether, as these plans would cramp his ideas.'[26] Both young men were aged twenty-three and seem to have regarded Lord Howe, quite mistakenly, as just another traditionalist. Knowles claimed that a year later he developed a system of numerical signals which he

> gave to Admiral Lord Howe on his arrival at Newport Rhode Island; and his Lordship afterwards introduced them into the Channel fleet. Sir Charles Knowles did not publish them out of compliment to Lord Howe.

He also claimed to have discovered tabular signals on his passage home to England, inspired by a chess board. Unlike Howe and Kempenfelt, Knowles was never in a position to give his reforms practical effect; his signals and fighting instructions were no more than projects. In view of his active and distinguished career as a naval officer, however, it would be wrong to class him entirely as a theoretician.

Knowles's signals were carefully designed for practical tactical purposes. He employed ten flags, together with a chequered red and white preparative flag to be hoisted where best seen. The general signals were divided into ten classes of fifty signals each, indicated by one of the ten flags flown either at the ensign staff, mizzen peak or jack staff. The number of the signal within its class was then signalled from one of the mastheads. If more than one signal were hoisted at any time, that class which was indicated by a flag flown on the ensign staff would be shown at the mizzen topmast. That indicated by a flag at the mizzen peak would be completed by a signal at the main topmast. The class begun on the jack staff would be completed on the fore topmast. In case two of the masts were shot away, the class signal was to be hoisted below the signal itself at the remaining position. When the number of the signal was greater than 10, it was made, not by a hoist of two flags, but by a pendant, red for tens, blue for twenties, yellow for thirties, and red, white and blue for forties. General action signals did not require a class flag.

The 'Class First for Action', not to be confused with the unclassed general action signals, included signals for the headmost or sternmost ships of the fleet, or of each division, to discontinue the action and go to the assistance of the centre and rear or of the van, centre or rear, and for the reserve corps to do the same. The second class was for evolutions or changes of sail, and for anchoring. The third class was for weighing and unmooring, and for divisions changing stations. The fourth class was for changes of direction; the whole fleet together or in separate divisions. The fifth class was for evolutions together or in succession. The sixth class was for tacking and miscellaneous purposes. The seventh and eighth were mainly for officers and masters respectively, and for pursers, boatswains, gunners, carpenters, etc, to repair on board the admiral, and orders for taking on wood and water. There were no signals allocated to classes nine and ten, but a separate 'Particular Signal Book by Day' was also provided, consisting of chasing and general signals to and from private ships, made with eight flags and four pendants flown in the old way in particular positions.

---

26 pp12, 23

27 Quoted in a letter from Sir Charles G F Knowles to Corbett, 20 March 1905. In 1929 a copy of the publication was discovered by Admiral Holland.

The night signals were divided into five classes. Guns were used to indicate the classes, lights and digits, and false fires the tens. The first class provided for general control and formations. The second was for evolutions of the whole fleet and divisions. The third was for battle, and included number 13, ordering 'the whole fleet to break through the enemy's line in succession'. The fourth and fifth classes were for battle and miscellaneous purposes. There were eighteen night signals for private ships.

The fog signals were made by a preparatory gun, followed by a two-minute interval and by minute guns for the classes, of which there were only three. After another two-minute interval, quick gun fire was used for tens and half-minute guns for the digits. Class two, signal number 1 was for 'The whole fleet to engage as they can most conveniently', which would certainly have been a hazardous operation in fog.

Whatever the technical merit of these signals, and their tactical implications, the book as a whole seems an astonishing achievement for a young man of twenty-four – even with Nelson's captaincy at the age of twenty in mind. His separation of battle signals, placed first in an unnumbered class, his separation of the other signals into classes, his use of the full programme of French evolutions, his care to provide for action by the whole fleet or its separate divisions, and his attempt to bring the night and fog signals into an ordered relationship with the day – all mark Knowles as a true innovator.

In 1798 Knowles published another signal book. In his file copy he wrote the following note:

> These signals were written in 1778 as an idea – altered and published, then altered again in 1780 – afterwards arranged differently in 1787 and finally in 1794, but not printed at Sir C H Knowles's expense until 1798 when they were sent to the Admiralty but were not published although copies have been given to Sea Officers.[27]

Knowles's claim to have discovered the tabular flag system with the assistance of a chess board is dubious. By that time the French system was well known in England. It is, however, difficult even to surmise what influence he may have had on Kempenfelt and others. By the time Knowles's tabular system was printed in 1798 it had undergone repeated revision to keep it from being outmoded by what Kempenfelt and Howe had done. It then provided the admiral with 295 numerical signals.

The most interesting features of Knowles's system are the tactical ideas which lie behind the signals. Battle signal number 6, which may have been original to the 1778 version, reads, 'Every ship in the fleet to break the line, follow the ship she is engaged with, and endeavour to take or destroy her'. This is a most uncompromising expression of free tactics. British ships were not only to break the enemy's line, but were to disregard their own. The assumption obviously was that in a true pell-mell British close fighting capacity would prevail. Against the tactically demoralised fleets of the French Republic and Empire this was not unrealistic, as was proven on several occasions. Against the fleets of the *ancien régime* it would have been a gamble. Knowles obviously also assumed that French determination to avoid a pell-mell could only be dealt with in this way, and that the French would not be able to prevent such an attack.

More controversial still are signals numbers 32 and 33. The first had been changed in manuscript to read, 'To break through the enemy's line together and engage on the other side'. Number 33 made provision for the circumstance in which, 'having the weather gage of the enemy, the Admiral means to pass between the ships in their line for engaging them to leeward, or being to leeward to pass between them for obtaining the weather gage'. In manuscript this signal had been changed to read, 'To break through the enemy's line in succession and engage on the other side'. The obvious difficulty of carrying out such manoeuvres is recognised by the addendum, 'NB: The different captains or commanders, not being able to effect the specified intention in either case, are at liberty to act as circumstances require'. It was not made at all clear whether Knowles envisaged in number 33 cutting the enemy fleet in line ahead, in bow and quarter line, or in two or more columns. Did he mean that regardless of the relative position of the two fleets at the moment the signal was flown, ships were to obey it as best they could?

Knowles's night signals for the Admiral were on a highly original system. They were divided into groups and were unnumbered. Signals were made by firing one or more guns, up to four, then a rocket 'with stars' or 'with rainfire', followed where necessary by one or two guns. Night signals for private ships, of which thirty were provided, were made by various arrangements of lanterns, one to four, 'hoisted where best seen', with false fires but no guns. A manuscript note indicates that Knowles derived his signals from the French *Tactique Navale*.[28]

Knowles's Fighting and Sailing Instructions consisted of fifty-five articles, the first fifteen being illustrated by diagrams and linked with signals from the signal books. He stated: 'These Instructions were written in 1780 and afterwards very much curtailed though the general plan is the same.'[29] In the first article he established a revolutionary order of sailing in parallel lines. If the squadron consisted of six to nine ships there should be two columns, three columns for ten to sixteen ships, and five or six columns for over seventeen ships. In all cases all lines would be as nearly equal as possible. If there were four lines, each sub-division of the centre would equal the whole van or rear. If five columns were required, each sub-division of the van and centre would equal the rear. On this basis squadron strength was to be determined solely by the needs of sub-divisional equality, a revolutionary departure from the whole tradition of squadron organisation. The commander-in-chief's station

was to be two points on the weather bow of the leading ship of the weather line. The distance between lines was to be not more than two miles and the ships half a mile or less from each other. The striking point about this arrangement is the shortness of the individual lines, presumably to ensure quick changes of course and order, and quick deployment for battle. On the other hand the distance between lines and between individual ships suggests a desire to give lateral spread over a wide sea area and to ensure enough room for any change of order required by a sudden shift of wind or appearance of the enemy.

In Article 34 Knowles wrote that, when the fleet was formed into two 'grand' divisions from the order of sailing in three columns, the centre would form one line to starboard, and the van with the rear astern would form the other line to larboard. If the fleet was sailing in four columns, the van and rear should still form the larboard division, and the starboard should be composed of the two divisions of the centre. If the order of sailing was in five columns, the starboard division of the van squadron should lead the larboard line. If six columns were the basis, however, the starboard division of the van should fall in behind the larboard division of the centre.

Article 13 was for a linear engagement with the enemy when to leeward of them and on the same tack. The British fleet should be parallel with the enemy 'but somewhat astern of them'. When the admiral hoisted a Dutch Jack 'where most easily seen', the whole fleet was to bear up four points together, thus closing the enemy in bow and quarter line. When close enough to engage, the admiral would haul down the flag, and 'each ship is then to haul up, and close engage her opponent in the enemy's line; pursue her, but not out of sight, without waiting for any other signal and endeavour to take or destroy her'. Knowles added in manuscript that 'the signal 29 may be made to engage the enemy to leeward – but with the exception of those three-decked ships that are leewardly who are to fight to windward'.[30]

Of very great interest is Article 14, which stated that the French had a manoeuvre to enable a line of battle to leeward to break off an action. The odd-numbered ships would bear away, covered by the even-numbered ships, which would subsequently bear away themselves covered by the odd-numbered ships bearing up and opening fire. Knowles wrote:

> It is, therefore, absolutely necessary, when a British fleet are to windward of an enemy, and have brought them to action . . . that every ship should (without waiting for any signal for that purpose) pursue her opponent, and endeavour to take or destroy her: for a line of battle is no longer useful after the enemy's ships are brought to action, as personal courage must then decide it . . .

This was one of Knowles's most forceful pronouncements, and seems to label him as an almost uncompromising opponent of the line, except as a means of approach. It has a real Nelsonic ring about it, all the more remarkable considering the year in which it was written. The fact that, so far as is known, the French did not practice the manoeuvre of retreating to leeward by alternate numbers, does not vitiate his idea. Any intelligent observer in the van of Byng's fleet at the Battle of Minorca would have been justified in drawing

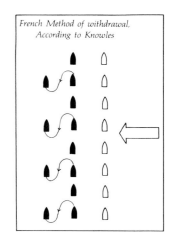

French Method of withdrawal, According to Knowles

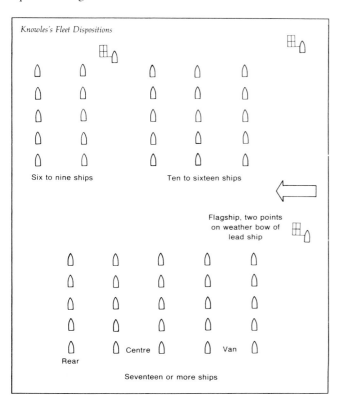

Knowles's Fleet Dispositions

Six to nine ships   Ten to sixteen ships

Flagship, two points on weather bow of lead ship

Rear   Centre   Van

Seventeen or more ships

28 Information to Corbett from Admiral Sir Charles Knowles, 20 March 1905; *NMM*, Corbett Papers, Box 14

29 Corbett's copy is inscribed at the top by Knowles in his own handwriting, 'Written in the year 1780 by Sir Charles Knowles'; *NMM*, Corbett Collection, TUN/41 (formerly S/P/Am/6). Information to Corbett from Admiral Sir Charles Knowles, 20 March 1905; *NMM*, Corbett Papers, Box 14.

30 Reprinted in Knowles's 1830 pamphlet, pp30–1

the conclusion, as Knowles did at the Battle of Grenada, that the French did indeed employ such a tactic.

Knowles was more conservative in Article 18, in which he required ships to keep the line unless specifically ordered to break it. 'NB: the line is not to be broke, if possible, when the fleet are much inferior to the enemy.' A handwritten Article 38, and the directions for the ships of the reserve corps, indicate that Knowles was aware of the value of the line of battle in most circumstances. He wrote that with a very large fleet, forty or fifty sail of the line, the most effective tactic would be action on opposite tacks to the enemy, because the successive broadsides of so many ships would be devastating. The *corps de réserve* should not be used to double the enemy, but rather to reinforce such weak points as might develop in the line of battle, and to provide tow for disabled ships. As an observation in Article 14 Knowles had written:

To double the enemy's line, that is, to send a few unengaged ships on one side, to engage, whilst the rest are fighting on the other, is rendering those ships useless; every [enemy] ship which is between two [of ours], has not only her two broadsides opposed to each of theirs, but has likewise their shot, which cross, in her favour. Therefore, if ships form a double line, it should be on the same side, whether it be to windward or to leeward of the enemy.

In Articles 16 and 17 Knowles expressed the belief that in the open sea it was always possible for a fleet to defeat an attack on its rear by tacking or wearing together, and engaging the enemy on opposite tacks. If the enemy should be caught close to the shore where there was not room to tack, however, it would be profitable to engage them from the rear in succession, inverting the order of battle. If their van and centre tacked to support their rear, 'then the battle will soon be reduced to a scramble, wherein British seamanship and courage will soon show their superiority, as it has ever done on those occasions'.

Knowles stated that when Howe succeeded Keppel at the Admiralty (presumably meaning the first time, in 1782), he wrote offering to show Howe his father's plans and papers, and mentioning his own 'work on naval tactics'.[31] Howe asked to see them all, and was then told by Knowles that the French 'felt the pulse of our Captains and avoided a close action', and that the way to force a close action was suggested by Hoste, 'and is in the French Naval Tactics'.

This Lord Howe executed at the battle of the 1st of June, 1794, when each ship followed her opponent, and thereby did not check the ardour of the men, which was done by the 21st Article of the old Fighting and Sailing [sic] Instructions; the line is not to be broken, unless by signal, until all the enemy's

ships are disabled or run; which Lord Howe changed to 'beaten or give way' though retaining the article.

Knowles wrote that Howe, 'on seeing the French manoeuvre [Article 14 of Knowles's instructions], struck it [Article 21] out of his own Fighting & Sailing Instructions, many of which his Lordship altered.' This is quite untrue. In Article 15 of his 1782 instructions, Howe changed the words to 'disabled or broken', and they remained as such until the Admiralty book of 1799 took over Howe's system, when the words were reversed to read 'broken or disabled'. Considering that Knowles's evidence involved an alleged French manoeuvre that did not exist, a fact of which Howe must at least have had some inkling, it can hardly be supposed that Howe acted on those grounds alone.

Knowles was an extraordinary figure, and his claims are bewildering. Yet if he really compiled his Fighting and Sailing Instructions in 1780, and there seems no reason in this case to doubt his written word, these alone entitle him to be regarded as a great tactical innovator. Written in that year they would have done credit to any naval officer, however high in the service. As the work of a lieutenant aged only twenty-five or twenty-six, they were a *tour de force*. In 1782 he compiled an excellent *Ship's Orderly Book*, which he had printed and sent to the Admiralty in 1783. He also wrote day and night signals for a convoy.[32] He was invalided from the navy after the Battle of St Vincent and does not appear to have taken any further part in tactical reform, although his health recovered and, at the age of seventy-six in 1830, he published his famous pamphlet attacking the claims that Clerk of Eldin had inspired Rodney's break of the French line at the Saints.

William Dickson's signals, 1780

Among a batch of papers preserved by Thomas Graves are four sets of signals issued or prepared by William Dickson, Captain of the frigate *Greyhound*, when serving in the West Indies under Sir Peter Parker in 1780.[33] The first is simply a private signal book in the traditional form, but incorporating Parker's additionals. With it, however, is a set of true numerical signals. This employed ten flags, including a green flag, the only known instance of this colour. The location of the flags, at the mizzen, foremast head, and mainmast head, indicated respectively units, tens and hundreds. The flags and most of the arrangements could be changed for security. This system was simple enough 'to be understood by the meanest capacity'. His night signal system was also numerical, but depended upon ten special arrangements of lights to indicate each number, 'by which means 12 Lanthorns only will refer you to the high Article 999'. The fact that these signals were not just a project is proved by their actual issue to Graves, then commander of the sloop *Savage*, by Dickson on 7 March 1780. Further evidence of their actual use is a note stating that 'for all other signals you are to refer to those you have received from Rear Admiral Parker'. Finally there was another numerical code of an even simpler form: ten flags numbered 0 to 9 which could be flown from anywhere in the ship, tens and hundreds being presumably shown by hoists of two and three flags respectively, one above the other. Again, as Dickson carefully noted, the signals could easily be changed by inverting the numbers.

---

31 Admiral Sir Charles Knowles, *Observations on Naval Tactics*, 1830, pp28–29 (*NMM*, Tunstall Collection, TUN/75)

32 All the Knowles instructions, etc, quoted, except the *Observations on Naval Tactics* of 1830, are from the autographed set given by Admiral Sir Charles Knowles to Sir Julian Corbett and 'bound together in 1910 with the MS notes from his own copies added' and with explanatory letters from Knowles inserted; *NMM*, Corbett Collection, TUN/41 (formerly S/P/Am/6).

33 *MOD*, NM/92 (formerly *RUSI* 92). Although the document is marked 'a copy' and the signature is not, therefore, in Dickson's own hand, its authenticity is undoubted. See *The Mariner's Mirror*, 40 (1954) no 2, p163, 'Answer', by Commander Hilary Mead.

CHAPTER SIX:

Tactics of the American Revolutionary War: Testing the New Systems

THE Channel and Atlantic operations of the British and French fleets under Admiral Angustus Keppel and the Comte d'Orvilliers, respectively, provide one of the few cases where the instructions, signals and actual battle tactics of rival fleets can all be considered together. The chief tactical event was the Battle of Ushant on 27 July 1778. This battle demonstrated the success of French efforts following the Seven Years' War to reform their tactical system. Its outcome, on the other hand, stimulated important developments in British tactics and signalling. After Keppel's retirement, Admiral Kempenfelt, as captain of the fleet to successive commanders-in-chief, was able to push ahead with tactical experiments, the effects of which gradually reached the more distant stations. The entry of France, and then of Spain, into the American Revolutionary War had led to the Western squadron, or Channel Fleet as it was increasingly being called, taking the primary role in British naval affairs, although very substantial forces continued to be sent both to the North American station and to the West Indies. New tactical ideas and signalling experiments tended to originate in the Channel Fleet and to pass from there to fleets in American waters.

Keppel's signals, 1778

At the time of the Battle of Ushant, Keppel was aged fifty-three and had a high reputation in the service as a fighting captain at Quiberon, a commander of the successful expedi-

Amiral *Louis Guillouet, Comte d'Orvilliers (1709–92)*. MM

tions against Senegal and Belleisle, and as assistant to Pocock at the capture of Havana. In 1778 his health was no longer good, however, and he was preoccupied with political concerns. He was not interested in tactical reform, but it must be admitted that he approached his command with great concern for efficiency in his own way. While his fleet was fitting out he drew up his own 'Signals & Instructions in Addition to the General Printed Sailing & Fighting Instructions', and on 28 June 1778, the day after returning from his first cruise, he issued a printed version.[1] This is not to be confused with the Admiralty work of the same title, nor with any of the Admiralty 'additionals' issued during the Seven Years War. Keppel's book was a selection from all these sources, with the articles reworded. There were twenty-four signals for day, sixteen chasing signals, nine for night and five for fog. The eighteen day fighting instructions are reminiscent of Hawke's issue of 1747 and in any case contain nothing new. There are four night fighting instructions.

On 21 August, two days before leaving Cawsand Bay for his third cruise, Keppel issued Captain (later Admiral) Adam Duncan of the *Suffolk* with a copy of the original work, with a handwritten reference to the incorporation of fighting instructions Articles 1 and 2 into a single article.[2] Keppel, apparently, considered it a necessary part of his duty to continue issuing the parent work. Nevertheless, Sir Charles Knowles stated that there was a re-issue of Keppel's own compilation in 1778, presumably for captains of ships newly commissioned for the war. Both books were in fact used during the Battle of Ushant. The schedule of signals which Keppel made from the *Victory* shows that, of the eleven general signals made, six were from the General Sailing & Fighting Instructions, and five from Keppel's own signals.

D'Orvilliers's signals, 1778

Across the Channel the situation was completely different. Unhampered by the British need for warships on the North American station, the Comte d'Orvilliers was free to prepare his fleet at leisure, while the Comte d'Estaing prepared another fleet for intervention in American waters at France's chosen moment. D'Orvilliers was a veteran of sixty-nine, but he had Pavillon as his chief of staff. His signal system for the campaign of 1778 was the most comprehensive and elaborate used until that time. It was not identical to Pavillon's project of 1776, but it is clearly derived from it and shared its strengths and weaknesses.[3]

The system started with a squared five-flag table of letters A to Z, less W. Sixteen of the lettered squares also had

1 *NMM*, DUF/31 (formerly Sm/167), a large folio manuscript dated 29 April, 1778, with a copy of Keppel's signature at the end, and CLE/2/9, printed version, signed and dated by Keppel, the *Victory*, St Helens.

2 *NMM*, DUN/4 (formerly S(D) 4)

3 M d'Antin, *Plans des Mouvemens et Positions, Respectives des Armées de France et d'Angleterre a la Journée du 27 Juillet, 1778*, Brest, 6 October 1778, *NMM*, REC/46 (formerly Sm/7)

Admiral Augustus Keppel (1725–86), by Sir Joshua Reynolds. NMM

(70)

ARMÉE NAVALE.

COMMANDÉE par M.r Le C.t D'ORVILLIERS lieutenant GÉNÉRAL des Armées navales Commandeur de L'ordre Royal et militaire De S.t Louis En 1778

SIGNAUX DE COMBAT DE L'ARMÉE NAVALE DU ROI EN 1778

Pavillon's tabular 'Signaux de Combat', employed by Admiral d'Orvilliers at the battle at Ushant, 1778. Published by d'Antin in Plans des Mouvemens et Positions Perspectivez des Armées de France et d'Angleterre a la Journée du 27 Juillet 1778, *Brest, 1778. NMM*

written on them standard, or 'primitive', orders of sailing, ending with the order of retreat. A hoist of two flags by the general enabled the co-ordinates to be read off, indicating the fleet's current order of sailing. From that table, reference had to be made to another squared table, alphabetical with the letters A to T, (less E, J, K and P) printed along the top and the whole alphabet (less W, and with J at the end) printed down the left side. In the squares of this table were numbers referring to the articles of the instructions. The general's next signal was a hoist of two flags, drawn from the previous table, indicating a letter. This letter could then be found in the left-hand column of the alphabetical table. The captains, by taking the co-ordinates of the two letters together, the first for columns and the second for lines, at last discovered the number of the article they were meant to obey. This article indicated the order the fleet was to form.

The art of this arrangement lies in its economy of flags; only five individual flags were necessary, though clearly a double set was required. The reason for the smaller number of columns than lines in the alphabetical table is that the letters indicating columns were those used in the previous table for existing orders of sailing, the remainder being unappropriated. The other letters, however, less W, were needed to produce all the references to the articles in the second table. Many of these numbered article references were the same because the same order of sailing could be reached in different ways from different starting formations. The whole affair seems ridiculously complicated, especially in making the existing order of sailing part of the signal. Nevertheless, with two hoists of a pair of flags each, drawn from a double set of only five flags, d'Orvilliers could move the fleet about much as he wished.

D'Orvilliers's signalling systems made elaborate provision for altering course at night according to a pre-arranged plan. Four separate square tables of eight flags each, controlled by a separate pendant, indicated the number of half hours which were to elapse before the alteration of course was to be made.

Signals to and from private ships were given in a ten-flag table of 'Signaux Particuliers', numbers 1 to 40 being signals made by the general or a *chef d'escadre* to a particular ship, and 41 to 100 being for signals from private ships. These latter were extremely detailed and well arranged. Although the squares were numbered, the flags were not.

Another ten-flag table was provided for 'Signaux de Combat', the 100 squares of which each bore the number of an article from the instructions, and a summary of it. As usual in these compilations, no attempt whatever was made to relate the flags to the numbers of the articles in a logical form. Article 21, for instance, was indicated by the third flag for the column in the table, over the first flag for the line. The articles themselves were not very progressive.

Yet another ten-flag table, with the flags differently arranged, was provided for signalling numbers. By the obvious and simple device of using different pendants in different positions for the hundreds, numbers up to 1200 could be signalled by a hoist of two flags and one or more pendants. This simple arrangement was capable of being made to serve the whole gamut of articles and signals required by the service, but in fact it was restricted to purely numerical statements such as latitude and longitude, soundings, etc. Instead of employing the numerical system for general communication, the system was further complicated

by the provision of a complex table for pass-words by day and night for each day of the week and with a hundred dual pass-words. So complicated had it all become that two further fifteen-flag tables were provided, incorporating many of the instructions already made, but expressing them with different flags. A separate table of signals was also provided for communication with the reserve corps.

The night signals were managed with equal complexity. There was first a nine-letter table giving seventy signals with some blanks. Next came a complicated 'general plan', then a large table of light and gun signals indicated both by letters and numbers. After all that came a set of sixteen separate tables, almost entirely numerical, entitled 'Signaux Généraux et Particuliers pour le Services des Armées', for day, night, fog, at anchor, sailing and coast purposes. These tables were intended to refer to all the previous tables, but the relationships are obscure.

The Battle of Ushant

On the British side the minutes of the two courts-martial which arose from the Battle of Ushant (27 July 1778) record an enormous amount of detail, much of it repetitive and very little of it having any value as a means of obtaining a clear tactical appreciation. On the French side the vast handwritten account of the operations of 1778–9 compiled by Lieutenant d'Antin, who was on d'Orvilliers's staff as *chargé des troupes*, includes large scale plans and views illustrating every phase of the battle. As these show the French fleet invariably well ordered and the British quite fantastically disordered, their bias is obvious, though they undoubtedly nevertheless provide a great deal of useful and reliable information.[4]

Having chased the French for four days without being able to force them to action, the British fleet sighted them once more at dawn on 27 July 1778. After discounting purely formalistic assertions, there seems no doubt whatever that Keppel wished to force an action, nor is it true to say that d'Orvilliers was trying to avoid one. The latter seems to have been quite prepared to fight on terms which gave him a distinct tactical advantage.

There was nothing at first to indicate more than a repetition of early manoeuvres, with the British to leeward pursuing and the French to windward holding off. The wind was south-west by west, and the two fleets were roughly parallel and at distance of about a league and a half. But whereas the French were in line of battle ahead, steering north-west, the British were apparently in a loose bow and quarter line formation and spread over a considerable distance. Though roughly parallel to the French, they were in fact on a converging course, at least to the extent to which this was possible without coming to the wind.

Even at this early hour, however, the British fleet was sufficiently out of alignment to effect the course of the coming battle. The order of sailing in the British Fleet was inverted, and Sir Robert Harland, second in command, had his squadron to larboard of Keppel in the centre and was well up to his station, in fact slightly on Keppel's bow; Sir Hugh Palliser's squadron on the other hand, stationed to starboard, was, according to Captain John Jervis of the *Foudroyant*, 'full three miles to leeward of the *Victory*, going under a slow sail'. Seeing this, Rear-Admiral John Campbell, Keppel's first captain, acting on his own initiative, signalled seven of Palliser's squadron and one of Harland's to chase. The effect of this seems to have been that, although these ships gradually made up their distance, Palliser's squadron as a whole became more disconnected than it had been before. Nevertheless, as Lord Longford, Captain of the *America*, one of the ships so ordered, stated at Keppel's trial, 'If the wind

---

4 The account of the battle which follows is taken from d'Antin, *op cit*, and the Proceedings of the courts-martial on Keppel and Palliser. No page references are given for these latter as there are four versions of Keppel's trial and three of Palliser's, quite apart from published excerpts and pamphlets.

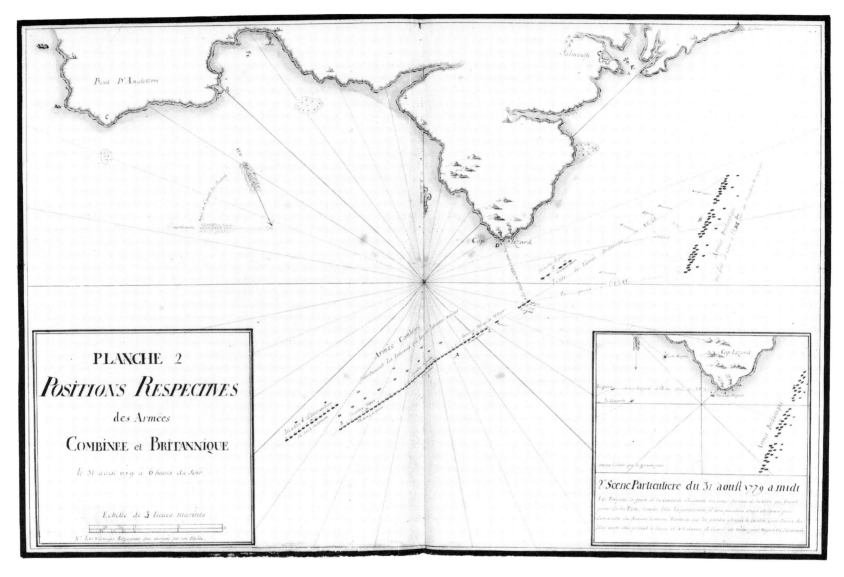

The positions of the British and French Fleets at noon on 31 August 1779, as published by d'Antin. NMM

had continued the same as it did, and the signal had not been made to chase, I do not think that any part of the Vice-Admiral of the Blue's division could have come into action at all.'

At about 8am d'Orvilliers signalled the French fleet to wear in succession. This had the effect of bringing them onto the starboard tack heading SW. According to d'Antin's diagrams, they were always well closed up in perfect order, but it seems more likely that by now their line was somewhat extended and not perfectly straight. As this stage, the wind shifted two points to the south against the French fleet, hiding them from the British by 'a dark squall'. The British were coming up in businesslike fashion, evidently intending to attack even though in poor order.

D'Orvilliers sensed a projected attack on his rear, and some of Palliser's Blue division were almost sufficiently advanced to weather him. Sometime between 9.30 and 10am, therefore, he signalled the French fleet to tack together, thus returning to his original course. Next he signalled his new van to run large four points while his new rear was getting into station. He then signalled the whole fleet to come to the wind. As seen from the British fleet, these movements were by no means perfectly executed. No doubt at this stage the French line fell into some disorder.

All this time the British fleet was continuing to work towards the French, but in no very good order; Keppel could

not afford the time to close his fleet in line abreast or bow and quarter line, if he was to be sure of forcing a battle. As Sir Robert Harland remarked (at Palliser's trial),

pursuing the French in the way we did that day, was a bold, and necessary, and allowable stroke of war, where nothing but risk, as it appeared, could stop the French fleet from getting off. The great decisive strokes of the day were to follow, by closing with the enemy and fighting it out.

When the French re-appeared from the squall about half an hour later they were much closer to the British, having fallen somewhat to leeward in the course of tacking. They were now steering north-west again, as when first sighted at daybreak. D'Orvilliers had signalled his van to keep their wind, so as to stop Keppel's Blue squadron from weathering them. Evidently this signal was not meant to apply to the whole van, so that, while the four leading ships turned to larboard in succession, the remainder held their course. Actually d'Orvilliers need not have worried. With his light squadron of three ships already stationed to windward of the van, he was well covered. Nevertheless, a few British ships were slightly to windward of the French van and were forced to give way. Captain Mark Robinson of the *Worcester* stated at Keppel's trial, 'If I had taken any wind I could have weathered more than half the French fleet, the wind having shifted more than two or three points to the westward.'

The British Red squadron was now some two miles from the French van and heading towards them at an oblique angle. It was 10.30am, and Keppel saw that if he was to fight he must do so at once, without attempting to close up his fleet and get it into reasonable order. He made the signal to tack together, which brought him into a rough line ahead formation on the starboard tack parallel to the French and to leeward of them, with Sir Robert Harland's Red squadron leading. The wind now shifted about two points to the west in Keppel's favour, giving him more freedom of manoeuvre. The two fleets were about to pass on opposite courses. The French, despite the partial withdrawal of their van, were in

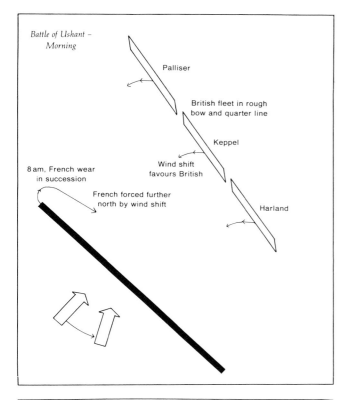

Battle of Ushant –
Morning

Palliser

British fleet in rough
bow and quarter line

Keppel

Wind shift
favours British

8 am, French wear
in succession

French forced further
north by wind shift

Harland

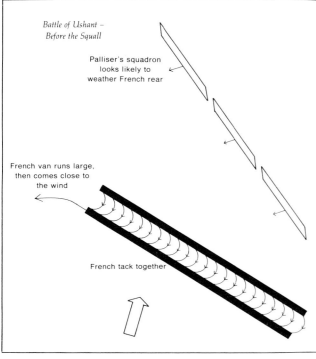

Battle of Ushant –
Before the Squall

Palliser's squadron
looks likely to
weather French rear

French van runs large,
then comes close to
the wind

French tack together

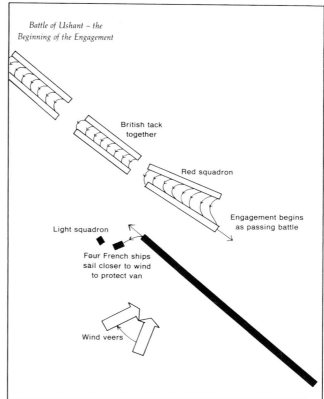

Battle of Ushant – the
Beginning of the Engagement

British tack
together

Red squadron

Light squadron

Engagement begins
as passing battle

Four French ships
sail closer to wind
to protect van

Wind veers

fairly regular order, but the British were strung out in a long irregular line, not even in their correct order of battle. This meant that when the two fleets began to pass each other, the British line had gaps in it of nearly half a mile and that some ships passed much nearer the French than others.

Firing began at about 11 am and at 11.05 Keppel signalled to engage, without having made any preliminary signal for the line of battle ahead after tacking. It says a good deal for the freedom of British tactics and the reliance placed on individual initiative that this omission was easily disposed of at Keppel's trial, Captain the Hon Robert Digby of the *Ramillies* remarking in his evidence that, 'A line of battle in my opinion, always retards. Had the signal been made, the Red division must have shortened sail.' Captain Samuel Goodall of the *Defiance* was equally firm on this point. Asked at Keppel's trial if action would have been possible if Keppel had formed a regular line of battle, he answered, 'No: the admiral had always offered them battle, and it lay in their breast whether they would bear down. Had he kept a line of battle that morning they could not have been brought to an engagement.'

The ensuing battle, with the fleets passing each other at a relative speed of about six to eight knots and with the opposing ships firing at varying ranges, could not lead to decisive results.[5] The battle was a desultory affair. The British ships engaged at widely varying ranges, and were separated from each other by wide gaps, some of the rear ships being up to half a mile apart. On this occasion there seems to be some truth in the standard British assertion that the French deliberately fired to damage masts and rigging. They were to windward, and thus heeling over in somewhat

5 John Bazely, Palliser's flag captain in the *Formidable*, stated at Keppel's trial that the maximum speed possible that morning for a ship with all sails set was 'between seven and nine knots.' As, however, the fleets were under the usual battle order of reduced canvas, their combined relative speed is unlikely to have exceeded eight knots.

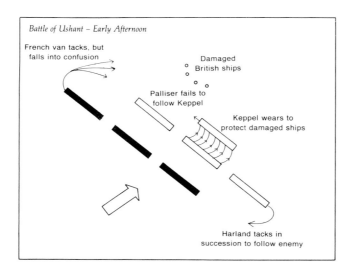

Battle of Ushant – Early Afternoon

French van tacks, but falls into confusion

Damaged British ships

Palliser fails to follow Keppel

Keppel wears to protect damaged ships

Harland tacks in succession to follow enemy

The Duc de Chartres, commanding the van, did not see the signal, but he wore his flagship and came under d'Orvilliers' stern to ask for orders. The signal was then obeyed, though somewhat slowly; the van ships which had kept to windward were now in bow and quarter line, having turned back onto the same course as the rest of the fleet, though remaining to windward. Eventually Chartres himself took the lead to save time and the rest of the French fleet was signalled to wear in succession. By now, what with the confusion over signals and damage to masts and rigging, their van ships were no longer in the original order of battle and they had already fallen away to leeward of the retreating British rear.

By about 12.30pm Sir Robert Harland had cleared the French rear with his van squadron and, on his own initiative, he signalled it to tack and so lead round on the opposite course to pursue the French rear. He still had six or seven of his ships with him, together with his flagship, out of a total of ten. The head of each fleet was now chasing the tail of the other. Harland was also followed by some ships of the centre, though owing to damage not all of his own squadron was able to come to the wind and tack. By that time the wind had dropped and the ships moved more slowly.

As Keppel's flagship emerged from the smoke, he could see Harland and part of the British van passing him to windward on the opposite course. This action he thoroughly commended but felt unable to follow, and at 1.02 pm he signalled the fleet to wear so as to close the French once more by approaching them on the larboard tack. By this time he was a mile or more beyond the French rear, but had refrained from wearing or tacking before because of the confusion it would have caused amongst his own centre and the ships astern him as they came out of the action, some being widely spaced but others close to one another; there was a clear danger that the *Victory* would have fallen aboard other ships. His choice to wear rather than tack in Harland's wake was, first, because it was obviously an easier manoeuvre for the damaged ships in the centre and rear, and second, because it enabled him to proceed direct to the help of at least four of his worst damaged ships, which seemed to be drifting away to leeward and across the line of the advancing French.

In less than ten minutes, however, the signal to wear was hauled down and at 1.26pm the signal to engage was also hauled down. At 1.40pm Keppel signalled for the line-ahead at a cable's distance. His immediate object now became that of protecting the damaged ships to leeward, which he imagined, quite erroneously, to be in some danger. In fact d'Orvilliers had little intention of attacking them, his own line by now being indifferently formed. Though his fleet was steadily drawing level with the British, he kept to leeward. Earlier, when the French had completed their wearing movement and formed again on the starboard tack, it had looked as if some of their leading ships had intended to weather the *Victory*, but soon afterwards they had headed away towards the damaged British ships to leeward and then still further to leeward, thus avoiding further action.

The British fleet at this stage was in some confusion. Furthest to leeward were the worst damaged ships. Next came Keppel with some of the centre and rear, while further to windward and astern was Palliser's *Formidable* with other ships of the rear, making no effort to obey Keppel's signal. At his own trial Palliser explained this by saying that, as his ship was end on to the *Victory*, he could not at first see the signal which was flying from her mizzen peak, which was the union flag and a blue flag with a red cross. Similarly he could not see the signal to wear, a blue pendant at the ensign staff. Meanwhile, Harland was still on the larboard tack pursuing the French rear in the opposite direction. His squadron had suffered the least damage and he had most of them with him.

squally weather, so that it would have required a deliberate effort to fire high. That they did so can be judged from the fact that many British ships were so damaged aloft that some were temporarily immobilised, though first-aid repairs were effected within a matter of hours. The British had some 500 men killed and wounded and the French nearly 700.

As the smoke rolled to leeward in clouds from the fire of something like a thousand guns, d'Orvilliers had a better view ahead than Keppel. As soon as he saw that his van had cleared the British rear he signalled them to wear in succession so as to follow the British round and attack their rear.

Captain Hugh Palliser (later Admiral Sir Hugh), (1723–96), by George Dance. NMM

At 1.50pm, therefore, Keppel signalled for the frigate *Proserpine* to come under his stern; he sent her with an order to Harland that he was to wear and take the rear of the line. By this means Keppel hoped 'to cover the rear' [should the French press home their assumed attack] and keep the enemy in check, until the vice-admiral of the Blue should come into his station, with his division, in obedience to the signal'. By the time the order reached him, Harland was already leading his division round to execute this manoeuvre on his own initiative.

In the process of wearing, Keppel had passed right round Palliser, who claimed later that the *Formidable* was too damaged to obey Keppel's signal immediately, as also were other ships of his division. Sometime about 3pm Keppel signalled to wear for the second time. This brought him onto the starboard tack and roughly parallel with the French, who had now drawn up about level, though some two miles to leeward. The damaged British ships which had drifted to leeward now seemed in much less danger. Keppel was greatly concerned that Palliser still lay to windward, and at about 3.15pm he signalled for the ships to windward to get into the admiral's wake. At 3.30 he signalled for the line of battle ahead at a cable's distance and at 3.45 for the frigate *Milford* to take an order to Harland that he was to resume his station in the van of the fleet with his division. At about 4.30 Keppel made the signal indicating ships out of their station, followed by signals to six particular ships lying to windward to call their attention to it. To have signalled the *Formidable* by flag would have been an open reprimand to Palliser, observed by the whole fleet; Palliser was to be treated 'with a delicacy due to his rank'. So at about 5.30pm Keppel signalled the frigate *Fox* under his stern to send his compliments, 'and to acquaint him, that he only waited for Sir Hugh Palliser and his division bearing down into his wake for him to renew the attack'. Captain Edward Windsor of the *Fox* received the message by megaphone from Keppel and repeated it to Palliser by megaphone, and Palliser answered in person that he had understood. Subsequently he claimed that he did not receive the message until 7.30pm, and that in any case he was still not quite ready to comply with it as he could not yet bend his fore topsail. By now the French were further to leeward, and from a strictly tactical standpoint the further movements of the British fleet had little significance.

Despite their heavier casualties, the French had had the best of the battle. They had caused much damage to the mobility of the British fleet, a point noted by Kempenfelt, who was a member of the court which tried Palliser. However, they made no attempt whatever to press their advantage. The confusion in the British fleet, due to lack of appropriate signals, misreading of and failure to see them, was compounded by the huge delays caused by sending frigates to deliver messages by word of mouth. Despite their voluminous signal system, however, the French communications were little better. The failure of the van to tack at once lost d'Orvilliers the chance to double the British rear, if, that is, he ever intended to attempt it.

Palliser, in enumerating the reasons for not shifting his flag from the damaged *Formidable* so as to keep up with the battle, mentioned the time which would have been wasted in moving 'my signal colours, necessaries and attendants'.

Just as at the Battle of Toulon and the Battle of Grenada (1779), Keppel was initially at a disadvantage because his fleet was not properly up with him at daybreak. As Admiral Campbell remarked at Palliser's trial, 'If Sir Hugh's ships had been weatherly enough to have been on the weather quarter of the *Victory* at daylight on the 27th [July], I think he ought to have been there; he should have endeavoured to be as near the enemy as his commanding officer was'. Only constant peacetime practice could ensure that, at the start of a war, an admiral could be sure of keeping his ships so well together during the night that at daybreak he would be able to work to windward immediately with his whole fleet properly aligned. Keppel showed himself throughout a vigorous officer inbued with the offensive spirit, but where tactical discipline was concerned he was certainly outmatched by his older and more cautious opponent.

The British courts-martial, 1779

In the subsequent quarrel between Keppel and Palliser, everything seemed to turn on one point: whether Palliser had acted in a mulish and bloody-minded manner in not coming more quickly to Keppel's support, or was his flagship too crippled (and his adjacent captains too unpractised in tactics) to obey his commander-in-chief promptly.

The whole affair was ridiculous. The French centre and rear in wearing had taken a wider sweep than necessary, while their van had made little effort to close the British by leading more to the wind. Thus the whole French fleet was already well to leeward of the British at the time that Palliser's sluggishness caused Keppel such concern, though Keppel naturally credited d'Orvilliers with having the same offensive purpose as he himself would have had if their positions had been reversed.

The quarrel between the British admirals, resulting first in Keppel's court-martial and then in Palliser's, produced a situation in which the real lessons of the battle were never properly considered, except by some of the officers forming the courts, of whom Kempenfelt was one. Keppel was accused, first, of failing to get his fleet into proper order before the battle began; second, that he failed to tack and double on the French after passing their rear; third, that he failed to renew the engagement; fourth, that he wore the fleet away from the French, thus giving them 'the opportunity to rally unmolested'; and fifth, that he failed to renew the action the next day. To all those charges Keppel had sufficiently convincing answers to cause the court to find that they were 'malicious and ill-founded'.

D'Antin gives detailed comments on the charges from the French point of view. They amount to saying that Keppel was a victim of superior French skill, and this, to some degree, was true. Keppel, so d'Antin said, acted sensibly enough; any disorder in his fleet at the start of the battle was due to changes of wind and thereafter his lack of initiative was due to his crippled ships and d'Orvilliers's masterly evolutions. He also played down Keppel's defence, except in so far as it reflected on his incapacity to defeat a better organised foe.

Palliser's court-martial, undertaken at his own request, enabled him to restate his case in full. He was acquitted, subject to the mildly worded rider that 'it was incumbent on him to have made known to his Commander-in-Chief the disabled state of the *Formidable*'. This seems a valid criticism, though as an issue it scarcely appeared in the trial.

Captain John Inglefield, then a lieutenant in Alexander Hood's *Robust*, and a witness at Palliser's trial, later characterised the battle as 'disgraceful to both nations – but certainly most to the French'.[6] By this he presumably meant that, whereas neither side showed either the intention or capacity to force a true pell-mell, the French, whose mobility at the end was least impaired, entirely failed to exploit their

---

6 *A Short Account of the Naval Actions of the Last War – by an Officer*, 1788 pp6–7 (*NMM*, Tunstall Collection, TUN/208)

success. This was an extremely relevant point and has been made subsequently by a whole line of British writers commenting on the American War as a whole. The British tended to employ younger admirals than the French, and no doubt sheer exhaustion at the end of a battle, following hours and hours of manoeuvring, tended to sap the initiative of the veteran French commanders, as in Rodney's case at the end of the Battle of the Saints. D'Antin, in giving the French view of the trials, stressed public uneasiness in England arising from memories of Admiral 'Binke' [Byng]. He naturally refrained from criticising the French command, though this does not necessarily mean that he failed in private to draw some chastening conclusions.

The Franco-Spanish fleet of 1779

The combined Franco-Spanish fleet which sought to gain control of the English Channel in 1779 consisted of twenty-eight French ships of the line and eighteen Spaniards, fairly

The Battle of the Dogger Bank, 5 August 1781, by Thomas Luny (see p 152). NMM

7 *NMM*, REC/46 (formerly Sm/7), pp117–9

evenly distributed throughout, in three squadrons.[7] A French ship was stationed at the head and tail of each squadron. D'Orvilliers was commander-in-chief, de Guichen commanded the van and Don Gaston Miguel the rear. Five other French and four other Spanish flag officers were also included. Don Luis Córdova, the Spanish commander-in-chief in the *Santissima Trinidad* of 110 guns, with two other Spanish flag officers, was in command of an entirely separate and all-Spanish squadron of observation, of sixteen and sometimes seventeen ships of the line. Separate from this was a composite light squadron of three French and two Spanish ships-of-the-line under a French *lieutenant-général*.

It appears that the French had practised special evolutions before the Spaniards joined them. Thereafter the combined fleet carried out evolutions in company with the squadron of observation and the light squadron, the two subsidiary forces taking up various positions and assuming various orders of sailing. Quite apart from more general questions of command, d'Orvilliers no doubt found it extremely difficult to achieve any tactical efficiency with this huge and strangely organised armada. Was it Spanish pride that produced Córdova's separate Spanish squadron of observation, or was it

that d'Orvilliers felt he could not effectively absorb any more Spaniards into his own line of battle, and certainly not the mountainous *Santissima Trinidad?*

Presumably experience in the campaign of 1778 lay behind d'Orvilliers' promulgation of a new set of signals in 1779. A sixteen-flag tabular system was provided for 'Signaux Généraux et Particuliers des Armées Navales à l'ancre et dans les Rades', designed to cover all aspects of harbour service; no doubt they saved much time and manpower in using boats to carry orders and messages. Another sixteen-flag table provided for 'Mouvemens Ordres et Signaux Généraux de l'Armée du Roy Commandée par Mons le Cte d'Orvilliers en 1779'. A new feature of both these tables was an additional column of signals written up outside the main tables. A special key flag was provided to indicate reference to this column rather than to the main table. In effect this provided an abbreviated signal system, but in rather a clumsy manner. A new type of table was also introduced, entitled 'Ordres et Signaux Particuliers aux Vaisseaux de Chaque Escadre en ajoutant ou signal la flamme'. It had sixteen flags along the top and eighteen down the side, and covered changes of course and changes of sail and chasing in very great detail,

together with soundings and means of dealing with prizes. Separate pendants were used to indicate which squadron was being addressed.

Two more tables completed the issue for the year. The first was a small table for single-flag general signals, and for single flag plus a pendant signals for fireships. The second was a sixteen-flag numerical table with the squares numbered 1 to 256, used solely, as far as can be seen, for referring to ships' numbers, a list of the fleet with their numbers being written opposite, including twenty Spanish ships. As in the numerical system of 1778, none of the flags themselves was given a number. Nor, of course, did any of the letters of the alphabet appertain to any particular flag. Only thirty different flags altogether, apart from pendants, were employed in the 1778 and 1779 system. Thus many flags appeared in several different tables, while in some tables the order of flags was partially the same as in others.

The mere recital of these two catalogues of signal codes might suggest to the reader that the whole system was utterly illogical and quite unworkable; this conclusion, however, would be wrong. We know from Kempenfelt's observations that he envied the smartness with which the French executed their signals. What, in fact, we have been examining is the internal structure and mathematical relationships of the different parts of the system. For working purposes the system could be condensed, simplified and rewritten in the form of a handy signal book, just as Pavillon had arranged for when drawing up his famous tables of 1776.

A manuscript signal book used in the French fleet in 1779 employed a lettered table with number references to the articles of the instruactions in the squares.[8] The book conformed to what had become the standard French pattern. Sixteen flags were used for the main signals. The first opening of the book showed the first flag in the series painted large at the top of the left and right pages. The sixteen flags were also painted in columns on each page, with the respective article and its number written opposite. At the top of the pages of the next opening appeared the next flag in the series, and so on. A signal could thus easily be read, the top flag indicating the double page on which it appeared, and the bottom flag the actual article or signal written opposite on that page. Making the signal demanded more skill, but as the signals were grouped in logical order, this was not too difficult, and an index with detailed headings could always be made.

In the library of the Royal Navy Staff College, Greenwich, is a vellum bound copy of *Tactique Navale*, as used by the fleet commanded by d'Orvilliers in 1779. It was issued from d'Orvilliers' flagship, the *Bretagne*, on 21 May 1779, signed by Pavillon. At the beginning is the name of the *Bienaimé*, one of the ships in d'Orvilliers's fleet in 1778–9, and with the date 1779 in a different hand. Inside the cover is an inscription stating that the book was captured in the *Ville de Paris* in the Battle of the Saints in 1782. It was evidently out of use by then, since de Grasse had his own signals.[9] It was not a book of instructions; over three-quarters is devoted to seventy-six evolutions in which the existence of the enemy is ignored. There is a sixteen-sided table, employing the usual sixteen flags, and at the beginning of the evolutions is a plate entitled 'Mouvements de Guerre'. Although this book seems disappointing to the historian, it impressed contemporary

8 *NMM*, SIG/C/3 (formerly Sm/9); manuscript signal book, 'Movemens, Ordres et Signaux Généraux de l'armée du Roi, commandée par Monsieur le Comte D'Orvilliers, Lieutenant-Général en 1779, Channel'

9 The book also bears [Prince] William Henry's signature, and was part of his collection. It was transferred from the Royal Library at Windsor to the Royal Navy Staff College in 1928.

opinion enough for a Swedish translation to be printed in Stockholm in 1787.[10]

The Spanish navy entered the war with a system of tactics devised by José de Mazarredo Salazar, *Teniente de Navío* [Commander] *de la Real Armada*, and expounded in *Rudimentos de Táctica Naval para Instruction de los Officiales Subalternos de Marina*, printed at Madrid in 1776 and dedicated to King Charles III.[11] Despite bearing some evidence of the influence of Hoste and Morogues, this is a text book for junior officers rather than a treatise intended for naval officers in general, though it could clearly have been read with profit by all alike. In common with the French writers, Mazarredo said very little about fighting the enemy. Broadly speaking, his tone was sophisticated and undogmatic.

Mazarredo did introduce a new idea, the use of fireships by the windward fleet, if threatened with doubling, as a means of covering its retreat to windward. Salazar also showed himself an innovator in his treatment of breaking the enemy line. He proposed that, when the fleet was to windward, the centre should break through the enemy centre. In the process of breaking through, the enemy's centre ships immediately astern of the break would be forced away to leeward, so disorganising the enemy rear and isolating it. Meanwhile, the enemy van would have no choice but to stand on to avoid being put between two fires, and it would thus become completely separated from the remainder of the fleet. Exactly the same movement might be executed from leeward, though in that case the enemy's rear would be forced to give way to windward, thus exposing itself to the fire of the centre and rear ships of the attacking fleet. The innovation lay in the details of the proposed movements, as shown in the relevant diagrams.

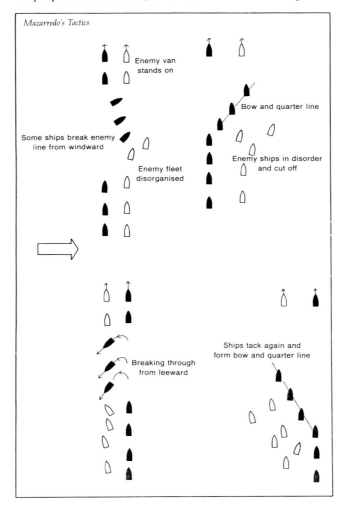

Mazarredo's Tactics

Enemy van stands on

Some ships break enemy line from windward

Bow and quarter line

Enemy fleet disorganised

Enemy ships in disorder and cut off

Breaking through from leeward

Ships tack again and form bow and quarter line

In case of a breakthrough from windward, the ships executing the movement would run to leeward in succession, the leading ship first, and break through in some manner not clearly defined, the implication of the diagram being that the enemy was inferior and that a gap had occurred in their line. Having broken through, the fleet would form a bow and quarter line on the same course as the enemy. This would place them athwart the gap in the enemy line, and on a bearing from each other of five degrees in relation to that line, the leading ship being the most advanced and thus furthest to leeward.

When breaking through from leeward the evolution became a little more complicated, since the ships executing it had first to tack in succession, the rearmost leading. Having broken through, again in a manner unexplained, they were required to tack again so as to get on the same course as the enemy. This would bring them into bow and quarter line athwart the gap on a line of bearing 25 degrees from that of the enemy, the leading ship being in this case furthest to leeward and nearest the rest of the fleet, since it began the movement last. Again there is the implication from the diagram that the enemy fleet was inferior, and that a gap was available to be exploited.

How any evolution such as this could have been carried out by the entire fleet, practising tactics according to the actual battle methods used at the time, it is difficult to imagine, particularly in the Spanish fleet of the time. Nevertheless, Mazarredo was, in this respect at least, an original theorist. His book is extremely well laid out and printed, references being to paragraphs or figures, never to pages. His proposals serve only to emphasise the wide gap which existed between theory and practice.

Mazarredo also drew up a signal book, specifically for Córdova's fleet, which was printed in 1781. It must have been used in the operations against Gibraltar, and it does not seem unreasonable to suppose that Córdova's signalling system was somewhat similar when he first joined d'Orvilliers in 1778. It is true that one of the signal books used in the Spanish fleet during the Seven Years War, a poor affair which only mentioned the enemy in seven articles out of 207, was reprinted in 1765.[12] Nevertheless, the Spanish navy appears to have shared, and more than shared, the theoretical side at least of France's naval renaissance. Mazarredo's signal book of 1781 is in some respects an improvement on Pavillon's. Like the latter, it employed a tabular system, but much less complex. It employed tables 20 by 20, each permitting 400 signals. The flags themselves were not specified on the tables, numbers being used instead, so that the flags employed could easily be changed. The whole business of reference to notional letters of the alphabet, and transposition of signals from one table to another, was avoided.

The signal book was evidently prepared for Franco-Spanish co-operation, as it begins with special signals for indicating Spanish and French squadrons, divisions, frigates, the reserve corps, etc. The 400 signals for use at anchor covered not only every feature of fleet administration, as in the manner of Morogues, but also shore bombardments and landings, no doubt with an eye to the siege of Gibraltar. Twenty special signals allowed for reporting the movements of ships, presumably to be made by private ships. The signals for use under sail by day, made with a combination of

10 *NMM* (formerly Tunstall Collection S/P/SW/1)

11 *Ibid*

12 *NMM*, SIG/C/13 (formerly Sp/177). *Senales que han de observer y practicar los Navios de la presente esquadra de Mando del Capitan General de la Armada, Marques de la Victoria . . . Duda en VIII de Octobre de Año de MDCCLIX, estando embarcado en el navio El Real Fenix*, reprinted 1765.

'cornets',which were swallow-tail flags, other flags, and flags from the table, included a series of battle signals. The series of 400 general signals covered the whole management of the fleet under sail, again in the manner of Morogues. Included, however, was another set of battle signals of the standard kind, but with some interesting extensions and refinements. Number 250 was to advise that the general intended to attack during the night if an opportunity should occur. Number 251 advised that the general did not intend to attack during the night unless a favourable opportunity should occur. Number 252 ordered individual ships to engage when able. Numbers 269–70 ordered the leading ships to steer so as to get to pistol shot of the enemy van, or rear, if possible. Number 271 ordered the fleet to prevent the enemy breaking the line, even at the expense of being repeatedly run aboard. The night signals were made from a table of lanterns and guns. Nineteen signals were provided for use at anchor, and three separate sets of nineteen signals for use under sail.

No-one studying this book could criticise the Spanish either for a lack of useful signals for battle and general purposes, or for over elaboration of signalling technique. Although still tied to the tabular system, their arrangement was brilliantly simple compared with that of the French.

Kempenfelt and the Channel Fleet

Although he was never commander-in-chief, Richard Kempenfelt played a major part in the tactical reform of the Royal Navy. He was the son of a Swedish Jacobite colonel who had accepted service under Queen Anne and had become lieutenant-governor of Jersey. He was highly regarded in the service, especially after being flag captain to Pocock and Admiral Sir Samuel Cornish in the East Indies, but he had only reached the Post List at the age of thirty-nine. He was thus junior in the service to Keppel, Howe, Palliser, Rodney, Byron, Barrington, Rowley, Graves and Hood, though senior to all of them in years. Following the Keppel-Palliser scandal, it had become difficult to persuade any of the leading flag officers to accept command of the Western squadron or Channel Fleet, as it was coming to be called. As a result, the large forces gathered in the Channel in 1779 to resist the combined fleets of France and Spain, comprising thirty-five to forty-five of the line and some thirty frigates, fire-ships, sloops, cutters and brigs in attendance were commanded in turn by Sir Charles Hardy, a resuscitated veteran of sixty-three who died within a year, and by his successor Francis Geary, aged seventy-one, who resigned five months later. His successor, George Darby, an able and vigorous man, nevertheless felt bound to resign when the Whigs returned to power in May 1782.

In May 1779 Kempenfelt was appointed as Captain of the Fleet, or chief of staff, to Keppel's successor, Hardy. He continued as captain of the fleet under Geary even after receiving his flag as a rear-admiral. Under Darby he hoisted his flag as third-in-command of the fleet and a year later held an independent command, during which he achieved a notable victory. After North's resignation he served as second-in-command of the Western squadron under Barrington and then under Howe, meeting his death when the *Royal George* sank in August 1782.

Throughout the spring and summer campaigning season of 1779 the Channel Fleet was on the defensive, despite its size and close proximity to good home supply and repair bases. It was inferior to the combined fleets of France and Spain, which were co-operating for the planned invasion of England. Kempenfelt, however, did not believe that British tactics had to be ruled by circumspection. On 27 July 1779 he advocated a more aggressive approach:

Rear-Admiral Richard Kempenfelt (1718–82), by Cornelis Ketel. NMM

'Tis an inferior against a superior fleet; therefore the greatest skill and address is requisite to counteract the designs of the enemy, to watch and seize the favourable opportunity for action, and to catch the advantage of making the effort at some or other feeble part of the enemy's line; or, if such opportunities don't offer, to hover near the enemy, keeping him at bay, and prevent his attempting to execute anything but at risk and hazard; to command their attention, and oblige them to think of nothing but being on their guard against your attack.[13]

Viewed in broad strategical terms there may be a good case for saying that, despite its relative inferiority to the combined fleets of France and Spain, the British Channel Fleet was larger than it need have been. The invasion threat was not realistically very great, and Franco-Spanish strategical co-operation was not at all good. Meanwhile, in the West Indies and North America, British naval forces were kept inferior to the French so as not to denude the Channel Fleet. It is even arguable that this fleet itself would have been more tactically efficient if it had been kept down to thirty of the line. Whatever tactical device Kempenfelt had in mind for use against a 'feeble part of the enemy's line', however, he was clearly making the condition that, for the weaker fleet to succeed in its harassing and opportunist role, it must be at least tactically equal to the enemy, if not superior. In fact, the reverse was the case.

Judging by his letters to the comptroller of the navy, Sir Charles Middleton, Kempenfelt was not only hamstrung by Sir Charles Hardy's lack of competence due to advancing age, but also by the tactical inefficiency and lack of practice of the fleet as a whole. 'It is with the greatest difficulty,' Kempenfelt wrote, 'I can ever prevail upon him to

13 *Barham*, I, p292

manoeuvre the fleet, he is always [so] impatient and in [such] a hurry to get to the westward, to the northward, or the southward, that he won't lose time to form a line.'[14] Benjamin Thompson, Count von Rumford, who was serving as a volunteer in Hardy's flagship during the summer and autumn of 1779 in order to supervise some experiments in ballistics, wrote a long account to Lord George Germain which gave a damning picture of tactical incompetence, in which Kempenfelt came in for his share of criticism.[15]

You must know we never as yet had attempted but one simple manoeuvre, and that is to draw the ships into a line of battle, one directly ahead of the other, and upon a signal to tack about all together. In this we never have succeeded. It has always been more than two hours before the ships ahead have got into their stations – the line has always been very crooked, and the ships at very unequal distances, and when we have come to put about confusion has commonly ensued, and we have been obliged to end our manoeuvring abruptly by making the signal for the ships to return to their stations in the order of sailing. This has arisen sometimes from one cause, and some-times from another, but I don't remember a single instance in which we have attempted to manoeuvre when we did not make at least one evidently wrong and contradictory signal.

In one incident reported with chagrin by Thompson, Kempenfelt could find no signal to order a manoeuvre by divisions. When a method was pointed out to him in the book, the wrong signal was flown, apparently one ordering the return of weekly accounts. A similar story appears in Sir John Barrow's *Life of Richard, Earl Howe,* about the period after command of the fleet had passed to Admiral Geary. Exasperated by Kempenfelt's attempts to manoeuvre the fleet, 'Geary at last grew impatient and going up to Kempenfelt, and laying his hand gently on his shoulder, exclaimed with good-natured earnestness, "Now, my dear Kempy, do for God's sake, do, my dear Kempy, oblige me by throwing your signals overboard and make that which we all understand – Bring the Enemy to Close Action!" '[16]

Undaunted, Kempenfelt continued throughout the rest of his life to experiment with tactical reforms.

Kempenfelt's tactical ideas at this time are best illustrated from his letters to Sir Charles Middleton. On 28 April 1779 he wrote:

The course of this [Palliser's] court martial has given me a clearer idea of the action between our fleet and that of France last summer than I had before conceived; and the different state the two fleets were in after the action confirms most strongly what I have always thought; that is, that the disabling of your enemy in his masts and rigging should have no small share of your attention and your fire . . . Owing to the different directions of their fire [our fleet] . . . except the red division [were]so totally shattered in masts, rigging, and sails that for the whole evening of that day they could not all form into a line. The French ships, on the contrary, were so little injured in these particulars that they had the perfect command of their yards and sails . . . There is no strength and force without motion and direction . . . 'Tis plain to me that our fleet, after that action, for all the first part of the afternoon, was at the mercy of the French. Unconnected to succour and support each other, what defence could they have made against the attack of a close, well-formed line of ships? Why the French did not profit from this advantage they had, I can't conceive.[17]

Like Howe, Kempenfelt was in favour of reviving the use of fireships.

I really don't see the necessity of waiting until a ship is disabled before you apply the fireship. I think the best time to use them is in the very beginning of the action; the smoke covers them as well then as afterwards; and is it not best to do

the greatest injury to your enemy as soon as you can to hasten his defeat?[18]

It might have seemed a bad time for experiments, but Kempenfelt felt he had no choice. On 18 January 1780 he wrote to Middleton:

I believe you will, with me, think it something surprising that we, who have been so long a famous maritime power, should not yet have established any regular rules for the orderly and expeditious performance of the several evolutions necessary to be made in a fleet. The French have long since set us the example. They have formed a system of tactics, which are studied in their academies and practised in their squadrons . . .

Indeed, 'tis too obvious to make any arguments necessary to show that fleets as well as armies require rules for the execution of their movements, and that the one stands in need of tactics as well as the other; without which both are unwieldy masses, where force is lost for want of form and order. Superior address in conduct may make up for the want of numbers, but what is to be expected when skill and address are wholly on the side of numbers?

The accounts Kempenfelt had heard of French seamanship at Ushant indicated that what they gained by good tactical control was not lost by any fault in shiphandling. In his opinion,

We should therefore immediately and in earnest set about a reform; endeavours should be used to find out proper persons, and encouragement offered for such to write on naval tactics, as also to translate what the French have published on that subject. They should enter into the plan of education at our marine academies.

But the most effectual way to obtain every wished for reform in your fleet is to find out a man to place at the head of it, who has genius to comprehend every requisite regulation, and has activity and spirit to enforce their execution.[19]

Kempenfelt attempted to create a new tactical system suited to a large fleet opposed by an even larger and better tactically organised enemy, to give this system expression through a combination of signals and instructions, and to introduce these signals and instructions to a body of officers likely to be both suspicious and contemptuous of sophisticated innovations. The best that could be expected of Hardy was that he might somehow be persuaded to sign what was laid before him. In default of personal direction by the commander-in-chief through a series of conferences with flag-officers and captains, Kempenfelt was forced to try to gain his ends by bombarding the fleet with new sets of signals and instructions on a purely *ad hoc,* improvised basis. Never in the history of any navy, at any period before or since, have so many signal books and fighting instructions been issued by one man in the space of only twenty-seven months, during at least ten of which the main body of the fleet was laid up for the winter.

Kempenfelt's tactical reforms

On 31 May 1779, Sir Charles Hardy issued Captain Adam Duncan of the *Monarch* with a large folio copy of the Sailing

---

14 *Barham,* I, pp294

15 W G Perrin, editor, *The Naval Miscellany,* vol III, London, 1928, pp127–31

16 Sir John Barrow, *Life of Richard Earl Howe,* London, 1838 (hereafter cited: Barrow), p141

17 5 September 1779; *Barham,* I, p296

18 *Barham,* I, pp309–13

19 *Barham,* I, pp290–1

& Fighting Instructions, reprinted and bound up with the Admiralty additionals in signal-book form, as previously issued by Keppel.[20] The signals, however, were differently arranged and at the end were a set of printed but unnumbered additional fighting instructions, which were in fact additions to the Admiralty additionals, with signals. They included a signal:

If at any time I perceive the leading ship in the Line of Battle does not steer down enough on the ship she is opposed to, in the Enemy's line, although, I may not be able from the distance, to direct the precise number of points I would have her bear away and would have her bear down on the nearest course she can steer to bring the enemy to action.

Who thought of this? A similar signal in Rodney's book might have been used to resolve Captain Robert Carkett's confusion at the Battle of Martinique (see Chapter Seven below).

So far, Kempenfelt seems to have been feeling his way by using the instructions bequeathed to the fleet by Keppel. A whole collection of instructions and signals, however, issued to Captain Lord Longford of the *Alexander* and bound up together, gives what must be a fairly complete picture of what was happening between May 1779 and October, when the campaign for the year was nearly at an end.[21] Here we find the full extent of Kempenfelt's early reforms revealed for the first time.

Articles 6 to 17 of the day signals, together with an Article 18 in manuscript, are word for word the same as articles selected from Howe's 1782 'Instructions for the Conduct of the Ships of War, Explanatory of, and Relative to the Signals contained in the Signal Book herewith delivered', the companion volume. The general purpose of these articles was, first, to give the admiral better control of the fleet, and second, to standardise methods for carrying out various evolutions. Especially important is Article 9 (p12 Article 4 in the 1782 instructions) for ships unable to keep their station in the line. They were to fall astern so as to avoid disordering the fleet, the gaps being filled by succeeding ships.

The captains of such ships [as have to fall astern] will not thereby be left in a situation less at liberty to distinguish themselves; as they will have opportunity to render essential service, by placing their ships to advantage when arrived up with the enemy, already engaged with the other part of the fleet.

Most of the sailing and fighting instructions and all the instructions for fog were the same as those hitherto thought not to have been issued together until 1781, so that they antedate much of Kempenfelt's work by a year.[22]

The nine night instructions, with headed references in the style of Howe, include Article 9, for captains, in case the enemy was suddenly encountered at night, to have 'discretion to prosecute a closer engagement with the enemy, or to suspend the continuance of their attacks, as they may have reason to think either proceeding most efficacious for the advancement of the public service'. This is entirely in the spirit of Howe's night battle instructions of 1782. It implies a high degree of professional skill and co-operation on the part of captains and a degree of confidence placed in them by the admiral, and is the complete antithesis of everything ordered and aimed at by Rodney. With so many of these 1779 instructions apparently anticipating Howe's work of three years later, it is tempting to make Kempenfelt the inspirer of Howe's innovations; it may, however, equally have been Howe who was supplying Kempenfelt with ideas. This seems in fact more likely, especially in view of a letter written, apparently to Kempenfelt, on 10 September 1779, in which Howe comments,

Admiral Darby's Additional Instructions of 18 October 1779 for the deployment of the fleet en echelon, *from Admirals Keppel and Hardy's Signal Book 1778–9. NMM*

20 *NMM*, DUN/6 (formerly S(D) 6) On 13 May 1779, Vice-Admiral George Darby, then acting as one of Hardy's divisional commanders, had issued Duncan with a printed copy of the Admiralty Additionals in the standard form with the index (DUN/5 (formerly S(D)), ie, not in signal book form as issued by Keppel a year earlier. To what extent Darby was acting on Hardy's ie, Kempenfelt's authority is uncertain. *NMM*, COR/61 (formerly Sp/116) is a slightly different printing but undated and unsigned. Darby, Kempenfelt and Duncan, the future Lord Camperdown, had all been members of the court-martial trying Palliser, which had ended only on 5 May.

21 *NMM*, PAK/3 (formerly Sp/135)

22 Barrow, p142

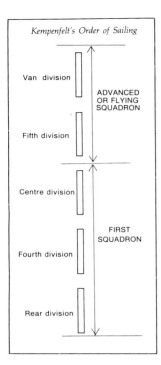

Kempenfelt's Order of Sailing

Van division

ADVANCED
OR FLYING
SQUADRON

Fifth division

Centre division

FIRST
SQUADRON

Fourth division

Rear division

I should be glad to know what part of our signals have been adopted; I should thence be able to form an opinion of the principle upon which that great machine (the Grand Fleet) is to be put in motion. Our signals were adopted rather for a single squadron; and, though most of the articles might be applied to a larger force, (fleets being composed of squadrons collectively arranged), the necessary continuation of the signals being different, the propriety of their use under different circumstances will vary also. In the disposition of them I had those objects in view. But the choice to be made of the expedients which any set of signals has provided for, will constitute, as we know, the ability of the flag-officers.'[23]

The words 'our signals' seem to refer to a joint production.

Bound in at the end of the volume are a number of additional signals and instructions issued in manuscript to Lord Longford by Vice-Admiral George Darby. Fifteen additional fighting signals, issued 1 September 1779, include: number 1, for attacking enemy's rear in succession; number 4, to break through the enemy line and attempt to cut off part of their rear, when fetching up to leeward on the contrary tack; number 5, for attacking the enemy rear and inverting the line; numbers 6, 7 and 8, for doubling; numbers 10, 11 and 12, to signal the admiral's intention to attack the enemy centre, van or rear; number 13, to extend the line to prevent being doubled. Directions for forming the order of retreat follow, and also slightly different directions, dated 18 August and erased. Further signals are dated 27 September. Directions for the order of sailing in three columns in echelon and for the order of sailing in five columns are dated 18 October. At the end are directions for the order of retreat and for forming the line of battle from it, dated 22 September.

With forty of the line under his admiral's orders, Kempenfelt saw that some new squadron organisation was essential, the traditional scheme of the van, centre and rear being inadequate for such a large fleet, and especially so for the kind of offensive-defensive tactics which he had in mind. His idea, as shown in the Order of Sailing, which was the last item in the volume, was to organise the fleet in five divisions under its five flag officers.[24] The centre under Hardy, together with the so-called Fourth division and the rear were then to form the First squadron, which would be the main battle force. The van, together with the Fifth division, would form the Second squadron, which would be regarded as an additional, reserve, detached, advanced or 'flying' squadron. Each squadron was given its own sailing order by divisions. When turning from a close-hauled line ahead to a line with the wind two points abaft the beam, the senior admiral's division of each squadron was to take station to windward of his squadron on either tack. When sailing large, the senior admiral's division was to be to starboard of his other division or divisions. The First squadron, when sailing by divisions in three separate lines ahead, might be signalled to proceed *en echelon*, thus:

Admiral's centre division

Fourth division

Rear division

The leading ship of the Fourth division was to be level with the centre of the Admiral's division and the leading ship of the rear level with the centre of the Fourth division. Divisional commanders might be signalled not to lead their respective lines but to keep station in the middle.

A further version of this cruising formation disposed the Second squadron to starboard of the First squadron, thus:

First squadron	*Second squadron*
Admiral	Van division
Fourth division	Fifth division
Rear[25]	

Following Howe, who followed Admiral Henry Osborn, Kempenfelt introduced the order of retreat for use against a 'much superior' enemy 'or after you have suffered much in action'. This order was also useful for covering a convoy of merchantmen, which could be sent ahead protected by the ships forming the two sides of the angle. The order was to be formed by the fleet being 'upon the sides of an obtuse angle of 135 degrees, formed of the two lines by the wind for the starboard and larboard tacks; the admiral is to windward at the point of the angle and in the centre of his fleet'. To change the order of retreat to the order of battle to starboard, the leading ship on the starboard wing hauled her wind to starboard, followed by the rest of the wing in succession as they reached her wake. Meanwhile, the larboard wing continued together on a course parallel to that originally steered by the starboard wing. On reaching the wake of the starboard wing, they turned to starboard together and the line ahead was formed.

As in Howe's signals, Kempenfelt dealt carefully with alterations of course both together and in succession. A

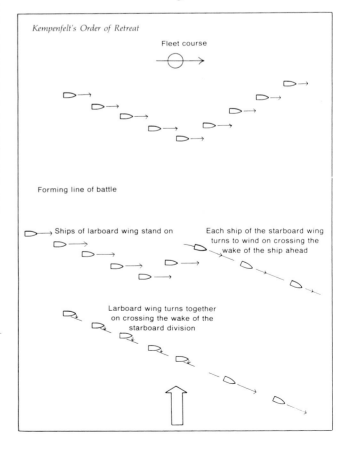

Kempenfelt's Order of Retreat

Fleet course

Forming line of battle

Ships of larboard wing stand on

Each ship of the starboard wing turns to wind on crossing the wake of the ship ahead

Larboard wing turns together on crossing the wake of the starboard division

23 *NMM*, PAK/3 (formerly Sp/135), issued by Darby to Lord Longford at Spithead 18 October 1779 (part printed in *Signals & Instructions*, pp125–31, together with the full list of the line of battle for the fleet).

24 *NMM*, PAK/3 (formerly Sp/135), issued by Vice-Admiral Darby, second in command of the fleet, to Captain Lord Longford, *Alexander*, 18 October 1779 (not printed in *Signals & Instructions*).

25 Kempenfelt to Middleton, 8 April [1780]; *Barham*, I, pp323–4

special feature of the Hardy-Kempenfelt instructions and signals is the use throughout of the word 'attack' in place of 'engage'.

In 1780 Kempenfelt devised yet another tactical organisation for the fleet, although it is unclear whether he persuaded Admiral Sir Francis Geary to adopt it. For a fleet of from fifteen to twenty-six of the line, he was content with the traditional arrangement of three squadrons. If the fleet exceeded twenty-six of the line, he wanted each squadron divided into two divisions, or for a fleet of forty-five of the line upwards, three divisions.[26] This proposed reorganisation is highly relevant to the instructions accompanying Geary's 1780 signal book, which include the order of sailing in three squadrons abreast as the standard formation, with a variation by which each squadron was itself formed into two columns. A large fleet was enabled to sail in six separate lines ahead of a few ships each disposed abeam of each other, or, if required, in pairs *en echelon*.

Kempenfelt's signal books

Meanwhile, Kempenfelt was trying to give better executive expression to his tactical reforms by means of more efficient signalling. Here he was in serious difficulties. It was always possible to make piecemeal issues of additional fighting instructions and orders of sailing, with the necessary signals attached, but it required a revolutionary act to introduce a new signalling system. It was likely to provoke resentment, confusion and disorganisation throughout the service.

Kempenfelt's signal book of 1779 was clearly influenced by Howe's of 1776 but, unlike Howe's, Kempenfelt's book was still tied to the old combined books of instructions. During 1779 Kempenfelt made use of a printed version of the kind of signal book Keppel had introduced, or at least made use of, in 1778. It was a book incorporating the main signals abstracted from the parent work and the Admiralty additionals, with page and article references to these books as bound together with continuous page numbering.[27] The form of the book was entirely traditional, with the flags having signification according to where they were flown, except for a very few special cases of 'where best seen'. All of Kempenfelt's new instructions, including the order of retreat and the fleet order in three divisions disposed abreast in echelon, were incorporated, in each case with explanatory directions and diagrams. Although there were still plenty of spare places in the ship for flying the existing flags, more flags were introduced, presumably to avoid confusion in distinguishing between different flags flown in the same place. Whereas Saunders could work his system of 1760, including all his additional signals and signals for private ships, with twenty-two flags and four pendants, the Channel Fleet signals of 1779 required forty-two flags and thirteen

pendants. Some of the flags were hopelessly confusing.

It is difficult to know the extent to which Kempenfelt really understood Howe's ideas. He wrote to Middleton on 17 October 1779, 'Signals pointed out by numbers, and according to that method observed in those you sent me, I have long been acquainted with. They don't require many flags; however there must be three of each sort, or you can't express a number which consists of three figures of the same rank, as 222 or 333, etc.' This seems scarcely a valid objection to the system as a whole. Either such signal numbers could be left out or a substitute flag used. Kempenfelt continued:

This numeral way of pointing out the signal is very convenient for those to whom they are addressed, for the ready finding out their meaning, and a very great ease to the compiler of the signals. The objections against them are that, for the most part of the signals, you have three flags to show instead of one in the common method; that ships may often be in a position not to see the flag at the mizzen topmast-head; that to ships at a distance it will be more difficult to ascertain the colour of two or three flags than of one; that when signals are simple, with only one position, that, assisted by circumstances, may lead to the knowledge of the signal, though the colour of the flag may not distinctly be distinguished.

Why Kempenfelt cited the difficulty of seeing flags at the mizzen topmast head as an objection to the numerical system, when that was a problem peculiar to the old one, is mysterious. Either the objections were not really his own, or he had failed to understand the numerical system. He concluded:

I think all these objections may be obviated; and in my opinion, the signals by numbers, or those by a superior flag, as used by d'Orvilliers the last two summers, are by much superior to any method we have. I could in a very few days (less than a week) arrange our signals according to both the above methods, could I have those days to myself without interruption.[28]

Were the objections his own or those of others? How could they be obviated? Surely it would have been better to keep the old system for the time being, rather than become committed to an all too obvious second best which could lead nowhere?

Writing to Lord Sandwich, First Lord of the Admiralty, on 9 July 1779, Middleton stated that it was agreed between Kempenfelt and himself 'to introduce the French system [of signals], if practicable, into Sir Charles Hardy's fleet'. 'That gentleman [Kempenfelt] has well considered the subject', he continued, 'but at this juncture it would be dangerous to exchange even a bad set of signals that are generally known for a better which the greatest part of the officers were unacquinted with'.[29] Middleton suggested that Kempenfelt should be brought ashore and that the two of them should, together with 'some other sea officers which I can name . . . simplify this very complex business and . . . reduce it within the practice of the most ordinary understanding. The materials for this purpose are lying by me, but engaged as I am in an office which I own is almost beyond my strength it would be madness to undertake anything new.' Middleton further suggested that Kempenfelt should be permanently withdrawn from Hardy's fleet to act as assistant to the comptroller for the whole administrative work of the Navy Board. Nothing came of either of these suggestions.

On 27 May 1780, Geary, now in command of the Channel Fleet, issued Captain Adam Duncan of the *Monarch* with a larger printed folio, 'Instructions for the Conduct of the Ships of War, explanatory of, and relative to the Signal Book herewith delivered', and followed by 'Signal Book for the Ships of War'.[30] The wording is exactly the same as that used

26 See *Signals & Instructions*, pp135-9

27 No printed copy of this signal book has been found. *NMM*, Tunstall Collection, TUN/2 (formerly S/MS/Am/1) is a folio MS copy with a loosely inserted note: 'Robert Callander, Clerk of His Majesty's Ship *Terrible* at Sea, August 12th 1779.' The *Terrible* (74) Captain Sir Richard Bickerton, Bart, was stationed in Sir Charles Hardy's Centre division of the fleet.

28 Kempenfelt to Middleton, 17 October [1779], *Barham*, I, p301

29 *The Private Papers of John, Earl of Sandwich, First Lord of the Admiralty 1771–1782*, G R Barnes and J H Owen, editors, London, 1932-8 (hereafter cited *Sandwich*), III, p42

30 *NMM*, DUN//14 (formerly S(D) 14). *NMM*, SIG/A/14 (formerly Sp/91) is a duplicate copy of the signals unsigned and undated, but with MS insertions, while *NMM*, PAK/4 (formerly Sp/137) contains the signals and instructions (less pp17-8) bound together.

by Howe. Hardy had died on 18 May, almost his last act being to refuse Kempenfelt permission to reorganise the fleet for the coming campaign.[31] Geary had hoisted his flag as commander-in-chief only on 24 May, and was no doubt prepared to sign anything which Kempenfelt put in front of him. It has hitherto been assumed that his particular form of signal book and the accompanying instructions were not issued at least until 1781, but the signed and dated copies issued to Duncan settle this point completely. By abandoning the parent work and the Admiralty additionals, and issuing a true signal book with accompanying instructions, Kempenfelt had become a revolutionary. His adoption of Howe's ideas was a very radical move because the Channel Fleet was the main fleet of Great Britain at the time, and contained five flag officers.

The signal book was on traditional lines but, like Howe's, the signals were numbered serially throughout. In the main they repeated the signals of the previous year, 1779, but also included many fresh ones of an administrative and general nature, up to a total of 397. There were fifty-six spaces for night signals, with some blanks, the signals themselves being shown under the headings number of lights, how disposed, place, guns, and false fires. In addition to the night signals for private ships was a separate section of eleven night signals for 'ships ordered to observe the motions of the enemy', thus

foreshadowing the great system of night reconnaissance used by Nelson with such effect on the eve of Trafalgar.

At the end was a ten-flag table for numerical signalling, arranged exactly like the numerical table possibly attributable to Howe in the manuscript copy of his fog and night signals of 21 August 1778.[32] There was, however, a very important difference in purpose, for whereas the 1778 table was stated to be for the use of ships not possessing the flags used in Howe's signal book, the 1780 table was for general usage. 'The Admiral to intimate his intention of directing the fleet by Numerary Signals will hoist the [Royal] Standard where best observed.' A Union flag hoisted over the Standard was to be the signal for reverting to the normal system of flags in particular positions. Are we to assume that this was Kempenfelt's way of trying to introduce true numerical signalling into the fleet, or did he actually regard the tabular system with all its inconveniences as a desirable half-way house between a true numerical system and the old one? The latter explanation seems unlikely; with so much difficulty being experienced in the recognition of signals, it may be that the numerical table was intended to facilitate distant signalling. Whatever his reasons, why did he introduce into his table a red flag pierced white not included amongst the twenty-seven flags used for normal signalling?

Soon after issuing Geary's signals and instructions of 27 May 1780, Kempenfelt seems to have produced yet another pair, which contained the same ten-flag table for numerical signalling. The printed versions have disappeared, but the Naval Library possesses a manuscript copy of both the signals and instructions, and signed: 'Completed this book, August 22, 1780, *Defence* at Sea – James Hardy'.[33]

The flags used for the signals were not the same as those in Geary's book, nor the same as those in Darby's book issued for the ensuing season's campaign in February 1781. They are, however, exactly the same as those in an unsigned and undated manuscript private pocket signal book in the Tunstall collection.[34] The dating of this latter book is fairly certain, as the chequered pendants which Kempenfelt told Middleton he had changed in his February 1781 book issued in Darby's name are struck out in pencil and marked 'striped'. On the other hand, the red-white-red and the blue-yellow-blue flags which Kempenfelt implies he had changed at the same time from horizontal to vertical are already in their new form without any sign of erasure or alteration. The explanation for this may well be that, as in the case of so many private signal books transcribed from an official copy which possibly bore heavy corrections itself, the copyist made a mistake which he later corrected.

The instructions which appear only in James Hardy's copy are incomplete, but include the bulk of the fighting instructions in Geary's book of 27 May as well as those issued by Kempenfelt for Darby in the following year. They include several from Howe's fighting instructions of 1776 and his additional fighting instructions of 1777. In fact, the whole trend of Kempenfelt's sailing and fighting instructions during his service in the Channel Fleet from 1779 to 1782 seems to be in the direction of copying Howe.

On 11 February 1781 Vice-Admiral Darby, in command of the Channel Fleet since Geary's retirement in September 1780, issued a 'Signal Book for the Ships of War together with 'Instructions for the Conduct of the Ships of War,

Admiral Lord Howe's 1778 manuscript tabular system for signalling numbers. MOD

FLAG.

Numerary Table.

	1	2	3	4	5	6	7	8	9	10
	Union	Red	White	Blue	half Red half Wh	half Blue half Yel	Dutch	Red Pend	Yellow Pend	Dutch Pend
Union	1	11	21	31	41	51	61	71	81	91
Red	2	12	22	32	42	52	62	72	82	92
White	3	13	23	33	43	53	63	73	83	93
Blue	4	14	24	34	44	54	64	74	84	94
half Red half White	5	15	25	35	45	55	65	75	85	95
half Blue half Yellow	6	16	26	36	46	56	66	76	86	96
Dutch	7	17	27	37	47	57	67	77	87	97
Red Pendant	8	18	28	38	48	58	68	78	88	98
Yellow Pendt	9	19	29	39	49	59	69	79	89	99
Dutch Pendt	10	20	30	40	50	60	70	80	90	100

31 Letter to Middleton 8 April 1780; *Barham*, I, p323

32 *MOD*, NM/34 (formerly *RuSI* 34)

33 *MOD*, Ec/50, together in a large folio vellum bound volume

34 *NMM*, Tunstall Collection, TUN/8 (formerly S/MS/Am/8)

explanatory of, and relative to the Signal Book herewith delivered'. These were clearly produced by Kempenfelt, and Thomas Graves significantly wrote the words 'Admiral Kempenfelt' on his copy of the instructions.[35] Nevertheless, in almost every respect they followed Howe's signal book and instructions of 1776. The signal book is a printed folio with many handwritten insertions. There were only twenty-seven flags and seven pendants, compared with the forty-two flags and thirteen pendants used in the summer of 1779. The flags, in which combinations of blue and yellow predominated, were a different selection from any used before. The structure of the book was much the same as that issued over Geary's signature in 1780. There were two flags on each thumbing referring to the left and right hand pages, with a written description of the respective flags at the head of each page. The main signals were numbered consecutively throughout. The fog signals were by guns, drums, bells and muskets. There were fifty-six night signals made with lights in different patterns, generally hoisted 'where best seen', guns and false fires, and there were separate night signals for the private ships. There was a table of compass signals and another for horary signals. The day signals included number 78: 'When fetching up with the enemy to leeward and on the contrary tack – to break through their line and endeavour to cut off part of their van or rear.'

The instructions issued with Darby's signal book were much the same as those issued by Geary in May 1780, and also those of unknown date issued later in the year.[36] As in the case of the signal book, the lay-out and printing were much the same as those of Howe's issue of 1776. Nothing in fact in the whole story of tactics and signalling at this time is more remarkable than the way in which Howe stamped his image on the work of all those who favoured his methods. To compare these with the printed issues of Keppel and Admiral Marriott Arbuthnot, even in terms of typography alone, is to realise that one is in a different world.

The Darby-Kempenfelt signal book included the same ten-flag table for numerical signalling, with exactly the same flags, as appeared in the signals of the previous year.[37] It must be asked again whether Kempenfelt really believed in numerical signalling in its true and unadulterated form. If so, was he prevented from introducing it by the opposition of the traditionalists in the service, of whom there always were a goodly number? The difficulty in answering these questions is not due to lack of evidence but rather to its inconclusive character.

Writing to Middleton on 9 March 1781, a month after the issue of Darby's signals, Kempenfelt said, 'Every article or order in a signal book should be numbered; and then, any ship possessed of eleven flags, different the one from the other, may express signals to any number.' Did he mean that the eleventh flag should be a substitute, the intention being to express signals from 100 upwards with a three-flag hoist and so avoid the use of pendants? Even so it would be impossible to signal three repeats, such as 111, 222, etc. The letter continued:

It has been a common saying that it is an advantage to go by signals that we have been used to, and when a new set comes out, to say we have our trade to learn again. This style was very proper with respect to the different signals used by different admirals formerly, when the signals were jumbled together without form or order, and when a long acquaintance with them was necessary to find out the meaning of any signal that was made in the chaotic state in which they were. But when the signals are formed upon a proper plan they require no study to comprehend them, and when a signal is made you can immediately turn to the article or order alluded to.

Had I remained ashore this winter, which my health very

much required, I think I should have been able to have rendered the signals much more perfect and useful by the helps I have received from others and my own observations. The plan I followed in the signals I made was not that I most approved of. That which I would have adopted – though most evidently the best – I could not get any of the admirals or officers of note to approve and countenance. I therefore followed in a great measure Lord Howe's mode, he being a popular character. The night and fog signals we use are almost entirely his, and both extremely defective. I would have used the French night signals, which are by much superior to anything of the kind that has yet appeared, but I was afraid of prejudice, for not an admiral I showed them to but started objections.

A scheme for fog signals has lately fallen into my hands, which is very simple and clear, which when I have leisure, I shall recommend for use.[38]

This is one of the most famous and most puzzling letters ever written about naval signalling. Did Kempenfelt mean that by introducing his tabular system he was copying Howe, and that by making it thereby an alternative to the normal system he was placating the traditionalists? If so, what evidence have we, apart from the manuscript project of 1778, that Howe ever countenanced the tabular system at all? Certainly it never appears in any of his printed signal books, except for the use of private ships. On the contrary, it seems to have been a Kempenfelt speciality, and copied by him from the French. What then was it that the 'admirals and officers of note' refused to countenance? Was it the true numerical system, with ten flags numbered 0 to 9 with pendants for the hundreds?

Some further light is thrown on Kempenfelt's views by an undated translation which he made himself of the whole of La Bourdonnais' system of numerical signals, numerical distance signals and night signals, as printed in Bourdé de Villehuet's *Manoeuvrier*.[39] This translation was sent to Middleton on about 12 March 1781.[40] No comment was added, but assuming this date to be correct, it was immediately followed by a letter of 14 March 1781 describing the night signals actually in use in the French Navy.[41] These were accurately described as being on

the same principle as their day signals, made with two flags, a superior and inferior. The small table, at the upper part of the page to the left, shows the guns and lights used to express the numbers of the flags in the large table: thus, one gun and one light signifies number 1. The flags in the large table are placed to help the conception by showing the affinity of these with the day signals, the only differences being that lights and guns are substituted to represent flags.

Kempenfelt's comment was: 'This method of exhibiting night signals is certainly very simple, easy to make, easy to be understood, and much less liable to be mistaken than any that have been yet in use amongst us.' Careful reference to the

35 *MOD*, NM/101 (formerly *RUSI* 101) printed folio. Another copy in manuscript is NM/85 (formerly *RUSI* 85), which declares them to be 'signals by Rear-Admiral Richard Kempenfelt'

36 *MOD*, NM/101 (formerly *RUSI* 101) printed folio matching the signal book

37 *MOD*, Ec/50, and *NMM*, Tunstall Collection, TUN/8 (formerly S/MS/Am/8)

38 *Barham*, I, pp430–41

39 First published in 1765. De Sauseuil's English translation did not appear until 1788.

40 *Barham*, I, pp343–8. The editor, Sir John Laughton, seems to have had some reason for tentatively assigning this date to the draft.

41 *Barham*, I, pp348–9

Vice-Admiral Sir Hyde Parker (1714–82) (detail), by George Romney. NMM

signals themselves arouses wonder that a man of Kempenfelt's extraordinary intelligence and technical ability could possibly have penned such a eulogy. Not only are these signals complicated in themselves, but they are based on a dead-end system of flag signals. Admittedly, the existing British night signals were inefficient. Nevertheless, they were amenable to improvement on lines far better than anything so far devised by the French. The only criticism Kempenfelt allowed himself was that the system required too many gun signals. On the credit side, it must be admitted that the French system did not require lights to be shown in particular positions but 'where best seen', and this was certainly an advantage.

Colour and size of signal flags and pendants

Kempenfelt had long been interested in the form of signalling flags. In his correspondence with Middleton he had a good deal to say about the most suitable sizes and proportions, and arrangements of their colours. In a letter of 20 March 1780 he suggested that the 'dimensions which seem to be most generally approved of' for flags was 9yd long by 5½yd deep. The next day he suggested that a depth of 5yd would simplify construction, and on 14 October he suggested that flags for repeating frigates should be 5yd by 2. On 19 February 1781 he changed his mind and suggested flags should by 7yd by 5yd, as 'most convenient . . . for large ships'; 6yd by 4 would be enough for frigates. Signal pendants for frigates he thought should be 6ft at the staff, 2ft at the end of the flag, and 45ft in length. On 2 March he added that pendants for flagships should be 8 or 9ft at the staff, and also 45ft long, but yet again changed his mind about flags, which he now said should all be made to a ratio of three to two. Signal pendants should not be swallow-tailed, and should be in the proportion of three at the staff to one at the end of the fly. Triangular flags should be 14½ft at the staff and 20½ft long.

Crosses on flags should be in width one-fifth of the breadth of the flag. Striped flags should have seven stripes with the darker colour on the outside sections. Chequered flags should have four chequers by three, with two light chequers and one dark next the staff. 'Pierced' or 'bordered' flags should have the pierced part in breadth half the total depth of the flag, and the border all the way round one quarter of the depth. Chequered pendants should have five chequers by two. Pendants striped vertically should have five stripes, the darker colour next the staff. Pendants striped lengthwise should have five stripes with the darker colour on the outside sections.[42]

Kempenfelt believed that every flagship and every repeating frigate should have a set of flags consisting of two of each flag and four of each pendant. This represented a considerable amount of cloth, and the expense evidently worried the Navy Board, which submitted 'to their Lordships whether some general mode should not be adopted to confine this Article within some limited Bounds'.[43]

The Battle of the Dogger Bank

It is evident from Vice-Admiral Sir Hyde Parker's Battle of the Dogger Bank, from the events of Kempenfelt's own engagement with de Guichen in December, and from Howe's relief of Gibraltar in 1782, that the new tactical ideas in the

fleet had very little effect at first on the conduct of battles. Very early on 5 August 1781, Parker, escorting a convoy bound for England out of the Baltic, encountered Rear-Admiral Johan Arnold Zoutman escorting a Dutch convoy intending to enter the Baltic. Parker ordered his convoy to make sail to the west for England, but Zoutman, keeping the Dutch convoy together on his lee side, drew his escorting force together into line of battle. With a north-easterly breeze, Zoutman formed his line under topsails and foresails at a cable's distance a point off the wind on the port tack. Parker was to windward and at once signalled general chase.

Neither squadron could be described as anything but second class. Parker had five of the line, a 50 and a 40. Two of his line of battle ships were good, but the rest of his squadron were old and crank. Zoutman also had seven fighting ships, three of the line, three 54s and a 40, a poor force and no doubt suffering from lack of fighting experience over a period of nearly seventy years. He was supported, however, by his five frigates, which, keeping very slightly to leeward, were ready to fire through the gaps in his line of battle.

Parker's general chase under a press of sail soon had his ill-balanced force strung out in order of speed, which led him a little after 6am to signal for the line abreast. Though the leading ships took in sail to let the rest come up, the squadron as a whole continued to press forward before the wind, Parker being anxious to get to close quarters as quickly as possible. Even so, with his ships in line-abreast at a cable's distance, he presented the ideal target, each ship steering for the enemy and bow to broadside. Nevertheless Zoutman held his fire until 7.55am allowing the British to turn into line ahead parallel to him on the port tack and open fire at what was said to be 'half musket-shot'. If this was an example of old-fashioned chivalry it certainly cost Holland the victory, as otherwise the British ships would have been severely damaged before being able to reply. Following tradition, Parker insisted on bringing his flagship, stationed fourth in the line, against Zoutman, who was stationed fifth. With only seven ships apiece, numbers two and three in the British line were opposed to numbers two, three and four in the Dutch, while the rearmost British ships had nothing to fire at.

Parker realising the unsatisfactory nature of his deployment, soon hauled down the signal for the line and hoisted the signal for close action; all ships then became heavily engaged. At 10.45am he rehoisted the signal for the line, and at 11 he carried his three stern ships with him to leeward,

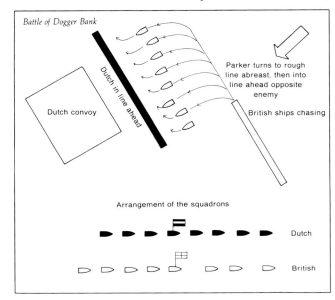

Battle of Dogger Bank

Dutch convoy

Dutch in line ahead

Parker turns to rough line abreast, then into line ahead opposite enemy

British ships chasing

Arrangement of the squadrons

Dutch

British

42 *Barham*, I, pp313, 315, 334, 337, 338

43 Navy Board Out Letters to Admiralty, 1 February 1781, *NMM*, ADM/B [?]

between his next ahead and the Dutch. At about 11.45 the action drew to a close, with both sides, somewhat unusually, severely damaged in their hulls. Neither was capable of renewing the battle and one of the Dutch ships sank the same night. Both convoys escaped entirely unscathed but, whereas the British reached their destination, the Dutch convoy retreated to the Texel.

Kempenfelt and de Guichen

Kempenfelt was instructed to intercept a French convoy of some 200 ships carrying troops and stores, partly for the West Indies and partly for the East, and escorted by nineteen of the line under de Guichen. He was given only twelve of the line, detached from the Channel Fleet, a force far too weak for such a task. When he sighted the French at daybreak on 12 December 1781, fifteen of their line of battle escort were to leeward, together with part of the convoy, steering west, with the wind at south-east, while the rest of the convoy, with only four of the line in company, were to windward roughly on the same course.

At the start Kempenfelt was to the north and east of the French, but by carrying a press of sail and steering south-west he managed to pass between the windward section of the convoy and the main French fleet. Three out of the four line of battle ships with the windward part of the convoy managed to slip past. The fourth, an 80 and the Marquis de Vendreuil's flagship, escaped only with damage. This very short encounter took place at 12.45pm, after which the British force turned to capturing the transports and merchantmen of the windward group. Two or three were sunk and fourteen captured. Others were struck but could not be secured. All this time, de Guichen was to leeward and though he formed line he was never able to approach nearer to Kempenfelt than five miles. Nor did he make any serious attempt to work to windward on the starboard tack until it was already too late.

Kempenfelt's signals on this occasion show a sense both of urgency and orthodoxy. At 9.42am he signalled for the line-ahead at two cables and to call in all detached ships.[44] Kempenfelt noted,

Had I waited for the slow sailing ships [to take their stations in the line] it would have been dark before we could have got near the enemy, and the convoy would have all escaped. . . . Had I that day known the superior strength of their squadron, and that it was such that I could not in prudence risk an action with, I should have made a general chase of it, and in that case very few of their transports would have escaped; for had any number of our ships been half an hour's sail ahead of us – and many would have been more than that, had the signal for the two-decked ships [to chase] continued out – they would have got in time between the convoy and the ships of war to have intercepted them.[45]

Howe's relief of Gibraltar

Though unimportant in the history of tactics, Howe's relief of Gibraltar in 1782 was a great convoy operation, since the Combined Fleet avoided action even though Howe was embarrassed by the presence of a very large convoy. Three events, however, are worth noting. It was on the way out,

Howe's Order of Sailing

according to Sir Charles Knowles, that Admiral Barrington persuaded Howe to adopt the Order of Sailing in six columns (six ships in each) instead of in three columsn. 'From the sailing in columns and the introduction and use of the signals by numbers', Knowles commented, 'may be dated the era of the improvement in the navy'.[46]

It was also on the voyage out, according to Barrow, that Howe called Barrington and the captains of his squadron on board and asked them to give their opinion in turn, beginning with most junior (as in a court-martial) on the merit of trying to force a night action. Each was in favour of this, supposing it to be the approved line, until Jervis's turn came. A night action, he said, invited confusion and firing into one's own ships, and 'would deprive the British fleet of the advantage of making use of his Lordship's admirable code of day signals, while those for the night were very imperfect'.[47] Barrington agreed with Jervis.

Despite his overwhelming success, Howe was not satisfied with the conduct of all his captains and flag officers. Station-keeping was poor, especially when the ships were caught in the current at the mouth of the Straits, while Rear-Admiral Sir Richard Hughes mistook his station and drew four other ships of his division off their proper course.[48] There was even difficulty in forming line 'advantageously' in the open ocean. Hughes was again at fault, and when he signalled a ship of his division to leave her proper station it was 'properly disregarded by her captain'.[49] Evidently there was still a considerable gap between theory and practice.

Kempenfelt's final signal books

So active was Kempenfelt during the year 1782, until his death in the *Royal George* on 29 August, that it is difficult to place his work in proper perspective. By now he seems to have become convinced that the best immediate course for the Royal Navy was to copy the French system of signals. On 2 February 1782 Sir Richard Bickerton, of the *Gibraltar* at Spithead, issued a printed signal book and instructions to Captain Lord Longford. Any doubts as to whether these were the work of Kempenfelt is dispelled by the evidence of later issues authoritatively attributed to him. Thomas Graves made a careful copy of the instructions, and added to the printed label a note that they had been issued by Kempenfelt.[50] Since the signal book and the book of instruc-

44 'Copy of the minutes of signals taken by Mr Davis Atkins, Acting Lieutenant on board His Majesty's Ship *Edgar* on 12th & 13th Dec 1781, in sight of a French fleet and numerous convoy'; correspondence of Admiral John Elliot, *NMM*, ELL/400 no 58.

45 *Barham*, I, p357

46 *Observations on Naval Tactics*, p25

47 Barrow, pp145-6

48 Not to be confused with Sir Edward Hughes, commander in chief in the East Indies, who fought five battles with Suffren.

49 Barrow, pp154-5

50 *MOD*, NM/53 (formerly *RUSI* 53). Graves's copy is also interesting as it contains a loosely inserted reading list with the following titles: *Manoeuvrier* [Bourné]; Bouguer; Fournier; Hoste; Morogues; *Naval Evolutions & Naval Tactics* (Steel) *Practical Seamship* (Hutchingson).

tions issued by Bickerton form a pair, Graves's attribution of the latter to Kempenfelt must apply to the signal book as well. A complete copy of both the signal book and the instructions was made by Admiral William Page in 1784 from a printed copy lent him by Captain (later Admiral) Peter Rainier. Page stated that these books were issued by Sir Richard Bickerton to the squadron to take out to India in February 1782, and that they were 'generally said, and no one doubted, to be the compilation and much the invention of the gallant and most able Admiral Richard Kempenfelt'.[51]

Kempenfelt's 1782 book was an exact copy of the signal system used by de Grasse, except that it had only sixteen flags instead of eighteen. A two-flag hoist was employed, and reference was by means of a book with the top flag of the hoist indicating the page, and the second indicating which of the articles on the page was intended. All the flags were the same as those used by Kempenfelt in 1779; none was numbered. To simplify using the book for sending signals, they were grouped according to subject. Howe's and Kempenfelt's new signals were all included, as can be seen in the summary below:

Summary of Kempenfelt and Howe's additional signals, included in Kempenfelt's signal book of 1782

No 218	Attack enemy's rear in succession. (Howe)
No 222	Van to double enemy's van. (1779)
No 223	Van to double enemy's rear. (1779)
No 224	Rear to double enemy's rear. (1779)
No 225	Particular ships to attack enemy ships seeming to be kept in a separate body. (Howe)
Nos 226–30	Engage enemy's centre, van or rear
No 233	Headmost or sternmost ships to signify when they can weather the enemy 'or if to windward of the enemy or on the contrary tack, for the sternmost ship to signify when she is far enough astern of their rear to be able to lead down out of their line of fire'.
No 234	Attack from astern to windward, the headmost ship to engage the sternmost enemy and each ship in succession to pass ahead and to windward under cover of the next ahead until the sternmost ship is engaging the enemy's headmost. (Howe's sixth additional, 1777).
No 235	'When fetching up with the enemy to leeward, and on the contrary tack, to break through their line and endeavour to cut off part of their van or rear.'
No 236	Ships to steer independently of each other to engage respectively ships opposed to them in the enemy line. (Howe)
No 243	'For the van ship to put upon the same tack as the enemy and engage his rear to windward, the rest to do the same as they arrive in his wake.'
No 244	Ditto, and engage his rear to leeward.
No 245	Van ship to steer so as to pass close to the enemy's sternmost ship.
No 246	Ditto, enemy's leading ship.
No 247	'For the van ship to keep no nearer than is requisite, to prevent the enemy's passing to windward of our fleet, and to be at the same

time within engaging distance, in case the fleet should come to action on contrary tacks.'

Kempenfelt's instructions provided four orders of sailing, assuming a fleet of fifteen ships in three squadrons. Order number 1 put the second squadron in the van, and the third in the rear. Order number 2 reversed the sequence of squadrons, but kept the squadrons themselves in the normal order. Order number 3 had the squadrons in normal order, but with each squadron in reverse order. Order number 4 was for both the fleet and each component squadron to be in reverse order. When the squadrons were directed to sail in two lines, the first half of each squadron was to form the weather or starboard line, and the second half the lee line. Five cables, or a good half mile, were to be kept between the lines, and two cables between the ships.

Article 9 instructed that

By day, when the fleet is sailing by the wind, a ship of the line from the centre squadron is to keep two points on the Admiral's weather bow, one from the leeward squadron two points on his lee bow, and one from the weather squadron directly to windward, or two points before the Admiral's beam. They are to keep six miles distant from the Admiral, but in thick weather to draw closer.

When sailing large, or before the wind, the ship from the centre squadron is to keep right ahead of the Admiral, that from the starboard squadron two points out to starboard of the fleet, and that from the larboard squadron two points out to port.

In the night each ship is to keep ahead of her squadron about one mile distant.

They are to examine such ships as may pass near them.

The ships are to take this duty in rotation (the three-decked ships excepted) beginning with the senior captain of the squadron.

Kempenfelt's Fleet Arrangement

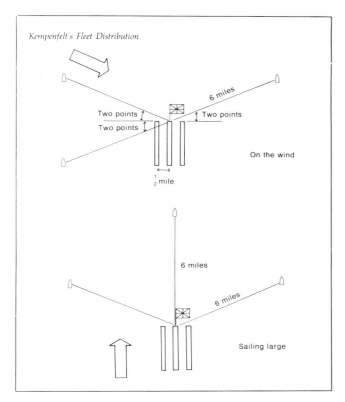

Kempenfelt's Fleet Distribution

The relief is to be in the evening one hour before sunset.

The fighting instructions, firmly based on early Howe, have only three important additions. The meaning of Articles 3 and 12 seems to be that, if a battle looked like being a close thing, the admiral should not necessarily try to weather the enemy. Instead he should be content, at any rate at the start, to engage them from leeward as the two fleets passed on opposite tacks. Article 13 also provided for engagement on opposite tacks:

When weathering the enemy on the contrary tack and signal is made to engage their van, the leading ship is then to bear down to the van ship of the enemy and engage, passing along their line to windward to the sternmost ship of their van squadron; then to haul off close to the wind, the rest of the fleet doing the same in succession.

The object of this last seems to have been to avoid losing contact with the enemy by passing too far astern of them. By hauling off close to the wind, the weather gage would be retained, and a second attack facilitated.

The difficulty in determining the dates of various changes in Kempenfelt's book is exemplified by a printed signal book signed 'John L Ross, *Ocean* at Sea, 2 June, 1782', and issued to Lord Longford as captain of the *Alexander*.[52] This book was clearly issued by Rear-Admiral Sir John Lockhart Ross Bt, who had commanded the Fifth division of the Channel Fleet under Hardy in 1779 and was now in command of one of its squadrons. Nevertheless the signal book is practically a replica of that issued over Geary's signature on 27 May 1780, and it has the same blank between the ruled lines at the top of the first page.[53] Geary's book, however, shows red-white-red and the blue-yellow-blue flags in horizontal stripes, while the Ross book shows them with vertical stripes. In this respect the Ross book is similar to the signal book produced by Kempenfelt at some time during 1780, but of which only manuscript copies survive.[54]

This, the very last Kempenfelt signal book, completely abandoned the French system, and reverted to the standard system favoured by Howe and used earlier by Kempenfelt

himself. It was a printed folio in the style introduced by Howe, thumb-indexed on his standard plan, that is with twenty-seven single flags flown in particular positions.[55] The signals can be described as strictly operational, including signals for use at anchor. There was a minimum of administrative signals, compared with those generally included by the French.

After his successful convoy action against de Guichen, Kempenfelt could expect to command a squadron of the Channel Fleet in the campaigning season of 1782. The position of Lord North's Ministry was becoming insecure, however, and following the loss of Minorca in February, the ministers resigned on 20 March. As a result, Keppel was ennobled and succeeded Lord Sandwich as First Lord. Darby, as a member of the Board, resigned command of the Channel Fleet. He was succeeded by Lord Howe, with Kempenfelt as second-in-command, and the latter transferred his flag to the ill-fated *Royal George*. Howe's orders to Kempenfelt of 30 April stated: 'while you remain on this service you will establish such signals and instructions for the government of the ships under your orders as you may think fit'.[56] Kempenfelt worked through the winter months, and on 4 May, the day after leaving Spithead, he wrote to the Secretary of the Admiralty that he had been 'constantly using all possible despatch for perfecting the necessary signals and instructions for the conduct of the squadron to be in readiness for putting to sea'.[57]

Whether he was referring to his own detached squadron or to the fleet as whole is uncertain. When the fleet was cruising earlier with Barrington, before Howe took up his command, its total force was only twelve of the line and it might well have been deemed a squadron. Anyhow, there is no doubt that Barrington issued Captain George Keppel of the *Fortitude*, one of his twelve ships, with yet another printed signal book drawn up by Kempenfelt.[58] The fact that Barrington's signature is undated raises the possibility that Captain Keppel already had the book, and that it was confirmed by Barrington on taking over the command. In addition, Barrington's name is erased, and 'R Kempenfelt' in his own hand written beneath. Some unimportant instructions issued and signed by Kempenfelt and dated 6 May 1782 are loosely inserted. This shows that, regardless of the date of its original printing and possible issue, it was used by Kempenfelt himself during his last cruise.

There was a flag at the top of each page of the day signals, two flags being painted on each thumbing for the respective double pages when open. Combinations of red and white, and blue and yellow, predominated. The chequered red and white, and blue and yellow, flags were retained. The red-white-red and the blue-yellow-blue flags had vertical stripes. The day signals were numbered right through consecutively from 46 to 412 and there were seventy-nine blanks. Although there was some attempt to keep similar types of signals together in small groups, the battle signals were widely dispersed, presumably to avoid confusion. They

52 *NMM*, PAK/4 (formerly Sp/137)

53 *NMM*, DUN/14 (formerly S(D) 14)

54 *MOD*, Ec/50, and *NMM*, Tunstall Collection, TUN/8 (formerly S/MS/Am/S)

55 *NMM*, SIG/A/14 (formerly Sp/91) is an unsigned, undated and apparently unissued signal book. It has flags exactly as above, though the signals are those of Geary's book of May 1780 (*NMM*, DUN/14 (formerly S(D) 14).

56 Quoted from *Signals & Instructions*, p43

57 Quoted from *Signals & Instructions*, p44

58 *NMM*, Corbett Collection, TUN/42 (formerly S/P/Am/7)

contained nothing new, being a repetition of Howe's of 1776 with such subsequent additions as have already been noted. They had different numbers from those used in the Bickerton book and the Page transcript.

The orders of sailing included: '105, Each squadron in two lines. Ships' distances from each other two cables.'; '103, Each squadron in one line, and the admirals commanding the squadrons to lead. Ships' distances one cable to one cable and half'. Number 104 was the same as 103, save the admirals were to assume their stations in the order of battle, and 105 was a variant of 104, the squadrons being 'ranged to windward and ahead of each other, the weathermost to lead. The sternmost ship of the two weather squadrons to be abreast of ship next ahead of the Admiral in the squadron immediately to leeward of them. Ships' distances one cable to one cable and a half.'

The familiar ten-flag numerical table was included with the same flags as before, including the red pierced white. The explanation given for its use was that,

It being under certain circumstances very necessary that the fleet be governed by signals, so framed as to convey the same signification from whatever part of the ship they may be shown, the following method will be occasionally adopted, whereby is expressed the number of the signal meant to be carried into execution.

It is possible that Kempenfelt was seeking an excuse for introducing quasi-numerical signalling to a skeptical fleet, but more likely that he had not, in fact, progressed beyond recognising that the numerical system could be useful if the flagship had suffered extensive damage to its masts, or was attempting to send signals to ships at very great distances.

In taking leave of Kempenfelt, we should note his boundless energy and his loyal subordination to men higher in rank though much younger than himself. He played an important role in the tactical reform of the Royal Navy. He had exploited Howe's tactical ideas, introduced useful ideas from the French, and provided many new sailing formations for a large fleet. In particular, he established the importance of the windward position in sailing, regardless of the traditional precedence of the respective squadrons. Evidence for the continued popularity of Kempenfelt's French version is found in a copy of the signal book and the instructions printed by Douglas & Aikman, Kingston (Jamaica). They are signed by Rear-Admiral Joshua Rowley, on board the *Monarch* at Port Royal, Jamaica, and issued to Captain Hotchkys of the *Shrewsbury*, 20 January 1783.[59] Nevertheless, it is difficult to regard Kempenfelt as a suppressed genius or as in any way the superior of Lord Howe.

Minor signals projects

An anonymous and unfinished manuscript experimental signal book of the Kempenfelt period makes use of a table of sixteen flags for all purposes. In table number 1 it supplies 256 signals. In table number 2 the same flags supply signals 257 to 512, with the help of pendants. The same sixteen flags were used for a special rendezvous table and again for a parole table.

The night signals are stated to be those 'of the late Admiral Kempenfelt', with an additional light. The directions for these as well as for the fog signals are quite unnecessarily obscure and long-winded. As neither the signals nor the index has been completed, it is impossible to judge of the character and quality of the signalling and tactical ideas. At the end, the anonymous author provided a set of eight flags, each of which could be used in reverse, to make sixteen substitutes for his original choices, 'as some officers seem to consider the number of flags in my plan as an insurmountable objection'. The objection seems unreasonable, as the flags, though doubtless too many, are well chosen. The substitutes, on the other hand, which include several thin rectangular stripes, appear to be entirely unpractical. The book is inscribed on an early blank leaf 'Cha[rles] Hen[ry] Lane' and at the end in pencil, 'Cop[ie]d b[y] Lt Daniell'.[60]

In 1787 Robert Liddel, a purser in the Royal Navy, published a guide to the construction by captains and officers of signal books for their own use, *The Seaman's New Vade-Mecum: containing a Practical Essay on Naval Book-Keeping*. The format he devised showed all the flags and pendants in 'the view'. He provided illustrative pages with coloured flags both for the old system and for a numerical code of his own devising, and he provided a tabular form for the latter employing a 'true' table, which was laid out so that, for example, the flag 6 over the flag 9 actually indicated the number 69.[61]

Captain Thomas West's *Naval Signals Constructed on a New Plan*, published in 1788, is one of the better quasi-numerical systems. Ten flags, 0 to 9, were used according to the French system to indicate a signal by referring first to the appropriate page in the signal book, and then to the article on the page. Four pendants could also be used to indicate one of four possible applications of the signal. For example, flag number 2 followed by flag number 3 indicated page 2, Article 3, which called for the fleet to form in line abreast. With a red pendant added, the meaning was changed to an order to the second grand division to form in line abreast on the starboard side. Book I, containing only ten pages, covered sailing formations. A preliminary signal could be given which would indicate that Book II, which contained general and administrative signals and a few battle signals, should be used. Other books were projected but not shown. Although the four-pendants scheme for 'particulars' is ingenious, the system was not a true numerical one and the flag designs were bizarre and inappropriate.[62]

Signals for Ships of war, compiled and arranged by William Goddard, late Secretary to Sir Thomas Palsey, Bart was published in 1794. The admiral's and private ships' signals were put together, forming a total of 499, arranged according to thumb-indexed subject headings. It contained nothing new and no flags were shown. A new arrangement was made in 1807, providing for 1999 signals and a numbered list of ships. The numerical flags were those of 1804 with a pendant to add 1000 to the number for any signal.[63]

59 *NMM*, DUC/48 (formerly Sp/10) and DUC/48 (formerly Sp/68) respectively

60 *NMM*, Corbett Collection, TUN/38 (formerly S/MS/Am/4)

61 *NMM*, Tunstall Collection, TUN/39 (formerly S/P/Am/5)

62 According to Tunstall's notes this book was once in the Mead Collection.

63 *MOD*, EC/107, undated, and Ec/49

CHAPTER SEVEN:

Tactics in the American Revolutionary War: Progress under the Old System

AMONGST Sir Julian Corbett's notable discoveries was that there was a dichotomy of thought between those indoctrinated with Kempenfelt's Channel Fleet reforms (themselves influenced by Howe) and those attempting to make piecemeal progress by elaborating still further the new 'Signals & Instructions in Addition to the General Printed Sailing & Fighting Instructions'. This dichotomy may possibly have been less clearly defined than it seems today, but there are indications that a real difference in tactical outlook existed. For a time it had seemed that Howe would create a new system on the North American station but, after his departure, tactics both there and in the West Indies moved along traditional lines. The result was little more than expansion and tinkering with the existing system. The reforms were generated in the Channel Fleet and only propagated when Channel Fleet officers were appointed to the North American and West Indian stations. Here they found natural conservatism, reinforced no doubt by station tradition and possibly by a genuine failure to understand what was going on in home waters. Rodney remained unreceptive to the end, even though he saw the new material, and spent leave in England between his two West Indian campaigns.

It would seem that both Rodney and Admiral Sir Samuel Hood, though not exactly contemptuous of Channel Fleet 'evolutions', had their own ideas of how to defeat the French. These ideas were based more on a well-formed line of battle and instant obedience to signals than on new developments in tactical approach. Unlike the post-Trafalgar period, reforms in tactics and signalling during the American War tended to go hand in hand. Those who tended to reject the new ideas about how to approach and attack the enemy also tended to ignore the need for a reformed signal book with an independent status. It was Byron, Barrington, Rodney, Hood, Arbuthnot and Hughes (still ignorant of the reforms) who did most of the fighting, rather than the Channel Fleet men. Kempenfelt's success against de Guichen may have been a tactical masterpiece in miniature, and Howe's relief of Gibraltar a tactical masterpiece on a very grand scale, but in each case they were denied success in actual combat with an enemy fleet. Meanwhile, Rear-Admiral Thomas Graves, the only Channel Fleet admiral to command in chief in American waters, lost a golden opportunity and suffered a severe tactical rebuff. In the end it was Rodney, the most inflexible and unreceptive of contemporary tacticians, who won the only substantial British victory of the whole war. In the history of tactics, however, the Battle of the Saints is something of a blind alley, except for its successful demonstration of Sir Charles Douglas's gunnery reforms. In the East Indies Sir Edward Hughes, a man with ideas somewhat similar to those of Rodney and Hood, laboured heroically and with some success against Suffren, reputedly a very great tactician, without having even the chance to scan a Channel Fleet signal book.

Even on the North American station, however there is some evidence that tactics had become a subject hotly discussed at wardroom or perhaps even gunroom level. The copy of the newly printed 'Admiralty Signals and Instructions in Addition' issued by Vice-Admiral the Hon John Byron on 3 March 1779 contains comments on tactical problems by Lieutenant Thomas Graves;[1] one is of special interest. At the very end of the fighting instructions Graves noted in pencil: 'Is it not wonderful that there are no signals for breaking the enemy's line which may be advantageously done by the fleet to windward bearing down on the rear of the enemy, or when on different tacks to cut off the rear of the enemy to leeward.'[2] Although the irony of this criticism has long been apparent, its tactical significance has scarcely been appreciated. Here was Thomas Graves, then aged thirty-two or thereabouts and still only a lieutenant, able to discourse for his own personal satisfaction, not only on 'breaking the enemy's line' but on how it should be done. Hitherto, so far as we know, this manoeuvre had been regarded mainly in a negative sense, as a penalty arising from an admiral's default or clumsiness, or as something consciously attempted by the enemy. In either case it was to be vigorously prevented. Yet here was Graves writing about it as something commonly assumed to be within a British admiral's competence, and as a sound method of attacking the enemy.

The sailing and fighting instructions issued by Byron on 25 October 1778, on taking command of the North American station, were in the same traditional thirty-four-page printed format with the additional chasing signals at the end, exactly as had been used since 1703.[3] The continued employment of the old book made it inevitable that only a very few new signals could be incorporated. Inside the cover of what appears to have been Thomas Graves's copy of the Signals & Instructions in Addition, issued by Marriot Arbuthnot before leaving England to take up command, is the following sagacious warning:

NB. Any signals which the Admiral may order to be reprinted in future, must be printed with exactly the same number of pages as are in this book and each page must contain the same number of articles – otherwise all the [private] signal books made for them will be useless.[4]

To give any continuity value to these private books it was essential that the official book, whenever reprinted, should retain the same pagination, additions being fitted in at the end of the relevant sections, provided there was sufficient room

---

1 *MOD*, NM/92, (formerly *RUSI* 92); an MS copy, together with MS signals in a separate volume, is in NM/87 (formerly *RUSI* 87)

2 All Graves's pencil notes are printed by Corbett in *Signals & Instructions*, pp215–29

3 *MOD*, NM/92 (formerly *RUSI* 92). Byron's signals, issued separately, are in NM/105a (formerly *RUSI* 105a)

4 2 April, 1779; *MOD*, NM/92 (formerly *RUSI* 92). The publication included three additional articles dealing with bow and quarter lines. After taking command of the North American station in August 1779, he made a number of additions to the new 'Admiralty Signals & Instructions in Addition . . .'

on the page. If there was not enough room, they were printed at the very end of the book regardless of their context.

On 25 August 1779, on arriving at New York, Arbuthnot issued the fleet with a printed book of his own devising, but probably arranged by his secretary, William Green. Although it had the same title as the new 'Admiralty Signals and Instructions in Addition to the General Printed Sailing and Fighting Instructions', it was in fact an entirely different work, printed in official folio form.[5] Ostensibly it was a printed signal book in the old form, each page being divided into three columns for flag, place flown and signification. Under signification, however, the relevant article was printed in full, while the order of the book was governed by the order of the articles and not of the flags. Its main value was for quick reference. Corbett described the contents of the book as 'distinctly reactionary' so far as the fighting instructions are concerned, on the grounds that some of the best articles in the Admiralty Signals and Instructions in Addition were left out.[6] These included Articles 6 to 11 for altering course and for various forms of bow and quarter line, and Article 21 to lead down against opposite opponents and 22 to engage the enemy in reverse order from the rear. On the other hand, Arbuthnot's book included four articles of long standing which had not previously found their way into the Admiralty book.

Sometime in 1781 Arbuthnot issued a printed folio of Additional Signals for the North American Station.[7] This took the form of a three-column page, headed signification, flags and place flown. There were forty-nine articles, including number 24, to engage at night, and number 25, to engage the enemy, if encountered at night, as best able and 'in the closest manner'.

The retirement of Byron from the West Indies command in 1779, and of Arbuthnot from North America in 1781, marks the end of the last stage but one in the attempt to develop tactics and signalling by continuous additions to the permanent Sailing & Fighting Instructions. Rodney issued no signal book at all in 1781. His so-called signals were no more than private manuscript books of the signals made by individual officers to give effect to his general corpus of instructions,

old, new, and personal-additional. These books, of which a number survive, were made, just as in the past, for the use of flag officers and captains, and were, as before, ruled, written and painted by their secretaries and other skilled assistants. At the Battle of the Saints in 1782 Rodney's fleet possessed no official signal book as such. On 18 December 1779, however, before sailing from England for his first West Indian campaign, Rodney had issued those ships accompanying him with his own Signals and Instructions in Addition. On 29 March 1780, two days after reaching Gros Islet Bay,

5 *MOD*, NM/92 (formerly *RUSI* 92), contains one copy of the Admiralty book and two of his own.

6 Corbett was wrong in stating that Article 15 of the Admiralty book was omitted. It appears as Arbuthnot's Article 6. On 17 June 1780, while at Staten Island, Arbuthnot issued three printed additional battle signals for the fleet to form in bow and quarter line, somewhat similar to the instructions previously left out. On 19 August 1780, while watching the French squadron in Newport, he issued a long, detailed and printed but unnumbered 'additional instruction for extending the line of search,' evidently to ensure contact with the French if they tried to slip out. Both sets are in *MOD*, NM/92 (formerly *RUSI* 92).

7 *MOD*, NM/83 (formerly *RUSI* 83).

8 *PRO*, Rodney Papers 15 and 19, and *MOD*, NM/83 (formerly *RUSI* 83), printed in *Signals & Instructions*, pp180–234

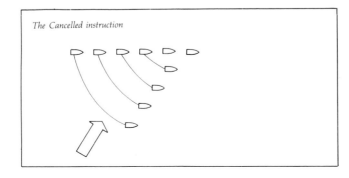

The Cancelled instruction

St Lucia, he issued another slightly different set.[8] Both sets had very considerable omissions from, and additions to, the Admiralty version, as well as several minor alterations as regards flags used. Rodney's fifteen additional sailing instructions were largely miscellaneous, including useful and semi-traditional items omitted from the Admiralty book. The most interesting is number 49, for when the admiral 'would have all the ships sheathed with copper chase on any point or quarter of the compass'. Experiments with copper sheathing had been going on for many years and during the course of

the American War a great number both of new and existing ships were so fitted. This was one of the very tangible achievements of the supposedly incompetent Board of Admiralty headed by Lord Sandwich. The French were doing the same thing, and only the Spaniards lagged behind.

Rodney also made some material changes in the fighting instructions section of the Sailing and Fighting Instructions in Addition. He cancelled Articles 3, 9, 10 and 11, dealing with the bow and quarter line. Article 11 was for forming the line in inverted order by 'the second ship passing the

The Moonlight Battle; the Battle of Cape St Vincent, 16 January 1780, by Francis Holman (see pp 164–5). NMM

headmost, and so on in regular succession until the sternmost becomes the leader', with the squadron in bow and quarter line. Graves's comment had been that it would only be satisfactory for ships to form on each other's weather quarter – which it appeared was the intention of the article – if the enemy were coming up with them to leeward. 'But if the enemy is coming up with them to windward they should be formed on the lee quarter of their leader.'[9] No doubt Rodney had himself noted the lack of clarity of this instruction as it stood. He also cancelled Article 26, which directed that, when a fleet were superior to the enemy, surplus ships should quit the line without being signalled, and rake the enemy's van or rear.[10] He probably had two reasons for making this latter deletion. It was against his principles to concede any initiative to his captains, and the instruction, though of some ancestry, was difficult to execute and had never produced much in the way of practical results.

Rodney's four additional fighting instructions contain one quite original article, number 29 or 71 (according to the continuous or non-continuous numbering of various copies), for 'all the three-decked and heavy ships [to] draw out of their places in the line of battle and form in the van [or rear] of the fleet'. Corbett divined in this article both the direct influence of Morogues and the idea later adopted by Nelson at Trafalgar for the three-deckers to lead. Morogues, however, never went beyond stressing the tactical value of heavy ships compared to two-deckers, a mere emphasis on the obvious. Nor did he provide any signal for their special employment. Rodney never made use of his own signal in any of his full or partial engagements, perhaps because he never had as many as five 'heavy ships' of 90 guns or over at his disposal except at the Battle of the Saints. At the Saints he adopted the traditional plan of having one for each of the three flagships and placing the remainder one ahead and one astern of his own flagship in the traditional position of the commander-in-chief's heavyweight seconds, thus giving additional strength to the centre. If flag officers were not to be deprived of their traditional claim on heavy ships, and there were only a tiny sprinkling of such ships, how could such ships be moved ahead or astern of the fleet without depriving the flag officers of control of their respective squadrons? Possibly, like Morogues, Rodney was thinking in terms of a possible future fleet including enough three-decked private ships to allow a much freer disposal of force. Even Howe's exceptionally powerful fleet of thirty-five of the line at the First of June included only seven three-deckers,

of which six were flagships whose admirals carried divisional responsibilities. Though apparently standing for everything retrogressive in tactics and signalling, Rodney voiced the demands of all his great successors when he wrote: 'On, my Lord, the three-decked ships are what must maintain the sovereignty of the sea; nothing can withstand them . . . I had rather have ten three-decked ships than eighteen two-deckers, and would answer with my head for their defeat.'[11]

Rodney's articles, variously numbered in different versions as 73 and 74, 31 and 32 with sub-division 33, or 26 and 27 with sub-division 28, which began, 'When the Commander-in-Chief means to make an attack upon the enemy's centre', or van, or rear, have become famous through an abortive application at the Battle of Martinique. This was clearly a signal of general intention and Rodney deliberately refused to commit himself on the means of execution. An attack on the enemy's centre might require a very different approach from an attack on the enemy's rear, quite apart from the state of the wind and weather and the relative disposition of the two fleets, especially if the fleets were on opposite courses. To have issued more elaborate instructions and signals would have been contrary to his general attitude, since any attempted sophistication of tactical approach would have necessitated at least giving his squadron commanders and possibly even his captains that kind of individual discretion which Howe favoured, but which he himself strongly opposed.

In November 1781, in temporary command of the British fleet in the West Indies, Hood issued a set of signals rather less extensive than those used by Rodney either in 1779–80 or in 1782, and with only thirty-seven flags in place of forty.[12] On 23 January 1782 he issued instructions with signals for his abortive attack on de Grasse's fleet at anchor in Basseterre Road, St Kitts. His plan was to attack in succession from windward, each ship engaging the first three ships only of the enemy; then to veer short round and fall onto the rear in succession.[13]

On 19 February 1782 Rodney arrived at Barbados to resume command of the West Indies station and the same day issued a corpus of printed instructions even more elaborate than Arbuthnot's.[14] Typical is his alteration of Article 27 of the 1703 Sailing & Fighting Instructions making it absolutely clear that 'no ship, or any particular part of the Fleet whatsoever, shall presume to tack, or wear, whilst in action, unless by signals, or order from the Admiral so to do'.[15] Article 31, for forming the line, which had read, . . . 'and if he would have those lead who are to lead with their starboard tacks aboard' or 'those lead who are to lead with their larboard tack aboard', were altered by Rodney in handwriting to: 'and would have him lead who commands in the second post', thus shifting the emphasis of the signal from ships to flag officers. Corbett links this change with the third charge against Keppel of having failed to make the signal in its old form.[16]

Article 32, the last, was given a long addition in handwriting to the effect that signals made to the fleet when in line of battle and duly repeated and acknowledged, should remain in force even though hauled down, until actually annulled 'by the annulling flag, or until it be naturally superseded by the purport of some subsequent signal'. Nevertheless, it was added, signals for the line-ahead or abreast or engaging would be kept flying as long as the commander-in-chief judged proper for the fleet to continue in line or engaging. Corbett saw in this an attempt to prevent a repetition of the misunderstanding about Rodney's intention to attack the enemy's rear at the Battle of Martinique.[17]

Rodney also issued in 1782 a new printed version of the Admiralty Sailing and Fighting Instructions in Addition.[18] The first twenty-three of the thirty-one printed fighting

9 *MOD*, NM/83 (formerly *RUSI* 83); *Signals & Instructions*, p220

10 *Ibid*, pp227–8

11 Rodney to Sandwich, 25 April 1782, *Sandwich*, IV, pp264–5

12 Manuscript copy of Hood's signals issued to Captain Cornwallis of the *Canada*; *NMM*, Tunstall Collection, TUN/10 (formerly S/MS/Am/10). Loosely inserted is a sheet of additional signals for landing and re-embarking the marines, seamen and the 69th Regiment used in the abortive attempt to relieve the British garrison of St Kitts.

13 *MOD*, NM/24 (formerly *RUSI* 24), Hood's signals issued to Captain Graves of the *Bedford*, with various additions, 12 January 1782.

14 *MOD*, NM/94 (formerly *RUSI* 94)

15 *Signals & Instructions*, pp274–5; from the copy in the Library of the United States Navy Department, Washington

16 *Signals & Instructions*, p279

17 *Signals & Instructions*, p280. Rodney made some useful additions to the original night sailing signals of the parent work: 5 and 6 for chasing NE, NW, SE, and SW and giving over the chase; 7 'If in pursuit of a number of the Enemy's ships, and the headmost brings to any of them, and passes on without taking possession of them;' signal follows; on which chasing ships further astern will keep a careful look-out and take possession of them. He also issued eight extra fog signals. See the copy in the US Navy Department Library but not quoted by Corbett.

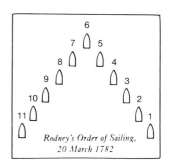

*Rodney's Order of Sailing,
20 March 1782*

instructions were the same as the first twenty-eight of the original Admiralty issue, less the five which Rodney had cancelled in 1779. Articles 24 to 28 contained his handwritten additions of 1779, for the heavy ships to draw out of the line, for close order in battle, and for preparing to attack the enemy van, centre or rear. Article 29 provided the signals for the second or third in command to lead the fleet with his divisions. Article 30 gave signals ordering the van or rear to close up on the centre, and forbidding individual ships to chase a beaten portion of the enemy fleet. Article 31 directed in detail how, should some ships have become separated from the fleet, the admiral would form a temporary line of battle ahead or abreast by signals to individual ships as to their temporary station. This article was originally devised for Rodney by Captain Walter Young.

Rodney's signals in 1782, judged by numerous examples, were arranged as before in the traditional manner with forty flags. Many of these were underemployed, so that he had a large number of positions in reserve for extra signals, though it is doubtful if this entirely offset the risk of confusing some of the flags.[19]

On 20 March 1782, only a few days before the Saints, Rodney issued an order of sailing by which the commander-in-chief in the *Formidable* was to lead the centre squadron and the fleet, with Hood in the *Barfleur*, followed by the van 'two miles broad on the weather quarter of the *Formidable*, that is about 3 points abaft her weather beam'. Rear-Admiral Francis Samuel Drake in the *Princessa* was to lead the rear 'two miles from the *Formidable* narrow on the lee quarter, that is to say to leeward and astern withall'.[20] Each squadron was to sail in arrow-head formation with the flagship leading, the remaining ships being disposed outwards from their own flagship. Thus in a squadron numbered 1 to 11 in order of battle in line ahead with the flagship number 6, the order of sailing would be as shown in the diagram on this page.

On 10 April 1782, the day after the abortive action with de Grasse, Rodney issued an additional Article 32 to his printed additional fighting instructions:

If at any time from baffling winds or other cause I think it expedient for my division to lead the fleet in line of battle for attacking the enemy . . . And if I would have the van [or rear] division become the centre . . . The Leading ship of my division is then to lead the fleet, and the leading ship of the division next to follow is, as above alluded to, of course to be next to the sternmost of mine and so on.[21]

On 25 April, thirteen days after the Battle of the Saints, the British fleet met off Haiti and then finally separated, Hood taking the main body to watch the French at Cap François while Rodney took his prizes and the badly damaged ships to Port Royal, Jamaica. On the very same day Hood gave out a list of additional signals. These included the following instruction: 'When fetching up with the enemy to leeward and on the contrary tack', if the admiral wished the fleet 'to break through their line and endeavour to cut off part of their van or rear', he would fly a chequered yellow and blue flag at the foretopmast head. In the main body of the signals, for which in the particular manuscript book cited there is neither date nor ascription, is the instruction: 'When standing towards the enemy in a line of battle on a contrary tack or being actually in battle', a red and white quartered flag would be flown at the fore, or the same with a red pendant over it at the mizzen topgallant, mastheads 'for any particular ship to lead through their line and all the rest to follow in due and close succession to support her for that purpose'. The same signals appear in two manuscript signal books, one entitled Hood's Signals of 12 January 1782. The additions are in both books dated 27 April.[22]

The signals did not provide for a general manoeuvre of breaking the enemy's line. They were solely concerned with the tactical situation presented by the Battle of the Saints, and the implication is that the admiral might signal the leading ship or any other ship to execute the manoeuvre if opportunity offered. The fleets would have to be passing close to each other and the enemy would have to be sufficiently close-hauled to allow for the necessary deviation off course to windward by the British. Neither the manuscript signal book actually used by Commodore Edmund Affleck at the Saints, nor any of the so-called signals 'by Rodney' contain either of these signals, which indicates that Rodney, alone amongst the leading admirals, possessed no signal for breaking the enemy's line at the time the battle of the Saints was fought.[23] When Admiral James Pigott arrived to supersede Rodney he was persuaded to take over Hood's two signals for breaking the line, and issue them to the whole fleet. The process by which Hood guided, or tried to guide, Rodney's successor can be read in his own letters to Jackson.[24]

Some attempt was made to improve the signalling system. In one of Hood's signal books of 1782 an additional left-hand column was ruled, and headed on every page 'N Signal' indicating an intention to number the signals right through in the manner of Howe's and Kempenfelt's earlier books, while leaving the traditional arrangement of flag, where flown and signification intact. This intention was never fulfilled but, in the same year, a West Indies signal book was produced, with Hood's and Pigott's additional signals included;[25] all the signals were numbered in the main text, though not in the index, with painted flags at the end. The signal for breaking the enemy line, with a flag chequered yellow and blue, was number 11, while the signal for any particular ship to do so, with a flag quartered white and red, was number 243, again ascribed to Pigott. The signal referring to additional fighting instruction number 34, for divisions in parallel lines in echelon, was given the signal book number 265, very near the end.[26]

Admiral Robert Digby on the North American station issued a signal book on 5 October 1782.[27] It was the work of John McArthur, later secretary to Lord Hood and joint biographer of Nelson, and it was used experimentally by a small cruiser squadron for three months.[28] It is a well arranged printed book, thumb-indexed, with the flags and

18 *MOD*, NM/94 (formerly *RUSI* 94): but see also NM/43 (formerly *RUSI* 43) in MS.

19 *MOD*, NM/28, 31 [wrongly dated in the catalogue as 17925] 43 and unidentified file (formerly *RUSI* 28, 31, 43 and 94); Tunstall and Corbett Collections, *NMM*, TUN/3 and 6, (formerly S/MS/Am/3 and 6)

20 *MOD*, NM/92 (formerly *RUSI* 92) and *NMM*, Tunstall Collection, TUN/3 (formerly S/MS/Am/3)

21 *MOD*, NM/94 (formerly *RUSI* 94) and US Navy Department Library, Washington; printed in *Signals & Instructions* pp300–1

22 *MOD*, NM/24 (formerly *RUSI* 24), and *NMM*, Corbett Collection, TUN/6 (formerly S/MS/Am/6). The signals in the second book appear twice over, the book being drawn in the usual tumb-indexed form but with a second part in which the signals are fully repeated, including coloured flags, thumb-indexed according to subject matter. In the first part the signal for a particular ship to perform the manoeuvre is attributed to Pigot.

23 *NMM*, Tunstall Collection, TUN/3 (formerly S/MS/Am/3); and *MOD*, NM/28, 31, and 43

24 David Hannay, editor, *Letters of Sir Samuel Hood (Viscount Hood) in 1781–2–3*, London, 1895 (hereafter cited *Hood*), pp138–61

25 *MOD*, NM/18a (formerly *RUSI* 18a), labelled 1790 but actually 1782

26 Instruction number 33 in *NMM*, Corbett Collection, TUN/6 (formerly S/MS/Am/6)

27 *MOD*, NM/76 (formerly *RUSI* 76) and *NMM*, KEI/S/9 (formerly Sp/9)

28 *Signals & Instructions*, p58, where, however, the description of the book towards the bottom of the page is somewhat misleading.

Captain (later Vice-Admiral) the Honourable John Byron (1723–86), by Sir Joshua Reynolds. NMM

positions on the old system and the signification columns partly left blank and partly completed in manuscript. Some of the twenty-four flags, however, are of unusual design, for example, a blue flag with the sinister chief quarter white; a flag divided diagonally blue and yellow, and the reverse; a white flag with a red lozenge and a white flag with a thin perpendicular pale blue stripe down the middle. Though the signals are not numbered in the main text, there is a separate index of the signals with numbers for use with a numerical table of ten flags selected from the main series of twenty-four. There is also a separate numerical table for private ships, which includes only the established flags, since only the admiral or senior officers would have had the experimental set. There is nothing very new about the main signals, except one for the van of the fleet to pass through the enemy line.

Amiral Jean-Baptiste Charles Henri Hector Théodat, Comte d'Estaing (1729–94), bust by Houdon. MM

Byron and d'Estaing

The Battle of Grenada (1779) is yet another example of a straggling British attack against a well-formed French line, though unlike the battles of Toulon and Minorca, the British mistake was due less to pedantic adherence to the line than to complete neglect of it. The British commander-in-chief, Vice-Admiral the Hon John Byron, aged nearly fifty-six, had a great reputation as a navigator and explorer, but Sir John Laughton was probably correct in asserting that 'It is very doubtful whether he ever saw a fleet extended in line of battle before he saw the French fleet on the morning of 6th July, 1779.'[29]

His opponent, the Comte d'Estaing, was aged fifty. He had been an infantry officer until the end of the Seven Years War, when he was promoted to *lieutenant-général*, and soon after the peace to governor-general of San Domingo. Then, by one of those acts of military metamorphosis of which only the French service was capable, he became simultaneously *Chef d'Escadre des Armées Navales*, and *vice-amiral*.[30]

The Battle of Grenada resulted from d'Estaing's capture of the island from the British on 4 July 1779, while Byron was clearing a home-bound convoy. By nightfall on 5 July d'Estaing had re-embarked most of his troops in transports in St George's Bay, covered by his battle fleet of twenty-four of the line. He knew that Byron was coming south to attack him and he had his frigates on the lookout. That evening they reported Byron's approach. As expected, he was steering south along the west side of the island with an off-shore breeze, E by N. D'Estaing kept his fleet at anchor until the last moment, chiefly because he had so little breeze at Georgetown that he was in danger of being carried out westward by the lee current if he put to sea before the dawn breeze took effect. At 4am, however, his fleet put to sea with orders to form line ahead, *par rang de vitesse* (in order of sailing speed), on the starboard tack, that is, as quickly as possible and without regard for stations in the official line of battle. Even so, he had some difficulty in getting his ships formed.

Byron meanwhile was coming along the western side of the island, helped by the freshening dawn breeze. He had with him eighteen of the line and astern a convoy of troops ready for the recapture of the island, guarded by three of the line under Rear-Admiral Joshua Rowley. Though Byron was well aware of the enemy's presence and indeed hoped to catch them soon after daybreak while still engaged in attacking the fortress, his fleet was still unformed, despite the fact that he had been prevented from sailing fast by the presence of his convoy. At 5.15am on sighting the French fleet to the south, apparently huddled in confusion and numbering only fifteen of the line, he signalled general chase, hoping to score a surprise, and at the same time he signalled Rowley to leave the convoy and rejoin the fleet with his three ships.[31]

This general chase was Byron's undoing, since it tended to disperse his ships instead of collecting them together. Both Anson and Hawke had been careful to form in line ahead before making their attacks against weaker French forces in their respective battles of Finisterre. Moreover d'Estaing was not to be caught so easily. By now the dawn breeze had reached him and his line was beginning to form on the

29 Source unknown

30 This happened as a result of the reorganisation of the French Navy by which new regiments were created to act as marine gunners and marine infantry. See Calmon-Maison, *L'Amiral D'Estaing*, 1910, Chapter ix.

31 Byron's own account is printed in *The Mariner's Mirror*, 30 (1944), pp88–90

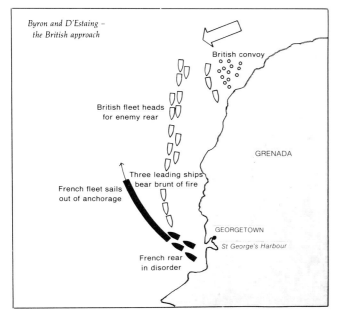

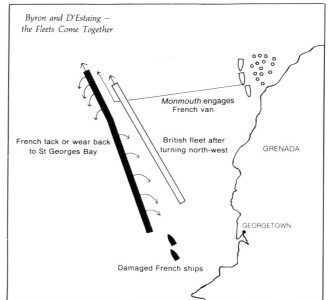

starboard tack on a roughly north-west course. Byron still did not realise the truth, and thought that he faced a distinctly inferior force. At 6.30am he signalled for 'the five or seven ships nearest the enemy to draw into a line-ahead in order to engage the enemy in their rear, endeavouring at the same time to get up with their van'. This was Hawke's old signal of 1747, intended for use against a fleeing enemy, and it was quite inapplicable here as the enemy rear was further away than their van by at least two miles. Hence not only was the element of surprise – in so far as it ever existed – completely sacrificed, but the five leading ships had to sustain the fire of a number of French ships to starboard before they could wear round and engage the French rear on a parallel course.

Had Byron attacked the enemy van he might have been more successful, as he would then have engaged the nearest French ships instead of the furthest, and would thus have been able to bring his whole fleet more quickly into action in support of his five leaders. It is true that he had no signal for attacking the van, but it would seem there was nothing to have prevented him from signalling 'general chase to the south-west', that is directly towards the leading French ships, followed by 'engage' and then 'engage closer'. Considering the kind of flag officers and captains he had, this should have produced the necessary pell-mell attack on the French van. Alternatively he could have signalled for the leading ships to lead more to starboard, followed by the signal to engage. More conservatively, he could have signalled his leading ships to shorten sail, and then have used Howe's additional signal of 1762, 'for the squadron when in chase to engage the enemy in line of battle according to the order in which they may successively arrive up with them'.

Byron's fundamental mistake seems to have been his inability to keep his fleet formed in reasonable order during the night. What was its formation at sunset on the night of 5 July? What orders did he send or what signals did he make immediately before sunset with regard to his night sailing formation? He was steering a steady course and he had the northern tip of the island of Grenada ahead of him to act as a check on his position. With a steady breeze and the need for not outstripping his convoy, it seems incredible that had his ships been in a fixed formation, they should not have been able to maintain it better during the short summer night.

It is much to Byron's credit that, when he realised that it was he who was outnumbered and not d'Estaing, he signalled

for closer action and did everything possible to force it. It is also to his credit that he was well supported by the whole fleet. Indeed the attack of the three leading ships, the *Sultan*, Alan Gardner, *Prince of Wales*, Vice-Admiral Barrington, and the *Boyne*, Herbert Sawyer, was so impetuous that they drove two of the French rear ships out of the line with severe damage and casualties. Having already worn round, Sawyer and Barrington were trying their best to get up with the enemy's van, Gardner's *Sultan* being entangled to leeward with the French ships.

At 8.30am Byron signalled the remainder of the fleet to wear in the wake of the leaders. The bulk of the British fleet was still straggling south, close inshore, and now began wearing round in succession, making what was practically a hairpin bend so as to follow the leading ships and get on a starboard course parallel with the enemy. The British fleet was thus steering in the opposite direction from the French, in order to lead round onto the same course and then try to draw level with the French van. Having lost the choice of heading the French at the start, they were now compelled to sail twice the distance to achieve the same result.

Three British ships were so out of position to leeward in the run south that they found themselves fired on by the whole French fleet as it passed on the opposite tack. After an hour the indefatigable Barrington and Sawyer came up from the south and passed between these ships and the French, still trying to get up with the French van. Meanwhile Byron had sent a schooner to order two ships to leave the line and pursue the two rear French ships which had been damaged at the start of the action. This shrewd attempt, however, proved abortive, and the schooner now passed on to the north to order Admiral Rowley's *Suffolk* to attack the leading French ship as it came north. Rowley obeyed this order with great bravery and skill and to the relief of the three British ships which had been too far to leeward, and which now lay crippled as successive French ships fired on them in passing.

Meanwhile, Captain Robert Fanshawe of the *Monmouth*, another of the ships recalled from guarding the convoy, acting without orders, cut right across the loop now formed by the British line and engaged the ships of the French van, suffering very severely in the process. At 10am Byron, realising that his whole attack was falling into disorder, cancelled general chase and signalled both for the line of battle ahead at two cables and for close action. As each ship tried to reach its proper station in the line there was much delay, some ships having to tack and then wear round again.

D'Estaing was now in a winning position. Except for the two crippled ships in rear, his line was intact. The British fleet, with three ships badly damaged and several hardly able to keep station, was still not properly formed in line. He thus had twenty-two effective ships against what now amounted to sixteen or seventeen. His van overlapped the British, and on his starboard bow was the British convoy. He could either cripple the British still further and then go for the convoy, or else make as if to attack the convoy and so force the British to fight him at a tactical disadvantage. As he drew further ahead, Byron kept close to the wind so as to cover his convoy and avoid being doubled. At 2.30pm he hauled down the signal for closer action.

D'Estaing was now clear of the British. His line was well formed and he could choose his method of attack. Instead, however, of attacking Byron or heading for the convoy, he put about – some of his ships tacking, some wearing. He then steered south-east back towards St George's Bay. Soon after, Byron signalled to tack together, putting the two fleets roughly on the same course but some distance apart. He could do nothing to save the three ships so severely damaged earlier through being too far to leeward; they were still further to leeward now and nearly in line with d'Estaing's approach. The French ships, however, held their course and contented themselves with firing at the British stragglers as they passed. The battle was over and no British ships were lost.

D'Estaing was strongly criticised by Suffren, captain of his van ship, and by other officers, for not pursuing his advantage. His aim seems to have been to prevent Byron slipping inshore of him, seizing the harbour of Georgetown, capturing the French troops lying there in transports, bringing in his own convoy, and landing the troops to recapture the fortress. Yet had he beaten Byron decisively there could hardly have been a Battle of the Saints. The decisive factors in his failure were probably three: his genuine preoccupation

with the safety of Grenada, the 'custom of the service' with regard to maintaining the safety of the fleet, and, perhaps, a cautious respect for the fighting qualities of the individual British ships.

The Moonlight Battle

On the British side, George Brydges Rodney stands out not only as the most controversial naval commander of the American War but also as the most controversial figure in the whole history of British tactics and signalling. In many ways he was unsuited to be a commander-in-chief, being violently contemptuous of those he disliked, and there were many. He had been a difficult man to serve under, even at the age of forty-four. Now, at sixty-two and broken in health, he was full of energy one moment, vacillating the next, highly suspicious of his subordinates, and inclined to treat them vindictively. Launched into the tropical West Indies, he proved himself both a tyrant and a martinet. Nevertheless, he achieved a very great victory. He was a tactical reformer and was popularly credited with 'breaking the enemy's line', although in fact he had actually refused to adopt the new signal specially devised for such a manoeuvre. When in good health and not exhausted he was still capable of driving a great fleet into action, assisted by his faithful assistants Sir Charles Douglas and Captain Walter Young, and his private physician Sir Gilbert Blane. He may well have studied Morogues, for he accepted the notion that a ship-to-ship action between equal forces offered little hope of success. To win, it was necessary to concentrate a superior number of ships against an inferior, leaving the remaining enemy ships disengaged.

On his way out to the West Indies via Gibraltar, Rodney scored an important tactical success. In the early afternoon of 16 January 1779, with his powerful fleet formed in good order, he was off Cape St Vincent when a Spanish fleet of eleven of the line and two frigates was sighted in some disorder. Rodney was ill with gout and his flag captain, Walter Young, directed the battle. Young wanted a general chase, to cut off the Spaniards from Cadiz.

This the admiral opposed and ordered the signal for a line of battle ahead at two cables distance. This I opposed in turn, conscious that a great deal of time and distance would be lost, and proposed the signal for a line of battle abreast at a cable's distance. At the same time (as he was confined to his bed) I kept out a stiff sail on the *Sandwich* that no time might be lost; and, as soon as I discovered the enemy making sail to get off, the signal for a general chase was made. When our ships got up the signal [was made] to engage to leeward and in succession, and the signal for close action at the same time.

The admiral's ill state of health and his natural irresolution, occasioned our shortening sail frequently, which prevented me from bringing the *Sandwich* so early into battle, as I could have wished; and it was with difficulty that I succeeded at last, as he attempted several times to have the ships called off from chase; but on representing to him that they were too far ahead to either see or hear our signals, and that it was our business to be at the head or near to them, that we went on and stopped one of the enemy who was prepared to evade us.[32]

32 Young to Middleton, 24 July 1780, *Barham*, I, pp65–6. The British ships gained on the Spanish either because they were coppered or because they set more sail, or both. The signal to engage to leeward in succession is Howe's signal for inverting the line of 1776 and earlier, number 18/61, *NMM*, Tunstall Collection, TUN/111 (formerly F1/5). The leading British ship was to engage the rearmost enemy while the second nearest ship, regardless of the order of battle, was to engage the next but one astern of the enemy, and so on. Text in *MOD*, NM/92 (formerly *RUSI 92*), and printed in *Signals & Instructions*, p223.

Admiral George Brydges, Lord Rodney (1719–92), first Baron Rodney, by Sir Joshua Reynolds. NMM

The battle lasted from 4pm until after midnight and seven of the eleven Spanish ships were sunk, taken or wrecked. Though Captain Young claimed to have directed the battle, Sir Gilbert Blane, in attendance on Rodney, suggests that it was a more equal partnership. In any case, Rodney took the entire responsibility for what was done and also the praise. Even if Young did not entirely direct the battle, the action itself was extraordinary, and praise was justified. A night action on a lee shore in a gale was a serious risk, as was discovered next day when the British fleet were in some danger of being driven ashore.

Young, in the same letter in which he described the battle, deprecated the use of general chase against an equal or superior enemy, particularly if formed in line of battle or in close order. In such cases he favoured a close line abreast, if the British fleet was to windward, as the best formation and one from which 'any evolution is instantly put in practice'. He considered one cable distance sufficient for this formation, as distances were invariably overjudged.

The Battle of Martinique

On the morning of 17 April 1780 the British fleet of twenty of the line and a 50 was to windward of the French fleet of twenty-three of the line, at four to five leagues distance. The wind was east and de Guichen was forming the French line on the larboard tack, that is, heading south. At 4.30am Rodney signalled for the line ahead on the starboard tack so as to close him, and at 5.45am for the line ahead at two cables. At 6.45 Rodney made the signal for Article 74 of his new additional fighting instructions, which read, 'When the commander-in-chief means to make an attack upon the enemy's rear he will hoist a flag half-blue half-yellow at the main topgallant mast head with a white pendant under it'.[33] There were separate signals for attacking the van and centre. In his own words, 'I gave notice, by public signal, that my intention was to attack the enemy's rear with my whole force; which signal was answered by every ship in the fleet.' But this is not the same thing, for whereas the instruction merely indicates the point of attack, Rodney's statement implies an actual concentration of force. Yet nothing was said about this in the instruction, nor provision made for how it was to be done (for instance, certain ships astern might be detached to double on the enemy's rear and so put them between two fires). It was no more than a signal of intention, requiring further signals to give it effect. In the manuscript signal book made for Commodore Edmund Affleck, the signal is condensed to read 'to attack the enemy's rear', as if implying an immediate executive purpose which, if baulked by the enemy, might automatically cease to operate. Here, therefore, was a further source of possible confusion. Rodney is known to have been exercising his fleet in tactics a fortnight before the battle, but there is no evidence that he had gone through his additional fighting instructions with his flag officers and captains, article by article, to see that they were properly understood. Sir Gilbert Blane, writing twenty-nine years later, said that all the captains had been informed of his plan of attack either orally or in writing; if this is true, Rodney's explanation of this particular point must have proved inadequate.

For the time being, however, the signal clearly held, whatever its interpretation. Rodney continued manoeuvring to give it immediate effect, his further signals being as follows: '7.00am line ahead at one cable distance' (this presumably was a closing-up movement on the centre at the

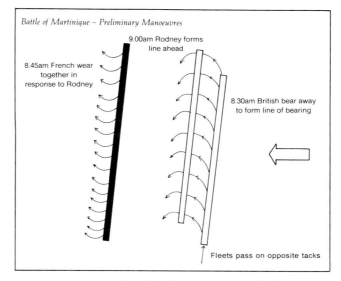

Battle of Martinique – Preliminary Manoeuvres

9.00am Rodney forms line ahead

8.45am French wear together in response to Rodney

8.30am British bear away to form line of bearing

Fleets pass on opposite tacks

moment when the bulk of his fleet was level with the French centre and rear); and, '8.30am, line of bearing, the ships to bear N by W and S by E of each other, at two cables' (the wind being then on the starboard beam). The extension of two cables distance was presumably to avoid bunching and possible collision should that course need to be altered again before reaching the enemy.

De Guichen seems to have divined the plan and sought to thwart it; at 8.45am he wore his fleet to the north, thus offering a van-to-van attack, which of course was not to Rodney's purpose. At 9am Rodney replied by signalling his own fleet to haul to the wind and to form line ahead at two cables on the port tack. This brought the two fleets on opposite courses once more, and at 9.42 Rodney's signal was repeated.

The fleets were now some three miles apart and the situation was again becoming favourable for Rodney's plan, the two leading ships in the British van having reached slightly beyond a point level with the rearmost ship of the French, with the fleets roughly parallel. So at 10.10am Rodney signalled for the fleet to wear and come to the wind on the starboard tack, thus coming once again onto the same course as the French. At 10.18 the signal was repeated, and at 10.19 the signal made for the rear ship to wear. At 10.36 Rodney signalled for the line-ahead at two cables. By this

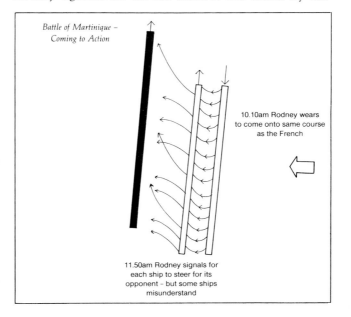

Battle of Martinique – Coming to Action

10.10am Rodney wears to come onto same course as the French

11.50am Rodney signals for each ship to steer for its opponent – but some ships misunderstand

33 *Signals & Instructions*, p230

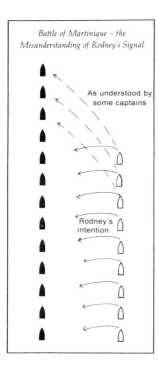

Battle of Martinique – the Misunderstanding of Rodney's Signal

As understood by some captains

Rodney's intention

time, the French were very strung out and it seemed only a matter of intelligent manoeuvre to bring the whole of the British fleet against their rear and centre, astern of the French flagship, thus representing a concentration of twenty ships against, possibly, twelve.

At 11am Rodney signalled for the fleet to prepare for battle and to alter course to larboard, thus closing the French, and at 11.28 for the rear division to close the centre. Then at 11.50 came the fateful signal for Article 21 of the additional fighting instructions:

'If the squadron should be sailing in a line of battle ahead to windward of the enemy, and the commander-in-chief would have the course altered in order to lead down to them, he will hoist a Union flag at the maintopgallant masthead and fire a gun. Whereupon every ship in the squadron is to steer for the ship of the enemy, which, from the disposition of the two squadrons, it must be her lot to engage, notwithstanding the signal for the line-ahead will be kept flying; making or shortening sail in such proportion as to preserve the distance assigned by the signal for the line, in order that the whole squadron may, as near as possible, come into action at the same time.'

Five minutes later Rodney signalled 'Engage', followed by 'Come to a closer engagement'.

At that moment the headmost ship of the British van was about level with the headmost ship of the French centre. From Rodney's point of view the situation was highly favourable. He could bring the whole of his fleet against the French centre and rear without any fear of collision, as the French line was strung out over a distance of some eight miles.[34]

This, however, was not the way in which the key man in Rodney's fleet saw the situation. Captain Robert Carkett, commanding the leading ship, assumed that Article 21 of the additional fighting instructions meant what it said, that is that each ship was to steer for her opposite number in the enemy's line, counting from the van.

Captain Carkett was not the only commander to make this interpretation. His next astern followed, and together they straggled north, striving to reach the head of the French van before engaging, which eventually they did. Rear-Admiral Hyde Parker, in command of the van, did the same, rebuking by signal two captains who seem to have divined Rodney's real intention.

The captain of the *Yarmouth*, the second ship in the centre squadron, finding himself two miles from his next ahead, a mile from his next astern and a mile and a half from the nearest Frenchman, who incidentally was a mile astern of his own next ahead, was so utterly perplexed that for a long time he did nothing at all, for which he was later dismissed from the service. Not only did the van ships ignore the original concentration signal but several of them engaged at too great a distance. This was due partly to their original efforts to make up distance on the French van and partly to de Guichen signalling his ships to make more sail, thus extending his line still further and sending the ships of the British van chasing

even further ahead of their centre.

Rodney himself in the *Sandwich* (90) made a spirited attack with the three ships astern of him on the ships immediately astern of de Guichen, and soon found himself right through the French line. But de Guichen wore with his seconds, gave the *Sandwich* a terrific hammering for an hour and a half, and then withdrew. There was also heavy fighting between the respective rear squadrons. The *Montagu* was heavily punished and her captain killed. Rear-Admiral Joshua Rowley, son of Sir William, forced the French rear ships to wear out of the line to leeward and did not rejoin the centre until later, for which he was severely reprimanded by Rodney. Soon after 4pm the battle was over, and de Guichen bore away to the west.

Had Rodney repeated his signal to attack the enemy's rear immediately before the signal to lead down, there would have been less cause for confusion. No doubt he relied on what might be deemed the standing order implied by Article 12 of his additional fighting instructions: 'In sailing in a line of battle, every ship is to keep at the same distance those ships do that are next the commander-in-chief, always taking it from the centre', with various signals for his seconds and the ships further ahead and astern to correct their distances.[35] Yet Hyde Parker clearly imagined that the French van must somehow be contained and so prevented from doubling. This was not only a traditional notion, as found expression at the Battle of Toulon in 1744, but had recently been re-emphasised by Howe. Thomas Collingwood, with a broad pendant in the leading ship of the centre, took the same view, his master's log recording, 'Bore away for our opponent, which bore NNW about two or three miles'. His opponent, according to Rodney's assumption, was roughly due west of him and about the fourteenth ship counting from the French van, whereas as eighth ship in the British line he assumed he should engage the eighth Frenchman.

De Guichen's account

De Guichen's official report on the Battle of Martinique is very well supplemented by the journal kept by Buer de la Charulière, his chief of staff throughout the campaign, who later wrote a book on naval tactics.[36] At 7am de Guichen signalled the French fleet to tack together and form in line of bearing to try and weather the British, but annulled it because,

pour se mettre en bataille tribord amures dans l'ordre renversé il oût faillu que les vaisseaux eussent arrivé pour aller prendre par un mouvement successif les eaux du vaisseau de tête, ce qui ne devait pes avoir lieu devant l'ennemi, en ce que ce mouvement aurait semblé éloigner de lui.'[37]

This is a little difficult to understand. How could a movement even remotely intended as aggressive be interpreted by a skilled opponent as anything else? And even supposing that it was, might it not also have been suspected of being a *ruse de guerre*?

Later in the morning, when the French were on the larboard tack and making sail to close their line, and the British rear was about level with the French centre, de Guichen signalled to tack together and form in line of bearing. His aim was to force an action from leeward. The whole British fleet, however, was now a little large to larboard, so de Guichen signalled for the line of battle in reverse order on the starboard tack. Then, on seeing his line too extended, he signalled the fleet to make sail, close up and prepare for battle.

At 12.30pm seeing that his line was still too far extended and that the British van was about to attack, he signalled his leading ship to shorten sail and the rest of the fleet to make

---

34 In the notes attributed to Rodney printed in the third edition of John Clerk's account of the action in *An Essay on Naval Tactics*, London, 1827, p95, it is stated that the Marquis de Bouillé, Governor of Martinique, was asked by de Guichen to tell Rodney that, had the latter's signal been obeyed, the French would have lost six or seven ships and de Guichen would have been his prisoner.

35 *Signals & Instructions*, pp220–1

36 Both accounts are quoted from at length in R Castex, *Les Idées Militaires de la Marine du XVIIIme Siècle*, Paris, 1911 (hereafter cited: Castex, *Les Indées Militaires*), pp81–8

37 p82

more sail to close the line. At 1.15 some of the van ships began to engage, followed soon after by the rear and, at 1.45, by the centre.

By 2.20pm de Guichen was convinced that Rodney intended to lead part of his centre through a gap caused in the French centre by a damaged ship falling to leeward. He therefore signalled his whole fleet to wear together so as to cut off Rodney and his seconds. When, however, he realised that he had mistaken Rodney's intention, he annulled the signal, which in any case was unseen by most of the fleet because of the smoke.

What clearly emerges from this dual account is that, when the battle began, the French fleet was too far extended and that, despite signals, this could not possibly have been remedied by the time at least some of the van ships began to engage. The French account thus confirms the British view of the situation. Rodney had a wonderful opportunity for his attack on the French centre and rear, and the straggling French van could not have taken any part in the battle for some considerable time. It should have been a miniature version of Trafalgar. If only Rodney had repeated his signal of intention, followed by 'Engage', and nothing else, it would have been just such a victory.

The May skirmishes

From 10 to 21 May the British and French fleets were in sight of one another, Rodney eager to engage under reasonable circumstances, de Guichen only willing to engage under favourable circumstances. To begin with the French had the wind and daily bore down in line abreast to within random shot and then brought to, refusing to close. On 15 May, when this happened as usual, Rodney crowded sail and pretended to retreat. The French pressed on and suddenly the wind changed sufficiently for Rodney to signal his van to tack and gain the weather gage. De Guichen promptly wore his fleet, placing himself on the same course as Rodney and only slightly ahead.

Rodney now had a chance to attack the French rear if only his ships could gain a little on the recently cleaned French vessels. He afterwards stated that 'though the wind would then have permitted the British fleet to get to windward, the moment the British van had got near the enemy, the signal was ready to take the lee gage close, had not the wind changed six points at once.'[38] In other words, Rodney was quite prepared to attack from leeward but was headed by the wind. De Guichen then tacked, sending his van across the van of the converging British. The three leading British ships were for a time engaged with the French centre, but were forced to bear away, and the fleets passed each other just out of range on opposite courses. Although Rodney had obviously had a chance to cut the French line from leeward as the two fleets converged, he did not attempt it. How could he possibly have done so, seeing that he had no signal?

On 19 May the fleets were engaged once more. While they were passing on opposite tacks, the wind shifted in Rodney's favour, so that his van seemed able to weather the French rear. However, according to a note he made in his own copy of John Clerk's account of the action in *An Essay on Naval Tactics*, Rodney did not seek to gain the weather gage. The result was that only the British van and the French rear were seriously engaged, de Guichen hauling his wind and avoiding further contact.

Rodney's reaction

Rodney's account to the Earl of Sandwich of his reaction to these tactical fiascos was a masterpiece of vituperative prose:

I can easily imagine the concern your Lordship must have felt at the dastardly behaviour of a fleet which called themselves British. The delinquents were too many for me to supersede, but I gave them public notice that no rank whatever should screen any person who disobeyed orders or signals; that I should hoist my flag on board a *frigate*, in the centre, from whence all signals should be made; that all the *others* should attend me, and be ready to carry [out] such orders as I might judge necessary for his Majesty's service.

It is impossible for your Lordship to conceive the infinite utility of this resolution. Men conscious of their bad behaviour in the battle of the 17th of April, and who would have been glad to have watched every opportunity of contributing to my disgrace as an officer, even at the risk of a defeat of the British fleet, were by this resolution thunderstruck, and found themselves under the absolute necessity of doing their duty and that my eye was more to be dreaded by those who betrayed their country's honour than the enemy's cannon.

Rodney was delighted when Rear-Admiral Hyde Parker was recalled, writing that he was

a dangerous man with a very bad temper, hostile in the highest degree to the Administration and capable of anything, if he thinks he is within the pale of barely doing his duty. His cunning and art, however, has failed him; and if it is thought advisable to call him to account for his conduct during the action on the 17th of April, his head will be at stake for palpable disobedience of signals and daring to carry the van and his squadron two leagues ahead of their station, when he himself ought to have been only one mile ahead of the *Sandwich*.

Rear-Admiral Rowley came off no better:

I don't know which is of the greatest detriment to a State, a designing man or a man without abilities entrusted with command. Had not Mr Rowley presumed to think, when his duty was only obedience, the whole French rear and the centre had certainly been taken.[39]

Rodney will always be a controversial figure in the history of naval tactics, as also in the broader fields of naval strategy and command. The fact must be faced that great admirals can be unpleasant people, and that great powers of strategical and tactical direction do not necessarily go hand in hand with modesty, friendliness and unselfishness.

Rodney incessantly drilled the fleet, and was so stimulated by the exertions that his health greatly improved. Captain Walter Young's account of these happenings, however, is less enthusiastic. During the 15 May encounter, Rodney had taken himself on board the frigate *Venus*, Captain John

---

38 Rodney's note in Clerk's *An Essay on Naval Tactics*
39 31 May 1780, *Sandwich*, III, pp215–7

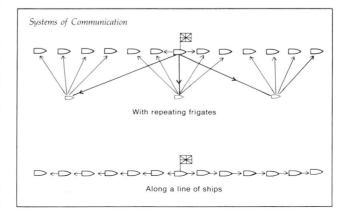

Systems of Communication

With repeating frigates

Along a line of ships

Douglas, so that he could better see the fleet, but Young said that tactical direction had actually been left to himself: 'His being in a frigate was of no service, as he always kept to leeward of our line. The enemy being to windward, he could never be a judge of it.'[40]

On signalling, Young was extremely interesting:

I have delivered the signal books to Sir George for his perusal, though I am apprehensive those are books he will pay little attention to; but I think the colour of the flags are well adapted, all but the chequered ones, which I would recommend not to be introduced. It would be attended with many good circumstances if you were to allow a complete set of signal colours to each captain, reducing the size of them, and let every ship repeat the signals made by the admiral; there would then be no excuse for any officer saying he did not see or understand the signal, and it would effectually put an end to that of captains stopping of signal colours to make a set for themselves; if they begin to put this in practice the warrant officer has a fine field for embezzling them, which few of them fail to do.[41]

This revolutionary proposal, which not even Howe was ready to concede, would have given every captain the same signalling facilities as that held by squadron flag officers and captains of repeating frigates. Signals by private ships back to the admiral would have remained the same, but every captain would have had both the right and the duty to repeat the

admiral's signals. The hint of unofficial means used to make up sets suggests that many captains saw themselves in the guise of possible senior captains in a sudden emergency. Young raised the same point again in a letter to Middleton of 31 July:

Indeed, I wish to see every captain in the navy who commands a line-of-battle ship to have a complete set of flags, and to repeat the signals through the line, as the French do; and I wish most sincerely that an established code of signals were adopted for the use of his Majesty's fleet. I am certain it would be attended with great advantage to our country and the service at large.[42]

In his next letter, 22 September 1780, Young suggested an important tactical reform:

in the whole of his [Kempenfelt's] code I do not see a temporary line of battle, which is absolutely necessary, as detachments may be sent from the main body, and ships separated by chasing. Enclosed with this I have sent one [a signal] that I have established, for your approbation. I had, on our coming upon the West India station, fixed chasing signals for every point of the compass, and night signals for the same purpose; but I found the captains were averse to them, therefore gave them up, though with much reluctance.[43]

The proposed 'signals for a temporary line of battle' were drawn up in the form of a lengthy fighting instruction. The admiral would first make the signals for the line of battle ahead or abreast as required, followed by 'the established preparatory flag'. Each captain would then be signalled in turn to take his place in the temporary line and would have to keep his position in the line recorded on 'a blank form'. When the signal for a temporary line was taken in, captains would 'fall into their stations with the commanders of the divisions, as established by the regular line of battle delivered to them'.[44] The purpose would seem to have been to allow the admiral to reconstruct his line of battle to the best advantage in the temporary absence of certain ships, instead of being compelled to form his temporary line by closing gaps and thereby placing less experienced captains in key positions, as for instance at either end of the line.

Walter Young died on 2 May 1781, but his instruction, word for word, was issued by Rodney on 19 February 1782, the day of his return to the West Indies to take over the command in chief from Hood.[45] Young was undoubtedly an officer of great ability and immense tact, his position as flag captain cum captain of the fleet to Rodney being one of the most difficult that any senior officer has ever had to fill. Middleton wrote: 'I believe there is not another man breathing so calculated to control and guide to fame a character [Rodney] that Nature never intended should be either a hero or a man of business.'[46]

Arbuthnot and des Touches

The comparatively unimportant action between Mariott Arbuthnot and Commodore des Touches clearly illustrates the kind of advantage the French were able to obtain through superior evolutionary control and prompt obedience to signals, together with their complete failure to exploit it. Arbuthnot with seven of the line and a 50 was trying to force an action on des Touches who, with an equal number of ships roughly equal in gunpower, was trying to carry 1000 French troops from Rhode Island to Virginia. The troops were embarked in the warships, so that it was not a convoy battle. The British ships were coppered and the French were not, so that when early on 16 March 1781 Arbuthnot sighted the French in hazy weather off Cape Henry, there seemed every

40 Young to Middleton, 3 June 1780, *Barham*, I, pp61–2

41 Young to Middleton, 28 June 1780, *Barham*, II, pp67–70. The signals, no doubt, were some of Kempenfelt's experimental signals sent out by Middleton; the flags, including the chequered ones, as described in Kempenfelt to Middleton, 20 March 1780, *Barham*, I, pp313–5. The 'Sea-Officer' [O'Bryen?] who translated Morogues in 1767 was also opposed to the use of chequered flags: see *Naval Tactics . . . by Mons De Moroques*, p38, footnote.

42 *Barham*, I, p72

43 *Barham*, I, pp75–6

44 *Barham*, I, pp76–7

45 *MOD*, NM/94 (formerly *RUSI* 94) and *Signals & Instructions*, pp290–300

46 Memorandum in Middleton's writing, *Barham*, I, p97

Admiral Marriott Arbuthnot (1711?–1794); engraving after a painting by William Hodges. BM

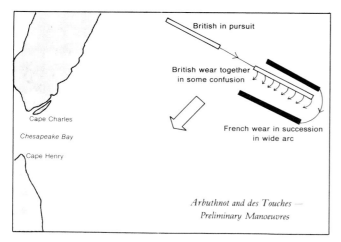

Arbuthnot and des Touches —
Preliminary Manoeuvres

chance of his forcing a pell-mell action. Shifting winds made things difficult for both admirals, but after passing the French on the opposite course and to windward of them, Arbuthnot saw his chance and tacked in pursuit, hailing Phillips Cosby, captain of his leading ship, to inform him 'I meant to engage them to leeward, as the weather was very disagreeable, hazy, and a large sea. Nothing could bear a more pleasing prospect than my situation.'

Cosby was now 'about two gunshots from their rear and one point to windward'. Finding that he was being gradually overhauled, and sensing the threat to his rear, des Touches wore his squadron in succession, coming right round in a big hairpin bend to leeward on the opposite course to Arbuthnot, thus completely sacrificing any hope he might have entertained of contesting the weather gage. On the other hand, he correctly judged that under the prevailing sea condition he would be able to use his lower-deck guns and the British would not.

According to Arbuthnot, Cosby at once wore the *Robust* round on the same tack as the French and began to engage without any signal to do either. Arbuthnot meant to wear anyhow on a tighter bend than the French so as to keep to windward of them. Cosby's action, however, 'obliged me to form under the fire of the enemy's line; and as the van was by this means put into confusion, I was single in bearing down to connect the line, exposed to the fire of the admiral and his seconds'.

Arbuthnot's letter to Lord Sandwich, from which these quotations are taken, is both vague and inadequate as an account of the battle.[47] Apparently des Touches, firing up to windward from his leeward position, damaged the two centre British ships considerably and then passed ahead to do the same to the three leaders. Later he wore his squadron round in succession, hit the three leading British ships once again as he passed them on the opposite course, and then withdrew and eventually retired to Newport.

Arbuthnot seems genuinely to have thought that he had won the battle. In his earlier despatch he had reported that

By 3 o'clock the French line was broken, their ships began soon after to wear and to form their line again with their heads to the south-east into the ocean [away from the American coast]. I cannot but regret the early flight of the enemy prevented the action from becoming general.[48]

During the fighting, however, he had kept the signal for the line flying and failed to signal close action. Contemporary British opinion rated his action a serious failure. Nevertheless, he had succeeded in stopping the French troops joining La Fayette's force in Virginia. This seems a farcical end to the battle, since des Touches had clearly won and, with five British ships practically immobilised, he could have done as he pleased.

De Grasse's signals and instructions

De Grasse's signal book for 1781 made no use of numerical signals.[49] None of his flags bore numbers, nor did his signals. His book had a page reference system by flags, of which he employed eighteen, using two-flag hoists. A substitute flag was provided, but he used only sixteen of the flags in the lower position. A total of 228 signals could be made. The book reveals no progress in battle tactics beyond the ideas of Hoste and Morogues, and, in consequence of its peculiar arrangement, left no room for signalling expansion to cover new tactical devices.

The accompanying 'Signaux pour les Mots d'Ordres' included a table of lettered flags, A to I, and K, evidently derived from Pavillon.Combined with another letter table it could be used for signalling both to warships and to merchant ships. Another table of twenty-eight signals made with single flags, pendants and cornets indicated landfalls and enemy ships sighted. There was also a table of battle signals by twenty-eight single flags, but it included nothing beyond those already mentioned. A table of administrative signals concluded the book.

Taking into account considerations of speed and handiness in sending and receiving, and the possible confusion of flags, it must remain open to question whether de Grasse's signal system was superior to the antiquated one operated by Rodney. Admittedly de Grasse scored heavily on logical order and the fact that the use of two flags, flown where best seen, provided a double check on possible confusion in reading. Yet he had no room for expansion like Rodney, who, by using more flags, had more signals as well as plenty of positions for flying particular flags still unappropriated. Moreover, the British use of pendants together with flags, apart from vanes and pendants to indicate particular ships, had already brought into existence something of a quasi double-flag signalling system.

The signals used by de Guichen in his battle off Martinique with Rodney were on the same plan, but he employed only sixteen flags for his general day signals.[50] The signals were numbered, but because they could only be used through the medium of the page system of flag hoists they were only a means of reference to a numbered list. Another sixteen-flag system, fourteen of the flags for which were of different

47 Arbuthnot to Sandwich, 30 March 1781, *Sandwich*, IV, pp168–9

48 Also written from Lynnhaven Bay (Chesapeake Bay) 20 March 1781.

49 *Movemens Généraux à l'Usage de L'Armée de Roi Comdée par Mons le Comte de Grasse, etc, Brest, 1781, de l'Imprimerie de R Malassis imprimeur Ordinaire du Roi et de la Marine*, MOD, 44 (formerly *RUSI* 44)

50 *Signaux Généraux de Jour, de nuit et de brume à l'usage de l'armée du Roi Commandée par Monsieur le Comte de Guichen. A Brest, de l'imprimerie de R Malassis . . . 1781*, MOD, Ec/3. There is a manuscript copy with all lower hoist flags on a single page in *NMM*, HOL/31 (formerly S(H)) 31.

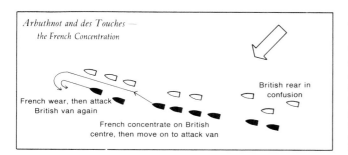

Arbuthnot and des Touches —
the French Concentration

French wear, then attack
British van again

French concentrate on British
centre, then move on to attack van

British rear in
confusion

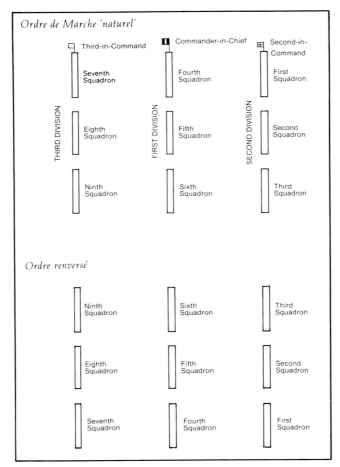

Ordre de Marche 'naturel'

Ordre renversé

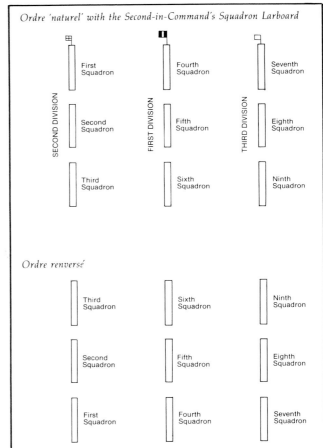

Ordre 'naturel' with the Second-in-Command's Squadron Larboard

Ordre renversé

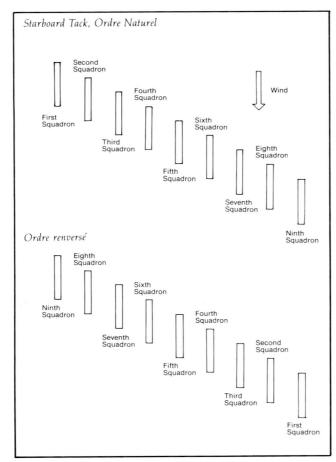

Starboard Tack, Ordre Naturel

Ordre renversé

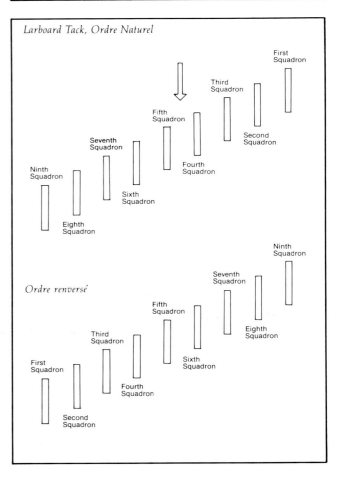

Larboard Tack, Ordre Naturel

Ordre renversé

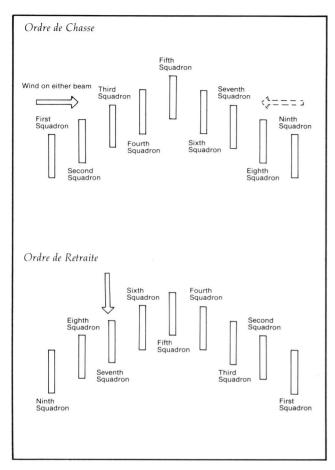

Admiral Samuel Hood (1724-1816), first Viscount Hood, by James Northcote. NMM

design, was used for combat signals; it contained some interesting sophistications, including an order for the leading ships to stay on the same tack as the enemy, and attack his rear 'au vent', to attack the rear 'sous le vent', to pass the enemy's leading ship 'de la vent', and for the leading ships to pass the enemy rear in close order as they were best able. There was another sixteen-flag system for 'Movements Généraux et Ordres Particuliers aux Vaisseaux de l'Armée', used mainly for sailing formations. A numerical table employing thirteen flags was provided for communication of numbers 1 to 169, presumably for ship numbers.

De Grasse also published in 1781 'Ordres de Marche et de Bataille', in which he gave the French sailing formations, with each of the three squadrons of the fleet again sub-divided into three divisions.[51] In none of the diagrams was any indication given of the wind, the direction of which must be assumed in each case.

Hood and de Grasse

The abortive actions between Hood and de Grasse off Martinique on 29 and 30 April 1781 are mainly of interest as illustrations of the extent to which preliminary dispositions influence fleet actions. Hood, on Rodney's orders, was blockading four French ships of the line in Port Royal, on the western side of Martinique. Early on 28 April one of his frigates reported a powerful French fleet of twenty of the line, including three 80s, commanded by the Comte de

Grasse, with his flag in the *Ville de Paris* (110), escorting a convoy of 150 merchantmen. They were coming round the south-west tip of the island and, to windward of Hood who was between them and Port Royal harbour. Most of Hood's eighteen ships-of-the-line were coppered and he was unencumbered by a convoy. Nevertheless, he was outnumbered, heavily outgunned and to leeward of the French. Next morning (29 April) the French fleet was sighted coming round the south-west extremity of the island, Hood having apparently worked to windward of them, though this seems a little uncertain. The fleets began to close,

51 *Ordres de Marche et de Bataille de l'Armée du Roi Commandée par Monsieur le Comte de Grasse Lieut-Général des Armées Navales. Commander de l'Ordre Royal et Militaire de Saint Louis, en 1781. A Brest*, 1781, same imprint as for *Movemens Généraux* (cf n49), MOD, 85 (formerly *RUSI* 85)

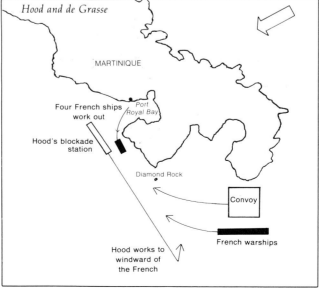

and meanwhile the four French ships-of-the-line which had been blockaded emerged from Port Royal. De Grasse had a difficult task. He was certainly in superior force but he had to pass his convoy into harbour and this meant working round Diamond Point and the Diamond Rock, a tiny islet beyond it. This he had to accomplish in the face of an extremely active though inferior fleet. He therefore fought a holding action, which continued in desultory fashion at long range for over four hours from about 11am, the fleets being sometimes on the same and sometimes on opposite courses.[52]

Hood at one stage tempted de Grasse to closer action by bringing to under topsails, 'but it was with Monsieur de Grasse the option of distance lay, and he preferred that of long shot. It was not possible for me to go nearer'. This statement is not at all clear. Did Hood mean that it was not possible because de Grasse had regained the wind, or because he himself could not risk forcing close action on a superior fleet? The upshot was that, with one of his ships taking much water, Hood had to retire to St Eustatius while de Grasse presumably collected the four French ships from Port Royal.

Light airs and calms during the night upset the formation of both fleets, but next morning (30 April) de Grasse was to windward and Hood was unable to weather him. Two more ships of the British van were now too leaky to keep their station in the line. With 1500 men sick, and short of complement, Hood 'judged it improper to dare the enemy to battle any longer'.

On the morning of 1 May there was another partial action, lasting about two hours. By now three more ships 'had their lower masts very badly wounded', so that Hood 'judged it right to bear away, and stood to the northward the next night'. Two of the three leaky ships at least were admitted by Hood to have been rendered so 'owing to the number of shot holes under water'. The captain and first lieutenant of one were killed. These facts tend to dispose of the idea that the French gunnery was completely ineffective and that 'never was more powder and shot thrown away in one day before' – even if this were true of the fighting on 29 April. On Hood's own admission, six out of nine ships in Rear-Admiral Drake's division were incapable of keeping their place in the

line after damage in three days of long-range skirmishing.

French accounts claim that it was Hood all along who avoided action and that Bougainville should have taken responsibility for securing four British ships seen to be foul of one another and separated from the fleet. The French also claimed that they pursued the British for three days, though they admitted that they failed to press their attacks with sufficient vigour.

Strategically, of course, this was an overwhelming French convoy success. The British had been outwitted from the moment they had missed the chance to observe de Grasse's departure from Brest and report it to Rodney; Hood was taken completely by surprise.

The Battle of the Chesapeake

In July 1781 Arbuthnot handed over the command of the North American station to Rear-Admiral Thomas Graves, pending the arrival of Rear-Admiral Robert Digby. Graves issued the squadron with a copy of the Signals & Instructions in Addition, together with Arbuthnot's handwritten additionals, dated 6 May 1781, endorsed 'The above signals given by Vice-Admiral Arbuthnot are what you are to observe and follow until further notice: London off Sandy Hook 6th July 1781, Thomas Graves.'[53] He also reissued Arbuthnot's additional instruction of 19 August 1780 with an identical endorsement, as well as other minor instructions and signals, including some dated as long before as 21 February 1780.[54] Next day, 7 July 1781, Graves reissued Arbuthnot's instructions for defence against fireships, originally dated 5 February 1781.

In all these ways Graves identified himself with the obscurantism of Arbuthnot's regime and made no attempt to introduce the new ideas germinating in the Channel Fleet, though he had persuaded Arbuthnot to accept a few Channel Fleet items in the previous year. On this occasion, however, he seems to have considered himself merely as a stopgap commander-in-chief.

De Grasse had arrived in Chesapeake Bay on 30 August with a powerful fleet from the West Indies, escorting a convoy of 3000 French troops from San Domingo, together with artillery and stores, and also a large sum of money borrowed from the Spanish at Havana. Cornwallis's army was already penned up in Yorktown by the concentration of land forces under Rochambeau and Washington. De Grasse's intervention helped to apply decisive pressure on land, while his fleet stood ready to prevent any attempt at a relief by sea. In addition there was a small French squadron at Rhode Island under de Barras, which might also be expected to join de Grasse in Chesapeake Bay.

Meanwhile Graves's fleet of nineteen of the line was heading south from New York. Only five of the line were his own, the remaining fourteen being part of Rodney's fleet brought north from the West Indies under Hood and Drake. These flag-officers and their captains were strongly imbued with their own special *esprit de corps*, and they had reached

Admiral Thomas, Lord Graves (1747?–1814); engraving by F Bartelezzi after a painting by James Northcote. BM

52 Hood to Middleton, 4 May and 24 June 1781; the first enclosing Hood to Rodney nd; *Barham*, I, pp109–19; Hood to Jackson, May 21 1781, *Hood* pp12–16; and 'A Journal . . . by the Chevalier De Goussencourt,' and 'Journal of an Officer in the Naval Army in America in 1781 and 1782', *Operations of the French Fleet under the Comte de Grasse, 1781–2*, New York, 1864 and 1971 (hereafter cited: *De Grasse*), pp45–6, and 140–3.

53 *NMM*, KEI/S/8 (formerly Sp/8), but lacking the additional signals after p24, and *Signals & Instructions*, pp 252–8 but numbered from 1 onwards instead of 12.

54 All in National Maritime Museum, KEI/S/8 (formerly Sp/8); the former printed in *Signals & Instructions*, pp246–8

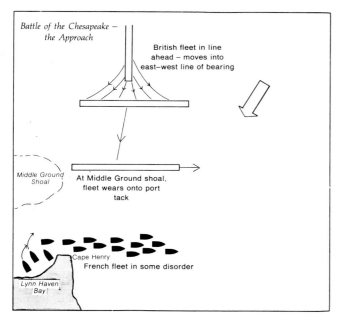

Sandy Hook only on 30 August, the same day that de Grasse reached Chesapeake Bay. The whole fleet, under Graves, set sail next day, and it looks as if there had been no time for a proper discussion of tactics, fighting instructions and signalling, nor was there any time for tactical exercises en route. Hood's and Drake's captains had presumably been issued with the additional signals given out by Arbuthnot earlier in the year and confirmed by Graves, but they seem to have received no proper explanation as to what the signals really meant, since Graves did not issue accompanying printed instructions.

At 9.30am on 5 September a frigate sighted the French fleet at anchor to the south-west between Cape Henry on the south side of Chesapeake Bay and the Middle Ground shoal to the north. The wind was NNE, very much in Graves's favour, and the flood tide was setting into the bay; he pressed ahead, and at 10.30am signalled to prepare for battle. At 11 he could see the French and realised for the first time that he was opposed by a much stronger force than that commanded by de Barras. He therefore signalled for the line ahead at two cables and stood on towards the south.

De Grasse had been badly caught. His two frigates stationed at direct signalling distance only had scarcely been able to give him an hour's warning, and even then he had hesitated, under the impression at first that Graves's fleet was that of de Barras. His fleet was anchored in Lynn Haven Bay, a small indentation just within Cape Henry and facing north towards the Middle Ground shoal. At that moment he had both the wind and tide against him and in addition he had 2000 men away in boats helping to land his troops and collect fresh water for the fleet. It says much for French efficiency that when, at 11.30am he signalled to slip cables and get to sea, he was promptly obeyed. He then signalled for the line-ahead on the port tack *par rang de vitesse*. As his fleet straggled eastwards out to sea, the rear ships, still practically

embayed, had to tack several times to clear Cape Henry. The van meanwhile was straggling and in very poor order.[55]

The two fleets were still some twelve miles apart, and at 12.45 Graves signalled for the line ahead at one cable, and at 1pm for an east-west line of bearing. This had the effect of bringing his fleet roughly parallel with the French, though converging slightly towards them, and at the same time running into the entrance of Chesapeake Bay.

My aim was to get close, to form parallel, extend with them, and attack all together: to this end I kept on until the van drew so near a shoal called the Middle Ground as to be in danger. I therefore wore the fleet all together and came to the same [larboard] tack with the enemy, and lay with the main topsail to the mast dressing the line and pressing toward the enemy, until I thought the enemy's van were so much advanced as to offer the moment for successful attack; and I then gave the signal for close action – the enemy's centre and rear at this time were too far behind to succour their own van.[56]

This sounds contradictory. Why should he wait for the French van to be 'so much advanced'? So slowly and so awkwardly were the French moving that Hood, writing next day, contended that their van might have been attacked direct as it emerged from Lynn Haven Bay,

which, I think, might have been done with clear advantage, as they came out by no means in a regular and connected way. When the enemy's van was out it was greatly extended beyond the centre and rear, and might have been attacked by the whole force of the British fleet. Had such an attack been made, several of the enemy's ships must have been inevitably demolished in half an hour's action, and there was a full hour and half to have engaged it before any of the rear could have come up.[57]

Graves meant to make his attack not in parade form but by a well-ordered concentration of firepower on the straggling French line. Following Howe's ideas of 1777, he was not prepared to allow his line to be overlapped by the more numerous enemy; the words 'extend with them' being a clear sign of his intention.[58] Nor was he prepared to take the risk of a general chase attack, especially with so many captains new to his command. Hood, it seems, would have taken the risk, but then fourteen out of the nineteen captains had been with him in the West Indies.

At 2.30pm Graves records that he signalled for 'the leading ship to lead more to starboard' in order to approach the enemy.[59] This was number 10 of Arbuthnot's additional signals.[60] The ships astern of the leading ship obeyed the signal in its obvious literal sense by turning to starboard in succession on reaching the point where the ship next ahead had turned. To Graves, however, the signal meant that the whole fleet should alter course in conformity with the movement of the leading ship, and not one after another in succession. By this means his whole fleet could have run

---

55 De Goussencourt, *De Grasse*, p69

56 *Sandwich*, IV, pp181–2

57 *Hood*, p31

58 See: 'Additional Instructions . . . by Lord Howe', *Signals & Instructions*, pp108–9

59 *Sandwich*, IV, p185

60 *Signals & Instructions*, p251

Battle of the Chesapeake –
Engagement

Middle Ground

Graves's intention

Actual course steered

Leading British
ships engage

Graves attempts to bear down

Cape Henry

French van ships damaged

down on a lasking course to engage, instead of making a slanting approach each directly astern of the ship ahead. In other words, Graves's signal was intended to produce the same result as a 'sailing' signal for forming a line of bearing. In view, however, of how Arbuthnot's signal actually reads – 'If at any time I would have the leading ship in the line alter course to starboard, I will hoist a flag half red, half white at the main topmast head' – the captains cannot be blamed for acting as they did. The similarity of his dilemma with that encountered by Byng at the Battle of Minorca is quite remarkable.

Seeing his whole plan of attack going awry, Graves repeated his signal twice, and also signalled various individual

ships to get into station and the rear to make more sail. At 3.34pm he made signal number 5 from Arbuthnot's additional fighting instructions of 1779, for the fleet then in line of battle to alter course to starboard and at 3.46 for the line at one cable.[61] 'The enemy's ships advancing very slow and evening approaching, the Admiral (judging this to be the moment of attack) made the signal for the ships to bear down and engage their opponents – filled the main topsail and bore down to the enemy.[62] This was followed by the signal for close action, and at 4.11pm the signal for the line was hauled down, 'that it might not interfere with the signal to engage close'. Nevertheless, the *Montagu*, twelfth in the line, luffed and opened fire at too great a range, thus forcing the ships of the centre astern of her also to luff so as to avoid overrunning and getting between her and the enemy. At 4.22 Graves 'hoisted the signal for the line-ahead and hauled down the signal for close action, the ships not being sufficiently extended'. But five minutes later the line ahead signal was hauled down again and the signal for close action rehoisted.

61 *Signals & Instructions*, p238

62 Graves's account, *Sandwich*, IV, p185. These words are from the official narrative endorsed in Graves's private letter to Sandwich; hence 'the Admiral'.

Graves tried to set an example by steering straight for the enemy, but soon found himself obstructed.

Meanwhile, the van of the fleet had begun to engage the four leading ships of the French van, which were very much exposed to windward. The French ships succeeded in beating the leading British ship out of the line, but were themselves heavily punished. De Grasse signalled to them to bear away to leeward but this proved expensive as it meant exposing their sterns to the British broadsides; the fourth French ship was badly hit in the process and her captain killed, and another ship was in danger of capture. Altogether only about eleven of the twenty-four French ships became engaged, the remainder being too far to leeward. On the British side only ten ships out of nineteen were seriously in action, and of these, ten suffered considerable damage to masts and rigging. They included the *Terrible,* now so crank that she leaked from the shock of discharge of her own guns and had to be burnt six days later.

All this time the British rear remained practically inactive,

Hood being under the impression that the signal for the line was still flying and that it had never been hauled down. There is much confusion of testimony here but, at any rate, the signal for close action was flying at 5.20pm, and the signal for the line was not, so that at 5.30 the British rear bore down – much too late. At 6.23 Graves finally hauled down the signal for close action and hoisted the signal for the line ahead at one cable. The French could be seen bearing away to leeward. It was nearly sunset, and at 6.30 all firing ceased.

The manoeuvres of the next nine days had little tactical significance and, when Graves sailed for New York on 17 September, Cornwallis's surrender became a certainty.

Reactions to defeat

Both Graves and Hood were greatly dissatisfied with their tactical fiasco, Hood especially so as he must have felt that he was partly to blame. Although he made no comment whatever on the failure of the approach signals to produce the

The repulse of the French in Frigate Bay, St Kitts, 26 January 1782 (dated 1783) (see pp 177–8). NMM

desired form of attack, he contended that the centre and van opened fire at too great a distance. This was true enough. When the leading French ships 'turned their sterns', the British van should have been signalled 'to make more sail to have enabled the centre to push on to the support of the van, instead of engaging at an improper distance'. This also was a reasonable criticism, and he elaborated it as follows:

Now, had the centre gone to the support of the van, and the signal for the line been hauled down, or the commander-in-chief had set the example of close action, even with the signal for the line flying, the van of the enemy must have been cut to pieces, and the rear division of the British fleet would have been opposed to those ships of the centre division fired at, and at the proper distance for engaging, or the Rear-Admiral who commanded it [Hood] would have had a great deal to answer for.[63]

In so far as this further statement was also justified, it clearly stemmed from one single fault; that the ships ahead of Graves opened fire too soon.

Graves's reaction was calmer and took the form of a memorandum issued next day to all flag officers and captains. It was clearly intended as a rebuke to Hood, and was accepted as such, though it was phrased entirely in professional and impersonal terms. From a man who had seen his whole plan go to pieces for lack of greater initiative on the part of a second-in-command with such a high reputation, it was mild indeed, perhaps too mild. One can imagine Rodney's phrases in similar circumstances.

When the signal for the line of battle ahead is out at the same time with the signal for battle it is not to be understood that the latter signal shall be rendered ineffectual by the strict adherence to the former, and the signal for the line of battle is to be considered as the line of extension, and the respective Admirals and captains of the fleet are desired to be attentive not to advance or fall back, so as to intercept the fire of their seconds ahead and astern, but to keep as near the enemy as possible, whilst the signal for close action continues out, and to take notice that the line must be preserved parallel to that of the enemy during battle without regard to a particular point or bearing. HMS *London*, at Sea, 6th September 1781.[64]

Hood, evidently strung by this implied rebuke and equally mortified by the British failure, wrote the following endorsement on his own copy when the memorandum was reissued by Graves at New York.

It is the first time I ever heard it suggested that too strict an adherence could be paid to the line of battle; and if I understand the meaning of the British fleet being formed parallel to that of the enemy, it is, that if the enemy's fleet is disorderly and irregularly formed, the British fleet is, in complement to it, to form irregularly and disorderly also. Now, the direct contrary is my opinion; and I think, in case of disorder and irregularity in the enemy's line, that the British fleet should be as compact as possible, in order to take the critical moment of an advantage opening and offering itself, to make a powerful impression on the most vulnerable part of the enemy. According to Mr Graves's Memo, any captain may break the line with impunity when he pleases.

---

63 Hood to Jackson, enclosure 1, 6 September 1781, *Hood* p32

64 Memorandum, 6 September 1781, *Signals & Instructions*, p260. The last word 'bearing' is printed as 'wearing'.

65 *Signals & Instructions*, p261

66 Rodney to Jackson, 19 October 1781, *Hood*, pp46–7

67 Graves to Sandwich, 14 September 1781, *Sandwich*, IV, p182; for editorial discussion see pp142–3

Of course there was never any question of breaking the *enemy's* line.

The captain of the *Bedford*, the Admiral's cousin Thomas Graves, commented in his copy of the memorandum:

NB. The foregoing order was given the day after the action with the French fleet of 24 sail of the line, in consequence of the second in command not bearing down and engaging the enemy, which was his duty; but kept his wind, by which means a most glorious victory was lost, and with it the loss of Lord Cornwallis's army in Virginia.[65]

The parallel with Lestock at the Battle of Toulon seems uncomfortably close.

Altogether there are four main points of criticism. The first is Hood's allegation that Graves wasted time in not attacking de Grasse much earlier. The second is his allegation that Graves failed to press ahead and so allow the British rear to get into action. The third is Graves's Memorandum, enjoining a more elastic notion of the line, the main object of which was to ensure that ships in close action should, nevertheless, have an unimpeded field of fire. The fourth is Hood's comment, that the Memorandum merely encourages individual caprice and hence disorder.

Rodney, now at Bath and 'very much out of order with a very violent pain in my stomach', misunderstood Graves's intention. Graves, he alleged, should have shortened his line so as to bring his nineteen ships against fourteen or fifteen of the enemy's, 'and by a close action totally disabled them before they could have received succor from the remainder, and in all probability have gained thereby a complete victory. Such would have been the battle of 17th April [1780] had I been obeyed'.[66] Graves himself, writing to the First Lord, made no direct allegations against anyone. He entirely glossed over the initial signalling failure when closing the enemy, and merely remarked obliquely of Hood, 'Unfortunately, the signal for the line was thought to be kept up until half after five, when the rear division bore down; but the fair occasion was gone.'[67]

Sir Charles Middleton, writing to Hood some months later, stated that Graves's private account, written to Sandwich,

appeared modest, with very little implication of censure. The not taking possession of the Chesapeake in preference to an action with a superior fleet seemed strange conduct when the object of the expedition was to succour Lord Cornwallis; but, until I received your account and observations, on the tenth [of January, 1782] I had not the least idea of your situation and the opportunity missed in consequence of it. This circumstance has been talked of by officers returning from America, but very little [was] known to the general public.

This raises yet another issue. Should Graves have steered at once into Chesapeake Bay, cut off de Grasse from the Franco-American land forces and captured the French troops still disembarking at Jamestown, or even have passed directly up the York River and opened communications with Cornwallis? This question has been much debated since, but it is unknown whether Hood actually raised it at the time, or only after Cornwallis's subsequent surrender revealed how critical his position had been at the moment when the battle was fought. The same criticism can be made of the similar opinion said by Middleton to have been expressed by officers returning from America. How many of them, had they been in Graves's place, would have been prepared to seize a position inferior to de Grasse without fighting, knowing as little of Cornwallis's plight as Graves then did?

New light is shed on this by additional signals issued by Graves at New York on 15 October 1781, four days before he sailed once more for Chespaeake Bay with all the ships and troops he could muster, being now aware of Cornwallis's desperate situation. Rear-Admiral Digby had already arrived to take over the command, but on the advice of the senior sea and land officers had agreed to serve under Graves until the expedition was over. These signals are of great strategic and historical interest, since they prove that Graves was now prepared to take risks which had seemed unjustified six weeks earlier when the conditions were much more favourable. The signals were:

1, 2, 3 For the van, centre or rear respectively 'to attack' the enemy's van centre or rear.

4, 5, 6 For the van, centre or rear 'to force through the enemy'.

7, 8, 9 For the van, centre or rear to retreat.

While numbers 1–3 and 7–9 were concerned with more precise tactical direction than had seemed possible on 5 September against de Grasse, numbers 4, 5 and 6 introduced the new Channel Fleet doctrine, but in extended form, the manoeuvre not being restricted to 'when fetching up to leeward of the enemy on the contrary tack'.

10 'To go into Hampton Road'.

11 'To push up to York Town'.

12 'To anchor as soon as the fleet gets above the enemy's fleet'

13, 14 'For the rear' ditto, 'For the centre' ditto.[68]

The really important signals are numbers 10 to 14, but coupled with numbers 1 to 9 they show a clear intention to fight if necessary in order to get into Hampton Road or 'push up to Yorktown', and so get 'above the enemy's fleet'. Evasion might possibly be achieved, but the whole purport of the signals was that a battle would and had to be fought.

Next day, 16 October, Graves issued four more signals, clearly designed to provide against another Chesapeake fiasco:

1 'For the van division to close the enemy's van and to engage as close as possible'.

2 Ditto the rear and the enemy's rear.

3 'To form a line parallel to that of the enemy's as near as possible'.

4 'For a particular ship to lead the fleet when in line or otherwise.'[69]

Finally on 18 October, the day before sailing, Graves reissued his post-Chesapeake Memorandum of 6 September, together with signals for regulating the line of battle.[70] Thus, on paper, he had done everything possible to provide for fighting his way through to relieve Cornwallis, for avoiding tactical misunderstandings, and for being prepared for the manoeuvre of breaking the enemy line without any qualifying restriction. Could any admiral have done better, at that particular moment, in making his paper preparations? Unfortunately Graves was too late. After the second return of the fleet to New York, Graves apparently issued his Memorandum for a third time[71] and on this occasion it provoked Hood's angry endorsement, noted earlier.

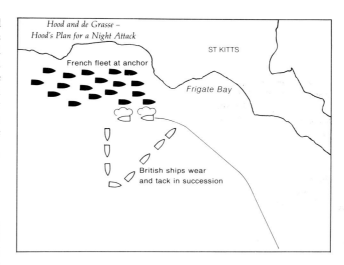

he sailed once more for Chespaeake Bay with all the ships and

68 *NMM*, KEI/S/8 (formerly Sp/8); signals issued in manuscript to the Hon George Keith Elphinstone, Captain of the *Warwick* (50)

69 *NMM*, KEI/S/8 (formerly Sp/8); signals issued in manuscript as before to Elphinstone

70 *NMM*, KEI/S/8 (formerly Sp/8)

71 6 November 1781, *Barham*, I, p127

72 'In attacking the enemy's ships to leeward at anchor, the ship to exchange her fire with the first and second ship of the enemy, stop at the third, and having given her fire to the third ship, to veer short round and fall into the rear, each ship following in succession.' *MOD*, NM/24 (formerly *RUSI* 24), printed in *Signals & Instructions*, p189

Hood and de Grasse at St Kitts

After the capitulation of Yorktown, the British garrison at St Kitts was besieged by French troops covered by the main French fleet under de Grasse. Hood, with a smaller force of troops embarked in his warships rather than in transports, advanced from the south to relieve the island. He intended to surprise the French fleet at anchor at dawn on 24th January 1782, and to adapt Howe's instruction for attacking the enemy's rear ships in succession by confining it precisely to the three rear ships.[72]

A collision delayed Hood's advance, and surprise was lost. At dawn on 25 January Hood was still about fifteen miles south of Nevis with the French about seven miles away to the north-west. Hood signalled for the line ahead at one cable, but it took about an hour and a half for the fleet to form and he was able to fill and stand on to the north. His plan now was to edge along the western shore of Nevis, across the narrow channel separating Nevis and St Kitts, and then pass inshore of de Grasse and occupy his anchorage in Frigate Bay. All this de Grasse himself could easily have divined. He had twenty-eight of the line to Hood's twenty-two, and at dawn the fleets were only seven miles apart. De Grasse had already formed his line and the chance to force an action against an inferior fleet was surely not to be missed. He had, however, already lost the first round by taking his fleet too far to the westward of St Kitts, and hence too much to

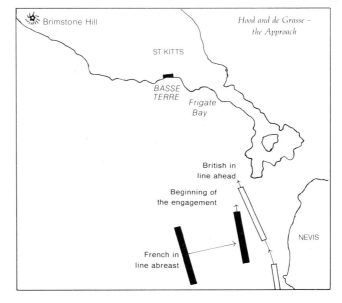

leeward. Nevertheless, he headed south to attack, and his slowness of approach in the light breeze was matched by Hood's slowness in forming a line.

As de Grasse approached, Hood altered course to larboard and westward to close him.

> Would the event of a battle have determined the fate of the island, I would without hesitation have attacked the enemy, from my knowledge of how much was to be expected from an English squadron commanded by men amongst whom is no other contention than who should be most forward in rendering services to his King and Country. Herein I placed the utmost confidence, and should not, I fully trust, have been disappointed.[73]

Nevertheless, Hood changed his mind at the last moment. Seeing de Grasse sufficiently to leeward, though still close enough for action, he decided to slip past him inshore and seize his anchorage. Having closed up his fleet and sent detailed orders to the leading ship, he 'pushed for it'.

All this time de Grasse was heading slowly south-east to attack. His van, or light squadron, which should have made every effort to intercept Hood, unaccountably bore away. De Grasse, thinking the van was becalmed by Nevis, signalled it to lie to, and soon after to fill and stand on. It was too late, however. At about 10.45am the north-east trade wind backed some four points towards the south, thus heading de Grasse and forcing him to bear away both from Hood and St Kitts. Conversely, Hood had now more sea room, the wind being on his starboard quarter instead of on his beam. Not to be denied, de Grasse tacked to the north onto the same course as Hood. By 2pm, and with more than three miles to cover before the British van could let go their anchors, de Grasse began to engage with his landing ships. Gradually his van drew up to engage the British centre while his centre engaged the British rear. Hood could not afford to set too much sail as he had to allow for precise anchoring which, if attempted too quickly, might have involved collisions and confusion in the face of the superior enemy fleet. His plan was for his leading ship to anchor first and as near as possible to a spit of land known as Green Point. The succeeding ships as they came up were to anchor in a line running out to sea to the west on a bank which lies on the west side of the island. This bank, Hood calculated, would accommodate about fourteen ships. When the depth of water became too great, the remainder were to anchor in succession, in line to the north-west and thus towards the shore, as soon as they found the water again sufficiently shallow. They would then lie in a curved formation.

The fleet would be near enough to stop any direct sea communication with the French fleet and their besieging forces. With springs on their cables, the ships could give their broadsides a wide arc of fire. In any case it would be difficult to attack them, as the normal direction of the wind would roughly bisect the angle made by the two arms of the formation. It was also practically impossible to double them by getting inshore at either end of the line.

Each of his ships, except the leader, had to pass her next ahead before anchoring, so that the line became inverted. But,

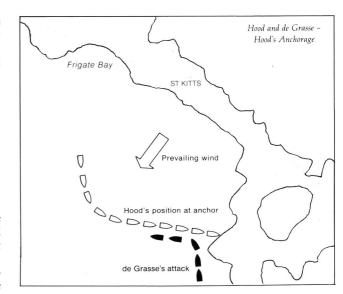

as Lord Robert Manners, captain of the *Resolution*, pointed out, this had certain advantages. The rear as they filed past covered the already anchoring van and centre. The centre, once in position, covered the anchoring movements of the rear. Nevertheless, Hood only just succeeded in getting his rear ships in under heavy fire. De Grasse made no attempt to lead round against the rear ships as they anchored furthest ahead; and, at 5.30pm, the first battle was over. Hood had achieved a complete success for the loss of one frigate and with one ship-of-the-line unable to find ground to anchor.

During the evening and the ensuing night Hood readjusted his line, moving four of his ships so as to make it impossible for the French to get round his easterly flank. When dawn broke on 26 January, the British fleet presented a respectable line with bowsprits and sterns almost touching, though at least four ships were under sail trying to correct their positions and anchor effectively. At 8.30am de Grasse attacked the Green Point end of the line. Though at first headed by the breeze, the French did better as they worked westward and seawards along the British line. De Grasse and his seconds backed their topsails so as to defile as slowly as possible past the British ships near the angular point. After this, he drew off, half his ships having never engaged at all.[74]

On 27 January the French made a further somewhat half-hearted attack, and thereafter they cruised before the island, keeping the British fleet under observation. On 28 January Hood landed the troops as a diversion to help the British garrison, but they were too few to make much progress and had to be re-embarked.[75] On 30 January de Grasse feinted to attack, but it came to nothing, though his fleet continued to remain in sight during the next fortnight.

On balance it can be said that the strategic victory had gone to the French. Had Hood been able to anchor at Sandy Point, right below the fortress on Brimstone Hill, he could have landed troops under his own guns and have kept up communication with the fortress by signal while under sail. Under these circumstances the situation of the French besieging force would have been difficult.[76] As it was, nothing more could be done to help the garrison.

When on 13 February the garrison surrendered, Hood felt it was time to be gone. He ordered the fleet to slip their cables simultaneously at 11pm that night. Each ship was to leave a lantern fixed to her anchoring buoy so as to give the appearance of a line of riding lights. At 9pm a lieutenant from each ship went on board the *Barfleur* to synchronise watches (an early instance of this procedure) and at 11pm exactly, the fleet began to make sail into the darkness. De Grasse was

---

73 Hood to Middleton, 7 February 1782, *Barham*, I, p144

74 Navy Board to Admiralty, 30 October 1786, *NMM*, ADM/BP/6b, contains a good contemporary sketch of the action.

75 See Hood's signals for November 1781, with a loosely inserted sheet of additional signals for landing and re-embarking marines, seamen and the 69th Regiment; *NMM*, Tunstall Collection, TUN/10 (formerly S/MS/Am/10)

76 See 'Journal of an Officer in the Naval Army in America in 1781 and 1782', *De Grasse*, pp168–9

completely deceived. He was some distance off, taking in provisions from victuallers, and did not realise the truth until at least thirty hours later. Hood was by then well clear, the extremely tricky night movements without lights having been executed without a single hitch.

The Battle of the Saints

The Battle of the Saints (9–12 April 1782) resulted from de Grasse's attempt to take a large convoy of merchantmen and 9000 French troops in transports to meet a force of Spanish ships and troops for a joint attack on Jamaica. As the French left Martinique they were sighted and pursued by Rodney's fleet from St Lucia.

At daybreak on 9 April, part of the French fleet and most of the merchantmen were still lying becalmed near Prince Rupert's Bay, Dominica. De Grasse, with the main body together with the transports, was catching the breeze in open water between the northern tip of Dominica and the small group of islands known as the Saints, lying roughly two-thirds of the distance between Dominica and Guadeloupe.

Although the British fleet lay becalmed to the south, the French still had a breeze. De Grasse, heading south on the larboured tack with his whole fleet in line-ahead in reverse order, was nicely placed to windward, between his convoy and the British. At about 8am Hood's van also felt the breeze, and the leading ships began to creep ahead, while Rodney's centre squadron and the rear were still becalmed. As soon as his rear squadron was abreast of Hood, the French fleet being in reverse order, de Grasse signalled his second-in-command, the Marquis de Vaudreuil, to break off and attack the British van. He himself continued running south with the van and centre to engage or at least threaten the main force of the British and cover the retreat of the straggling merchantmen. Vaudreuil led away to leeward and then wore round to the north, so as to get on the same course as Hood. He was also joined by three ships from the French centre, presumably detailed by de Grasse, so that he now had fourteen ships of the line against an immediate opposition of only eight, one of Hood's squadron being away to leeward and three more being still becalmed in rear. Hood was under easy sail so as not to draw too far ahead, and the French ships began to sweep past him at long range. At 9.48am one of them opened fire at the leading British ship, which replied, and at 9.50 Rodney, though still becalmed some miles astern, made the signal to engage. By 10.30 the first cannonade was over. Vaudreuil's force, having passed to the north, were tacking in

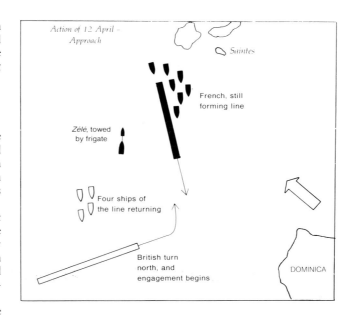

succession so as to run south, wear round in succession and repeat the cannonade, thus performing a circular movement. At 11am Hood backed his main topsail so as not to run too far ahead of the supporting ships eastern of him, and at 11.15am Rodney signalled to close the whole line.

By 12.14pm Vaudreuil was round again heading north and firing on Hood in the *Barfleur*, this time at closer range. In this action three British ships were badly damaged aloft and a French ship was damaged by a bursting gun. At 1.45 firing ceased, Vaudreuil having now passed Hood a second time without attempting any further attack. Indeed the remainder of the British van and some of the centre were beginning to creep up and exchange shots with the French van and centre. At 2pm Rodney hauled down the signal for the line ahead and at 3 signalled the whole fleet to wear to the south.

De Grasse had certainly secured the safe retreat of his transports and merchantmen. Nevertheless, it seems to be generally agreed by both French and British writers that he had lost a unique opportunity in not attacking Hood's leading eight ships with at least twenty of his own while keeping the remainder to windward so as to block Rodney from coming to relieve them.

De Grasse was provoked into further action against Rodney on 12 April by trying to save the *Zélé* (74) from capture after she and the *Ville de Paris* had collided during the night. The *Zélé* was taken in tow by a frigate towards Guadeloupe, but would clearly be captured unless de Grasse intervened, which he did by forming his line in reverse order on the larboard tack, heading south.

Captain du Pavillon was flag captain to Vaudreuil in the *Triomphant* and is said to have been so disturbed by de Grasse's signal that he had it verified twice before repeating it. Turning to Vaudreuil, he prophesied disaster.[77] He himself was killed in the battle.

With the trade wind blowing east between Dominica and Guadeloupe, the large island to the north, the French were steering SSW and to windward of the British but with their fleet not properly formed, several of those in the van being still too far to windward. Both fleets were in reverse order, the British line led by Drake's rear squadron and the French by Bougainville. Rodney had only five three-deckers, and, following the usual practice, three carried the flag officers

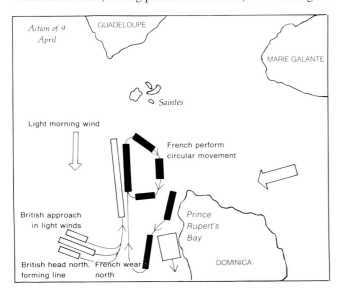

77 See de Goussencourt, *De Grasse*, p120

while the remaining two were seconds to Rodney himself in the *Formidable*. De Grasse had the same arrangement except that he had one extra three-decker for his extra, and fourth, flag-officer. The British fleet outnumbered the French by thirty-six to thirty and, although this was not in itself a decisive element in the battle, it marked the decision of the government to give Rodney local superiority.[78]

Although the four ships sent from Hood's squadron to chase the *Zélé* were not yet properly back in their stations, in the rear because the line was in reverse order, the rest of the British fleet was extremely well formed in line ahead steering about ENE on the starboard tack at a cable's distance. The French were cutting obliquely across the head of the British line on the opposite course and at 7.45am they were near enough for a ship at the tail of the French van to open fire at the leading British ship. Sir Charles Douglas recalled the approach to action:

> Thus standing towards each other on the contrary tacks, the wind moderate, the weather clear and the water perfectly smooth, the *Marlborough* being the leading ship of Admiral Drake's division fetched in with the sixth and seventh ship, counting from the headmost of the Count de Grasse's line and at half past seven [am] was fired upon; whereupon the signals for battle and close battle were made, our leading ship being supported by the quick and well directed fire of her followers sailing in due and close succession, now leading large, sliding slowly and closely down along under the enemy's lee.[79]

Rodney hoisted the signal for close action and for the leading ships to alter course to starboard so as to close the enemy. His signal was properly obeyed, each ship luffing as close to the wind as possible. De Grasse also altered course to starboard so as to keep his van in action with the British centre and rear, and to prevent the British from concentrating their fire on his own rear without having first received full punishment.

When the British van began to pass beyond the rear of the less numerous French fleet, Drake held his northward course, partly to give himself sea room to work to windward of the French and partly to give his squadron time for immediate and necessary repairs to the masts and rigging.

At the southern end of the battle the leading French ships were still not properly formed and were in danger of running under the lee of Dominica and being becalmed. As it was, the dying breeze was now coming more from the south of east and forcing them to edge away still further to starboard and towards the British, thus further upsetting their formation.

De Grasse was now in a most uncomfortable position, which he tried to remedy by hoisting the signal for the whole fleet to wear together, with the intention of putting himself on a parallel course with the British, heading north and on the same tack. By this means he could not only avoid being becalmed but also avoid losing the weather gage, since he could prevent the British van, now clear of his rear, from working round to windward of him. To execute this order properly the rear ship of all should have worn first, but her captain, being closely engaged, did not dare risk swinging his bows across the British broadsides. He therefore disobeyed the signal. Nor did Vaudreuil, de Grasse's second-in-command, now in the rear, repeat the signal. He also was closely engaged at musket shot, and deemed it impossible to execute the signal safely. Seeing that his signal had not been obeyed, de Grasse hoisted the signal for the whole fleet to haul to the wind again, thus avoiding the potential confusion caused by some ships wearing and not others. He then tried to extricate his fleet by making the signal to wear in succession. This would have had the effect of bringing his fleet parallel with the British and level with their centre and rear, with the leading ship wearing first and the remainder following round in what would have amounted to a U-turn.

Here again each ship would have had her bows raked in turn, and this time it was the leading captain who refused to obey and still kept his wind. Soon after this, the wind at the southern end of the battle, off Dominica, changed to southeast, taking some of the leading French ships aback. Those obstructed in their forward movement became bunched together, causing gaps to appear in the line. Those still able to proceed were turning away to starboard, while those still held up remained more to windward.

Rodney's *Formidable*, having engaged all the French van and centre in turn, was now approaching the *Diadème*, the leading ship of the French rear, which was taken aback and blocking the movements of the three ships astern of her. With the wind now round more to the south and in his favour, Rodney was close enough to the French to pass through the gap in their line and so to windward of their rear squadron. Five ships followed him, and each in turn, as they passed, poured a heavy fire from their starboard guns into the stern of the *Glorieux*, the last ship of the French centre. At the same time, Rodney's next ahead, the *Duke*, was taken by her captain, Alan Gardner, through the gap between the three ships bunched round the *Diadème* and the remainder of the French rear. When Rodney and those following him came round on this bunch from windward and opened fire with their larboard side guns, the four French ships were very severely handled.

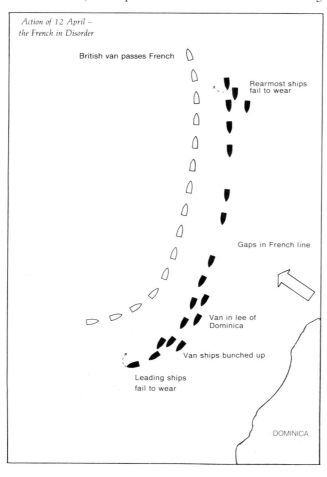

Action of 12 April –
the French in Disorder

British van passes French

Rearmost ships
fail to wear

Gaps in French line

Van in lee of
Dominica

Van ships bunched up

Leading ships
fail to wear

DOMINICA

---

78 Piers Mackesy, *The War for America, 1775–1783*, London, 1964, pp453, 457 and 459

79 Letter from Sir Charles Douglas to Sir Samuel Greig, Admiral in the Russian Service (private collection)

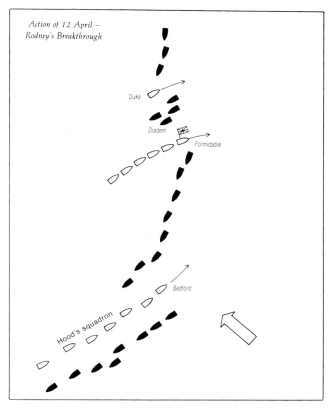

*Action of 12 April –
Rodney's Breakthrough*

Duke

Diadem

Formidable

Bedford

Hood's squadron

So far everything connected with the breakthrough seemed to have come about by accident. Later controversy centred on whether Rodney had planned the manoeuvre in advance, or whether, supposing it was a snap decision, the real credit for advising it should go to Sir Charles Douglas. It was almost certainly a snap decision, much as Gardner's was, since smoke obscured everything until the last minute, and those following Rodney did not know that they had cut the French line until a puff of wind blew the smoke away and showed the French to larboard and to leeward. Further in the rear, an exactly similar manoeuvre took place; Commodore Edmund Affleck's *Bedford* (Captain Thomas Graves) passed through yet another gap in the French line, possibly without realising it, and was followed by Hood with the whole rear squadron of twelve ships. As they cut through the French centre they gave the two leading ships, the *César* and *Hector,* a concentrated battering with their starboard guns, while firing their larboard guns into the ships round de Grasse.

Ostensibly all that had happened was that the two fleets had passed through each other and changed places, with the British now to windward. In reality their relative situations were totally different. In the breakthrough the French had lost their formation and sustained major damage. Gardner independently, Rodney and the five ships astern of him, and finally Affleck with the whole of the British line astern of him, had been able to fire their starboard broadsides into the sterns of the French ships ahead in the line, and their larboard broadsides into the clusters of ships trying to close the gaps. When, therefore, the French gradually emerged from the main engagement, they were in three groups with their damaged and disabled ships lying uncovered to windward.

By about 10.30am the first part of the battle was nearly over. Drake's squadron was already a long way to the north

and out of action. Rodney had ceased fire much earlier, though some of the centre was still in action. Some of Hood's ships were still engaged, their bows being towed round by boats so as to bring them back again towards the enemy when the breeze failed. Near these and somewhat to the south and leeward lay the worst damaged of the French ships. A little further to the south of these again were the ships under de Grasse. De Bougainville, with some of the French van, was still further south and to windward of de Grasse on the larboard tack. Vaudreuil with the French rear was to the west and leeward of Rodney and to the north-west and leeward of de Grasse. He was heading south-west and still further from the battle area. De Grasse tried to re-form his line, but it seems that he signalled the fleet to form on the larboard tack and on the leewardmost ships on Vaudreuil's squadron. In this way he lost all chance of supporting his worst damaged ships, and set in motion a general retreat. Had he signalled to form the line on the *Ville de Paris,* and thus closer to the British, Rodney would have had to act with more circumspection.

As soon as Rodney himself ceased fire, at about 9.30am, he sought to rally his fleet so as to renew the action. He made the signal to tack, since the bulk of the French fleet was now to the south-west and leeward of the British. Drake could not manage to tack in the light wind, but he wore round with his less damaged ships, though meanwhile Rodney had signalled for the van to tack and gain the wind of the enemy and for closer action. About 10.30 Rodney hauled down the signal for the line, repeated the signal to engage and for the van to make more sail. But with no wind and much smoke, little could be done or seen. At 12.30pm an easterly wind dispersed the smoke and made further action possible. All but about six British ships were able to renew the action. Rodney signalled the van to close the centre and later for closer action. The French were retiring to the north-west and gradually the British took up the chase under the direction of the divisional commanders, with Hood especially active.

During the afternoon and early evening four French ships, already damaged, were captured, though Rodney did not press the chase with any great vigour. Just as the sun was setting Hood in the *Barfleur* came up with the *Ville de Paris,* which was already engaged with three British ships. He raked her bows and de Grasse surrendered. It was Hood's great coup. Soon after, Rodney signalled the fleet to bring to, though some ships continued to chase and did not rejoin the fleet until later. Hood was furious, on the grounds that Rodney had failed to press the chase with vigour even at the risk of a night action and dangerous dispersal next morning. Without attempting to be dogmatic it is at least safe to say that Hood had a good case. Rodney's apparent refusal to reap the fuller fruits of victory, summarised in his famous complacency, 'Come, Come, we have done very handsomely', is clearly accounted for by physical and nervous exhaustion.

Order of battle in every form [wrote Sir Howard Douglas on the effect of cutting the line] depends essentially upon keeping distance. It is only by well directed cannonade, that impressions are to be made, and effects produced, that may disturb the enemy's order, and it is upon the talent with which this is watched for, and the promptitude with which advantage may be taken of such an effect, that success will depend. This was magnificently and splendidly done on the 12th April.[80]

This was the first time in the century that fleets passing on opposite tacks had been so closely engaged along their total lengths, in circumstances to which de Grasse contributed as well as Rodney. Though Rodney had no doubt brought station keeping and general fighting efficiency to a high level in a very short time, it was Hood who had commanded the

80 Sir Howard Douglas, *Naval Evolutions; A Memoire by Major General Sir Howard Douglas, Bart,* 1832 (NMM, Tunstall Collection, TUN/74 and 76) p63

bulk of the fleet in the West Indies since Rodney's last campaign. Between them they had set up a new standard in line of battle tactics.

Could the French have countered the cutting through process by bringing their disengaged larboard guns into action quicker and more effectively than could the British?

When Captain Alan Gardner carried the *Duke* through the French line, she 'received the fire from three ships more of the rear which cut our masts and rigging with sails all to pieces, after passing up raked us very much' (Master's log). Soon after, her main topmast and part of the top crashed down on the quarterdeck. This shows that some of the French rear ships used their larboard guns effectively, though in theory the *Duke* should have taken a terrible beating from all the seven ships of the French rear. On the other hand, the *Duke's* excellent gunnery enabled her own larboard guns to be used very effectively. In passing through the French line, moreover, she had heavily raked the three ships held up astern of the *Diadème* with her starboard guns.

The seven ships of the French centre ahead of the *Diadème*, when cut off from their van by Commodore Affleck leading through their line, could have put the British between two fires by engaging them with their larboard guns while the French van fired with their starboard guns. The *Ville de Paris*, de Grasse's flagship, was the third ship in this group. It appears, however, that the French ships ran right out of the fight to leeward, leaving the damaged *Glorieux* astern.

Though both fleets were taken by surprise by their passage through each other at three different points, the results were very different. The separate segments of the British fleet were less surprised in the sense that their lines were well ordered, while the French line was both disordered and taken aback by the wind. When, therefore, the French ran to leeward through the British in separate groups it was clearly a further stage in their tactical collapse – not a tactical remedy.

Sir Charles Douglas

Apart from their superior tactical cohesion, the British fleet is said to have been superior in gunnery as a result of improvements introduced in three of the most heavily engaged ships by Sir Charles Douglas. Ships' guns were always a danger to those who fired them. If the breeching ropes broke in the middle of a battle, as they often did, the gun might go crashing to and fro from one side of the ship to the other. This was especially dangerous in a heavy sea with a slippery deck, and had proved so both to Mathews's flagship at the Battle of Toulon and Hawke's at the Second Battle of Finisterre. Douglas's remedy involved the attachment of large steel springs to the breeching ropes, counter-weighing the breeching with heavy shot and fixing corrugated inclined planes in the rear of the gun carriage wheels.[81]

Flash from the touch hole and sparks from the muzzle were also dangerous, both to the gunners and the ship. Tin priming tubes had been available for some twenty years as a means of avoiding the use of loose powder scattered on the touch hole, but Hawke had reported these as 'very pernicious things', apt to fly out and wound the gunners.[82] Douglas introduced goose-quill priming tubes, filled with the best mealed powder, mixed with spirits of wine. Instead of the 'match' twisted round a linstock or more riskily held in the hand, Douglas applied a simple flint-lock striker with a lanyard attached for firing the powder in the goose-quill. Dangerous sparks were stopped by damping the wads inserted between the 'cartridge' and the shot, and between the shot and the muzzle. For greater ease and rapidity of fire Douglas introduced flannel cartridge cases instead of silk ones. These left less debris in the barrel, which avoided the need for frequent wormings in the middle of the fight.

These devices proved very successful at the Saints.

> Not a single goose-quill tube failed [wrote Douglas] nor did a gun require being wormed, so long as the flannel-bottomed cartridges lasted on board the *Formidable* or *Duke*, nor of the 126 locks on board the latter (every lower deck [gun] having two) did a single one fail. . . . the *Duke*, from the improvements alluded to, fired sometimes fully from both sides, and even with as much ease as if they [the guncrews] had been exercising; nor did a single atom of gunpowder catch fire by accident on board of her, she having, as usual, and as now is becoming the practice, as well as the *Formidable* and divers other ships, used wetted wads.[83]

Research by Dudley Pope, however, reveals that as a result of experiments at Woolwich, locks were ordered for all quarterdecks by an Admiralty letter of 21 October 1755.[84] Tin priming tubes were also to be used with locks and it was these to which Hawke objected as being dangerous. The same letter ordered the use of flannel cartridges instead of the current paper ones, presumably for all guns. Nevertheless there seems no evidence to support the view that all quarterdeck guns were fitted as ordered, and it may be that after Hawke's objection to the tin primers the whole plan was dropped and that Douglas's use of goose-quill tubes facilitated the re-introduction of the locks. The issue of the flannel cartridges still needs clarification.

Tactically, Douglas's most effective improvement was in the difficult process of traversing the guns:

> By the means, then, of bolts placed in the side, right in the middle between every two guns, into which we occasionally hook their tackles, we are able to point all of them, without using a crow or handspike, where knees called standards do not interfere, full four points before or abaft the beam, which I presume is to a degree of obliquity until now unknown in the Navy.[85]

---

81 Sir Charles Douglas to Middleton, 5 September 1779 and 17 September 1781, *Barham*, I, pp269–72 and 274

82 *Barham*, I, p282n

83 Douglas to Middleton, November 1781, *Barham*, I, p282. Douglas had been captain of the *Duke* since 1779 until taken into the *Formidable* as Rodney's First Captain.

84 *The Mariner's Mirror*, 51 (1965), no 1, p70

85 Douglas to Middleton, 12 July 1779, *Barham*, I, p268. The knees or standards were angular sections of timber set between the side of the ship and deck above to give extra strength, especially in view of the curvature created by the ship's tumblehome. The effect in some cases was to confine the guns within what amounted to semi-compartments.

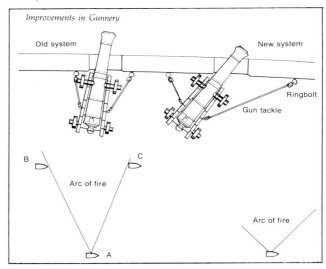

Improvements in Gunnery

Old system   New system

Ringbolt

Gun tackle

B   C

Arc of fire

Arc of fire

A

This meant that instead of having to bring the ship directly parallel to its opponent so that the guns, unable to fire except at right angles, might hit the mark, they could now fire at an angle up to 45° from each side. Each gun now was therefore capable of a total arc of fire of 90°. A ship so fitted could simultaneously engage two enemy ships, each lying at a distance of a quarter of a mile and in line-ahead at a distance of two cables from each other. This marked a tremendous tactical advance from the traditional situation where ships lying as little as two points on the bow or quarter of an enemy ship were practically immune from attack, though themselves able to rake the bow or quarter of the enemy, provided their broadsides were at right angles to the target.

Douglas wrote to Middleton that the ability to direct fire obliquely, as well as with greater rapidity, enabled the British ships so fitted to sustain a devastating fire against the French ships passing them on opposite tacks.

The *Ville de Paris* [de Grasse's flagship] . . . thought secure in being, as was supposed by them, out of the *Arrogant*'s line of fire, because four points on her bow, did, to their infinite surprise, in that direction receive such a broadside as had wonderful effect . . . if every ship in the fleet had been so appointed, and without standards, like the *Arrogant*, fewer – possibly very few – of the enemy's ships would have escaped. . . . I only allude to probable consequences, presuming that, had such oblique fire been general, few, if any, of the enemy's masts had been left standing. Lieutenant Butler . . . says that from the middle deck of the *Formidable* he never fired less than two, sometimes three broadsides at each passing Frenchman before such Frenchman could bring a gun to bear on him – from such guns excepted as standards are in the way of.[86]

To Lord Sandwich, Douglas wrote,

Sir George [Rodney] has so much confidence in the obliquity and quickness of fire of this His Majesty's ship [the *Formidable*], her guns being fitted and appointed as the new approvedly tremendous *Duke's* are as far as time would permit, that the *Formidable* penetrated the enemy's line of battle between the 2nd and 3rd ship astern of the gallant Grasse, almost totally silencing their fire by pointing her guns – where the untoward knees called standards do not interfere with the tackles – as far forward as possible, according to the new exercise, before they [the French] got so far aft as to be upon her beam.[87]

As far as is known, only the *Formidable* (90), *Duke* (90) and *Arrogant* (74) possessed all the necessary fittings on this occasion; their improved gunnery undoubtedly helped both to defeat and demoralise the French during the middle and end of the battle.

De Grasse's narrative

In October 1782 de Grasse published his own account of the battle with a schedule of signals certified by his chief staff, the Chevalier de Vaugirauld.[88] At 5.45am he had signalled for the line of battle on the larboard tack in reverse order, and at 6.15 to make more sail. This placed his third squadron, under de Bougainville, in the lead. He was worried about the possible capture of the damaged *Zélé* and felt compelled to fight to relieve her. The fleets were approaching on opposite courses, and he intended to get on the same course as the British in order to engage their van with as much advantage as on 9 April, that is if they continued to stand towards his centre. The British by this stage had stopped chasing the *Zélé* and were forming line on the starboard tack. When the wind backed a little to east, de Grasse was able to weather the oncoming British and felt that his own line could no longer

be cut. The situation was also more favourable for his own attack and for enabling his van to avoid being becalmed under the lee of Dominica, which was certain to happen if they continued on their immediate course. At 8am therefore, he signalled for a change of course to starboard, steering SSW. The engagement had begun at this point and his turn to starboard enabled de Grasse to keep his van in contact with the British and at the same time to carry it away from Dominica.

It seemed to de Grasse that the moment had come to put his fleet on the same course as the British and to concentrate his attack on their van, an attack they would find it difficult to relieve from their centre and rear. So at 8.15 he signalled the whole fleet to go about together. This signal was not carried out. He was now undecided as to the best method of going about, as his ships were already under heavy fire, including the unfortunate *Ville de Paris*, which went through several manoeuvres, but found her sails so riddled with shot as to prevent her from tacking. Yet every ship in the fleet could, like the *Ville de Paris*, have worn if unable to tack. The signal was repeated, but with no effect, and de Grasse was forced to fight on the opposite tack sooner than he would have done, and less advantageously, since his third squadron [under Bougainville and leading the fleet] was too far ahead to support him.

After this the wind veered to SSE, thus heading the French and favouring the British. Finding his signals unexecuted, de Grasse gave up his plan of attack on part of the British line and concentrated on stopping his third squadron, in the van, from being becalmed. He still felt it was possible to wear the fleet onto the opposite course. In this way his third squadron, now back in the rear in *l'ordre naturel*, would have been able to engage that part of the British fleet which had meanwhile sailed past the second squadron and the centre, and would have enabled the second squadron and centre to undertake temporary repairs before a second attack.

At 8.45am de Grasse signalled the whole fleet to come to the wind on the larboard tack and then to wear together. This signal was not carried out, although it was seen and repeated by the frigates. The commander of the third squadron held on his course, and the van followed his lead, rather than de Grasse's signal.

The centre and the second squadron (still in the rear) were forced to pass the enemy at close range. The *Ville de Paris*, heavily damaged aloft and with yards lacking braces and sheets, moved more slowly than the third squadron. There was now no proper order on which de Grasse could form a line. The third squadron, together with some ships of the first (the centre) which had followed them, were cut off from the rest of the French centre, which had now been cut through by the British centre. Meanwhile, the British van continued to overlap the French rear. The British centre outsailed the French centre, but made no attempt to double. The British rear, however, doubled part of the French centre and the third squadron, having cut through the French line.

The breeze died away and the fleets were now becalmed, with smoke covering the whole battle area. The centre and that part of the second squadron now with de Grasse had ceased fire. Firing continued, however, between the British and the ships of the centre and third squadron which had been doubled. The calm lasted three-quarters of an hour, until a breeze dispersed the smoke, and the battle then continued. At

86 4 May 1782, Barham, I, p284

87 13 April 1782, *Sandwich*, IV, pp256–7

88 *Mémoire du Comte de Grasse sur le Combat Naval du 12 Avril 1782, avec les Plans des Positions Principles des Armées respectives. MOD, EG. 161*

Amiral *Pierre André de Suffren Saint Tropez (1729–88)*. MM

Admiral Sir Edward Hughes (c1720–94), by Sir Joshua Reynolds. NMM

10.45am de Grasse signalled his frigates to close him to repeat signals, but only the *Richmond* obeyed. He could now see the second squadron (under Vaudreuil) re-forming and heading SSW. The rest of the centre, as well as the third squadron, were making a long detour to get to leeward of the six ships still following the *Ville de Paris*. The *Glorieux* was one of this number, and de Grasse was not too heavily engaged to signal the frigates to take her in tow. The *Richmond* passed a cable under fire, but when the wind freshened and other ships came under fire the cable was cut.

At 1.15pm, seeing his signals so little attended to, de Grasse made a general signal for the whole fleet to close. The third squadron did not obey and there seemed no hope of saving the *Glorieux*. The battle was, however, not yet lost, and action of a kind continued. De Grasse signalled to close in reverse order of battle on the larboard tack and for the second squadron, still in the rear, to come to the wind together. The British van now overlapped the French rear to such an extent that the British ships would have been out of the battle completely if de Grasse's signal had been obeyed. At 1.45 he signalled the second squadron, in the rear, to shorten sail, and this at last was obeyed. The third squadron did not shorten sail, though the second was within range of de Grasse to leeward and well enough formed. At 2pm he repeated this signal for the line in reverse order on the larboard tack. This was again ignored by the third squadron and by the ships of the centre already in retreat. At 2.45, therefore, de Grasse signalled the second squadron to shorten sail as the only means of closing them, and this signal was finally obeyed when it was repeated at 3.07.

At 3.30, seeing that so many of the fleet were running large, away from the enemy, instead of obeying his signals, de Grasse signalled the whole fleet to come to the wind together on the starboard tack, and thus onto his original course at the beginning of engagement. He thus hoped to take advantage of the easterly breeze, which would enable the fleet to rally and come to the wind. This signal was again not obeyed by the third squadron and those ships of the centre with it.

Meanwhile it was necessary to prevent the *César, Ardant* and *Hector* being cut off, and to give the *Ville de Paris* a chance to recover her position. It was also necessary to slow down the British pursuit, which was being executed without much regard for order in the belief, no doubt, that little further resistance was to be expected.

At 4pm de Grasse repeated the signal for the whole fleet to close and form in *l'ordre naturel* on the starboard tack. The second squadron obeyed, and, had the third done so, together with the ships of the centre accompanying them, the line could have been formed. Meanwhile the *César* had been forced to surrender and the *Hector* and *Ardent*, in danger of being cut off, signalled for help. The second squadron did all they could to disengage the *Ville de Paris*, but the rest of the fleet continued to keep the wind astern of them. At 4.15 de Grasse repeated his signal in the hope that the example of the second squadron might influence the rest of the fleet to come to the wind together on the starboard tack. He repeated it once more at 5.45. It was still ignored, and the retreating ships crowded sails – most of them setting studding sails as well.

The second squadron, faced by the whole British fleet, which had abandoned the pursuit in order to concentrate on them, gave up the attempt to detach the *Ville de Paris*, and the flagship was soon surrounded on all sides. As every boat in the ship was holed, de Grasse was unable to shift his flag.

Suffren and Hughes

The five battles fought in the East Indies in the years 1782–3 between Pierre André, Bailli de Suffren-Saint-Tropez, and

Admiral Sir Edward Hughes exemplify the deadlock in tactics which characterised so much of the American War. In each of these battles, internal circumstances severely handicapped the rival admirals at different times and in varying degrees: leaky and barnacled ships, crew shortages, devastating sickness amongst crews and shortage of water and provisions. External, and in the long run more important, circumstances were the strategic implications overhanging each battle. Upon each action depended the capture or loss of the naval and commercial bases Negapatam, Trincomalee, Cuddalore, Batticaloa, Madras and Porto Novo. The battles were dominated by the relationships of the admirals with their respective political and military masters and colleagues; French relations with Hyder Ali and Tippoo Sultan were particularly important. The pervasive influence of the policy and trade of the rival East India Companies could always be felt. Preoccupations with the great seasonal movements of the rival merchant fleets, as well as of the merchantmen in use as transports for military stores, were never absent.

The two admirals were well contrasted. Suffren was reputed a man of genius, a tactical innovator, careless of his personal appearance, a great fighter, a man capable of inspiring great efforts; above all, a man determined to make the defeat of the enemy's forces his chief object as a means of winning the war for France in the East Indies. Hughes, a picturesque though less sensational figure, was without a flicker of tactical originality. Sir Herbert Richmond describes Hughes as a 'safe' man, and he was recommended by Palliser to Lord Sandwich as such: 'He will not wander out of the path that may be prescribed to him, and follow any schemes [sic] or whims of his own, nor never will study to find fault with his orders, but always how he may best execute them for His Majesty's service.' [89] The implication is that, tactically, Hughes would not attempt anything new. To call Hughes dimwitted, docile or conformist, however, seems more than unfair to a man who, by sheer hard fighting, checked the reputedly greatest tactician of the age. In the execution of his entirely standard orders, he was able to rely on the firm support of his captains. This was an advantage which in the end was to prove fatal to Suffren's visions of absolute victory.

In the Battle of Sadras, 17 February 1782, Suffren, having the wind and also the advantage of twelve ships to nine, sought to crush the British rear by leading his coppered ships, which were less hampered by marine growth, along the British line only as far as Hughes in the centre, while his disengaged ships doubled Hughes's rear. Hughes, sensing nothing of this plan, calmly awaited the French in a well-formed line to leeward. Suffren issued the clearest orders to Captain Tromelin, his second-in-command, both verbally and in writing, together with the over-riding Nelsonian exhortation to do the best in the circumstances to assure success. Nevertheless, the plan failed.[90] Five French ships did at one time or another fire on the *Exeter*, the rearmost British ship, and three of them made some attempt to work round her stern – one of them, the *Brillant*, with some success – but five of the French rear ships were never properly in action at all. There was just enough wind for these ships to have worn to leeward and got into action, but there was not enough for Hughes's disengaged ships in the van to tack and double Suffren from windward. When night and a squall ended the battle, therefore, Hughes had survived what might have been a major disaster, but he had been unable to exploit the weakness and confusion of the French attack.[91]

In the Battle of Provedien, 12 April 1782, Suffren again attacked from windward, with twelve ships to eleven, coming down with his coppered ships in the lead, intending a ship-to-ship engagement and without any plan for concentration. His attack was somewhat ragged and, though a fierce action developed in the centre, the two leading French ships held off once they had run the length of the British line, while the French rear failed to close in at all. Nevertheless, the concentration of French fire in the centre drove the *Monmouth* out of the line, disabled. With his fleet uncomfortably near the shore, Hughes now wore in perfect order and engaged the French on the port tack. Suffren also tacked in response, and the battle continued with the two fleets on the same course. Suffren's flagship soon lost her foretopmast and had to leave the line, with the admiral transferring his flag. Squalls and darkness ended the battle and Hughes anchored, with the *Monmouth* in less than four fathoms. Suffren drew off with his flagship in tow of a frigate and two other ships badly damaged aloft.

Next morning Hughes re-anchored his ships in a well-sited line with springs on their cables so as to be able to align their broadsides on any bearing from which the French might attack. Meanwhile, carpenters, smiths and riggers were sent from the whole fleet to help repair the *Monmouth*, anchored inside the line.[92] The two fleets remained in sight of each other for a week, Suffren also carrying out repairs and trying to tempt Hughes to abandon his defensive position and fight. When this failed he withdrew. Two of Hughes's ships had scarcely been able to fight at all owing to scurvy, and a week after the battle he landed 1000 sick men at Trincomalee.

In the Battle of Negapatam, 6 July 1782, Suffren started with twelve ships to eleven but, this time, he was to leeward, and had lost the use of the *Ajax* the previous day, when her maintopmast and mizzen topgallant mast broke in a sudden squall, so the numbers were therefore equal. On 6 July Hughes tried to close the French from windward in line of bearing, ordering a ship-to-ship engagement except that his next astern was to join with him in attacking the same French ship. Suffren tacked to pass astern of Hughes and seize the weather gage. Hughes was nearly caught, and by the time he

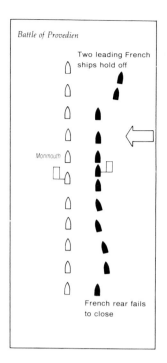

Battle of Provedien

Two leading French ships hold off

Monmouth

French rear fails to close

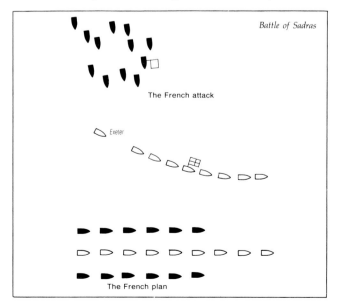

Battle of Sadras

The French attack

Exeter

The French plan

89 Quoted by Sir Herbert Richmond in *The Navy in India*, pp88–9

90 Castex, *Les Indées Militaires*, gives an extract from Suffren's instructions to Tromelin, p306

91 Suffren in describing the battle states that except for the *Brillant* which actually doubled the British rear, no ship had been as isolated as his own or had 'essayé autant de corps'; Castex, *Les Idées Militaires*, p308

92 See *Le Balli de Suffren: Documents Inédits . . . par Gustave Lebat*, 1901, pp33–43

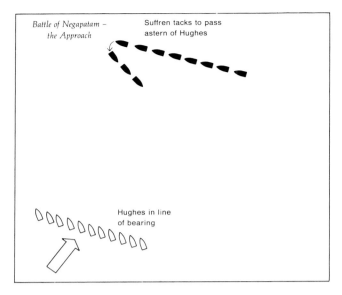

Battle of Negapatam – the Approach

Suffren tacks to pass astern of Hughes

Hughes in line of bearing

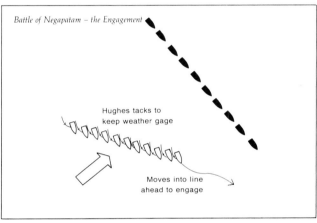

Battle of Negapatam – the Engagement

Hughes tacks to keep weather gage

Moves into line ahead to engage

Battle of Trincomalee

Four French ships engage *Exeter*

Exeter retreats

Some French ships fail to engage

Vengeur on fire

had also tacked and begun to modify his order from line of bearing to line ahead, his ships were somewhat in disarray. Nevertheless, he made the signal to engage. The attack was poorly executed, and the two leading ships collided, but righted themselves in time to make a joint attack on the leading French ship, which suffered heavily. The four rear British ships, when they came up, failed to close the five rear French ships and kept up only a distant fire. As before, the main fight was in the centre between the admirals and their seconds.

After an hour and a half, a change of wind accompanied by a squall brought confusion to the battle. Suffren and Hughes both tried to re-form their lines, but the French had two ships dismasted and another which was thought to have surrendered, while three British ships were ungovernable. So, with the wind dying away again, the battle came to an end, and the two fleets gradually separated. Despite this inconclusive end, British casualties were twice those of the French.

In the Battle of Trincomalee, 3 September 1782, Suffren had the weather gage and was again superior in numbers, with fifteen ships to twelve. He had, moreover, replaced some of his more unsatisfactory captains by younger and supposedly more reliable men. His only special order was that his two rear ships were to concentrate their combined fire on the rearmost ship of the British.

Like Hughes in the previous battle, he found it impossible to keep his fleet in proper order when closing the distance in line of bearing, or a lasking course. As his ships were not all capable of the same speed under the same sail, some outdis-

tanced others and several lost their correct alignment.[93] Seeing that his leading ship was now within range, and that nothing better in the way of formation could be hoped for, he signalled her to stand down direct for the British. He fired a gun to attract attention but this was taken as a general signal to engage, so that the action began with several of the French firing from too great a distance.

The four leading French ships engaged the leading British, the *Exeter*, and drove her out of the line to leeward, but they failed to capture her. The two rear French ships failed to obey the signal to double on the rearmost British ship, and one of them caught fire and retreated. In the centre a tremendous battle ensued for a full hour, at the end of which Suffren's flagship and her next astern were dismasted and the ship astern had lost her mizzen topmast. Meanwhile five of Suffren's ships had hardly opened fire at all, despite a general signal to engage at pistol range. In the British line three ships, including Hughes's flagship, were practically unmanageable and three more were in danger of sinking altogether, having received many wind and water shots – a tribute to the much-maligned French gunners, often accused of firing only to bring down masts and spars.

Suffren now signalled his van to tack in succession and support him in the centre, but as the wind had dropped, they had to have their heads towed round by their boats, while being fired on by the invincible *Exeter*. Then the wind changed and Hughes signalled to wear. Those of his ships which were still capable of making way came round in good order, though punished by the French as they turned their sterns. Suffren meanwhile contrived to get his manageable ships to tack, so that the active units of the two fleets were again on the same course, though not in any proper order. As night fell, firing gradually ceased and the fleets separated.

Neither admiral felt capable of trying to renew the action next day. Hugh survived, despite the severe damage inflicted by the smaller number of French ships which had actually engaged him at close quarters in the centre, because at least five French ships were never fully engaged at all. Suffren's comments on three of these – 'Mal', 'Très mal', and 'Mal comme toujours' – sufficiently indicate the mortification he must have felt in being deprived of what might have been an outstanding victory.[94]

In the Battle of Cuddalore, 20 June 1783, the last of the series, Hughes finally had numerical superiority over Suffren, with eighteen ships to fifteen. He had, however, less than two-thirds of his full complement of seamen, and in addition his fleet suffered from over 2000 men sick with scurvy and dying daily in considerable numbers. He was also short of water. Suffren, with fewer ships, outnumbered Hughes in men by about 8500 to 5500. On the other hand, two of his ships were leaking badly, eight were uncoppered, and nearly all had been uncleaned for more than three years, whereas several of Hughes's had been recently docked in Bombay. Neither admiral, therefore, had anything like a fully efficient force.

Suffren had decided to stake everything on a concentrated attack on Hughes's rear, more radical in concept than the attack he had intended in the first battle. Seven only of his ships were to fight in the usual way, and these were to space themselves out along the British line so as to contain the thirteen ships of Hughes's van and centre at long range. The five French 74s were to engage the five rear ships of

93 Confirmed by letter from the *Flammand; Le Balli de Suffren: Documents Inédits*, p55

94 'Je ne puis attribuer cette horreur qu'à l'envie de finir la campagne, à la mauvaise volonté, et à l'ignorance, car je n'oserais soupçonner rien de pis'; Castex, *Les Idées Militaires*, p314

Hughes's line in combination with his three remaining ships, which were to double Hughes's rear from leeward. Suffren would thus have a superiority of eight to five in the rear. On the day of the battle, however, despite having the weather gage, he completely abandoned his plan and ordered the traditional coterminous linear action. He merely warned his rear ship, a weak 40, to avoid close action with the British 74s and 64s.

The French fleet closed the British in the afternoon. Suffren, in somewhat grudging obedience to a royal order, dated 6 May 1782, flew his flag in a frigate, an experience which led him to remark after the battle, 'en sera la première et dernière fois'. In a fierce battle lasting two hours and ending at dark, neither side gained any advantage and neither admiral felt able to fight again when opportunities occurred in the next few days. Suffren's lack of ships had been offset by his stronger guncrews, and he had in no sense been worsted.

John Clerk of Eldin wrote that Suffren

has given us something new, not only by obliging Sir Edward Hughes to act on the defensive, but by having, in his masterly seamanship, attempted a change, and put in practice a new mode of attack from to windward. He is also the first of an enemy, for this century at least, who will be allowed the honour to have made an attack upon a British Squadron.

Suffren was praised for concentrating, in the first battle of 17 February, a superior number of ships against an inferior. In the second battle it was said that 'his attempt upon the van, equally well concerted on this occasion, evidently proves him to be an officer of genius and great enterprise'.

John Clerk of Eldin

It is fitting to conclude the study of tactics during the American War by looking at the work of John Clerk of Eldin. Clerk is distinguished, not only as the author of the first complete and original work on naval tactics written in English, but also as the first to attempt any serious study of naval tactics in themselves, as something apart from naval evolutions. His interest in tactics stemmed particularly from his friendship with Commissioner Edgar, who had been present in Byng's flagship at the Battle of Minorca and in Boscawen's at Lagos Bay. The friendship had begun in 1770, so that by the time of the Battle of Ushant, Clerk, aged fifty, was already a skilled civilian critic and could analyse, or attempt to analyse, the tactical aspects of each new battle as news of it arrived. He went about his investigation systematically, examining past examples and asking fundamental questions. Considering the four trials of Keppel, Palliser, Byng and Mathews, he regarded it as remarkable that 'not one single hint has escaped, from anyone concerned, that it was possible anything could be attributed to the system of the attack itself, or that any kind of improvement should be attempted'. When noting that the French had invariably avoided attacking from windward, he wrote, 'Shall we not have reason to believe, that the French have adopted, and put in execution, some system, which the British either have not discovered, or have not profited by the discovery?' What Clerk was saying amounted to a condemnation of the whole British system of fighting. Reading his book today one can at once sense a well argued case.

Clerk was a well-to-do Scottish merchant and a dilettante of considerable distinction both in arts and science. Prevented from going to sea by his family, he became passionately interested in sea life and ships, using every opportunity to visit ports and talk with naval officers, merchant seamen and shipbuilders. In this way he acquired a great mass of information about shipbuilding, navigation and naval gunnery. He was led on to study tactics, using model sailing ships which he rigged himself, and making small models which he could carry in his pocket and set out for demonstration and discussion on a table. He moved in intellectual circles in Edinburgh, and was at the same time eager to put his ideas at the disposal of the Royal Navy of King George III. In this respect he formed part of that highly distinguished body of Scotsmen who, after the eclipse of active Jacobitism, felt no longer inhibited from taking service under the English Crown. It was no mere coincidence that at the very moment that Clerk's book began to appear, names such as Dalrymple, Rose, Douglas, Duncan, Hope, Elphinstone, Patton, Murray, Mackenzie, Cochrane and Carnegie were being mentioned on the flag list, in most cases for the first time.

Clerk had one particular advantage over all his highly distinguished French predecessors. Not only was he a civilian and, therefore, entirely uncommitted by training or profession to the support of any particular system, but also, unlike Hoste and McArthur, he had no official connection with the sea service. He was a rich man in an independent position, holding no office that could be prejudiced by the expression of inconvenient views. He could be neither intimidated nor pigeon-holed by official superiors. He did not depend in any way on the financial success of his publications.

Fifty copies of the first part of *An Essay on Naval Tactics* were printed in 1782 to be 'handed about among friends'. It was reprinted for sale to the public in 1790.[95] The second, third and fourth parts were published in 1797, together with a reprint of the first still bearing the date 1790, in a single collective volume. All four parts were republished in the second edition of 1804. The third edition, published in 1827, after Clerk's death, includes notes originally made by Rodney in the margin of his privately printed copy of 1782.

Unlike Hoste and many of his successors, Clerk wrote with only one object in view: the defeat of the enemy. He makes no mention of evolutions or orders of sailing. His book was a tract, not an instructional treatise. He was not

Battle of Cuddalore – Suffren's Plan

The Hon John Clerk of Eldin (1728–1812); plate from Cow's 'Scott Gallery'. BM

---

95 There is a rare copy in the library of the Royal Navy Staff College, Greenwich.

concerned with defence, which is why the book does not discuss the relative advantages and disadvantages of ships and fleets fighting from windward or leeward, so much as their relative positions for attacking from windward or leeward.

His main theme is advanced in the first few pages. The attack from windward was the wrong method of attack. If made at right angles in line abreast, it exposed the attackers to a devastating broadside of raking fire. If made on a converging course, either in line ahead or line of bearing, it meant that the leading ships would be heavily punished. As in the case of Byng's action, this would be certain to throw the ships astern into disorder unless they immediately turned away and headed perpendicularly towards the enemy, in which case they would be heavily raked. They could never support the van effectively if it was held up and crippled. Unlike the French theoreticians, Clerk continually reminded his readers that the opposing fleet were unlikely to remain idle spectators of any particular form of attack. Although it might suit their purpose, for the moment, to receive the attack in line ahead with only steerage way on, they were always free to act to the best advantage.

Using Pocock's first battle with d'Aché, 29 April 1758, as an example, Clerk showed what he described as the Curve of Pursuit from windward.[96] If the ships of the attacking fleet approached on a lasking course, each aiming at her opposite number in the enemy's line, it might be possible by crowding sail to bring every ship into action simultaneously at musket shot. If, on the other hand, they began by approaching in line abreast they would have to keep correcting their course to conform with that of the enemy, who would all the time be drawing ahead, even if only with steerage way. As a result, the attackers' line of approach would take the form of a curve, thus increasing the distance to be covered and gradually leading into a line-ahead. Eventually they would come within musket shot of the enemy and onto a parallel course, but somewhat astern. To draw up level and engage ship for ship, they would have to endure the fire of each of the enemy ships as they passed along the enemy line.

The leeward fleet was in a position to turn away from an attack, and form a new line further to leeward, thus challenging the windward fleet to another expensive onslaught. In any case, the leeward fleet would be in a far better position to support its own leading ships if they were hard pressed, since they could be signalled to turn away to leeward in succession while the centre and rear made sail to cover their retreat. A further sophistication was open to the leeward fleet, in which alternate ships might be ordered to quit the line and start forming a fresh line further to leeward. Meanwhile 'the intermediate ships left behind them in the line would be sufficient to amuse even the whole' of the attacking fleet.

Clerk deployed his evidence, beginning with

engagements, where the British fleet being to windward, by extending their line of battle, with a design to stop, take, destroy, or disable, the whole of the line of the enemy's ships to leeward, have been disabled before they could reach a situation from whence they could annoy the enemy; and, on the other hand, where the French, perceiving the British ships in disorder, unsupported, and thus disabled, have made sail, and, often throwing in their whole fire upon the van of the British fleet, ship by ship, as passing in succession, have formed a line to leeward, to be prepared if another attack should be made.[97]

The particular examples Clerk employed were the battles of Minorca, Grenada, Arbuthnot and des Touches, the Chesapeake, and Martinique. By analysing each battle carefully with the help of the best sources then available and with copious diagrammatic illustration, Clerk easily established his thesis. He was not concerned with criticising the plans or conduct of any particular admiral but merely with showing that the accepted British practice of attacking from windward was unremunerative. To contrast British tactics with those of the French, Clerk chose examples 'of engagements where the French, by keeping their fleet to windward, have clearly shown their dislike, as well of making the attack themselves, as of suffering the British fleet to approach them while in this windward situation'.[98] The particular examples employed were Rodney's two abortive actions on 15 and 19 May 1780, Hood and de Grasse off Martinique on 29 April 1781, and the Battle of Ushant. Here again Clerk had no difficulty in showing that, when the French were themselves to windward, they never initiated a serious attack.

Clerk's central idea about how the British fleet should be handled first found its way into his work as a short section entitled 'Other Observations', placed at the end of his comments on Ushant. It had originally been intended for inclusion in his privately printed edition of 1782 but was eventually omitted as being premature.[99]

In a case where two fleets approaching on opposite courses were contending for the weather gage, the losing fleet would invariably turn away to leeward, and the fleets would pass on opposite tacks. Clerk contended that the fleet which lost the

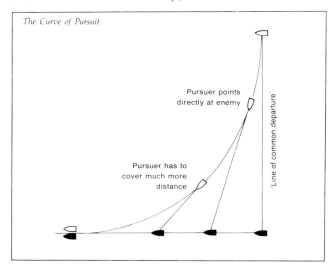

The Curve of Pursuit

Pursuer points directly at enemy

'Line of common departure'

Pursuer has to cover much more distance

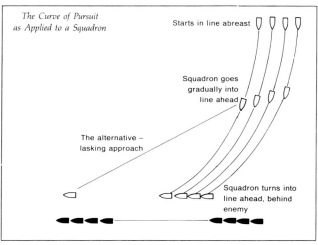

The Curve of Pursuit as Applied to a Squadron

Starts in line abreast

Squadron goes gradually into line ahead

The alternative – lasking approach

Squadron turns into line ahead, behind enemy

96 p160 (second edition, 1804)

97 p43 (second edition, 1804)

98 p86 (second edition, 1804)

99 p103 (second edition, 1804)

weather ought to cut through the enemy line instead of giving way, this action being more prejudicial to the fleet which was cut than to the cutting fleet, which would retain its original formation and be able to use both broadsides. The ships of the cutting fleet could rake the sterns of the enemy's centre or rear ships, these last mentioned being thrown into confusion and forced to leeward because they could no longer proceed straight ahead. 'The very first time ever we shall have the spirit to make the experiment, the success will

be sufficient to justify the attempt, by convincing us, that the risk or damage to shipping in making the attempt, will be found to be of less moment than in any one other mode of attack whatever', was Clerk's prediction. If the leeward fleet could not cut the enemy line, the next best thing to do would be to double the enemy's rear immediately an overlap was evident. Clerk wrote that the court martial of Mathews after the battle of Toulon was the reason the Royal Navy never learnt the value of breaking the enemy line. Only those ships that failed to follow Mathews in attacking the enemy should 'be considered as breakers of the line'. 'And hence that Sentence of the Court Martial which broke Mr Mathews', concluded Clerk, 'ought virtually to be considered as the source of all the many naval miscarriages since'.[100]

In Part II Clerk developed his plans for the attack from leeward.[101] He demonstrated that a fleet to leeward which was determined to press an attack would eventually get their chance when the wind changed. Indeed, they might get it sooner if a ship in the windward fleet which sustained damage to her rigging through carrying too much sail in an effort to keep up to windward, had to be relieved. Since an engagement with the fleets passing on opposite courses had been proved unremunerative, the leeward fleet, if headed, should cut the enemy's line as soon as they could get near enough.

Clerk's second method for an attack from leeward was for the cut to be executed by the fifth or sixth ship in the leeward fleet. If the cut were made near the rear of the windward fleet the attackers would be able to concentrate on the separated

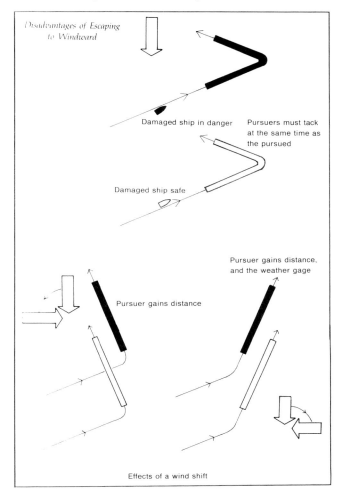

100 p117 (second edition, 1804)
101 p179 (second edition, 1804)

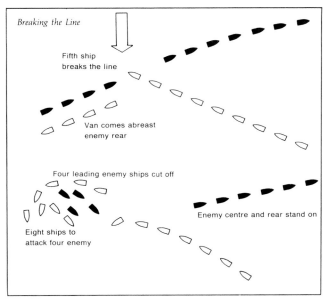

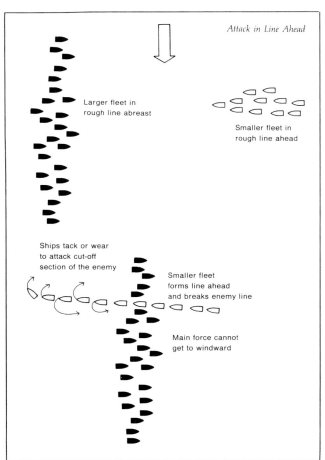

enemy rear ships; if towards the centre, the cut would still force the rear of the windward fleet down to leeward. Clerk admitted that it would be dangerous for a cut to be made by this method towards the enemy van.

In a remarkable foreshadowing of the Battle of Cape St Vincent, Clerk postulated a situation in which a large fleet was advancing in an irregular line abreast with the wind abeam. At daybreak and in foggy conditions, the fleet might be firing signal guns for getting into their proper order. A much smaller fleet, warned by the gunfire, might appear suddenly ahead on the opposite course and about level with the centre of the larger fleet. The commander of the smaller fleet could at once form a line ahead and cut right through the larger fleet. He could then tack together or in succession, concentrating on that part of the larger fleet to windward of him and leaving the leeward part unable to intervene. The parallel with the Battle of St Vincent is extraordinary, though St Vincent himself said that the claim that Clerk had 'any merit' in that action itself was 'totally out of the question'.

If an admiral was determined to make an attack from windward, Clerk's proposed method was for concentration on the three rear ships of the enemy line.[102] The particular ships detailed for the attack would have to be the best sailers in the fleet; the remainder coming up well to windward of the attacking group, in order to give support, should be in loose order rather than in line-ahead. Clerk argued that, once the three rearmost enemy ships were engaged, and their speed checked, their loss would be inevitable. Whatever the rest of the enemy fleet did in an effort to save them would fail. If the fleet tacked in succession or together, the supporting ships of the attacking fleet to windward would be there to deliver raking broadsides. If they wore in succession or together, the attackers could also wear and drive the leeward fleet into flight. The three victims would remain in the clutches of the attackers. Indeed, by this time the fourth and fifth rearmost ships might well have been engaged and surrounded by the attackers coming down *en masse* from windward. Clerk considered various methods of cutting off the enemy's rear ships, but not the two specially advocated by Howe: attacking in succession, each ship then taking station in the rear; and attacking by inverting the line, the first ship attacking the

enemy's rearmost, and each succeeding ship passing the ship ahead of it on the disengaged side.

It was possible that the fleet whose rear ships were in danger might react more quickly by wearing together before the attack was launched. In that case, the same ships, now the headmost of the enemy fleet in reverse order, could be heavily attacked. The leeward fleet might indeed extricate itself, though the windward fleet had still a chance to harry their new rear. Clerk considered the possible changes of wind, as well as partial breezes, and concluded that, once the three rearmost ships had been heavily engaged, even the most favourable wind would not enable the rest of their fleet to extricate them without becoming involved in a pell-mell.

An easy objection to all these schemes for cutting the

102 p123 (second edition, 1804)

Attack on the Three Rearmost

Other ships come up when possible

Three fastest ships engage three enemy rearmost

Enemy tacks in succession to relieve three rearmost

After tacking, has lee gage and is in danger of being raked

Enemy tacks together

Enemy in disorder, and in severe danger of being raked

enemy's line from leeward is that Clerk never allowed sufficiently for vigorous counter action by that part of the windward fleet which had not been cut off. In general, he was far too much inclined to dismiss them as out of the battle, demoralised, and making off after realising the difficulty of helping their friends. To this, however, he had an even easier reply, although he never made it specifically. If the ships of the windward fleet did succeed in intervening, the result would be a real pell-mell which is exactly what the attackers, assuming they were the British, most desired.

Clerk undertook a lengthy analysis of the amount of fire given and received in action fought by fleets passing each other on parallel and opposite courses. At very slow speeds, no individual ship could possibly fire or receive more than one broadside at or from any other individual ship. At higher speeds their guns would bear on each other for less than a minute and it might not be possible to reload before the next

ship in the opposing line came directly abeam.

Clerk's bold statement on cutting the enemy's line as a deliberate tactical procedure must not be confused with anything that took place at the Battle of the Saints, on which occasion the three movements were all 'accidental and unpremediated'. In Part IV, which was first published in 1804, Clerk reviewed the Battle of the Saints at great length. Rodney's accidental cutting of the French line was shown to have reaped all the advantages that Clerk had prophesied. He did not, however, suggest at any point, in either his narrative or observations, that either Rodney's or Affleck's or Gardner's cutting through the French line was in any way inspired by anything that he himself had written. In everything he wrote, Clerk was extremely modest, especially in view of his apparently heterodox teaching. It was his enthusiastic and less cautious followers who made exaggerated claims on his behalf.

The Battle of the Saints, 12 April 1782 (detail), by Nicholas Pocock (see pp 179–84). NMM

CHAPTER EIGHT

Tactics of the French Revolutionary War

THE stimulus of defeat, and an empire lost, ensured that tactical reform continued in the British fleet following the American War. Howe was at the Admiralty from 1783 to 1788, with a single break of eight months. His administration was by circumstances one of demobilisation and retrenchment, but tactics and signalling continued to develop, stimulated also by the feeling that unrest in France and the alarming progress of the Revolution were inimical to any real peace settlement. Reform was inevitable with such men at the head of the service as Howe himself, Hood, Alan Gardner, and Jervis.

On 18 May 1785 Vice-Admiral Mark Milbanke, then port admiral at Plymouth, issued a set of numerical signals with 145 signals listed.[1] On 27 June 1786, Commodore Alan Gardner, commanding on the Jamaica station, issued another. The signals Gardner devised were not, it must be admitted, of a very progressive kind. There were no battle signals at all, except that to engage. No signals for fleet formations were included. The day signals for private ships were on the old system, as also were the signals for calling officers. Fog signals were by a table, employing guns first to indicate the column, and then after a pause of two minutes, the horizontal line. Despite its limitations, however, this set of signals was a pioneer code by a man whose reputation had so far rested on his fighting qualities alone. On 16 October 1787, at Spithead, Lord Hood issued a set of ninety signals worked on a true numerical basis, but accompanied by a squared table apparently used to express the same numbers.[2]

Admiral Philip Patton's system, 1787

Not all British tactical progress was in the right direction. Philip Patton, a highly capable and experienced officer who had been constantly at sea since 1755, also produced a new system. He began his work well before 1784 (when his first draft appeared[3]) and his book was finally printed at his own expense in 1787. By that time, events had already passed him by; Howe had begun to move towards a true numerical system. Patton got no further than attempting a compromise between the old and new systems, presenting a scheme for operating both at once. In his letters to Sir Charles Middleton (now Lord Barham) advocating his own scheme of signals, and castigating those of Howe and his ex-flag captain, John Leveson Gower, Patton seems hardly to have understood current trends in tactical thinking, quite apart from his admission that he found it difficult to obtain access to their latest codes, or to those of Hood. Even in that age of free discussion, security was beginning to be an issue.

Middleton, whatever he might have said to Kempenfelt by way of encouragement, was a complete reactionary who knew nothing of tactics, and had never seen a battle in his life. In a splendidly blimpish effusion, with a veneer of apparent common sense, he asserted that seamanship was enough to win battles.

The more I consider the situation of the fleet, the more I am convinced of the necessity of confining your movements to the old Fighting Instructions, and which are certainly better adapted to fleets who fight on an equality than the more modern craze. Simple movements with the assistance of good leading ships are within the reach of the meanest capacity: but when additional signals are given out without system or order, as I am afraid is the case with all those I have yet seen, it must only bewilder the judgment and increase the confusion . . . A facility for forming and tacking the line seems to comprehend the whole of what is necessary at present. . . All I fear at present is the confusion of our signals; but however bad they may prove, they will do no harm if left asleep, and however ignorant our commanders may be at setting out, their seamanship will give them great advantages in forming and keeping the line as soon as they have been practised so as to understand it.

Perhaps Jervis was not being too severe when he wrote with characteristic violence of 'that damned fellow Sir Charles Middleton . . . the utmost extent of his abilities having gone no further than . . . compiling with the aid of that dull dog Patton the most voluminous stupid code of signals that has been exhibited by the signalmongers of the present age'.

As time went on, Patton became more and more worried about his failure to obtain official recognition. He abandoned the idea of sending a printed copy of his book 'to every admiral upon the list', because of the expense, but seems to have distributed widely both the book and an eight-page pamphlet on naval signalling.[4] He toyed with the idea of seeking the patronage of his young ex-shipmate the Duke of Clarence, but foresaw difficulties.[5] Nothing came of a grandiose plan of his that the Admiralty should appoint a jury of flag officers to select the best signal system amongst the many projected.

Patton's book is a large and cumbersome folio weighing over 6lbs. It contains 352 numbered pages, many left blank, together with many unnumbered blank pages. It was fit for use only on a cabin table.

1 *NMM*, HOL/39 (formerly S(H) 39). The book is in manuscript and is unfinished.

2 *NMM*, HOL/39 (formerly S(H) 39); see also *NMM*, ADM/BP/7, Navy Board to Admiralty, 1 October 1787. Hood evidently requested signal colours with a printed set of ten flags, quite unlike those used for his signals of 16 October.

3 *Barham*, II, pp368–85

4 *A System of Signals combining the method commonly used in the British Navy of making the signal from fixed places of the Masts or Rigging with a Numerary Method By which the Flags are showed where they may be best seen. So as to gain the Advantages of both these Methods: at the same time, Securing to the Commander in Chief, a Certain proof, that his signals have been seen and understood, in every Ship under His Command, NMM*, Corbett and Tunstall Collections, TUN/38 and 43 (formerly S/P/Am/2 and 8) [Tunstall notes other copies in the Mead Collection and the National Maritime Museum, which have not been identified]; and *A Short Sketch of the use and importance of Naval Signals* (*NMM*, Tunstall Collection S/P/Am/3). There is a clear reference to the *Sketch* in *Barnham*, II, p375.

5 Prince William Henry, Duke of Clarence, was given or obtained a copy; *NMM*, Corbett Collection, TUN/43 (formerly S/P/Am/8), signed 'William Henry' and corrected in manuscript by the Prince's hand.

The system of day signals it contains depended on the use of ten flags, numbered 0 to 9, and four pendants, which allowed for fourteen groups of signals, up to a total of 925. Flags hoisted at the foretopmast head expressed units, at the main topmast, tens; and at the mizzen topmast, hundreds. Signalling by the old method was facilitated by having the two methods printed on the left and right openings of each double page. By using the three topmast heads and the mizzen peak, together with one or two of the four pendants hoisted either above or below, or both above and below, the flag in question, a large number of options was possible with a single flag. In the case of the groups signified by the pendants themselves, another, and even a third pendant, was hoisted below the indicative one. In all but three of the groups, Patton confined himself to forty-two signals, and in all but two groups there were seven signals on each page. With maddening eccentricity, however, he omitted numbers from his series, without any discernible reason. An outstanding virtue of the system, however, is that every ship in the fleet was envisaged as possessing all ten flags and pendants, which would permit two-way signalling, and repeating of signals.

Each group of signals was preceded by a set of general orders, which provided a useful and necessary codification of service practice, the majority of which presumably were already established. Group 4, for evolutions in the order of sailing and for bringing to, tacking and wearing, included the standard movements, and most of Kempenfelt's signals and instructions for the period 1779–81. Group 7, line of battle and evolutions, was thoroughly in tune with the Howe-Kempenfelt developments. The general form and tone of the signals were to give the admiral flexibility of control, and to ensure proper station keeping and tacking and wearing procedure by individual ships. There was no hint of evolutions for their own sake.

The night signalling system allowed for seventy signals, divided into six classes of ten signals, except for Class 3, which had twenty signals and was restricted to private ships signalling the admiral. The admiral could make his signals in one of three ways. Lights could be employed to indicate by their number the class of the signal, and by their arrangement the number of the signal within the class. The admiral would probably also use gun signals to indicate the number of the signal. A third possibility was that the admiral might not use lights at all, but would signal entirely by guns, one to six to indicate the class, followed after a pause by one to ten to indicate the number.

Every ship was required to acknowledge the admiral's signals by showing at the mizzen peak the correct number of lights for the class, and by firing musket shots to indicate the number of the signal in the class. If the admiral had made his signal entirely by guns it was to be answered similarly with musket shots. To make their own signals, private ships employed three lights in different arrangements to indicate the twenty signals of Class 3. They could also signal by means of guns for the first ten, and false fires for the second. The admiral acknowledged the signal by showing a single light at the ensign staff. There were only ten fog signals, made using guns and acknowledged by muskets.

By combining the old and new systems, Patton evidently pleased nobody, quite apart from the fact that the tide was now flowing strongly with the new. His night signals were confusing in themselves, quite apart from the mad-hatter explanations he provided. The tactical signals did little more than reproduce the progressive ideas of 1779–81.

'An Essay on Signals', 1788

By contrast with Patton's work, *An Essay on Signals by an Officer of the British Navy, 1788*, is a neatly bound volume measuring only 7 by 4½ in.[6] The author proposed a true numerical code using ten flags, three triangular flags, and a cornet. He provided 1051 signals, arranged in alphabetical order according to the first word of the signal. The normal procedure was to fly the flags one above another at the foremast, but for long-distant signalling the hundreds could be flown at the mizzenmast head, the tens at the main, and the units at the fore. A flag at the mizzen peak denoted thousands. Triangular flags were used for substitutes. The yellow cornet indicated that the number of the signal was to be read as a numeral. A really distinctive feature of the book was the device employed for night signalling. This was a square signal lantern with numbers printed on the four rectangular crown-glass windows. Gun signals were used for fog signalling. Up to six guns fired at different rates, from quick firing to six-minute intervals, made possible thirty-two different signals.

In the main the book is simply a signal book, without an instructional element. One of the best features is the very large number of interrogative signals, requiring some kind of numerical answer, which enabled the admiral to gather information from the fleet about numbers of sick and casualties, latitude and longitude, soundings, how many days supply of various stores on board, etc.

Unfortunately the author spoiled his effort by devising an additional set of thirty-two quite unsuitable flags to be used as single-flag battle signals. Supplementary flags for a single-flag signalling had already been introduced by Kempenfelt, but the object here was to provide special and distinctive flags to be used in battle only, and preferably with the admiral in a frigate. Their complex design and colouring made them hopelessly indistinct. Some were not merely confusing but bizarre: the sun rising from a strip of blue sea against a red background; red with a blue border and with a red cross on a yellow fly; yellow with a red horizontal stripe through the middle thickened near the fly and with a blue oval over all; red and with the Man in the Moon in yellow and white. Nevertheless, the book is a notable contribution to signalling literature, even if it never received encouragement and came too late to have practical value.

The National Maritime Museum possesses an ingenious experimental numerical signal system in manuscript, in which all the signals could be made by a hoist of two of the ten numerical flags, which were so designed that they could be flown upside down to double the number of options.[7] In 1788 Rear-Admiral the Hon John Leveson Gower, who had been Howe's first captain at the relief of Gibraltar in 1782, and had come to command of the Channel Squadron, issued another set of numerical signals. However, his system was not at all progressive as it depended upon an eight-sided table.[8]

Lord Howe and Sir Roger Curtis, 1788–9

The close association between Howe and Sir Roger Curtis dated from the American War, when Curtis was appointed to be Howe's second captain on the North American Station. Subsequently Curtis had played a distinguished part in the defence of Gibraltar, and he was destined to be Howe's First

---

6 *NMM*, Corbett Collection, TUN/37 (formerly S/P/Am/1) with the book plate of Prince William Henry, Duke of Clarence and William IV. The plates are marked 'Published 25 Aug 1787 [sic] by S Hooper.'

7 Source unknown

8 *MOD*, Ec/5

Captain at the Glorious First of June. There is no evidence to suggest that, apart from being a good chief of staff, he had any wide experience of tactics or had made any serious study of signalling.

Howe resigned from the Admiralty for the second time in July 1788, and sometime during that year Curtis devised a new set of signals and instructions, of which a copy in the form of a general index to the signals, initiated by Curtis and dated 1789, still survives.[9] This book is frequently quoted in a set of handwritten discussion notes dealing with draft signals and instructions, written jointly by Howe and Curtis in September 1789, which throws much light on the Howe-Curtis partnership.[10] There are also frequent references to Howe's signals and instructions of 1782. In many cases, criticisms and objections appear in the same hand as that of the originating item, suggesting that these had been copied by the author of the draft from a paper already received from the other partner.

It is impossible to imagine Howe at any moment in his career, and certainly not at this particular moment, proposing anything as retrogressive as Curtis's 'Admiral's Signals', which employed a table of sixteen flags, along with ten

9 *NMM*, Corbett Collection, TUN/7 (formerly S/MS/Am/7), a marbled covered notebook 7¼in × 4¼in.

10 Seven double foolscap sheets and four smaller ones, some headed 'Sr R C' and one dated 'Sept 1789'; *NMM*, Tunstall Collection, TUN/72 (formerly HC/MS, once in Lord Sligo's collection of Howe Papers)

11 Barrow, pp142–3

12 MS line of battle and order of sailing, signed by Howe, 10 July 1790; *NMM*, Tunstall Collection, TUN/19 (formerly S/MS/R/4); *NMM*, DUC/1.2; and the former *RUSI* 146, which may be in the *MOD*, but has not been identified. Tunstall identified *NMM*, 37 MS 1792, as another undated copy, but it cannot be traced. *NMM*, DUC/2 (formerly Sm/91) has order of sailing signed by Howe 19 October 1790; squadrons to be ½ mile apart, divisions 'nearly at the same distance;' ships three cables apart in fair weather.

13 *NMM*, MKH/A/2/n/4, signed by Howe, *Queen Charlotte* at Spithead, 25 July 1790, and issued to Captain Keith Stewart, *Formidable*

supplementaries, presumably to signal numbers. As usual, the numbers assigned to the tabular flags had only an indirect relationship to the number of the signal. The fact that they had any kind of relationship at all makes them all the more confusing. For example, flag hoist 1 over 1 equals signal number one, but flag hoist 5 over 9 equals signal number 73. The same flags could also be used to provide single-flag signals of importance, such as number 3, 'to alter course together', and number 7, 'to engage the enemy'.

The notes provide a wealth of information about the progress of Howe's and Curtis's thoughts about tactics. For example, Curtis proposed a signal, no 70, which would require the fleet 'to cannonade the enemy's ships within reach, aiming at their masts – the guns to be pointed by the most expert gunners'. Howe queried it, wondering whether 'under *any circumstances* this appointment in a signal book would be advisable? And, if not productive of an *evident* effect, it might not tend to relax the Ardor of the men?'

Curtis dealt very fully with night signals, using Howe's method. Howe had always stuck to his original system of 1776, classifying signals according to the number of lights shown, followed by 'how disposed', (ie, in a square or triangle, etc.), followed by where shown in the ship, and finally by guns and false fires if any. In fact, however, Curtis provided no indication of where the signals were to be shown. There are numerous references in the notes to Howe's signal book of 1782 and Curtis's of 1788.

These documents show that Curtis had little appreciation of what Howe was aiming at. Far from feeding him with new ideas, he was merely offering outmoded, pedestrian devices. In October 1789 Howe wrote to an unknown correspondent that 'the looseness of our present system of tactics in the navy, if any system may be properly said to exist, is such that I cannot say I have quite made up my mind upon the plan that I would recommend for publication'.[11] The following year, however, Howe was given command of the Channel Fleet formed as a result of the Nootka Sound dispute with Spain. In responding to the occasion, Howe grasped the nettle and issued to the fleet a true numercial signals system.

Howe's numerical signals, 1790

The Channel Fleet was an impressive force of thirty-one ships-of-the-line divided into three squadrons, each of two divisions. Admiral Samuel Barrington, with Jervis, commanded the van, Howe with both Lord Hood and Sir William Hotham commanded the centre, and Sir Alexander Hood with Sir Richard Bickerton commanded the rear. The order of sailing was still in three parallel columns in the same order as in line of battle, but with Howe supernumerary and leading the starboard or weathermost column.[12] The fleet's tactical efficiency failed, again, to reach Howe's exacting standards.

The *Signal Book for the Ships of War*, issued to the Channel Fleet, marks the final development of Howe's signalling and tactical system.[13] It was the long-delayed masterpiece for which the more progressive British sea officers had been waiting. It was re-issued by Howe himself as well as by Hood, Jervis and others, and remained the standard form for the navy until the amalgamation of his signals and instructions in the first official Admiralty signal book, which was issued in 1799, and remained in effect until 1816.

Twelve special-purpose flags were supplied, three of which could be inverted, and these could be employed for a number of miscellaneous single-flag signals. Otherwise, all inhibitions were cast away, and a true numerical system established. Five columns were provided in the signal book for numeral signals. In the first was entered, in writing, the number of the signal, the second gave the 'purporte', the arrangement of the signals being determined by the alphabetical order of the words in this column; in the third column was entered a brief 'signification'; in the fourth and fifth columns the page in the instructions, and the relevant article number were noted. All signal numbers and many signals were inserted in handwriting, especially the battle signals, possibly for security reasons. There were 160 numbered signal spaces in all, with many blanks.

Howe's introduction of numerical signals caused the Board of Admiralty, on 26 June, to request the Navy Board to supply flags and other necessary signal colours, provided they should judge them 'fit to be adopted'. It is interesting that the practicality of the design was referred to the expert judgment of the Navy Board. On 28 June the Navy Board gave their approval, and on 2 July they supplied the Admiralty with a complete list.[14]

Important and otherwise new signals in Howe's 1790 'Signal Book for Ships of War'

Article 19 (MS) 'Ships of the fleet are at liberty to fire upon the enemy in passing them, though not prepared to bring them to a general action immediately.'

Article 20 (MS) 'Particular ships or divisions of the fleet to attack or harass the rear of the enemy or such part thereof as the detached ships may be of competent force to engage for giving opportunity to bring on a genereal action.'

Article 22 (MS) 'To quit and withdraw men from ships captured after having destroyed or disabled them so that the enemy cannot carry them off if time permits, and to join the fleet.'

Article 40 (MS) For passing through the enemy's line in either direction [as in Signal 75 of 1782].[15]

Article 41 (Printed) 'The ships to take suitable stations for their mutual support, and engage the enemy, as arriving up with them in succession.'

Article 42 (MS) 'Each ship of the fleet to steer for independently of each other the ships opposed in situation to them in the enemy's line.'

Admiralty Flags of 1790 designed by Lord Howe
1 Red
2 White with narrow red border
3 Blue, white, blue vertical
4 Yellow, with narrow black edging top and bottom
5 Quarterly white and red
6 White and blue diagonally divided
7 Blue with a yellow saltire
8 Yellow and blue vertical
9 Blue, white and red, horizontal
10 Blue with a white centre
11 White
12 White with a blue cross
13 Red with a white cross
14 Chequered yellow and blue
15 Union
16 Red and white horizontally divided
17 Blue and yellow vertically divided
18 Blue and white striped horizontally

19 Blue cornette
20 White and black divided horizontally
21 Yellow
Flag sizes: large ships 12ft broad by 14ft long;
frigates 10ft broad by 12ft long.
Pendants sizes: large ships 8ft at staff, 36ft long and 2⅔ft at end of fly; frigates 6ft at staff, 27ft long and 2ft at end of fly.[16]

Howe was in a dilemma when it came to signals from private ships. Despite all his urbanity and his willingness to seek the opinion of his subordinates on grave tactical matters, he was still an autocrat, for whom it was unthinkable that private ships should have the full set of ten flags, plus the substitute flag and the hundred pendants. As a result, sound principles adhered to with such praiseworthy persistence were cast aside and the private ships given an eight-flag table providing sixty-eight numbers, together with substitute and annulling flags, a pendant signifying numbers only, and an interrogatory pendant as well. To make themselves understood the private ships had a parallel set of twelve flags and two pendants. Of these, only one was contained in the

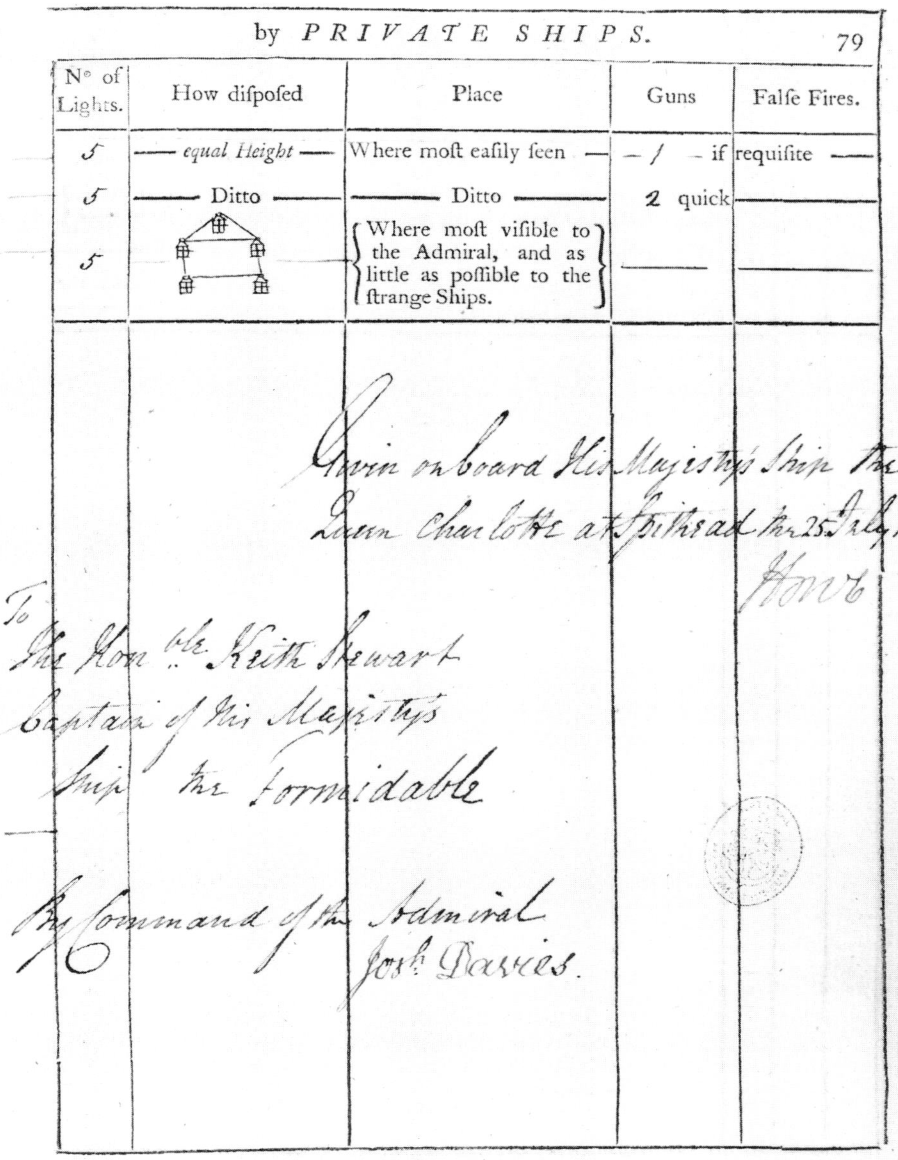

Admiral Lord Howe's signature, dated 25 July 1790, at end of his Signal Book for the Ships of War, issued to Captain Keith Stewart. NMM

Nº of Lights.	How disposed	Place	Guns	False Fires.
5	equal Height	Where most easily seen	— / — if	requisite
5	Ditto	Ditto	2 quick	
5		{ Where most visible to the Admiral, and as little as possible to the strange Ships. }		

14 *NMM*, ADM/B.P./10 (1790)

15 Sir Julian Corbett was under the impression that this form of the signal was 'for an entirely new manoeuvre' (*Signals & Instructions*, p255) and based a good deal of his argument about Howe's tactical ideas on this erroneous supposition.

16 *NMM*, ADM/BP/10 (1790)

admiral's numerical series, the rest being drawn from the special-purpose flags. At the head of the table it was stated that it might 'be used by the admiral occasionally'. The night signals – both for the admiral and for the private ships – were much the same as those in the signal book of 1782. In view of these complications, it seems unlikely that Howe was dissuaded from allowing private ships the numerical signals because the Navy Board refused to issue more flags. The choice seems to have been his own.

For many years there was some doubt as to the actual authorship of the accompanying 'Instructions . . . Explanatory of, and Relative to . . . the Signal-Book herewith delivered', issued by Howe in 1790. Luckily there is a copy in the Mead Collection, signed 'Howe, 9 October 1790, *Queen Charlotte* at Spithead'.[17] This was after the fleet had finished its cruise but was prepared if necessary to put to sea again. In layout, general content and actual text, it is practically a reprint of Howe's instructions of 1782, and thus of 1776 as well. So, once more, there is evidence that the system of tactics devised for the Royal Navy, under which all the great victories of the Revolutionary and Napoleonic Wars were won, was already promulgated before the American War of Independence had properly begun.

The Instructions are divided up in exactly the same way as before: 'General Instructions'; 'Instructions . . . Preparatory to, and in Action with the Enemy'; fog; sailing by night; battle by night. There are twenty-four battle instructions in place of the previous twenty-two. Numbers 23 and 24 are for exercising particular squadrons or the fleet – additions which would have been useful in 1790 no doubt. There is also number 25 added in writing, to the effect that, if a detachment of the fleet was ordered to attack part of the enemy fleet but was so far ahead that it risked being 'overpressed by a superior force', it should confine itself to 'retarding the progress of the enemy' until the admiral arrived with the rest of the fleet, and brought on a general action. There seems to be a hint here of the 'advance squadron'. The fog, night sailing and night battle instructions are the same as before.

Howe also reissued his printed 'Additional Instructions Respecting the Conduct of the Fleet Preparatory to and in Action with the Enemy', dating from 1 July 1777, and almost identical.[18] It seems curious that by this stage he should not have thought of incorporating his additionals into his main book of instructions. No doubt keeping them separate was the way of least resistance, the Royal Navy having been long accustomed to additionals of every kind. In view of the amount of time he had been devoting to tactics and signalling during the previous year, it is hardly likely that his action, or rather inaction, was due to laziness. A more likely reason is that he was already in the process of producing a complete volume, entitled 'Additional Instructions for the Ships of War', which appeared in 1793. In the meantime, he was experimenting with the separate sections of this volume by issuing them piecemeal.

On 10 July 1790, while the fleet was still preparing at Spithead, Howe issued a 'Separate [printed] Instruction respecting the signal with a white flag with a blue cross, page 9 of the Signal Book.'[19] This is a preparative signal (Article 1) which ushers in a highly ambitious series of instructions for night battle. Article 2 states:

If the preparation so required refers to a fleet or squadron of the enemy present, either greater in force and number of ships than the British fleet, or which the Admiral may for other reasons mean to defer bringing to action until at, or after, the close of day, when he judges it may be done with greater advantage in two or more divisions acting independent of each other; and it then happens that the circumstances of the case do not admit of a more leisurely communication of his plan; he will signify the form (if to be different from any before delivered) in which he would have the ships drawn up for conducting the attack after dark; by showing the flags expressive of the different divisions, in succession at proper intervals and putting abroad with each divisional flag the signal pendants distinguishing the several ships of which he would have (in such succession of place) each of the two or more divisions to consist. The commanders and captains are thereupon to take their stations accordingly; and to keep their ships in readiness for proceeding to engage the enemy in the order so pointed out when the proper signal is made for engaging the enemy at or after the close of day. But the divisions of the fleet, for the purpose of this signal, will be to remain as in the established form of battle expressed if no such change is to be made in the disposition thereof as in this article provided.

This article is quoted in full not so much as a further example of Howe's prose style, which unfortunately did not improve with age, but as evidence of the lengths to which he was prepared to go in order to ensure a satisfactory form of attack under hazardous conditions. Article 3 seems to suggest a ship-to-ship engagement, though Sir Julian Corbett saw in it Howe's intention, like Nelson's at Trafalgar, to contain the enemy with his own centre division.

On the same day, 10 July, Howe issued two further documents. Of the first, 'Instruction respecting the formation of the Order of Battle', the main purpose of the most important article, number 1, was to insure correct leading by the head or tail ships of the respective squadrons when the fleet tacked or changed formation.[20] 'Of the Order of Sailing', consisted of ten articles and was his first full-dress attempt to regularise the cruising formations adopted by the fleet at sea.[21] It was the original version of what appeared in his Additional Instructions three years later. Article 1 reads: 'When the fleet is formed in Order of Sailing, in three columns or parallel lines, it is meant that the Commander in the Second Post should keep with his Squadron or Column, to Starboard when Sailing from the Wind, and to Windward

17 Mead Collection, *NMM, TUN/127* (formerly Mead/Ins/4) Another copy in the Mead Collection, *NMM, TUN/128* (formerly Mead/Ins/5), is inscribed 'Rear-Admiral Kingsmill' as the cover label.

18 *NMM, Sp/80*, signed Howe 10 July 1790; see also *Signals & Instructions*, pp319–26, reprinted from the Bridport Papers, *BM*, Add Ms 35194. Compare pp108–16, Howe's additionals of 1777. In the 1790 reissue the last four paragraphs of Article 9 are in manuscript.

19 Issued to Vice-Admiral Sir Alexander Hood (Lord Bridport) and printed, in *Signals & Instructions*, pp327–31

20 *NMM, DUC/1/3*; issued to Captain John Duckworth, *Bombay Castle*. The same instructions, issued to Sir Alexander Hood, are printed in *Signals & Instructions*, pp331–3

21 *NMM, DUC/1/4*; issued to Captain Duckworth

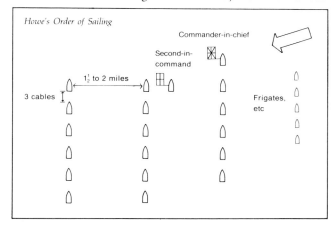

Howe's Order of Sailing

of the Centre Column by the Wind, on both Tacks.' Article 2 established that the columns were to be one and a half to two miles apart, with the ships at about three cables distance, and that the fleet would steer no nearer than a point from the wind to enable the most leewardly ships to keep their stations. Article 3 directed the fleet to keep station on the leading ship of the weather column. The admiral would himself usually take station about three cables ahead of the leading ship of the weather or starboard column. In Article 6 captains were warned not to fall out of station even if the ship immediately ahead of them did so. Article 8 ordered that when the fleet sailed in two columns the same regulations were to apply as when sailing in three. In Article 9 frigates, fireships and smaller vessels were ordered to keep to windward when the fleet was sailing by the wind, and to take station outward when the fleet was sailing large. And in Article 10 it was ordered that leading ships which became the sternmost, and vice versa, as a result of tacking, wearing or altering course, were 'to continue in this performance of the services incident to such changes in their situation, until replaced by some subsequent evolution in appointment'.

On 24 August 1790, a week after sailing from Torbay, Howe issued a 'Memorandum for the more particular explanation of the Admiral's intentions respecting the order of sailing'.[22] It was an exhortatory document on station keeping, drawn partly from Articles 6 and 7 of his Order of Sailing. In particular, it was noted that ships were often seen to fall to leeward for want of sail aft. The summer's cruise proved once again the tactical inefficiency of the British fleet, at least by Howe's own standards.

The fleet returned to Spithead in the middle of September and, in anticipation of an autumn cruise, Howe issued on 19 October a new order of sailing and a new order of battle.[23] The order of sailing was for thirty-seven ships divided into three squadrons each of two divisions. This time the admiral was usually to lead the starboard, or weathermost, column. In the order of battle, Barrington commanded the van and Sir Alexander Hood the rear, and the admiral in the *Queen Charlotte* was supernumerary to the centre. On the same day Howe issued two further instructions, based no doubt on the experiences of the recent cruise. The first, entitled 'Instruction respecting the disposition of the fleet in order of Sailing by divisions, denoted by the signal flag, half black, half white', was very long and confused, no clear distinction being drawn between divisions and squadrons, if indeed a distinction was intended.[24] The second instruction, headed 'Observation', sought to define the respective functions and responsibilities of squadron and division flag officers.[25]

Lord Hood and John McArthur, 1791–3

Howe was ordered to strike his flag in December 1790, once the dispute with Spain was over. Such was his zeal that, although he was already aged sixty-four and far senior to all

flag officers still able or willing to continue facing the 'rigours of the service', he described this event as 'my professional annihilation'. The initiative for tactical reform passed to Lord Hood, who for the next two years had command of the Channel Fleet, which continued on a war footing because of the 'Russian Armament' scare. A plan of Hood's fleet while lying at Spithead in two lines, shows, in the southern line, three divisions each of six ships commanded by Leveson Gower, Admiral Samuel Granston Goodall and Cosby; and in the northern line also three divisions, commanded by Hotham, Vice-Admiral Jonathan Faulknor and Vice-Admiral Sir Richard King. Hood in the *Victory* is sixth in the centre division of the northern line.[26]

Hood's position as tactician and signaller is difficult to determine. He was certainly a bitter critic of Rodney for his self-seeking at St Eustatius, and for not pursuing the French after the Saints. Yet he seems to have accepted Rodney's general ideas of tactics and fleet discipline, though he was quick to insert a signal for breaking the line once he had the West Indies fleet under his own command. Like Howe, he was a stickler for fleet discipline and station keeping. It is doubtful, however, if he was really interested in advanced tactical ideas.

In signalling, he was much influenced by his new secretary, John McArthur, who since 1782 had been working on the signal system he had presented to Admiral Robert Digby for use on the North American station.[27] The code which he eventually produced and submitted to the Earl of Chatham, then First Lord of the Admiralty, in 1790, was a much more ambitious but far less satisfactory affair.[28] It was in fact an adaptation of the French scheme, earlier adopted by Kempenfelt. Twelve flags were used and though no actual list is given, they were referred to as parti-coloured and quartered, white apparently being always one of the colours. The flags 'in their natural order' have the red, blue and yellow colours uppermost or next the mast. Each flag, however, could be inverted or reversed, which seems to imply reversing the flags divided vertically as they cannot be changed by being turned upside-down. The treatment of the quartered flags is uncertain.

Four tables of signals were provided. In Table 1 each of the twelve flags is used singly, after which the twelve flags are used in pairs, number 1 at the top, and each of the twelve under it in turn, followed by number 2 with the rest in turn under it as before. This produces 144 signals which, with the twelve used singly, make 156 in all. Table 2 has the top flag each time in its 'natural' form and the bottom flag inverted or reversed. No single flags are used, so that the total is 144. Table 3 has the flags singly, reversed, making twelve signals, and they remain reversed at the top when the flags are used in pairs, thus providing 156 signals. Table 4 has all the flags reversed and no singles; total 144. The grand total is thus 600.

Highly elaborate security measures were offered for changing the flags monthly, weekly or even daily, by combining the Dominical Letters and Golden Numbers with Lunar and Solar Cycles, 'changes might be made every lunar month for thousands of years'. Since the success of the scheme depended on the accurate use of the twelve flags and avoiding mistakes in reversing and inverting, changes in the flag values would have been most inadvisable.

When we come to the signals themselves, however, the confusion inherent in the scheme becomes fully apparent. Instead of the four tables being used to group similar types of signals, each table includes a good deal of material common to the rest; signals for attacking and engaging appear in all four tables with the reference phrase, 'Fighting Evaluations' at the bottom of the page. Confusion also exists in 'Line of Battle',

22 *NMM*, DUC/1/6; issued to Captain Duckworth

23 19 October, 1790; *NMM*, DUC/1/7 & 8; issued to Captain Duckworth

24 *Signals & Instructions*, pp333–5; issued to Sir Alexander Hood

25 *Signals & Instructions*, pp335; issued to Sir Alexander Hood. It seems possible that this Observation was only issued to flag officers and not to Captains.

26 Cecil King, article in *The Mariner's Mirror*, 24 (1938), pp176–83

27 *MOD*, NM/76 (formerly *RUSI* 76), and *NMM*, KEI/S/9 (formerly Sp/9)

28 Manuscript presentation copy to Lord Chatham, *BM*, Add Ms 38812; and manuscript presentation copy to the Duke of Clarence, *MOD*, Ec/143

which often includes what are clearly battle signals. There is little of tactical value that is new. Number 89 'For the van division to force through the enemy, break their line and double upon them', is a repeat of his signal of 1782. Private ships' signals are included in the main body.

It had apparently been Hood's intention to give McArthur's signals a trial during 1791, but the Admiralty refused to accept them. Whatever the real reasons may have been, the decision seems sound. McArthur therefore undertook to rearrange Howe's signals of 1790 for immediate use by Hood, which, in theory, proved acceptable to all parties. The signals were re-arranged alphabetically and were more fully indexed, but the result was too complicated, and the book seemed likely to defeat its most obvious purpose – simplicity in handling.[29] In the inflated and egotistical style which has damaged his reputation, McArthur wrote in his *Memorial*,

> Finding that some scruples of delicacy intervened in the adoption of any new plan of signals which would supersede that of Earl Howe's numerary code, the memorialist, therefore, made a new arrangement of his Lordship's day signals by simplifying the form of the Indexes . . . and engrafting many new ideas and instructions of his own. In this new arrangement plates or engraved plans are originally introduced by the memorialist explanatory of the principal evolutions, together with instructions illustrative of them; and the whole, with a paper containing the memorialist's observations and reasons for so making material alterations, were transmitted by the late Admiral Sir Hyde Parker to Lord Howe, who was pleased in the most flattering manner to approve of the alterations made *in toto*.[30]

McArthur also compiled new night signals from some 'undigested manuscripts' left by Kempenfelt. 'These day and night signals were approved by the Lords Commissioners of the Admiralty and were ordered to be printed under the inspection and superintendence of the memorialist, and were accordingly issued for the first time to the ships belonging to Lord Hood's squadron.' If his claim to have introduced the printing of diagrammatic plates of the standard sailing and tactical evolutions is correct, his contribution to the literary presentation of tactics is important.[31]

The 'Instructions and Standing Orders with Separate and Additional Instructions . . .', which Hood issued to the Channel Fleet in 1791, were firmly based on Howe. All the items in Howe's Additional Instructions appear, except the 'Instructions respecting the Disposition of the Fleet in Order of Sailing by Divisions . . .' and the unimportant rendezvous item. The Order of Sailing, however, is rearranged and cut down to eleven articles (the last in handwriting) with an added Memorandum, incorporating some of the material

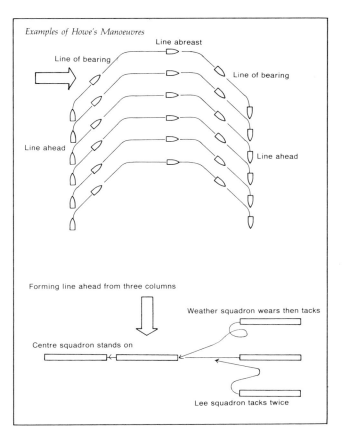

Examples of Howe's Manoeuvres

Line abreast

Line of bearing

Line of bearing

Line ahead

Line ahead

Forming line ahead from three columns

Weather squadron wears then tacks

Centre squadron stands on

Lee squadron tacks twice

29 Manuscript of 'Lord Howe's Day and Night Signals classed under General Heads, Alphabetically sub-divided and adapted for a Commander-in-Chief's ship. The whole arranged for Lord Hood in the Russian Armament, summer 1791. By John McArthur, Secretary', (formerly *RUSI* 133, present location may be in *MOD*, but the piece has not been found).

30 John McArthur's *Memorial* to the Admiralty, 9 November 1807

31 *PRO*, Admiralty Papers, Secretary's Department, Promiscuous M.4 (notes by Sir Julian Corbett). Some confusion about the useof signals at this time is due to the reliance placed by Corbett on the manuscript private signal book made by Lieutenant John Walsh of the *Marlborough* (formerly *RUSI* 33 but present location unknown). The scheme of painting of this book is that each flag is at the head of a separate page to represent the top flag of a two-flag hoist. The ten flags are then painted below to indicate the bottom flag of the hoist. Pendants are added for the hundreds. The flags are those of Howe in 1790.

32 *NMM*, Mead Collection, TUN/128 (formerly Mead/Ins/5), issued '*Victory*' at Spithead, 13th June 1791 to Captain Robert Kingsmill, the *Duke*', signed 'Hood' and countersigned by McArthur; *NMM*, MKH/4/2/ n/5 is another copy issued on the same day.

from the missing articles. In addition, Howe's Standing Orders of 1776 are reprinted as the first item in the book.[32] There are the same twenty-two battle instructions with the extra two already printed by Howe in 1790 and the twenty-fifth added in handwriting. In view of the great tactical ferment stirred up by the American War it seems extraordinary that Howe's early tactical work should have remained pre-eminent. In the field of signals it was, perhaps, less surprising; he himself had made the crucial change in 1790 which took the British service ahead of the French, and the lay-out and general arrangements of the signal book remained much the same as for 1776.

The following year, 1792, Hood had command of an evolution squadron which he exercised by signal during July and August, sometimes keeping his flagship out of the line. Starting with fifteen ships disposed in two and then in three columns, he practised them in forming line ahead on the lee column by the weather column proceeding to leeward in line of bearing, thus taking station ahead. From line ahead, with the wind west, they altered course together four points to starboard, thus coming into line of bearing, and a further alteration together of four points to starboard brought them into line abreast; and, with a further and similar alteration, they were in line of bearing again on the starboard tack. With the wind again west and the fleet in line ahead, they tacked together into line of bearing, altered course together four points to larboard, and then hauled to the wind together on the starboard tack. With the fleet in three columns in line of bearing, they formed line-ahead, the leeward column taking the lead. With the wind WNW, and the fleet in order of sailing in three columns slightly *en echelon* on the larboard tack, line ahead was formed to leeward by the two weathermost columns wearing to starboard, and then coming to the wind while the lee column passed ahead of them to lead the fleet. A whole day was spent in wearing in succession, the sternmost ship first, an evolution likely to leave large gaps in the newly formed line unless carefully executed. It was necessary to begin the evolution with the rear ships if the line was well closed up, as otherwise there was a risk of collision.

From the order of sailing in three columns, line-ahead was formed on the centre column, the weathermost column wearing and then tacking and the lee column tacking and re-tacking so as to get into their respective stations astern; this took forty-three minutes. With the fleet in line-ahead the line was inverted from van to rear by the leading ship shortening sail and the second ship making all sail and passing to windward, then herself shortening sail when at a proper distance ahead, the remainder running ahead similarly in succession. This took an hour and forty-five minutes. With the fleet again in three columns, line-ahead was formed on the weathermost column by the other two tacking and then retacking on reaching their position astern. With the fleet sailing in two columns on the starboard tack the larboard (and leeward) column tacked in succession, rear first, and then retacked, so as to form a larboard line of bearing. The columns would then be at an obtuse angle and ready to form the order of retreat, which apparently took only twelve minutes.[33]

None of those evolutions was very difficult for ships with good officers who were trained to work as a fleet. They were also severely practical as a means of enabling the admiral to form his fleet as advantageously as possible for an attack, either to get the wind of the enemy or to station his stronger or weaker ships at one or other end of the line. The timing of performance is of little value as a gauge of competence without an exact knowledge of the prevailing weather conditions and tidal currents. Nevertheless, they provide a valuable and possibly unique record of the kind of training attempted in time of peace.

On 7 May 1793, a fortnight before sailing for the Mediterranean, Hood, then at Spithead, issued his *Signal Book for the Ships of War*, with day and fog (but not night) signals, countersigned by McArthur.[34] This book is labelled on the cover 'New Arrangement', indicating that it was a variant of Howe's 1790 book, now regarded as standard for the whole navy. The main difference is that it is thumb-indexed for the signal numbers, which dispensed with the general-purpose page headings. More important is the set of eight diagrammatic plates illustrating the meaning of particular signals, just as claimed by McArthur in his *Memorial* (see above). It seems certain, therefore, that McArthur was actually responsible for this useful innovation, which ensured that captains had a standard reference for the practical interpretation of particular signals. Where necessary, the diagrams include a scale of cable lengths. Of course, the idea was taken from the French, but its value to contemporary British naval thought is proved by the incorporation of this set of plates in the Admiralty book of 1799.

On the same day, Hood issued his accompanying book of instructions.[35] Its aim was to consolidate Howe's Instructions and Additional Instructions, together with new material, in a single volume.[36] The arrangement must be McArthur's: it starts with a good index to all the articles and contains Howe's 'General Instructions', 'Separate Instruction . . . close of day' (Additional), 'Formation of the Order of Battle' (Additional) and fog instructions; it ends with a complete reprint of Howe's original Standing Order, issued on the North American Station in 1776. There are, however, three items which characterise the book as a Hood-McArthur production. The 'Order of Sailing' of seventeen articles is not the same as Howe's, though it incorporates a good deal of his material. To some extent the articles overlap the General Instructions but at least the wording is more precise than some of Howe's vaguer drafting. Article 8, which warned that when tacking in succession captains were not to be

misled if the captain ahead failed to put his ship in stays, is virtually the same as Article 9, 'ships not to be thrown out of their station by the inattention of their seconds ahead and astern' (marginal summary). Article 17 covers much the same ground as Howe's Additional 'Instruction respecting the disposition of the Fleet in Order of Sailing by Divisions', which is omitted as a separate item.

A clear innovation is the 'Instructions for the Conduct of the Fleet in the Execution of the Principal Movements of the Evolutions'. This Howe-like heading covers detailed references to the diagrammatic plates in the Signal Book, and provides a direct link between the two volumes, in addition to signal references which could be inserted against each article in the Instructions.

Another innovation is the form given to the thirty-seven 'Instructions . . . in Action, with the Enemy'. Howe's original twenty-six instructions and nine additional instructions were amalgamated with a certain amount of rearrangement and rewriting, especially towards the end.

Hood also issued on the same day *Night Signals & Instructions*,[37] which is by far the most interesting of the three volumes. It combines Howe's night signals of 1790 with a shortened form of his 1790 night instructions, extracted from his main instructions, and prints them in a single volume. Howe's principle of signals in one volume and instructions in another, with an additional volume, is abandoned in favour of a single volume incorporating all signals and instructions releating to the night. The fact that in 1799 the Admiralty not only applied this principle to the night signals and instructions but to the day and fog signals as well, shows how far-reaching was the influence of the Hood-McArthur partnership.

On Christmas Day 1793, when the allied naval forces were evacuating Toulon, Hood reissued his three volumes without alteration.[38] These seem to have continued in use after his recall,[39] and were certainly used by his successor's successor, Sir John Jervis.[40]

McArthur's contribution to all this has never been explained, but it is reasonable to assume that he influenced Hood very considerably. As admiral's secretary, he countersigned all copies of signals and instructions issued to the fleet over Hood's signature, so that these matters must have been

---

33 Formerly *RUSI* 154a; present location may be the *MOD*, but it has not been identified. *Signals & Instructions*, p68n, gives details including the times for certain evolutions, as taken from McArthur's plans.

34 *NMM*, SIG/B/64 (formerly Sp/17), issued to Captain Keith Elphinstone [the future Lord Keith], *Robust*. Admiral Holland, apart from confusing his footnote references, seems to have thought this to be a Howe book; see *Development of Signalling*, p13

35 *NMM*, KEI/S/10(c) (formerly Sp/19), also issued to Captain Elphinstone.

36 Although the first dated copy of Howe's Additionals is 5 July 1793, (*NMM*, Tunstall Collection, TUN/109 (formerly FI/3)), they must previously have appeared as individual items.

37 *NMM*, SIG/B/63 (formerly Sp/18), also issued to Captain Elphinstone. At the time it was first discovered, Corbett regarded this copy as unique (see manuscript notes inserted in cover pocket). There is now a duplicate copy, *NMM*, KEI/S/10(b) (formerly Sp/18 (1)).

38 *NMM*, KEI/S/11(a) (formerly Sp/20), a day and fog signals labelled 'New Arrangement'; KEI/S/11(b) (formerly Sp/21), night signals and instructions; KEI/S/11(c) (formerly Sp/22), instructions identical to KEI/S/10(c) (formerly Sp/19) of 7 May 1793, but a different printing.

39 *NMM*, DUN/20 (formerly S(D) 20), Night Signals & Instructions issued by Sir William Hotham to Captain Bazely, *Blenheim*, 18 March 1795.

40 *NMM*, Mead Collection, SIG/B/68, Signal Book for Ships of War, Day and Fog, 1795, issued to Robert Calder by Admiral John Jervis, Gibraltar, 19 November 1795, copy; and *NMM*, SIG/B/68 (formerly Sp/133).

constantly under his supervision. He was certainly a very able man who had a varied and distinguished career, but as far as his reputation went he was his own worst enemy.[41] His pompous style and the exaggerated claims made in his *Memorial* (1807) and his 'Thoughts on several plans combining a system of Universal Signals' (1799) have greatly prejudiced him in the eyes of historians. McArthur's fate was so closely linked with that of his master that when Hood was recalled from the Mediterranean in 1794, he lost the opportunity for furthering his signalling schemes by practical experiments at sea, though he continued to serve as Purser to Admiral Robert Man and as secretary to Sir Hyde Parker.

McArthur's 'Thoughts on several plans combining a system of Universal Signals' covers a wide range of topics. In Section 4 he proposed a system of day signals which could be used both afloat and ashore.[42] Using the current numerical flags in the signal book he provided for 9999 signals, by one or two flags 'hoisted forward', representing digits and tens, and one or two flags 'hoisted aft' representing hundreds and thousands. Two flag staves fixed in the ground would permit signalling from the shore. For spelling out words he used four flags: a red and blue horizontal and a red and yellow horizontal, both invertible, a red and blue vertical and a red and yellow vertical, both reversible. By hoisting these eight arrangements either forward or aft he could get sixteen letters of the alphabet, the remainder being supplied by pairs of flags forward and aft. The scheme was, however, a backward move towards flags in particular positions. So also was his scheme in Section 5 for signalling numbers with only five flags. 'Any Merchant Vessel possessed of these flags only, can substitute two table cloths for the other two, and make signals to a vast extent.'

In Section 6, after explaining an existing scheme for keeping signal flags stretched taut in light airs or calms, he proposed a new scheme for night signals. Here at last is something really worth while. Powerful lights were to be set forward and aft, with cylindrical covers on topping lines for obscuring them quickly. By displaying one or more of these lights in vertical or horizontal combination, fore or aft, or fore and aft, twenty-six signals could be made – the entire alphabet. Using single figures only, numbers up to 10,000 could be signalled quite easily. But the real innovation here was in having the lights kept covered, ready for quick uncovering and covering, thus facilitating quick changes of signal. Visibility is assured in clear weather by the greater power of the lights, each consisting of four oil burners cased with talc instead of horn or glass. No guns or false fires were needed for any of the signals.

Although it has been possible to work out a fairly coherent

account of Howe's tactical and signalling publications in relation to those officially emanating from Hood, there is good reason to suppose that this is not the full story. The mere fact that in his instructions of 7 May 1793, Hood appears to have anticipated the first publication of Howe's Additionals by two months, suggests that these instructions should be dated earlier, possibly as far back as 1790. More conclusive evidence of this is provided by a set issued by Rear-Admiral Alan Gardner on 19 March 1793, when fitting out a squadron for the West Indies.[43] These instructions are identical to Hood's of 7 May, which they thus anticipate by nearly two months, while anticipating the assumed date of publication of Howe's Additionals by nearly four months.

The interesting point here is not so much Gardner's anticipation of Howe as the identity of his outlook with Hood's. Gardner had commanded a 74 in the Channel Fleet in 1790 and since then had been a member of the Board of Admiralty. Like Hood, he appears to have had prior knowledge of Howe's Additionals and to have disapproved of the practice of issuing them in a separate volume. By what process the two men came to issue identical selections and rearrangements of Howe's work is unknown. It is quite possible that the Hood-McArthur Instructions, consolidating Howe's work, were first issued in 1792, during Hood's second year with the Channel Fleet – that is, assuming that Howe's Additionals were already current. One thing, however, is clear: the strong desire on the part of Hood and his followers to have all relevant and necessary instructions printed in a single book.

Howe's additional instructions, 1793

The outbreak of war with France brought Howe back to command of the Channel Fleet. Such changes as he made between 1793 and his practical retirement from command after the spring cruise of 1795, were designed more to secure proper execution of his tactical system than to improve either it or the Signal Book. It can be argued that Howe's reforming zeal had begun to wane. It can equally be argued, however, that until captains and subordinate flag officers could be relied on to carry out straightforward evolutions, there was no point in giving them freer initiative. The famous signal for 'the ships to take suitable stations for their mutual support, and engage the Enemy, as arriving up with them in succession', so highly approved of by Sir Julian Corbett, was worth exactly what the squadron and division commanders and captains were able to make of it. 'Suitable stations' and 'mutual support' are relative terms and imply a highly sophisticated capacity both in tactics and seamanship. Howe's deliberate caution, if indeed that was what he was now displaying, seems to have been justified by the Battle of the First of June, fought the following year, when his signal for the whole fleet to pass through the enemy's line was not properly executed, despite the fact that the French fleet waited to receive his attack.

On 28 May 1793, the day after resuming command of the Channel Fleet, Howe issued a slightly amended version of his 1790 Signal Book.[44] As before, several of the battle signals were issued in handwriting, again, presumably, for security reasons and not because they were added as afterthoughts. This version continued in use throughout the year and was used at the Battle of the First of June, 1794.[45]

On 5 July 1793 Howe issued a completely new volume of forty pages, entitled 'Additional Instructions for the Ships of War'.[46] The advantage of such a volume was that it served as a general appendix to Howe's main book of instructions and could be changed or added to without disturbing his own parent work.

---

41 He wrote a standard work on naval courts martial, edited the famous *Naval Chronicle* with James Stanier Clarke and wrote the standard life of Nelson.

42 *NMM*, Mead Collection, TUN/147, presentation copy to Sir Evan Nepean, then Secretary of the Admiralty. The signalling schemes included in this book are entirely different from the scheme he prepared in 1790.

43 *MOD*, Ec/110, issued to Captain Hutt, *Duke*

44 *NMM*, SIG/B/62 (formerly Sp/123), issued to Rear-Admiral George Bowyer from the *Queen Charlotte*, Spithead, and signed by Howe; also *MOD*, Ec/106, issued to Captain Collingwood, the *Prince*, same date. The two books are not quite the same. In Bowyer's copy the numerical signals start at Number 10 and in Collingwood's at Number 15. This was probably due to subsequent alteration.

45 *MOD*, NM/52 (formerly *RUSI* 52), printed signal book of the *Trompeuse*. A manuscript version used by Captain Schomberg, *Culloden*, was printed in T Sturges Jackson, editor, *Logs of the Great Sea Fights*, two vols, London, 1898 and 1900 (hereafter cited *Logs of the Great Sea Fights*), I, pp9-20. *MOD*, Ec/34 and *NMM*, Tunstall Collection, TUN/16 (formerly S/MS/R/1) are the same signals. In all these the signal for passing through the enemy line is numbered 34.

The first and most important new section, 'Of the Order of Sailing', consisted of sixteen articles. Articles 1–6 are a slightly expanded version of the first six articles of his Order of Sailing of 10 July 1790.[47] Article 7 is a warning against losing station by falling too far to leeward or astern. Article 8 warns that particular care was necessary for the leading ships. Article 9 is an order that, when tacking in succession, ships were to be put in stays before passing the wake of the respective leading ships, although they were to give place to ships 'accidentally retarded in so doing'. Article 10 is a warning that columns might have to be closed to a nearer distance than that given in Article 2. Articles 11–13 and 16 are much the same as Articles 7–10 of 1790. Article 14 contains instructions for wearing in succession, and for forming in the order of sailing or battle; when wearing together the ships were to come into a line of bearing. Article 15 orders that the fleet was to be ready at all times to meet the enemy at the shortest notice.

These instructions continued to govern the general conduct of British fleets throughout the Revolutionary and Napoleonic Wars. Though the whole book of Additional Instructions did not appear until the summer, the 'Order of Sailing' seems to have been current earlier in the year.[48] 'Instructions respecting the Dispositions of the Fleet in Order of Sailing by Divisions' were mainly concerned with directions for tacking and wearing together and in succession and for forming the order of sailing in three columns. 'Instructions respecting the formation of the Order of Battle' is a reprint of Howe's 1790 issue. So are the 'Additional Instructions respecting the Conduct of the Fleet preparatory to and in action with the Enemy', though in this case they date back to 1777. 'Separate Instructions respecting the Signal flag' are the same as issued in 1790 but with a different flag. 'Instructions for the Conduct of the Frigates' together with the 'Observations' date back to 1778 via the re-issue of 1782. The rendezvous instructions have no tactical importance. Thus, out of the seven sections dealing with tactics, only the first two are new. The remainder date either from 1790 or from the earlier days of the North American Command.

In issuing this volume, Howe might well seem to have fallen into the same error as the Admiralty when they issued Signals & Instructions in Addition, giving it collateral authority with the parent work. The difference would seem to be that, whereas the Admiralty's Additionals were not intended to be changed frequently, Howe's most probably were. This is the most likely reason why he failed to take the opportunity to effect consolidation into a single volume.

---

46 *NMM*, Tunstall Collection, TUN/109 (formerly F1/3), presentation copy, bound in red morocco, inscribed 'Given under my hand on board the *Queen Charlotte* at Spithead the 5th day of July 1793. Howe [signature]. To His Royal Highness The Duke of Clarence, Rear-Admiral, etc etc By Command of the Admiral. Geo Purvis.' The book, however, was retained by Howe and was until recently in Lord Sligo's possession. Instructions for the Conduct of Frigates was reissued separately, 4 December 1795, *NMM* Sp/29.

47 *NMM*, DUC/1/4

48 The whole set, apart from the partial issue of 1790, was issued by Rear-Admiral Alan Gardner to his West Indian squadron as early as 19 March 1793; *MOD*, Ec/110.

49 *NMM*, DUC/2 (formerly Sp/23)

50 An English translation of these signals is in the *MOD*, NM/95 (formerly *RUSI* 95) bearing the bookplate of Prince William Henry.

51 *Livre des Signaux de Jour, à l'Usage des Vaisseaux de Guerre Français*, 1819, pp210–3

52 *NMM*, SIG/C/4 and 5 (formerly Sp/69 and 69 bis), and Mead Collection, *NMM*, TUN/121 (formerly Mead/F/2).

'A Short Exposition of the Defect in our Present Naval Signals'

In 1793 'A Short Exposition of the Defect in our Present Naval Signals' was published by 'A Naval Officer'.[49] It complained of the general ignorance of signals amongst naval officers. The British system, it held, was bad, and that of the French only a little better. There were far too many flags both in the numerical, and the 'local', particular position, system. Under the 'local' system it was impossible for ships ahead of the admiral to see a signal at his mizzen peak. Plain flags, especially yellow and white, were hard to distinguish when between the sun and the observer, while red and blue had the disadvantage of being themselves squadron flags. Although the author did not recommend a system of his own, he did say that signals should be made with two flags, and not with pendants which were difficult to see. The general presentation of the book was muddled, and the book was also out of date. By the time it was published, a great many officers, other than Howe, had tried their hand at devising a true numerical signalling system.

French tactics after the American War

The ferment of tactical ideas in the British fleet was not shared by the French, who had been satisfied both by the outcome of the American War, and by their tactical performance at Ushant and the subsequent battles. The Brest Fleet, having adopted Pavillon's signals as developed for d'Orvilliers, had continued to use them throughout the American War. As it had become the premier fleet of France, its signalling system tended to prevail in other theatres of war. The signal book used by de Guichen was an adaptation of Pavillon's work by Buer de la Charulière, his *major d'escadre*. So also were de Grasse's signals as devised by de Vaugirauld, who had also acted as *major d'escadre* to du Chaffault.[50] The Treaty of Versailles put a virtual stop to tactical development in France, because Pavillon's system appeared to have justified itself.

Following the Treaty of Versailles, Vice-Admiral the Comte de Missiessy produced, in 1786, a signal book using twenty flags, five cornets and eight pendants, which appears to be no more than a further adaptation of Pavillon's system. Its publication led the government to appoint commissions at all the naval bases where naval commands were established to consider the future of signals. Without exception, all reported in favour of retaining the existing system 'jusqu'à présent'. The main reason for rejecting the true numerical system appears to have been the inconvenience of having to hoist three flags on one halyard to indicate any number in the hundreds. The size of flags suitable for fleet signals no doubt made a three-flag hoist difficult, but only a few years later the British were able to get around the problem by using a pendant and two flags below for the tens and units. The French also foresaw difficulties about compass signals.[51]

'Tactique et Signaux'

The rejection of the true numerical system was a retrograde decision, which resulted in the publication in 1787 of *Tactique et Signaux, de Jour, de Nuit et de Brume, à l'Ancre et à la Voile*. In its earlier form, it was printed at Brest for 'l'usage de l'Escadre d'Evolutions commandée par M le Mis de Nieul', and it was reprinted in 1790 for the squadron commanded by François-Hector d'Albert.[52] It thus became the book with which the navy of the Republic went to war in 1793.

Tactique et Signaux was a treatise packed with observations, disquisitions, comments and explanations, rather than a textbook. From a contemporary British point of view it would have been considered too long and repetitive, even by the leisurely and discursive standards of Howe. It began with a definition of tactics, an explanation of sea terms, a description of ship and fleet organisation and definitions of the various orders of sailing, particularly in three columns. To the usual list of advantages and disadvantages associated with the windward and leeward positions, a new one was added. Ships in the windward position, if they had the misfortune of having a mast go by the board, would be in danger of their guns setting fire to the sails, which would have fallen to leeward. A section on 'Principes pour la Chasse' was very much in the style of Hoste. The second paragraph of a section on 'Mouvemens de Guerre' was devoted to avoiding combat. On the other hand, far more space was given to boarding, described as 'une action de hardiesse et de vigueur', than might have been expected. Cutting the enemy line was described as 'très delicate et hardie, et ne peut être entreprise que par un Général consommé dans le métier, et qui a sous ses ordres une Armée très-exercée et leste dans ses mouvemens'. The signal technique was a straightforward variant of the tabular system, using a numerical table which could represent a double set of any nineteen flags.

In order to judge the practical value of this book, it is necessary to understand its background. In comparison with Morogues's book, which must have been rejected from the start as quite unworkable, *Tactique et Signaux* is a monument both of practicability and common sense. All ordinary necessities were provided for in a simple and efficient manner. Signals could be made by reference to the numerical table and could be read in the same way, though the table is printed in comparatively small type, – the squares containing three figures measuring only 7mm by 7mm. Real doubts, however, arise from the feeling, admittedly prompted by hindsight, that many of the tactical signals were never likely to be used, and perhaps were never even expected to be used. In the hands of a vigorous admiral, determined to defeat the enemy, they could easily have proved their superiority over those in the contemporary British signal book; but the situation, as we know, was very different. The signals inherited the elaboration bequeathed to the French navy by Hoste. The subtleties of attack postulated as 'très delicate' as well as 'hardie', provided for over-direction by the general, and encouraged him in this course. If French tactics were a reflection of the book, the book also reflects French tactical thinking, the kind of thinking which had often led to success but never to decisive victory.

When the Revolutionary war began, the French had one great asset, a true fighting spirit, as shown by their unflinching acceptance of Howe's attack on the First of June. This spirit was frittered away as time went on, by admirals who failed to draw the conclusion – obvious enough to the soldiers – that the quadrilles and pirouettes of the eighteenth century were no longer what was required. Of course the sea service was not able to benefit from the concept of *levée en masse*. Nevertheless, the Revolution offered a real chance to revise the whole system of signals in the light of pressing naval requirements. By tying themselves to the tactics of the *ancien régime* without possessing the technical skill to exploit them properly, the French found themselves at a serious disadvantage when facing the British, now tutored by Howe's reforms. By continuing to overload many of the signals with so much instructional matter, they failed to differentiate sharply enough between orders and explanations. By failing to adopt the true numerical system of signals, even after careful consideration, the French navy placed its commanders in a situation of serious tactical disadvantage.

It may perhaps be argued that these decisions were taken by the officers of the *ancien régime* in 1786 and 1787, and that the navy of the Republic was almost bound to accept them, in view of the executive and administrative confusion then prevailing. This is true but there was plenty of opportunity to make changes after that, especially during the peace of Amiens. The truth seems to be that the French navy retained all that was least valuable in the tactical and signalling system of the *ancien régime*, without acquiring any of the advantages which might have flowed from the acceptance of the revolutionary spirit and revolutionary techniques.

An attempt was made in 1798 to give the French navy a proper signal book when *Signaux Generaux, de Jour, de Nuit et de Brume, à la Voile et à l'ancre, à l'usage des Armées Navales de la République Française* was published at Lorient.[53] The book, however, was constructed on the plan first used in the American War; it was a complete throwback. The signals were the same as those given in *Tactique et Signaux* (1787), though the numbering and arrangement was slightly different. The usefulness of the book was, however, increased by omitting all preliminary matter to the actual signals.

In 1799 a new plan was adopted of printing *Tactique et Signaux* in two separate sections. The main part of the work appeared as before under the title of 'Signaux Généraux', but omitting all the tactical diagrams used to illustrate Articles 1–15 and 28–65.[54] The introduction was slightly altered, and there were new preliminary instructions for the reporting frigates. The numerical table numbers and flags were increased from nineteen to twenty, thus enabling the 'Ordres Particuliers' to be increased from fifty-seven to seventy-seven, though the new ones are unimportant. The single-flag signals were slightly different, and again overloaded with observations. The main day signals under sail were differently numbered and arranged. Articles dealing with the light squadron were extracted from the main series and put in a supplementary section at the end of the book; these were increased from six to thirty-eight by transferring articles previously shown in the chapters dealing with the duties of frigates.

The missing diagrams and general tactical introduction were printed in the second section of the book, which was entitled *Tactique à l'usage des Armées Navales de la République Française*.[55] The diagrams illustrate the attack from windward, doubling the enemy's rear from windward and leeward, doubling the enemy's van from leeward and cutting the enemy line from leeward. In each case the doubling or cutting ships were to resume their original course as soon as they had executed the main manoeuvre. This meant the same course as the enemy, since none of the manoeuvres was contemplated with the fleets on opposite courses.

A new printing of the Signaux was issued in 1801, in which the material remained practically unchanged. The Tactique was reprinted in 1807 without any substantial alteration. What seems somewhat surprising is that the words in the title 'a l'usage des Armées Navales de le République Française' were still retained after the formation of the Empire.[56]

53 *MOD*, NM/60 (formerly *RUSI* 60). It is signed 'Jean Bart,' presumably the great-grandson of the famous admiral.

54 *NMM*, SIG/C/6 (formerly Sp/90). This copy is inscribed 'Vice-Admiral Latouche à St Domingo' and thus seems to have been used by La Touche de Tréville when commanding in the West Indies in 1803–4.

55 *NMM*, SIG/C/6 (formerly Sp/90). The title page is dated 1805, but since the *Signaux* were published separately as early as 1799 the *Tactique* was probably published as a separate work at the same time.

56 *NMM*, SIG/C/7 and 8 (formerly Sp/41 and Sp/41 {bis}). The second copy, printed at Lorient, is inscribed 'Donné à la Maj. . .[?] le 26 Avril 1807.'

The Vicomte de Grenier, 1787

A striking contrast to the sterility of the official thinking on tactics is a book published in Paris in 1787 by the Vicomte de Grenier. *L'Art de la Guerre sur Mer, ou Tactique Navale* is a short work of about fifty pages in which Grenier set out an entirely new and revolutionary system of naval tactics.[57] Based on his own war experience, he asserted that two tactical objectives were fundamental: first, to concentrate the whole force against part of the enemy's, the remainder of the enemy being meanwhile rendered 'nulle', and defeated later; second, never to expose any part of the fleet to the enemy unless it were 'flanquée', so that an enemy attack could effectively be defeated. 'Never', wrote Sir Julian Corbett, 'had the fundamental intention of naval tactics been stated with so much penetration, simplicity and completeness.'[58] Grenier condemned Hoste, Morogues and Pavillon in revolutionary style, as men concerned only with setting ships in order of battle, 'et non celle d'attaquer avec advantage un ennemi et de s'en déffendre le mieux possible'. Their so-called tactics amounted, in Grenier's view, to no more than changing from the order of sailing in three columns into line ahead and back again, while the only fighting formation they offered was 'une seule ligne au plus près'.

Grenier believed that experience had shown the close-hauled line ahead to have several disadvantages for battle. A change of wind could easily break the line, as could poor station-keeping, or an enemy attack which disabled ships. During the necessary evolution between the order of sailing in three columns to the order of battle, the fleet was exceptionally vulnerable. Once in line the enemy was free to concentrate against one part of it. Hawke attacked Conflans's rear at Quiberon 'sans aucun ordre'. At Ushant and Grenada similar attacks were frustrated by the skill of d'Orvilliers and La Motte-Piquet and their captains, combined with the inefficiency of the British, rather than by any virtue arising out of the French line ahead formation. A further disadvantage of the single line of battle was that a commander could not assist part of his line which happened to be disabled without exposing another part to danger. In any case, the commander, being in the middle of his line, could not communicate efficiently with either end of it. Under these circumstances, battles were generally indecisive, with the commanders remaining in sight of each other until night, when they would slip away unobserved.

The idea of attacking from windward, Grenier wrote, was nothing but a legacy from galley warfare, when the wind aided the rowers in driving home their main form of attack, the ram. For broadside sailing ships, the windward position only helped the weaker fleet refuse action and retire; it was also easier in the windward position to ensure that smoke did not accumulate between decks. The leeward position was better because it enabled the lower tier of guns to be fired; this was especially the case when ships had been on a long cruise and lacked stability, because of the consumption of large quantities of victuals and ammunition. The leeward position also facilitated protecting disabled ships, which could drop away yet further to leeward. Grenier's idea was that every effort should be made to keep part of the enemy fleet to windward, and then to concentrate to attack it.

Grenier replaced the single close-hauled line of battle with a 'nouvel ordre de bataille', which consisted of the fleet in three divisions disposed along three sides of 'un losange régulier'. The order of sailing of the three separate divisions could be varied considerably, but must always result in one of two forms. Either one of the divisions would be in line ahead with the other two in line of bearing, 'en échiquier', or two divisions would be in line ahead, and the other on line of

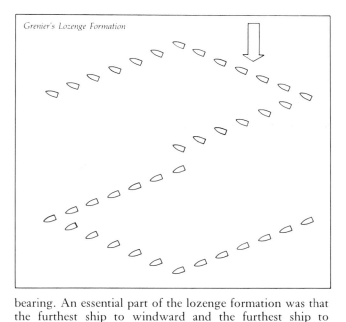

Grenier's Lozenge Formation

bearing. An essential part of the lozenge formation was that the furthest ship to windward and the furthest ship to leeward must be on exactly the same relative bearing to the wind. All movement in the face of the enemy must be calculated from the position of the leewardmost ship.

In defence, the lozenge formation did not leave any ships exposed to attack unsupported, and any attempt by the enemy to double either end of the division nearest to it or to cross its 'T', either ahead or astern, could be met by one of the other two divisions. By an ingenious system of battle tactics the lozenge could always be handled so as to put the enemy at a serious disadvantage. A and C divisions could always come to the help of B by tacking, or come up to take station on either side of B. A and C could also be used to cut the enemy line in two places. Alternatively A and C could pass together either ahead or astern of B so as to concentrate against the enemy, or to protect B against a concentration.

The real difficulty about accepting Grenier's lozenge formation as the best solution for the tactics of his time is that he never really developed it as a form of attack. Despite his criticism of Morogues, he was concerned with attack only in a theoretical manner. His concept of concentration against part of the enemy fleet was clear and effective, but how did the lozenge help to bring it about?

Grenier's willingness to accept the leeward position gives away his own essential defensiveness. He really regarded the lozenge as a formation which might give an inferior fleet a chance to counterattack, should the superior fleet show signs of weakness; whatever happened, the lozenge could never be doubled. Admittedly Grenier was correct in claiming that the lozenge formation would facilitate control. Signals would be seen more easily and quickly, and the admiral would be in a better position to gauge the situation faced by his more distant ships. Nevertheless, Grenier's failure to give details of other than defensive benefits of his formation, except for some very general remarks about the chase, suggests that his lozenge should be regarded as no more than one of a number of methods for avoiding the rigidity of the line.

The virtue of Grenier's work is primarily that it was genuinely concerned with naval warfare, not naval parades. His influence on the Dutch must have been considerable; at the time of the battle of Camperdown in 1797 the lozenge

57 Source unknown

58 Corbett, *Fighting Instructions*, p286. Corbett, however, misquoted Grenier's second dictum.

formation appears to have been one of their orders of sailing. Within a year of its publication in Paris, Grenier's work was translated into English by the Chevalier de Sauseuil, a French officer who described himself as a 'Member of the English Society for the Encouragement of Arts, Manufacturers, and Commerce.' [59] In the same year, 1788, he also answered Kempenfelt's call for a study of French tactics by the publication in London of his complete and accurate translation of Bourdé de Villehuet's 'Manoeuvrier' of 1765. The translation was dedicated to Prince William Henry, later Duke of Clarence and King William IV.

The Chevalier de Sauseuil said in his preface to Bourdé's work that he had been helped by a British naval officer with the technical terms, and that he had also incorporated notes made by another British officer who had himself translated most of Bourdé for private use. With these notes were five drawings by the same officer, reproduced as plates together with the original illustrations. The result was a useful work from which the English reader could obtain the fullest appreciation of Bourdé's whole attitude. It included, as in the original, the system of numerical signals invented by La Bourdonnais.

D'Amblimont's tactics and signals, 1788

A year after the appearance of Grenier's book, in 1788, the Comte d'Amblimont published *Tactique Naval, ou Traité sur les Evolutions, sur les Signaux et sur les Mouvemens de Guerre.* *Chef d'Escadre* d'Amblimont had commanded the *Vengeur* in the Battle of Ushant, the *Hercule* in the Battle of Martinique in 1780, and the *Brava* in the Battle of the Saints. He was an original thinker, who, starting from an entirely different position, anticipated Nelson in breaking up the fleet into *pelotons*, or tactical groups with different functions.

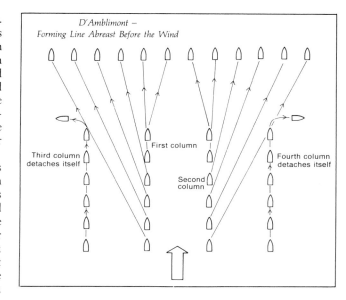

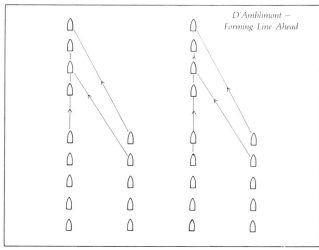

Claude François, Comte
d'Amblimont (1736–97). MM

D'Amblimont rejected the order of sailing which divided the line of battle into three similar columns, and replaced it with an order of sailing in four columns, only the centre two of which would form the line of battle. To form line abreast before the wind, the third and fourth columns would detach themselves, the main line being formed by the two centre columns, called the first and second columns. Those two would fan out successively so that the two rear ships would become the extreme starboard and larboard flanks. To form line of battle ahead the leading ship in the windward column would lead, and the leading ship on the leeward one would follow. Then the second ship in the windward column would take station in their wake, followed by the second ship in the leeward, and so on. The outer columns formed respectively a reserve for filling gaps, or effecting a concentration, and a light squadron.

The Glorious First of June

The Battle of the First of June, in 1794, was one of the greatest convoy actions in naval history. Conditions in

59 *The Art of War at Sea; or, Naval Tactics reduced to new principles: with a new order of battle – Translated from the French of Viscount de Grenier, Rear-Admiral in the French Navy, by the Chevalier de Sauseuil,* London, 1788

60 *NMM,* DUC/1/11. Corbett printed the diagram without any explanation in *Signals & Instructions,* p341

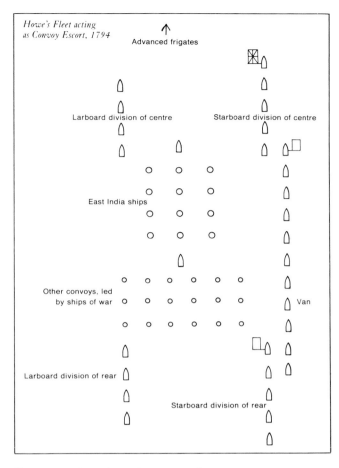

Howe's Fleet acting as Convoy Escort, 1794

Advanced frigates

Larboard division of centre   Starboard division of centre

East India ships

Other convoys, led by ships of war   Van

Larboard division of rear

Starboard division of rear

and one a boatswain. One can well believe that they were exceptional men,[62] or how else could Villaret de Joyeuse have brought them to such a state of tactical efficiency in so short a time?

Unlike the French army, the French navy derived no benefit of any kind from the Revolution. The flag officers had all been murdered or expelled, or had found it prudent to retire. After the revolutionary outbreaks at the naval bases, professional reliability had to be matched by loyalty to the new regime, which was made still more difficult by the counter-revolution in the Vendée. Though the sea service continued to be recognised as of major importance, its political and strategic direction was never sufficiently intelligent or sympathetic to give it the place it could and should have held both in Revolutionary and Napoleonic strategy.

On the other side, Lord Howe's fleet was the best with which Britain had ever entered a war, even though by his own standards it lacked both tactical knowledge and experience. His line of battle was studded with famous names and with names soon to be famous. Graves and Sir Alexander Hood were his second and third-in-command and in addition he had four junior flag officers. Each of the seven admirals was in a three-decker, of which the French had only four altogether. Howe's captains included James Gambier, Cuthbert Collingwood, Robert Calder and John Thomas Duckworth. Against this overpowering collection of talent, Villaret de Joyeuse offered the French fleet for attack without any sign of evasion or retreat. He had his orders. He must have guessed after three days skirmishing that Howe would attempt to break his line. Indeed his ordering of the French fleet seems to have been a counter-measure deliberately contrived and in part successful.

Howe had always been in favour of encouraging personal initiative and of giving his flag officers and captains as much

61 Norman Hampson, *Le Marine de l'An II*, 1959, pp194–6, 201–4

62 Norman Hampson, *Le Marine de L'An II*, chapter VI, where the whole question of the naval personnel is discussed at length.

France required the safe arrival of over a hundred supply ships from America. Any risk was deemed acceptable in their defence. The particular risk involved was that of using the Brest fleet as a diversion until the convoy was safe in French ports. This was a risk the *ancien régime* would not have accepted lightly.

For the British, the First of June was also a convoy battle to the extent that at the start of the operations Howe was concerned with clearing a large outbound convoy from the Channel. His order of sailing with the whole fleet acting as escort as far as the Lizard was as shown in the accompanying diagram.

When sailing large the starboard division of the centre was to be directly ahead of the convoy and the larboard division flanking the India ships.

The French fleet, consisting of three squadrons, each in three divisions, was commanded by Rear-Admiral Louis Thomas, Comte de Villaret de Joyeuse, an officer of exceptional ability.[61] Though promoted from lieutenant to captain only on 1 January 1792, he turned out to be just the man the Revolutionary navy needed, a member of the nobility, willing to serve the new regime, and firmly in favour of discipline, tactical training and hygiene. In May 1793, he was given command of a small squadron on the Vendée coast, and immediately set to work to give his force tactical efficiency and cohesion.

His second-in-command, Rear-Admiral François Joseph Bouvet, had entered the French navy as a lieutenant in 1786 after previous service in the French East India Company. His third-in-command, Rear-Admiral Joseph-Marie Neilly, of non-noble origin, was thirty-eight and still a lieutenant at the time of the Revolution. Of his captains, three are said to have been promoted lieutenants, eleven sub-lieutenants, nine captains or mates of merchant ships, one a seaman in the navy

Amiral *Louis Thomas, Comte de Villaret de Joyeuse (1750–1812), engraving by Forestier after a painting by Ambroise Tardieu.* MM

freedom of action as was consistent with efficient action by the fleet as a whole. At the same time he did everything possible to ensure efficient station-keeping and the prompt and efficient execution of all the complicated evolutions which his three- and six-column cruising systems entailed.

A good example of his careful balance of initiative and command control is an addition to his additional instructions which stated that when the signal number 166 was made for 'the ships severally to take such stations as most convenient at the time, without regard to any established order of sailing,' flag officers and other divisional commanders were neverthe-less to keep their appointed stations with reference to the commander-in-chief, and to make such signals as were necessary to their own squadrons and divisions.[63]

63 Issued to Sir Alexander Hood and printed in *Signals & Instructions*, p340. There is no date, but Corbett assigns it to 1794. The reference to p12, Article 17 of the Additional Instructions is puzzling, as Article 17, already on that page, relates to the stationing of fireships to windward of the line. An additional reference is made to signal no 166: 'the ships severally to take such stations as most convenient at the time, without regard to any established order of sailing.'

His tactical object was to devise some means of forcing a supposedly unwilling enemy to action. The traditional device, ordering ships in excess of the number in the enemy line to harry the enemy van or rear, was inefficient. First, it could only be undertaken when the enemy ships were near enough to be counted. Second, the choice of ships to perform the harrying depended entirely upon which tack the fleet was sailing on at the time. D'Orvilliers had had a light squadron at the Battle of Ushant, and had formed part of Cordova's Spanish fleet into a squadron of observation, but it is doubtful whether the tactical function of these groups was

the same as that which Howe now required for his fleet.

Howe revealed his plan on 20 April 1794 in a letter to Sir Alexander Hood, with a copy of the instructions he was issuing to Rear-Admiral Pasley, who was to command the advance squadron. This squadron was to consist of four 74s, three, including Pasley's, from the first division of the van and the fourth the last ship in the whole line of battle. When the signal was given, Pasley was to take station to windward of the weather column of the fleet and to intercept any enemy ships crossing ahead of the fleet, or as directed by the rear-admiral. In the presence of any considerable force of the

The Battle of the First of June, 1794, by Robert Dodd (dated 1795). NMM

enemy, he was again to take his orders from the rear-admiral, but also to act to the best advantage if the latter was 'not in a situation so as to advise you'.[64] This squadron played an important part in the operations of 28 May.

Doubt has sometimes been expressed as to which printed signal book Howe used at the First of June. This can now be resolved by reference to the signal book and accompanying instructions issued to Captain William Johnstone Hope of the *Bellerophon*, one of the ships of the Advance squadron, on 28 March 1794.[65] These signals are the same as those issued to Admiral Sir George Bowyer in 1793. The instructions included a new article, number 27, which established that, when the fleet was formed in an irregular line of battle, and it was wished that those ships which were accidentally leading each division should assume the usual duties of divisional leaders, the admiral would fly a triangular white flag with red fly (signal 6, page 4 of the signal book). These temporary arrangements of the fleet and this institution of the divisional leaders were also to take effect when the regular succession of the ships in each column in the Order of Sailing had been altered in the same manner. Divisional admirals were not to confine themselves to controlling their own divisions, but were to exercise authority over all the ships they should find to be out of station.

This instruction illustrates Howe's concern for correct leading of divisions when one or more changes of formation, combined with tacking or wearing, might have changed the original order of divisions and perhaps also have inverted the order of the individual ships in their respective divisions. Those familiar with the old infantry drill of the British army will recognise some similarity with the problems arising from getting the battalion 'inside out'. The really interesting point is the power given to divisional commanders to correct the station-keeping of ships in divisions other than their own.

When the fleets first made contact early on 28 May, the wind was SSW and remained much the same until 1 June.[66] The French were to windward in an irregular and reduplicated line abreast, and the British on the starboard tack in two columns, with four of the line under Admiral Sir Thomas Pasley acting as an advanced squadron and detached to windward.[67] The French began to form a very irregular line on the starboard tack, and Howe, after skilfully tacking the British fleet twice, managed by evening to bring his advanced squadron into touch with the French rear and partially to windward of them. To do this, he had made full use of his large repertoire of numerical signals both to the

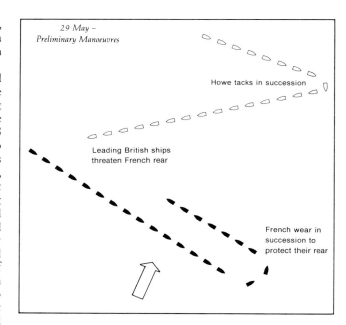

29 May –
Preliminary Manoeuvres

Howe tacks in succession

Leading British ships
threaten French rear

French wear in
succession to
protect their rear

fleet as a whole and to individual ships. The captain of the rearmost French ship, the *Revolutionnaire* (120), the only three-decker private ship, misjudged his capacity to beat off the British attack practically single-handed. As night came on, he received a terrible hammering and eventually struck to the *Audacious* (74), but was not boarded. Both ships withdrew independently and reached home ports in safety. On the whole, the British attack had been somewhat muddled.

Howe's frigates 'appointed to observe the motions of the enemy during the night', correctly interpreted both the letter and spirit of his reconnaissance instructions, so that at daylight next morning, 29 May, the British were well up with the French, and at 4am Howe was able to signal for the line of battle ahead and astern of the admiral 'as most convenient, without regard to the established form'. The two fleets were now parallel, both on the starboard tack, with the French considerably to windward. Howe at once decided to force an action, and signalled to tack in succession, headmost and weathermost ships first (Signal number 78). This had the effect of turning his van obliquely towards the French rear. Soon after this, he signalled to pass through the enemy line to obtain the weather gauge (number 34). Villaret de Joyeuse, sensing the threat to his rear, signalled to wear in succession, thus bringing the French fleet round in a hairpin bend, his van moving towards the British on a converging course as his rear continued to move away.[68] There was 'a great head sea', and the leading British ships took some time to approach the French rear and open fire on them. Howe meanwhile signalled that his ships had liberty to fire in passing the

---

64 *Signals & Instructions*, pp341–3, from the Bridport Papers. The term 'Rear-Admiral' could not refer to the commander of the whole van or the starboard division of the fleet when in two columns, in each case Admiral Thomas Graves. It might refer to one of the divisional flag officers.

65 *NMM*, Mead Collection, TUN/130 (formerly Mead/Ins/7) Both books signed 'Howe, *Queen Charlotte* at Spithead'.

66 Howe's line of battle was substantially the same as that issued on 19 April, *NMM*, DUC/1/10, and not as shown in Barrow p224, as apparently issued on 2 May. The fleet was organised both for three squadrons, each with two divisions, and for two 'grand' divisions only (starboard and larboard) the centre squadron thus being split. In each case the *Queen Charlotte* was supernumerary.

67 There are three separate sets of pictorial and diagrammatic descriptions of the operations: fourteen coloured plans by John Urquhard, Master of the *Bellerophon*, in *NMM*, DUC/1/14; nine coloured plans by S J Ballard, third Lieutenant of the *Queen*, In *NMM*, JCD/12 (printed in *Logs of the Great Sea Fights*, I); and three sketches by Nicholas Pocock, who was allegedly on board the *Pegasus*, Howe's repeating frigate.

68 The effect of the tacking and wearing was different. Tacking, from a course close to the wind, involved a turn of about 14 points. Wearing involved a 16-point turn, thus sending the French fleet in exactly the opposite direction on a parallel course.

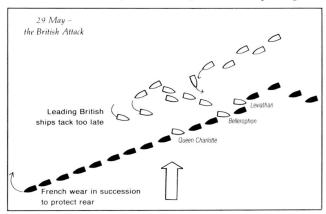

29 May –
the British Attack

Leviathan

Leading British
ships tack too late

Bellerophon

Queen Charlotte

French wear in succession
to protect rear

enemy, though 'the Admiral does not mean to bring them to general action immediately' (number 28). For the time being, his grand design had to be abandoned. As the French van began to appear to larboard, covering the retreat of their rear, the British engaged them at a distance of nearly three miles while continuing to fire on the French rear as it disappeared.

Howe now took a very important decision and one which is seldom mentioned in popular accounts of the battle. At about 11.30am he signalled the fleet once more to tack in succession, which would have had the effect of driving his van into the French van at an angle of about 40 degrees. Five minutes later he annulled the signal, but then he repeated it at about 12.15pm. His object now seems to have been to break through the French line in line ahead rather than simultaneously from a position parallel with them. His earlier signal had evidently been premature, the signal to tack having not yet taken full effect. When the leading British ships had at last tacked (together with the *Caesar*, out of station), they engaged the enemy's van and centre as they passed them to leeward on opposite courses, but without making any attempt to break through. Only the *Queen Charlotte*, about tenth in the improvised line, her next astern the *Bellerophon*, and later the *Leviathan*, actually succeeded in breaking the French line, the *Queen Charlotte* passing between the fifth and sixth ships of the French rear. The three British ships were then put on the larboard attack to come parallel with the French, who were now to leeward of them.

The fighting became confused as the main body of both fleets, having been heavily engaged on opposite courses, turned inwards towards each other once more. When the British ships had passed the end of the French rear, those capable of tacking did so. This had the effect of giving the British the weather gage. Howe now turned the *Queen Charlotte* back to rejoin his oncoming van. There had meanwhile been a good deal of hard fighting and a number of ships on both sides were temporarily disabled.

Three French ships were seen falling away to leeward, but were saved from capture by Villaret de Joyeuse's prompt action in wearing his fleet once more, and passing between them and the British. It was probably for this reason that he was content to surrender the weather gage. Moreover, damaged ships could more easily and quickly wear than they could tack. With a heavy sea running and his fleet disordered, Howe did not feel justified in pressing the action further; having obtained the weather gage, he could afford to wait.

During 30 May and the early part of the 31st, fog prevented an engagement and, although it had cleared by the afternoon of the 31st, Howe again decided to wait. His patience and deliberation throughout the entire operation is a remarkable, but also a typical, example of his methods. He sought a victory as crushing as that at Quiberon, but by different methods, attuned to different circumstances.

On the morning of 1 June the wind was south by west, and the French lay about five miles to leeward in an extended line ahead on the larboard tack. Howe was at last in the optimum position. Nevertheless, it took him five hours from his first general signal at 3.50am to bring his van ships to action, the intervening time being spent in aligning his fleet carefully at the proper intervals and then going down on the larboard line of bearing. At about 7.30am he signalled his intention to pass through the enemy line (number 34) and, at about 8.30am for each ship independently to steer for and engage her opponent in the enemy line (number 36). This seems to have

led to some misunderstanding, partly because of the second signal and partly because there was still uncertainty as to the mandatory force of number 34. Only five ships in addition to the *Queen Charlotte* succeeded in breaking the enemy line, heavily raking the respective bows and sterns of the French ships to starboard and larboard of them as they went through.

The battle of manoeuvre which had begun with the two vans at about 9.30am had lasted about three and a half hours, and the French were now thoroughly defeated. The majority of the British ships had succeeded in coming to close quarters without being immobilised or thrown into confusion by the French high-angle gunfire. They now turned to larboard, parallel with the French line, and opened devastating cannonades. Gradually the battle line broke up into a series of two-ship and three-ship engagements, drifting steadily to leeward.

French accounts testify to the tremendous impact made on their line by the *Queen Charlotte*.[69] Captain Gassin of the *Jacobin*, next astern of Villaret's flagship the *Montagne*, anticipated Howe's intention by pushing up very close to his admiral's stern. This left a gap astern of his own ship, which the succeeding French ships also managed to fill. However, in trying to clear his admiral's stern, following orders shouted from the stern gallery of the *Montagne* by the flag-captain, Gassin fell to leeward and did not play a vigorous part in the battle. Nevertheless, his earlier action saved the *Montagne* from being even more badly raked by Howe. As it was, the *Jacobin* and the *Achille*, her next astern, took a terrible pounding while the *Queen Charlotte* worked her way between them and then engaged the *Juste*, next ahead of the *Montagne*.

For some time, dense smoke obscured the battle area, but by about 1pm fire had slackened all round. Nine French ships were seen to be totally dismasted and two French three-deckers to retain only their foremasts standing. Other French ships and a number of British were also seen to be in various states of disablement. Having begun the battle more or less in

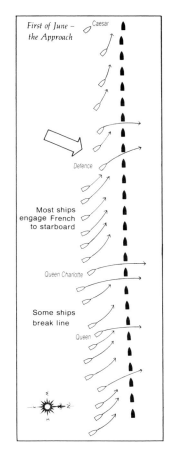

First of June – the Approach

Most ships engage French to starboard

Some ships break line

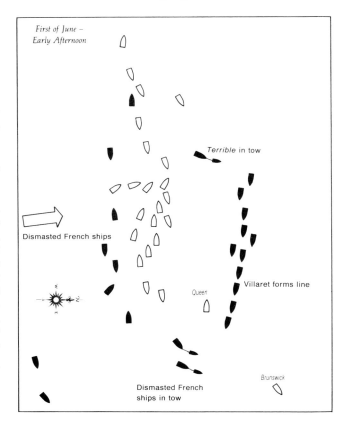

First of June – Early Afternoon

Terrible in tow

Villaret forms line

Queen

Dismasted French ships

Brunswick

Dismasted French ships in tow

First of June – the Queen Charlotte's Attack

Jacobin closes up on *Montagne*

Jacobin

Montagne

Queen Charlotte

Queen Charlotte attacks between the two

Jacobin

Montagne

69 *Sommaire de la Journée du Premier Juin*, 1794, printed in *A Narrative of the Proceedings of His Majesty's Fleet under the command of Earl Howe*, by Captain William Bentinck of the frigate *Phaeton*, 1796.

pairs, the two fleets were now in a confused and oval-shaped mass, the British ships still mainly to windward.

A critical situation had now been reached: the pell-mell was over, and the question now was one of collecting the spoils of victory and minimising the results of defeat. Villaret de Joyense seems to have realised this more quickly than Howe. He signalled his van to tack and join him, and for the rest of the fleet to wear. Neither signal was obeyed, partly no doubt because they were not properly seen, and partly because both manoeuvres were too difficult to execute, but eventually Villaret formed a line of eleven ships two miles to leeward on the starboard tack. This was to cover his disabled ships, which were working their way to leeward away from what had been the rear of the British line, some being towed by frigates and corvettes. All this time, both sides were making emergency repairs, the newly formed French line itself lying to. Further to windward seven French ships which had already surrendered lay cut off from the main body of their fleet. One of these, the *Vengeur*, sank before she could be boarded, but the remaining six were eventually towed to England. Eight more French ships in various states of disablement were to leeward, some being towed.

While Howe was considering pursuit, Sir Roger Curtis strongly advised him to signal the fleet to close. Howe agreed, and even recalled two ships which were about to collect two more prizes which they had just forced to cast off their respective towing ships. The reason for this sudden caution was that Curtis thought that Villaret de Joyeuse might use his newly formed line of not more than a dozen ships to capture the damaged British ships which had also fallen to leeward. The danger was illusory; Villaret's action was purely defensive. This failure to pursue was a great lapse on Howe's part but, as in Rodney's case, it was due to physical and nervous exhaustion. When, after five days and four nights of fighting, chasing, searching and anxiety, he at last forced and won the battle, it is hardly surprising to learn that he was in such a state of collapse that only the physical support of his personal staff prevented him from falling exhausted on the deck.

Tactically the battle was a great British victory. To sink a French ship-of-the-line and carry six more triumphantly to Spithead was a feat unparalleled since Barfleur and La Hogue. Howe had won partly by tactical virtuosity, but mainly by better gunnery and by being able to deploy a better trained and more disciplined fleet than had the French, although the disparity between the two in this last respect was not as great as some accounts would suggest.

Less than a quarter of the British fleet had managed to break the French line. Howe had made no attempt to concentrate on a particular part of the enemy line, as the French ships were too well closed up, nor had he wasted time by doubling. By dressing his line very carefully, he managed to bring about the pell-mell which was his main object. British gunnery had then won the battle, and this also was certainly due to Howe, though when congratulated by a formal deputation of petty officers and seamen of the *Queen Charlotte*, he replied with the modesty of the truly great, 'No, no, I thank *you* – it is *you* my brave lads, not I, that have conquered.'

Villaret de Joyeuse had also good reason to be satisfied. He had saved the convoy (and perhaps his own head), and had shown a very creditable degree of tactical skill against the greatest tactician in Europe then in high command. His skilled and resolute bearing at the end of the battle had saved a very heavy defeat from becoming a total disaster.

Howe's tactics after the First of June

As far as we can tell, Howe saw no reason to change either his signals or instructions after the First of June. Such lapses as had occurred – and there seem to have been quite a number – could be safely attributed to the professional failure of particular persons and not to the system. The evidence for this is that when Rear-Admiral Sir William Cornwallis joined the Channel Fleet immediately after the battle, he was issued with the same signal book and additional instructions of 1793 as had been used for the First of June operations.[70]

The arrangement for a Portuguese squadron to join the Channel Fleet caused Howe to issue preliminary Orders for Combined Fleet on 11 August 1794. The Portuguese were to form a 'reserve or advanced squadron' and to be stationed about three miles to windward of the 'body of the Fleet'. Their role was 'to act on such occasions of general engagement against the unoccupied Ships of the Enemy's Rear or to attack any of their separated Ships as circumstances render advisable', in the judgment of their admiral, unless otherwise directed by the Commander-in-Chief.[71] No doubt all this was done for form's sake, especially the use of the term 'Combined Fleet'; Howe probably regarded the Portuguese merely as an encumbrance. Meanwhile, he reconstituted the original advance squadron on 28 September 1794 and called it the Reserve.

Howe's line of battle for his autumn cruise consisted of thirty-two of the line, exclusive of the Portuguese. His normal order of battle consisted of the van, centre and rear squadrons, each in two divisions and each division under a flag officer. In the second order of battle the fleet was formed into two equal 'Grand divisions', Starboard and Larboard. This meant dividing the centre squadron and giving its starboard division to the Starboard Division of the fleet and its larboard division to the Larboard.[72] In each order of battle Lord Howe's *Queen Charlotte* was supernumerary. For the order of sailing in three columns the usual three squadron organisation was used. For the 'order of sailing by divisions', the fleet lines formed in a double line, each squadron having its two divisions in parallel columns. The *Queen Charlotte* was again reckoned as supernumerary in both orders of sailing.[73]

By the end of September, detachments to protect convoys had reduced the fleet to twenty-seven of the line. The two orders of battle were as before, but in the order of sailing in two columns the new Reserve squadron of four of the line was listed separately. After making various detachments and suffering such damage from autumn gales, Howe's fleet was then reduced to twenty-one of the line. Two orders of battle

70 *NMM*, COR/65 (formerly Sp/119) is the signal book, signed 'Howe, *Queen Charlotte*, Spithead, 17 June 1794'. *NMM*, Tunstall Collection, TUN/115 (formerly FI/9) is the additional instructions, signed 'A S Douglas' who was Howe's flag captain, same date.

71 *Signals & Instructions*, pp344–5. The instruction is also printed in W G Perrin, editor, *The Keith Papers*, London, 1927, I, pp199–200. *NMM*, Tunstall Collection, TUN/17 (formerly S/MS/R/2), MS signal book of Captain John Elphinstone, gives a list of the Portuguese squadron, their position in the order of sailing and some signals concerning them. There were only five of the line, a 40, a fireship and two frigates. Corbett suggests that the confusion between an advance and a reserve squadron was settled by the Portuguese becoming a 'true reserve'. But this, as he himself points out, was more to avoid incorporating them in the line of battle, remembering d'Orvillers's experience with the Spanish in 1779. See *Signals & Instructions*, p74.

72 These and the following orders of sailing and battle are given in the MS signal book of Captain John Elphinstone, flag captain to Rear-Admiral George Keith Elphinstone, later Lord Keith, in the *Barfleur* and *Monarch* 1794–7, *NMM*, Tunstall Collection, TUN/17 (formerly S/MS/R/2).

73 When at sea on 8 September 1794, Howe directed that, when he himself was not leading the centre squadron, it was to be led by the senior divisional flag officer. Normally, Howe himself would be ahead of the van; *NMM*, DUC/1/16.

exist, in the second of which the van and rear were reduced to six ships each, and were not subdivided.[74]

At the beginning of 1795 the strength of the fleet was up once more to thirty-seven of the line, organised as in the previous summer. In place of the advance squadron, two frigate squadrons were established, each of four ships and designated 'Advanced' or 'Wr [weather?] Quarter' respectively. This was the organisation used by Howe for his last appearance in command.[75]

When Sir Alexander Hood, now Lord Bridport, took over the command, the total dropped to nineteen and then eighteen ships of the line and there appears to have been no squadron organisation for the order of battle, only the Starboard and Larboard divisions. When, in March or later, the total number of ships of the line rose again to twenty-one, full subdivisioning with van, centre and rear was restored; the *Royal George*, Bridport's flagship, was kept a supernumerary.

In the spring of 1795 Howe took the Channel Fleet to sea for the last time, although the fleet was actually in Bridport's hands. On 9 June, before putting to sea, he reissued Howe's Signal Book, Instructions and Additional Instructions.[76] The same three books were issued on 19 October by Vice-Admiral Cornwallis, then commanding the rear squadron of the Channel Fleet.[77]

So far the development in signalling and instruction seems simple enough, but the existence of all three books with 1795 printed on the cover label indicates a different printing during the course of the year.[78] Study of the available evidence suggests a ferment of tactical and signalling development in the Channel Fleet during the years 1793–5, comparable with the ferment of 1779–82 during the American War, though with the difference that a general system had by this stage been established. Much of the evidence is puzzling and contradictory, and earlier attempts at elucidation have been unsatisfactory.[79] Anomalies also appear, thus adding to the confusion.[80]

It is certain, however, that a new edition of Howe's instructions was printed in 1796.[81] It was word for word the same as before, except for an important addition to Article 7 of the night sailing instructions, directing, 'as a fixed principle', that in fog or at night the fleet was to continue on the same tack, regardless of a change of wind, unless danger or a signal from the admiral dictated otherwise.

Cornwallis's retreat

Cornwallis's famous action off Belle Ile on 16 and 17 June 1795 illustrates the lack of tactical initiative displayed by the French even when led by Villaret de Joyeuse. Like so many French admirals of his time, Villaret did not invariably refuse a pounding, but he was incapable of exploiting a clear advantage.

On 16 June Cornwallis was cruising off Brest with five of the line and two frigates when he sighted what amounted to the main Brest fleet to leeward, consisting of thirteen of the line and fourteen frigates. With the help of a land breeze and a change of wind, Villaret de Joyeuse was able to divide his force, and by next morning to bring his ships in two roughly equal divisions on each flank of the British. Cornwallis formed his ships in a wedge formation, two and two with his flagship leading at the angular point. This, incidentally, was the exact reverse of the 'order of retreat' in which the wings were pushed forward, thus placing the flagship at the angular point furthest astern.

The French attacked throughout the day, their line of battle ships coming up on either flank of the British in succession, firing broadsides and then dropping to the rear again, in the manner recommended by Howe. One frigate

drew up level with the British larboard wing and opened fire, though the others kept to windward. One of the British ships was a heavy sailer and seemed in real danger of capture, until Cornwallis himself put about and relieved her, steering back towards the French fleet. This seemingly astonishing conduct convinced the French that he was expecting some

---

74 *NMM*, Tunstall Collection, TUN/17 (formerly S/MS/R/2)

75 *NMM*, (formerly Sm/108, present location possibly in Grey collection) is an exceptionally fine manuscript copy of Howe's signals for 1795.

76 *NMM*, COD/43, set of three books, signed 'Bridport', dated from the *Royal George*, Spithead, and issued to Captain Codrington, *La Babet*.

77 *NMM*, Tunstall Collection, TUN/62, 116, 117 (formerly S/P/R/9, FI/10 and FI/11) all dated from the *Royal Sovereign*, Spithead, and issued to Captain George Martin, *Magicienne*.

78 *NMM*, DUN/21 (formerly S(D) 21) Signals, DUN/22 (formerly S(D) 22 Instructions, DUN/23 (formerly S(D) 23) Additional Instructions, all with manuscript additions, but unsigned and apparently unissued. *NMM*, KEI/S/12(a) (formerly Sp/26), three copies of signals, KEI/S/12(b) (formerly Sp/27), three copies of instructions, KEI/S/13 (formerly Sp/28), one copy additional instructions. A curious feature of the additional instructions is that in both series noted above 'the Conduct of the Fleet . . . in action with the Enemy' and 'Separate Instruction respecting the Signal' are omitted. *NMM*, Tunstall Collection, TUN/112 (formerly FI/6) 'Additional Instruction 1795' on the cover label has the same omission. This copy, recently in the possession of Lord Sligo, seems to have been originally in Lord Howe's papers. Another copy of the same Additional Instructions, labelled 1795 and with the same omissions is in the Mead Collection, *NMM*, TUN/134 (formerly Mead/Ins/11). It was issued on 9 April 1796 by Vice-Admiral George Vandeput, *St Albans*, at Spithead to Captain Henry Darby, *Adamant*.

79 *NMM*, DUN/25 (formerly S(D) 25) is an unissued and undated printed signal book with DUN/30 (formerly S(D) 30) Additional Instructions. An accompanying note by Commander Hilary Mead reads 'Probably printed 1794, copies issued 1794, sometimes 1796, thereby straddling 1795 (dated) edition'. Admiral Holland's attempted elucidation of the signal book of this period, *The Development of Signalling in the Royal Navy*, p13, can no longer be fully accepted, quite apart from printer's errors.

80 *NMM*, HSR/Z/5 and COR/65 (formerly Sm/118 and Sp/119), two manuscript sheets of numerary signals apparently for private ships, but set on a squared table with flags entirely different from Howe's. COR/65 is signed Robert Stopford, *Phaeton*, 3 August 1794, then employed in Cornwallis's Bay of Biscay Squadron. S(D) 24 corresponds with HSR/Z/5.

81 *NMM*, Tunstall Collection, TUN/114 (formerly FI/8), '1796' printed on the cover label; an unissued copy.

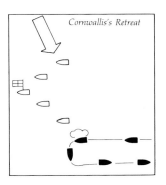

Cornwallis's Retreat

immediate reinforcement, especially as his frigates could be seen some way ahead making bogus signals to a non-existent fleet on the horizon. When, towards evening, some distant sails did in fact appear, the French were convinced that these were British warships and withdrew without even sending any of their numerous frigates to investigate.

Admiral Sir Charles Ekins commented,

> I never have been able to view this retreat in so high a light as it has been so often placed. Admiral Cornwallis deserves every credit for doing all he could, and no brave man would have done less. Had the French attacked the squadron seriously, it would have sold itself dearly, but it must have been overpowered. As in Barrington's affair at St Lucie [sic], I take more credit from the French than I am disposed to give to the English.[82]

Richard Hall Gower

A man of intense intellectual curiosity, Richard Hall Gower (1767–1833), served as an officer in the East India Company, retiring from the sea in 1802 at about the age of thirty-five to devote himself to naval architecture. Apart from designing new types of hulls and sail plans, Gower made a number of navigational and mechanical inventions. In 1793 he published a small *Treatise on the Theory and Practice of Seamanship*, which included a description of a number of fittings invented by himself. In the second edition of 1796 he outlined a scheme for a naval telegraph. This consisted of a large frame into which could be fitted letters of the alphabet followed by numbers painted black on thin white boards or sheets of tin. A signalling dictionary was to be made which could give up to 999 words for each letter of the alphabet; the signal would appear, for example, as A453.[83] By using the alphabet alone, any additional word could also be spelt out letter by letter. While each signal was being read, the symbols for the next – could be fitted into the grooves at the back of the frame, which only required rotating on a central pin to be immediately visible. This, Gower claimed, would overcome the inconvenience of making detailed flag signals, which could be especially difficult in windy weather.

In the third edition (1808) of his book, Gower claimed that his telegraph anticipated that of Sir Home Popham, a claim which seems entirely justified, since Popham's first practical experiment at sea did not take place until 1800. Gower expanded his ideas on signalling in the third edition, so as to cover all forms and contingencies. For day signals he proposed a simple numerical system of ten flags, numbered 0–9 with a substitute flag and a yellow pendant. His flags were white, red and simple combinations of red and white, and blue and white. He assumed that it was difficult to distinguish even between such sharply contrasting colours as red and blue and therefore arranged his flags so that the relationship of white to either red or blue was always different. The pendant signified 100 if hoisted above any of the flags and 200 if hoisted below.

As an alternative to the flags, Gower introduced 'shapes' in the form of two sets of cones and two sets of cylinders, to be made of light wicker or basket-work, painted black, or else of canvas set out with collapsible hoops to save stowage space. They were installed for hoisting at the yardarms. The shapes were very simple, and the triangular ones were reversible, while the small cones could be combined.

The list of signals is extremely simple: 0–Signal

acknowledged and understood; 1–Annul; 2–36 General; 37–49 Horary; 50–81 Compass bearings; 82–110 Private ships' signals, but in the form of ships of a convoy signalling the commodore, (as in an East India Company fleet); 111–145 Signals to particular ships from the commodore; 146–162 Questions from the commodore to particular ships; 163–175 Answering signals; 176–185 Signals at anchor; 186–203 Battle signals; 204–232 'Regular manoeuvring significations' (orders of sailing and changes of course); 233–247 Paroles. The signals, though primarily designed for East India Company fleets, could easily have been adopted and expanded for purely naval purposes.

The night signals were worked on the system by which one to four lanterns equalled 1 to 4; a blue light equalled 5; two guns equalled 10; three guns equalled 20; and four guns equalled 30. Thus 19 was expressed by two guns, one blue light and a hoist of four lanterns. The fog signals were made entirely by guns; at intervals; quick; a minute apart; and two minutes apart.

None of Gower's ideas was accepted, though he was willing to provide experimental signal shapes at his own expense. Being denied direct access to the Admiralty, he was unaware until some time later that Popham's methods, copying his own, were already being used by the navy. He frequently complained that his various inventions in ship design were copied and used without acknowledgement, particularly large sections of his *Treatise on the Theory and Practice of Seamanship*. It is easy to treat Gower as a disgruntled inventor who made exaggerated claims and had a persecution complex; it is probably more just to put him in the same category as La Bourdonnais.

Nelson and Hotham, 1795

Horatio Nelson's first tactical experience, apart from that of combined operations in the West Indies and Corsica, was in the Mediterranean Fleet under Sir William Hotham in 1795. Two partial actions with the French fleet, not very important in themselves, served to establish his position as a man who aimed at absolute victory.

On 13 May the *Ça Ira*, a powerful French 80, being damaged in a collision, fell out of the French line of battle with her fore and main topmasts and yards trailing over her larboard broadside. The British fleet was at that moment in general chase, and Nelson, taking the *Agamemnon* astern of the *Ça Ira*, began a highly successful attack. The wind was a little west of north and the two ships were steering west, the *Ça Ira*, being towed by a frigate. Nelson was able to approach

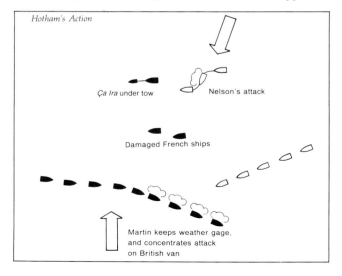

Hotham's Action

Ça Ira under tow

Nelson's attack

Damaged French ships

Martin keeps weather gage, and concentrates attack on British van

82 Charles Ekins, *Naval Battles*, p211

83 The example given in the book.

quite close, turn away to larboard, fire his starboard broadside into her stern, continue to chase, and then repeat the process. Although the *Ça Ira*, replied accurately and energetically with her handful of stern guns, Nelson managed to keep roughly within a hundred yards of her for over two hours. When a number of powerful French ships of the line began to close in to relieve her, however, he was glad to obey Hotham's signal of recall. Mahan comments that Nelson's courage and skill, though remarkable enough on this occasion, seems less remarkable than the fact 'that Nelson, here as always, was on hand when the opportunity offered; that after three days of chase he, and he only, was so far to the front as to be able to snatch the fleeting moment.'[84]

Next day a contest took place to secure the damaged *Ça Ira*, now being towed by a 74, the *Censeur*. Two British 74s were signalled to attack them, but were taken aback by the wind and heavily pounded. The two fleets now began to approach each other in line ahead on opposite courses, with the *Ça Ira* and the *Censeur* somewhat to leeward of their projected point of intersection. To save his two isolated ships, the French commander, Rear-Admiral Pierre Martin, would have had to resign the contest for the weather gage, not a difficult expedient, one would think, given French tactical doctrine. At the last moment, however, he kept his wind, thus allowing the British to pass between him and the two ships he was trying to save. By this manoeuvre he managed to bring the fire of practically the whole of his fleet against the two leading British ships, simply by filing past them closely and then drawing out of range by keeping to windward. This was no real solution, however, for after two hours' fighting the *Ça Ira* and the *Censeur* were both forced to surrender after Martin had continued on his course and abandoned them.

During this episode Nelson was stationed third in the line, immediately ahead of Hotham, and it was this proximity which enabled him to go straight on board the flagship and urge immediate and vigorous pursuit of the retreating French. Hotham had excellent reasons for refusing – both administrative and political – as well as having his two leading ships partially dismasted and two other ships, which had made an independent attack soon after dawn, much cut up in their sails and rigging. All this Nelson knew well enough, but he was in favour of accepting a calculated risk. Hotham's reason for not pursuing, given admittedly on the spur of the moment, is very similar to Rodney's after the Saints: 'We must be contented, we have done very well.' 'Had we taken ten sail,' commented Nelson, 'and allowed the eleventh to escape, when it had been possible to have got at her, I could never have called it well done.'

Sir John Jervis, 1795–9

Sir John Jervis once declared, 'Lord Hawke when he ran out of the line [at the Battle of Toulon] and took the *Poder* sickened me of tactics'. There is good reason to believe, however, that Jervis studied tactics assiduously. At the beginning of the war with France, he had command of the fleet operating against the French islands in the West Indies. He made few alterations from Howe's 1793 Signal Book or Additional Instructions when he issued his own in 1794, although in his main instructions he omitted battle Articles 15 to 26, as well as night sailings and Articles 12 and 13.[85]

---

84 A T Mahan, *The Life of Nelson, The Embodiment of the Sea Power of Great Britain*, London, 1898, 2 vols, I, p165

85 *NMM*, SIG/B/65 (formerly Sp/82) Signal Book, SIG/B/66 (formerly Sp/83 instructions, SIG/B/67 (formerly Sp/84) Additional Instructions; all issued from the *Boyne*, Carlisle Bay, Barbados, 6 January 1794.

Jervis's Secret Instruction

SECRET

[1]This is on a Supposition that seven or a less Number of the Enemy's Ships are separated to Windward, but should the Number exceed seven, *Our leading Ship is nevertheless* to Tack as soon as our *Seventh* Ship shall have passed thro' the Enemy's Line.
[2]No 8, 9, & c
[3]Nos 8, 9, 10
[4]No 1, 2 & c
[5]No 17, 16 & c
[6]No 14, 13 & c
[7]No 8
[8]No 1, 2 & c
[9]No 8, 9 & c
[10]No 17, 16 & c
[11]No 8, 9 & c
[12]No 12 & c
[13]No 10
[14]No 8, 9 & c
[15]No 9, 8 & c
[16]No 8, 9 & c
[17]No 7, 6 & c
[18]No 8, 9 & c
[19]No 8 or 9 & c
[20]No 1, 2 & c
[21]No 1, 2 & c
[22]No 1, 2 & c
[23]No 1, 2 & c
[24]No 17, 16 & c
[25]No 8, 9 & c
[26]No 1, 2 & c
[27]No 7 ½No 1
[28]No 1, 2 & c
[29]No 1, 2 & c
[30]No 1, 2 & c
[31]No 8, 9 & c
[32]No 8, 9 & c
[33]No 1,2 & c
[34]No 1, 2 & c
[35]No 8, 9 & c
[36]No 17, 16 & c
[37]No 17, 16 & c
[38]No 17, 16 & c
[39]No 1, 2 & c
[40]No 8, 9 & c
[41]No 1, 2 & c
[42]No 1, 2 & c
[43]No 8, 7, 6, 5,
[44]4, 3, 2, 1
[45]No 8, 9, 10, 11,
[46]12, 13, 14,
[47]15, 16, 17

When the Signal 50 is made, the weathermost and headmost Ships are to decrease Sail, and to edge down so as to collect a strong Body of Ships which are to form as they arrive up with each other, and to force thro' the Enemy's Fleet in the Direction, *where they judge the Body of our Fleet can fetch thro'*. When seven Sail (or an equal number of our Ships to that of the Enemy as shall be separated from their Centre & Rear) have passed through the Enemy's Fleet, Our headmost Ship is then to Tack[1] and a like number of Ships are to Tack (in succession) as there are of the Enemy's Ships so separated, and are to fetch up with, and attack each a Ship of the Enemy's Van, as they are able. The Remainder of our Centre and Rear Ships being intended to act against the Centre and Rear of the Enemy to Leeward.

It is very possible a great part of the Enemy's Centre and Rear may be separated from their Centre and Van, when this Evolution is effected; And as it is most likely that the nearest of the Enemy's Ships[2] which our Fleet has weathered *where our Ships first cut thro'* must be disabled in our passing, and will be obliged to bear up under the Lee of our Centre[3] and Sternmost Ships: In order to prevent their rejunction with their Van[4], the Rear[5] of the Fleet, is either to Tack or Wear, and the rest in succession on (according to the Conduct of the Enemy's Van and Centre) when the fourth[6] or fifth from our Rear Ship has nearly advanced up abreast of the Enemy's Ship[7], where our Ships first cut thro' and weathered; And should the Enemy's Van[8] bear up in order to form a Junction with their Centre[9] and Rear astern of our Fleet, *An equal Number of our Rear*[10] Ships (to those of the Enemy's Van who have been cut off from their Centre[11] and Rear, are to Tack, the Rear Ship first and the rest in succession, in order to oppose their Van[12], and the next[13] Ship of ours to that which last Tacked, is to Wear, and will become our present Van Ship, and she is to attack the headmost Ship of the Enemy's Centre[14] and Rear, and the rest of our Centre[15] and Rear are to Wear in succession, and to attack each her Opponent in the Enemy's Line. But if those Ships of the Enemy's Centre[16] and Rear which have been cut off from their Centre[17] and Van should have Tacked, our Ships are to continue on the same Tack they were at first standing, and each Ship of ours[18] is, in succession, to close with, and engage her Opponent[19] (or the nearest Ship to her at this Time) while our headmost [20]Ships which have passed thro' the Enemy's Fleet, and have Tacked upon the Enemy's Van[21]. are engaging them, The rest of the Fleet are to act in like manner by engaging each her Opponent, as they are then situated, and as they are able so to do.

If the Enemy's Van Ships[22] should Tack after our Van[23] Ships have Tacked, then our Van is to edge under their Lee in succession with a very easy Sail, at the same Time our Rear[24] and Centre are to Wear in succession, and to attack the Enemy's Centre[25] and Rear who are passing under the Lee of our Centre and Rear: And when the headmost Ships of our Van are nearly closing up with their headmost[26] Ships, they are to engage them as close as possible in passing; and when our headmost Ship[27] has passed their Sternmost[28], our headmost Ship must wear, and the rest in succession, and attack the Enemy's Van[29] Ships under their Lee in the best manner they are able always keeping in mind that the English Van Ships[30] are ever to keep to Leeward of the Enemy's Van[31] Ships in order to prevent the Enemy from making a rejunction with their Centre[32] and Rear, that have been cut off and are engaged with our Centre[33] and Rear. But should the Enemy's Van[34] Ships *Wear* after our Van[35] Ships have Tacked, then our Van Ships must lead large force Sail and form astern of our Centre[36] and Rear and engage the Enemy's Rear[37] Centre as they are best able. The Rear[38] of the English Fleet having now tacked in succession, our Line will become *reversed*, and of Course *our present Van*[39] will close with the Enemy's Van[40] (*who have bore up*) and prevent their Van from making a Rejunction with their Centre[41] and Rear taking Care to keep to Leeward of the Enemy's Van[42] as before directed to our former Van[43], in Case the Enemy's Van Ships should have Tacked; And the English Centre[44] and Rear at first our Van will wear in succession, the Sternmost Ship first, and each Ship is to attack her Opponent in the Line of the Enemy's Centre[45] and Rear to Leeward.

N.B. It is always to be understood by the Fleet, that if in the Execution of any of these Methods of Attack, any Ship should be disabled from fulfilling her part thereof, or from keeping her Station in the prescribed Mode, she is to give place (either by hauling her Wind or bearing up) so as to enable the first Ship astern of her (which is able) to supply her place, and that Ship is to take it with all possible Dispatch. The disabled Ship is to unite with the Main Body of the Fleet, and there to exert her utmost against the Enemy that her Crippled State will allow.

Given on board the *Victory* the
31st day of January 1796
[signed] St Vincent

TO
Captain Duckworth
Commander of His Majesty's
Ship *Leviathan* [issued to Duckworth in 1798]

By Command of the Admiral
Geo Purvis
Exd R C[alder]

Rear-Admiral (later Admiral of the Fleet) John Jervis (1735–1823), first Earl St Vincent, by Sir William Beechey. NMM

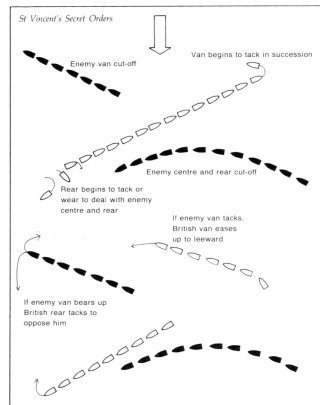

When in the following year he succeeded Hotham in command of the Mediterranean Fleet, he took over Hood's system employed on that station, presumably because the fleet was familiar with it.[86] He made a most important addition, however, by inserting as signal number 50 a tactical order: 'When approaching the Enemy's fleet on opposite tacks and the Admiral finds he cannot weather it, he may judge it expedient to make this signal . . . Vide separate Secret Instruction.' The Secret Instruction itself was an entirely separate document issued ten weeks later, and it opened up a dazzling series of tactical options.[87]

The purport of the Secret Instruction was that, in the event of the enemy succeeding in gaining the weather gage, the British fleet was to decrease sail, form into a strong body, and to force through the enemy line from leeward. It was then to tack and engage those enemy ships thus cut off from their centre and rear, and to prevent them reuniting. Meanwhile, the British centre and rear would engage the enemy centre and rear, doing everything possible to prevent them from reuniting with their van. Jervis had an answer for any manoeuvre the enemy attempted in an effort to reunite as a single line. Largely these British movements would have to depend upon the initiative of individual captains.

The Secret Instruction alone is sufficient to label Jervis as an incisive tactical thinker. Some observers might, however,

criticise it as far too complicated and theoretical. Jervis would undoubtedly have accepted this criticism, but would have added, no doubt, that he proposed to raise the fighting efficiency of his fleet to such a pitch that the instruction would no longer seem impracticable. There is no suggestion anywhere in the instruction of containing part of the enemy fleet with a possibly inferior number of ships in order to effect a superior concentration on the remainder. Jervis's idea was, evidently, that by breaking through the enemy's line, he would separate and disorganise the enemy ships, and that in the general pell-mell British fighting efficiency would prevail.

Jervis's objectives were illustrated by a set of seven extremely interesting tactical diagrams, apparently prepared for him by Captain Robert Calder, his Captain of the Fleet.[88] Number 1 shows the British fleet of seventeen ships in line ahead, having broken through the enemy line. Seven enemy ships (presumably the enemy van) are shown retreating to windward, and eleven isolated ships to leeward, the fleets having apparently met on opposite courses. The leading British ship is seen tacking, and the rearmost as having tacked or worn – in each case, presumably, to head off the isolated enemy van ships to windward from helping the remainder of their fleet to leeward. This looks very like what actually happened at the Battle of St Vincent, but with the starboard and larboard positions reversed, the rearmost ship imitating Nelson's action. The date of this diagram is uncertain. It is given in a different form from the rest, and is the only one which even hints at breaking the enemy line.

Number 2 shows the British fleet in line ahead about one and a half miles to windward of the French fleet, also in line ahead and parallel with the British. When the signal is made for the British fleet to bear down and attack, the leading ship will have to sail nearly a mile and a half on a diagonal course before she can open fire on her opposite number. This will take about twenty minutes, during which the leading French ship will have sailed about one mile. The ships in both fleets

86 *NMM*, SIG/B/68 (formerly Sp/133), day signals issued from the *Lively* 19 November 1795 to his Captain of the Fleet, Captain Calder, *Victory*, 'or of any other ship or vessel in which I may hoist my flag.' The book is thumb-indexed along the top and bound in blue morocco. The Mead Collection contains an exactly similar copy issued to Captain Thomas Wall, *Defence*, as well as a copy of the Night Signals, in Hood's manner, both similarly dated.

87 *NMM*, DUC/1/18. This copy, issued to Captain Duckworth on joining the Mediterranean Fleet in the middle of 1798, is dated 31 January 1796, showing that it was first issued ten weeks after the signal book.

88 *The Naval Miscellany*, II, pp301–7

are shown at two cables distance. The implication is that the French will continue on their original course at 3 knots and the British go down to attack at about 4½ knots.

Number 3 is signed Robert Calder, *Victory,* 19 August 1796. It shows a fleet of thirty ships in line abreast, in wedge formation and in the order of retreat and in line of bearing on either tack 'comfortable to the signal 32'. That signal was 'for the ships of the fleet to be kept on the larboard line of bearing from each other, though then sailing on the starboard tack or the contrary on the starboard line of bearing, though then sailing on the larboard tack'. Jervis issued a general order, '*Victory*: at sea. 9th May 1796', relative to signal 32. In case of a shift of wind forward, 'the leading ship becomes the axis' (instead of the flagship) and the other ships of her squadron or division were to make sail and hug the wind in succession as much as possible, the sternmost ships most, so as to retain their relative positions. If the wind came aft, the sternmost ship was to become the axis 'and those ahead of her are to make sail and cling to the wind, the headmost ship most.' [89]

Number 4, signed 'Robt Calder', as also were numbers 5–7, is explained as showing the British fleet attacking in line of bearing from windward and the French fleet turning away four points to leeward and in succession. 'The English fleet must then force sail and bear up also in succession, one point: the English fleet will then be eight and the French six points from the wind. Each ship to bear up instantly her opponent

does, by which they will preserve their bow and quarter line', and so be able to engage ship for ship 'as before directed'. The British fleet, having borne up four points in order to begin their approach, would need only to bear away one point more and force sail to reach round the French line as it turned away, and so force an action. There was no suggestion of breaking the enemy line.

Number 5 shows the situation arising should the French turn away four points together so as to avoid action. The British fleet, already approaching in line of bearing, 'must then alter their course together, one point to starboard, steering SSE; and force sail'. Number 6 shows a further attempt by the French to avoid action by wearing together. The British fleet, still shown as following their original mode of attack, 'must then alter their course together to starboard six points, steering SW by S and increase sail'.

Number 7 shows an extension of the French avoiding action taken in number 4. This time, instead of merely turning away from the British in succession, the French wear in succession, thus forming a new line on the opposite tack. The British answer to this was highly sophisticated. They

must then alter their course together ten points to starboard, when the fourth or fifth of the enemy's van [is] in the act of wearing and steer W by S, shorten sail and keep the main topsail shivering, by which manoeuvre the rear and centre of the French fleet will be greatly annoyed in wearing and their van unable, from their position, to fire upon our ships, until they have opened astern of their rear ships. The English fleet will be enabled to close up with them in the same manner on their starboard quarter in lieu of the larboard, as if they had not wore the lines being reversed.'

A tactical diagram by Admiral John Jervis's flag captain, Robert Calder, showing the means by which, should the French Fleet wear together, they might still be brought to action, about 19 August 1796. NMM

This means that the French centre and rear would have to receive the fire of the whole British fleet while passing on the opposite tack preparatory to wearing round in succession following their van. Thereafter the British would be able to engage their new line now formed on the starboard tack.

On 3 November 1796 Calder issued Jervis's instructions for the look-out frigates to keep two points on the admiral's bow. By day they were to be one to three leagues distant, and by night one to three miles, according to the weather. They were to make sail ahead of the fleet at 4am, and shorten sail at 4pm, so as to attain their proper distances. 'All night signals should be avoided as much as possible by these frigates, as they always cause confusion, more or less, in a fleet.' Instead, one of the frigates was to report back. On 28 March 1797, immediately after the battle of St Vincent, Jervis issued an order of sailing for the fleet in two parallel columns one mile apart, with the ships at intervals of two and a half cables.[90]

Admiral John Jervis's Order of Sailing in Two Squadrons, issued to Sir Charles Thompson, Vice-Admiral of the Blue, 19 August 1796. NMM

The leading ship of the larboard column was to be level with the second ship of the starboard column. The admiral was to be two points on the weather bow of the starboard column and three points forward from the beam of the leading ship of the larboard column. He issued an identical order of sailing on 12 April 1798.[91]

Jervis's thinking about cutting the enemy fleet from leeward, shown in Calder's first diagram, evidently should not be taken as indicative of any tendency towards radical ideas generally in tactics. His night signals and instructions of 1798 followed very closely on Hood's New Arrangement of 1793.[92] Two days after Nelson had won the Battle of the Nile, 3 August 1798, Jervis, now Earl St Vincent, issued a copy of his 1796 instructions.[93] These were exactly the same as those he issued in 1795, themselves a copy of those issued by Admiral Alan Gardner in 1793.

On 17 December 1798 St Vincent issued further signals and instructions, to come into force on 1 January 1799.[94] The day and fog signal book resembled the Hood-McArthur New Arrangement of 1793, except that the signal numbers were printed throughout. There were 181 instead of 162 signal spaces, the squared table was for nine instead of eight signals, and there was a table of horary signals.

The Battle of Cape St Vincent

At dawn on 14 February 1797, Jervis had his fleet well closed up in two parallel columns. The wind was WSW, and

---

90 *MOD*, NM/35 (formerly *RUSI* 35)

91 Copy sold at Hodgson's Rooms, 3 April 1908

92 *MOD*, Ec/164, bound in blue morocco. *NMM*, SIG/B/72 (formerly Sp/134), unsigned, is similarly bound and was possibly St Vincent's personal copy. *NMM*, Tunstall Collection, TUN/22 (formerly S/MS/R/7) is an undated manuscript copy of the whole of St Vincent's Night Signals, with explanations, directions and instructions, in pocket book form.

93 *NMM*, SIG/B/69 (formerly Sp/30) and *MOD*, Ec/165

94 *NMM*, SIG/B/70 (formerly Sp/95), issued to Captain Henry Savage, *Warrior*. It much resembles *NMM*, Sp/17, but appears to be a fuller and later printing of KEI/S/11(a) (formerly Sp/20). The Mead Collection has a copy similarly worded about coming into force in the New Year.

Order of Sailing in Two Squadrons
Larboard or Lee   Starboard or Weather

		11		
		0 *Victory*		
		1 0 *Blenheim*	0 *Repeater*	
Culloden	12 0	2 0 *Diadem*		
Orion	13 0	3 0 *Prince George*		
Colossus	14 0	4 0 *Irresistible*		
	15 0	5 0 *Britannia*		
Barfleur	16 0	6 0		
Excellent	17 0	7 0 *Captain*		
	18 0	8 0		
	19 0	9 0		
Goliath	20 0	10 0 *Egmont*		
Namur	21 0			

Each Squadron to be 8½ Cables or One Mile asunder and the Ships to be 2½ Cables from each other

Given on board the Victory
Off Cape St Vincent 6 Febry 1797

To
Charles Thompson Esq
Vice Admiral of the Blue
&c: &c:
By Command of the Admiral
Geo. Purvis.

(Sign'd) J Jervis

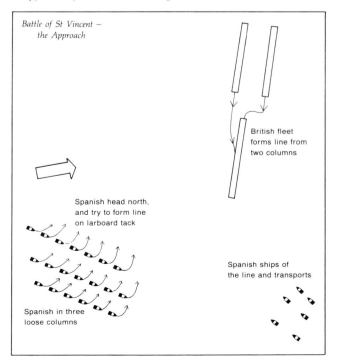

Battle of St Vincent –
the Approach

British fleet forms line from two columns

Spanish head north, and try to form line on larboard tack

Spanish in three loose columns

Spanish ships of the line and transports

HMS Victory *raking the Spanish at the Battle of Cape St Vincent, 14 February 1791, by Robert Cleveley.* NMM

he was steering a little west of south. He was warned of the Spanish fleet's presence and had cleared for action twenty-four hours previously. His look-out frigates soon reported the Spanish fleet in what turned out to be complete disorder. It was a hazy dawn, but gradually he made out to starboard (and therefore to windward), twenty or more Spanish ships-of-the-line and frigates, while very slightly to larboard, and to leeward, a smaller group of apparently eight or nine of the line. Jervis had fifteen of the line with three flag officers and a commodore, Nelson, under him. Realising his tactical advantage, he held on his course until 10.57am, when he signalled (number 31) 'To form the line of battle ahead and astern as most convenient', thus allowing his fastest sailers to draw to the head of the line. No signal was made for carrying a press of sail, since the instructions enjoined all ships to carry sail equivalent to that of the admiral. After altering course only one point to larboard and then back one to starboard, Jervis at 11.28am signalled (number 40) 'The admiral means to pass through the enemy's line'; two minutes later he signalled 'to haul the wind on the starboard tack'. So well was

he placed at dawn that his line of advance had scarcely changed in the course of nearly three hours. Considering night movements and the hazy weather, this was largely luck. Nevertheless he risked much: had the windward section of the Spanish fleet shown anything like an indication to form a real line and engage, their superior firepower could, on paper, have enabled them to immobilise his leading ships at long range and to have overwhelmed the remainder by sheer weight of metal.

The situation in the Spanish fleet, however, was chaotic. The fleet was short of men, short of supplies, short of training and in every way deficient, though, as before, the flag officers and captains possessed excellent signal books and detailed tactical manuals. At dawn on 14 February the Spanish Admiral, Don José de Cordóva, had his fleet in the order of sailing in three columns, steering ESE with column commanders leading their respective columns. This part of his fleet consisted of twenty of the line. To the east, and hence to leeward, were the smaller group of ships already sighted by the British and judged to be a detached division. In

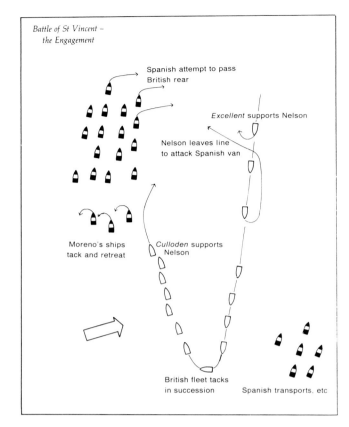

*Battle of St Vincent –
the Engagement*

Spanish attempt to pass
British rear

Excellent supports Nelson

Nelson leaves line
to attack Spanish van

Moreno's ships
tack and retreat

Culloden supports
Nelson

British fleet tacks
in succession

Spanish transports, etc

fact they consisted of four large merchantmen loaded with mercury and protected by two of the line and some frigates. Had Jervis known this he might well have captured the whole convoy and so seriously have damaged the Spanish economy.

When various Spanish ships began to report enemy ships astern of them, Cordóva took a quick and, as it turned out, disastrous decision. Realising that he was heading away from the enemy, and wishing to keep the weather gage, he signalled the fleet to form a line of battle 'as convenient' on the larboard tack. For a British fleet of the same size, well trained under Jervis, to have attempted such an evolution, suddenly and in hazy weather, would probably have led at least to temporary confusion, but it is not likely that Jervis, Howe or Hood would ever have attempted it, since it involved reversing the order of sailing in columns and at the same time inviting a general scramble to form a single line on the opposite tack. In addition, it gave no opportunity to the admirals leading the three columns to get to the heads, or even the centres of their columns in the reversed order. In fact all the Spanish ships seem to have been left astern, including, of course, Cordóva. Three of his ships – Admiral Juan de Moreno's flagship, together with another 112 and a 74 – failed to wear smartly enough and fell to leeward. Cordóva signalled to them to wear again and take station in the rear so as to avoid the advancing British.

At 11.31am, only five minutes after signalling his intention to pass through the enemy line, Jervis's leading ship, the *Culloden*, opened fire on one of Moreno's group. All three tacked independently and then withdrew to leeward towards their convoy. As the British line continued on its course, firing to starboard, the weather part of the Spanish fleet began to pass them on the opposite tack, pursuing a slightly diverging and north-westerly course.

At 12.08pm Jervis signalled the British fleet to tack in succession. Having forced the enemy fleet to separate, his first object was achieved, and he was now free to pursue the

retreating Spanish force to windward. Meanwhile his rear ships could continue to keep off the convoy group to leeward by firing their larboard broadsides as they passed. At 12.15 he repeated the signal to pass through the enemy line, presumably to reassure his rear ships that they were doing the right thing in leaving Moreno to leeward as they came up, and then tacking in succession. To have tacked earlier might have prevented this essential separation and have allowed the Spanish rear and the convoy group to rake the British ships as they came round in stays.

Moreno and the two ships with him did in fact attempt decisive intervention from leeward and succeeded in disabling the *Colossus*, fifth ship in the line, so that she missed stays and fell away to leeward. This caused a considerable delay and though Moreno's ships were eventually checked, their intervention certainly hampered the British pursuit of the Spanish weather division.

Cordóva realised that, despite his initial setback, he could still make use of his vastly superior strength. He decided to double on the British rear by getting his van ships to wear round on the opposite tack. By this manoeuvre he might capture the rearmost British ships and frigates and save the mercury convoy from capture, as well as protecting Moreno's three ships. Unfortunately, he could obtain no effective response from his flag officers and captains, despite signalling and hailing. Nevertheless, he succeeded in initiating a movement to close the British rear with a number of ships, his own included, though no proper line was formed.

Jervis saw that the Spaniards were slipping away from him, partly because of his delay in making the signal to tack and partly because of the delay caused by the disabling of the *Colossus*. Although the *Culloden, Blenheim, Prince George* and *Orion* were well round on the larboard tack and heavily engaged, the leading Spanish ships had reached far ahead of them, while others were bunching up further to windward. Nor was it any longer possible, in view of the delay caused by the *Colossus*, for the British line to continue its hairpin bend movement in smooth, uninterrupted fashion. Nor was there any further need to worry about Moreno who was now in retreat, with one of his ships working right round the British rear to leeward in order to rejoin Cordóva. The British rear ships, therefore, had an opportunity to turn directly to starboard and attack the main body of the Spanish fleet by the quickest route. At 12.51pm Jervis made the famous signal, number 41, 'The ships to take suitable station for their mutual support and to engage the enemy as arriving up with them in succession'.

With the line of battle thus abandoned, flag officers and captains were free to act to the best advantage. This did not mean that they were free to do just as they pleased, but rather to engage the enemy in a new line or part of a line as well and as quickly as they could.

Nelson may or may not have seen the signal, but in any case he had already anticipated it. Being the sternmost ship but two, he could see the need for cutting across the Spanish line of advance or, more accurately, retreat.

. . . perceiving the Spanish Fleet to bear up before the wind, evidently with an intention of forming their line, going large – joining their separated division – or flying from us, I ordered the *Captain* to be ware, and passing between the *Diadem* [next astern] and *Excellent* [rearmost ship], at ten minutes past one o'clock, I was close in action with the Van, and, of course, the leewardmost of the Spanish Fleet.

In this way he engaged the *Santissima Trinidad*, supported by two other 112-gun flagships as well as an 80. The *Excellent* (Collingwood), however, executed the same manoeuvre and was later followed by the *Diadem*. In the meantime, the *Culloden*, still leading the original line, was coming up together with the *Blenheim* and other leading ships. Thus an improvised line or quasi-line was formed along the leeward flank of the Spanish fleet. Further to the rear the remainder of the British fleet, having tacked, were now formed in a line somewhat to windward of the Spanish and were thus able to force their way amongst the bunched ships of the Spanish rear. There was, however, a period, which no doubt seemed long to those concerned, during which the *Captain* was insufficiently supported by the van ships as they came up behind her after tacking. Admiral Thomas Taylor made the interesting point that those ships should not merely have engaged the Spaniards once they were round on the larboard tack, but should have followed Howe's established doctrine (Additional Instructions . . . in Action with the Enemy, Article 6) that when the leading ship came up with the enemy's rearmost ship the second British ship was to pass ahead of her and engage the next enemy ahead, with the remaining ships doing the same in succession.

The result of the pell-mell round the *Captain*, and then ahead and astern of her as well, was that four Spanish ships were captured, two by Nelson in person, though they mounted a total of 382 guns. Córdova tried to set his fleet a personal example of stout resistance, and the *Santissima Trinidad* was so severely damaged that she actually surrendered, though she eventually escaped.

Córdova observed that when the improvised line was formed by Nelson and others 'along our line on the starboard side, their ten ships [the rest of the fleet] remained in consequence upon the other side [ie, more to windward], firing at us, in a regular and powerful line, by which manoeuvre they obtained a decision of the combat in their favour'. The *Santissima Trinidad*, he continued, was reduced to a wreck. Four ships were hailed and signalled to give her support, and two of these were amongst those eventually taken. 'In this situation – that is, against four times their force, counting besides numbers the superiority of their fire over ours – I being doubled on and cut off from the greater part of our said ships . . .' Córdova felt that he ought to have been relieved by the disengaged van or centre ships. 'Whoever considers the state of things, and moreover the rapidity and accuracy with which the English handle their guns, can imagine what must have been our condition by four o'clock.' Admissions such as these show why the victory went to the British fleet, despite the fact that they were outmatched by more than two to one in guns; and by much more in weight of metal.

'A System of Naval Tactics', 1797

Published anonymously and subtitled 'Combining the established theory with general practice, and particularly with the present practice of the British Navy',[95] *A System of Naval Tactics* was an extremely useful and well set out text book, very handy in form, with coloured figures distinguishing the different columns or squadrons in every evolution in red, white and blue.

Part I purported to be a translation of Morogues, though it was only partly so, as it left out the battle tactics, and rearranged and paraphrased the rest. The sailing diagrams were very well shown. There were tactical and battle sections at the end, which, though still purporting to be from Morogues, were actually drawn from Hoste and Bourdé de Villehuet. These sections are only relevant in so far as they reveal the interest shown at the time in those authors.

Part II gave summaries and extracts from Bourdé de Villehuet, with special reference to his so-called 'order of convoy'.

Part III was described as the 'Tactics of the British Navy', and purported to be 'compiled from official documents with the assistance of several experienced officers'. In the space of twenty pages the anonymous author covered the order of sailing in three columns, six columns and two grand divisions, changes of course and changing from various orders of sailing into the line of battle. For these he provided a summary of current practice as regards distances between ships and squadrons in the various orders, and rules to be observed when changing course or formation. This part was, in fact, a summary of Howe, re-written so as to make it more easily understood.[96] Nothing whatever was said in this part about fighting, no doubt for security reasons.

The Battle of Camperdown

Tactically, Camperdown (11 October 1797) was a Howe battle in the sense that on the British side it was fought with Howe's signal book and accompanying instructions. On the Dutch side it was fought with a numerical signal book containing 795 signals, and of unrivalled simplicity in arrangement – unrivalled even by Howe's.[97] It was based on ten flags, numbered 1–10, with pendants for the hundreds, but the book was thumb-indexed so that the relevant pendant, together with the ten flags, could be seen on each page, with the signal opposite the flags, the flags for the tens being shown along the top.

The battle signals provide for every imaginable contingency though, unlike the French, they are directed entirely towards the actual defeat of the enemy.

Dutch Battle Signals of 1797

No 715	Attack the enemy's van to windward with a superior force of the heaviest ships in the line of battle; light ships to support the attack in lozenge formation on perpendicular of the wind [the lozenge suggests Grenier's influence].
No 717	Attack the enemy's *corps de battaile* to windward with a superior force of the heaviest ships in the line; light ships to support their heavy ships at the head of the line and the rest at the rear.
No 726	Van to crowd sail to attack to windward and engage enemy's van while the centre and rear engage to leeward so as to double.
No 728	Van to cut through the enemy's line and the rest to turn [parallel] with the enemy and engage.
No 729	All ships to cut through the enemy's line individually.
No 730	Enemy's fleet being at anchor in a port or road, to double their van and engage.
No 735	Fleet to retreat fighting.
No 739	Squadron or ships signalled to cut through the enemy's line.
No 741	Admiral sees that he has the advantage; ships, therefore, which are able, to cut through the enemy's line and put them between two lines of fire.

95 *NNM*, Corbett Collection, TUN/96 (formerly T/P/R/1), bookplate of Philip Carteret

96 Sir Julian Corbett, in his *Signals & Instructions* pp75–8, quotes at length from Part III, noting the absence of any mention of breaking the enemy line. The two diagrams which he reproduces, however, are merely copies of those introduced by McArthur on behalf of Hood and later adapted for the Admiralty Signal Book of 1799.

97 *NMM*, DUN/31 (formerly S(D) 31)

No 744   Ships which have doubled enemy to rejoin the
          fleet.
No 748–9 Ships at the head of the line to prevent enemy
          doubling to windward: ditto at rear.
No 750–1 Ditto, to leeward, ditto.
No 752   Enemy is trying to cut our line; ships to close
          up and stop them.
No 754   Close up and stop enemy ships which have cut
          our line from rejoining their main body.
No 755   Enemy has cut our line; ships to form lozenge
          formation to leeward with the wind.

It is astonishing to discover that, although all the ordinary

words and standard terms are in Dutch, the particular evolutions in the battle signals are in French. This would seem to show a desire to graft tactical virtuosity, seen as a French accomplishment, onto the traditional phraseology of Dutch seamanship. There is no sign whatever of French wording in the Dutch signals at the start of the American War.[98] Either they were dazzled with the new French tactics and signalling (as Kempenfelt was, though in a different way) or else they had come under French influence as a result of the Revolution. The signals themselves are in the true Dutch spirit and seem to be almost in contradiction with those produced by Morogues and Pavillon. Nor do they read as mere notional flourishes – paper signals never to be used. On the contrary, their number and consistency seem to breathe a truly offensive spirit. In the lozenge formation employed in signal number 715 we see the influence of Grenier, whose book, first published in 1787, was no doubt read with great

The Battle of Camperdown, 11 October 1797, by Philippe-Jacques de Loutherbourg (dated 1801). NMM

98 *NMM*, Tunstall Collection, TUN/13 (formerly S/MS/D/1) signal book of Rear-Admiral Andreis Hartsinck, Mediterranean, 1775–6, old system with flags in particular positions.

interest in Holland as showing a real breakaway from convention.[99]

Except for the Dogger Bank in 1781 (itself a very second-rate affair), the Dutch had not fought a fleet action since Malaga in 1704, so the signals of 1797 were really intended for the kind of fleet the Dutch no longer possessed.

Admiral Jan Willem de Winter put to sea from the Texel with fifteen ships-of-the-line, very weak in firepower. He made no attempt to manoeuvre when approached by the British fleet but simply formed in line ahead on the larboard tack on a north-easterly course with the wind NW by N, within sight of the Dutch coast. Three heavy frigates were used to support his line. Admiral Adam Duncan had sixteen ships-of-the-line. Neither admiral had any three-deckers. Nevertheless, Duncan's fleet, although badly strained and worn by continuous North Sea cruising and only just recovered from the disruption of the Nore mutinies, was in good fighting form.

Being signalled by his advanced ships that the enemy were in sight to leeward, Admiral Duncan

immediately gave chase, and soon got sight of them, forming line on the larboard tack to receive us, the wind at NW. As we approached them I made the signal for the squadron [fleet] to shorten sail, in order to connect them; soon after I saw land between Camperdown and Egmont, about nine miles to leeward of the enemy, and finding there was no time to be lost in making the attack, I made the signal to bear up, break the enemy's line, and engage them to leeward, each ship her opponent, by which I got between them and the land, whither they were fast approaching. My signals were obeyed with great promptitude.[100]

In his letter to Lord Spencer he wrote: 'For the particulars of our victory [I] shall leave your Lordship to my public letter, and shall just say I was obliged to lay all regularity and tactics aside, we were so near the land, or we should have done nothing.'

Recent writers have been at pains to contradict the idea that Duncan's method of attack anticipated Nelson's at Trafalgar. They need not have worried, for Sir Julian Corbett settled the point more than half a century ago.[101] Duncan had his 'squadron', as he calls it, in two divisions, eight ships in his own division and the other eight under Vice-Admiral Richard Onslow. He saw at once what ought to be done and what in practice could be done. He must get between the Dutch and the land by steering for them as quickly as possible. His ships were in two loosely formed groups; his own division, less the *Adamant* which had previously been detached to observe and was now with Onslow, to the north, and Onslow's division, plus the *Adamant*, to the south. Yet speed must not produce confusion. To break through the Dutch line successfully, each ship must steer for her opposite number in such a way as not to block the fire or movement of flanking ships. To describe Duncan's approach, therefore, as headlong is inaccurate, especially as his speed could in any

Admiral Adam Duncan (1731–1804), first Viscount Duncan. NMM

case scarcely have exceeded three knots in the fresh or moderate breezes recorded in the ships' logs.

In the space of three hours – that is from the beginning of his tactical approach to the start of the battle – Duncan made some thirty signals.[102] He was making fresh signals at roughly six-minute intervals; Keppel made nineteen signals in the eight and a half hours of approach and battle of Ushant. Yet no one would suggest that Duncan was being unduly fussy or trying to exercise an unnecessary degree of control over his subordinates. On the contrary, he was trying to speed up the battle right from the start, his signals being mainly by way of making adjustments of distance and bearing so as to avoid the need for bringing to and dressing his line during the approach. His main signals, apart from a

99 A Dutch translation of Grenier by J J Melvill was published at Leyden in 1799 under the title of *Zee-Tactique of Konst des Oodogs te Zee*, but no doubt his work was well known in Holland before that.

100 Julian S Corbett, (and Herbert W Richmond, editor of vols iii and iv), *The Spencer Papers*, London, 1913, four vols, (hereafter cited *The Spencer Papers*), II, 197n. See also *Logs of the Great Sea Fights*, I, editorial comment by Admiral Sturges Jackson.

101 *NMM*, DUN/19 (formerly S(D) 19) is an manuscript notebook of Duncan's signals dated 1795. It contains nothing new.

102 The number is uncertain, owing to discrepancies between the log of the *Venerable*, Duncan's flagship, and the logs of other ships. See also the list of signals made, as recorded by the *Triumph*, NMM, SIG/B/75 (formerly Sm/32).

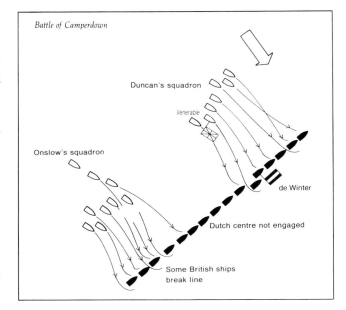

Battle of Camperdown

Duncan's squadron

Venerable

Onslow's squadron

de Winter

Dutch centre not engaged

Some British ships break line

number of general ones and those to particular ships to make more sail, appear to have been as follows:

11.08	No 48:	To form on the starboard line of bearing.
11.11	No 81:	To come to the wind together on the starboard tack.
11.17	No 95:	To take station in the line as pendants thrown out.
11.30	No 36:	Each ship independently to steer for and engage her opponent in the enemy's line.
11.35	No 14:	Bear up and sail large.
11.40	No 41:	The van to engage the enemy's rear.
11.53	No 34:	To pass through the enemy's line and engage from leeward.
11.53	No 41 repeated (to regularise the actual approach).	

In Duncan's order of sailing, which was the same as his order of battle, he placed Onslow's 'larboard or lee division' in the van, the *Monarch*, Onslow's flagship, being fifth ship. His own 'starboard or weather division' formed the rear, the *Venerable* being second ship, which placed her tenth in the line as a whole. When the Dutch were sighted, Onslow's division was roughly opposite their rear and Duncan's opposite their van. Hence the British van was in fact Onslow's division. The terms 'starboard and weather' and 'larboard and lee' were terms of title and precedence, used without reference to actual tactical situations and, therefore, extremely confusing. Had the battle been fought in line ahead with the British fleet in reverse order, Duncan might possibly have found his own position inconveniently remote from his leading ships, as Byng did at Minorca.

Duncan's signal lieutenant had been on sick leave ashore,

suffering from scurvy and only resumed his full duties on the morning of the battle. When ordered to hoist signal number 41, he ordered the signalman to hoist number 39 by mistake, which was for the van to engage the enemy's centre. He at once noticed his mistake, and had it hauled down. Number 41 was hauled up but not before number 39 had been noted in Admiral Onslow's flagship, which probably caused subsequent talk about Duncan changing his plans at the last minute and confusing the squadron.

When Onslow opened the battle at about 12.40pm, he succeeded in breaking through the Dutch line, raking the bow and stern respectively of the ships he passed between. Roughly half the British squadron succeeded in executing this manoeuvre in some measure; two separate pell-mells resulted, with the Dutch ships in the centre not fully engaged. The Dutch fought with all their accustomed bravery but were heavily worsted by the superior British gunnery and mutual support tactics of the pell-mell. The Dutch did what they could to assist ships nearest to them but the British showed marked superiority in their in-fighting changes of position. They also showed superiority in preventing the Dutch doing the same thing. By 3pm the battle was over and many Dutch ships had struck, with nine of the line and two other ships eventually secured. The prisoners included Admiral de Winter and another flag officer. The fleets had been carried to leeward in the fighting and many of them were now close in shore so that Duncan had great difficulty in getting off the coast with his prizes.

In terms of percentage of captures to total enemy force, this was the most complete victory achieved by British arms in the whole of the Revolutionary and Napoleonic Wars with the exception of the Nile.

The Battle of the Nile

It is customary to criticise the French commander at the Battle of the Nile (1 August 1798), Admiral François Paul Brueys d'Aiguïlliers, for taking up a weak position, but was it really so weak, except against an opponent like Nelson? He had anchored his ships in the western recess of Aboukir Bay on a line roughly NNW and SSE with a slight angular projection in the middle. They had rocky shoals immediately ahead and to the north west of them, and shoal water to the west and south. It would have been impossible or at least extremely difficult, to have attacked them with anything but a wind ranging between NW and SE. The only possible hope for the attackers would seem to have been for them to make a wide sweep into the shallow and uncharted bag of the bay, and then to approach from the south. The direction of the wind, however, made that approach impossible.

Brueys's worst mistake was undoubtedly his failure to maintain any reconnaissance, though this is seldom mentioned. With large parties of seamen ashore fetching water and engaged in quasi-military duties, it was especially necessary that he should have had good warning of Nelson's approach. As it was, he kept his four very useful frigates tucked away on the inside of his line of battle and seems to have relied solely on look-outs from the mastheads of his anchored fleet. Nelson, apart from his ships of the line, had just one brig. Hence the two fleets sighted each other at roughly the same moment.

After a long discussion with his flag officers and chief of staff, Brueys decided to fight at anchor, no doubt wisely. Observing, moreover, that the two British ships first sighted were still a long way off, he persuaded himself that the attack would be delayed until next day. This was at some time well into the afternoon. When, however, the British squadron, 'favorisée par une jolie brise de nord-nord-ouest, s'avançait

Amiral François Paul Brueys d'Aiguïlliers (1753–98), engraving by Leguay after a painting by A Lacauchie. MM

rapidement,' Brueys realised that the attack was coming that evening.

By the judicious use of springs, Brueys's ships should have been able to adjust their broadside positions so as to be able to develop a crushing fire against any force making a direct approach either in line ahead, line abreast or line of bearing. Villeneuve, who was third-in-command and stationed last but one in the line, says that he had 'deux grosses ancres, une petite et quatre grelins,' [cables]. What use he made of them is uncertain but apparently Brueys showed no effective initiative here, and in the battle the French ships seem to have continued lying anchored by the bows only. They were thus in the line of the wind and in line of bearing to each other corresponding to the line, slightly curved, on which they were anchored.

Apart from giving no orders to ensure that his ships could be worked round so as to develop the maximum fire face to the enemy, Brueys also ignored the danger of leaving wide gaps between the ships. He should have anchored his ships closer together, in which case the two wings could have met at a sharper angle, instead of leaving about three-quarters of a cable of clear water between each ship. Nor did he ensure that in case of emergency the guns on the larboard side of his ships were loaded and ready for firing. All these errors helped to lose the battle, but they were only symptoms of a more major defect, the general lack of fighting efficiency in the French fleet.

Sir Edward Berry, in his well-known pamphlet, says that while at sea Nelson had frequently had his captains on board the *Vanguard* to hear his plans for attacking the enemy.[103]

There was no possible position in which they [i.e. the French] might be found, that he did not take into his calculation . . . Upon surveying the situation of the enemy, they could ascertain with precision what were the ideas and intentions of their commander, without the aid of any further instructions; by which means signals became almost unnecessary . . .

It is almost unnecessary to explain his projected mode of attack at anchor, as that was minutely and precisely executed in the action which we now come to describe. These plans, however, were formed two months before an opportunity presented itself of executing any of them, and the advantage now was, that they were familiar to the understanding of every captain in the Fleet.

Berry was careful not to reveal any details of Nelson's tactical plans. Luckily, however, two of Nelson's tactical memoranda survive, both dated on the day he was joined by the ships detached by Lord St Vincent to reinforce him under Captain Thomas Trowbridge.[104]

General Order. *Vanguard*, at sea, 8 June 1798. As it is very probable the enemy may not be formed in regular order on the approach of the squadron under my command, I may in that case deem it most expedient to attack them by separate divisions; in which case, the commanders of divisions are strictly enjoined to keep their ships in the closest order possible, and on no account whatever to risk the separation of one of their ships. The captains of the ships will see the necessity of strictly attending to close order: and, should they compel any of the enemy's ships to strike their colours, they are at liberty to judge and act accordingly, whether or not it may be most advisable to cut away their masts and bowsprits; with this special observance, namely, that the destruction of the enemy's armament is the sole object. The ships of the enemy are, therefore, to be taken possession of by an officer and one boat's crew only, in order that the British ships may be enabled to continue the attack, and preserve their stations.

The commanders of divisions are to observe that no consideration is to induce them to separate in pursuing the enemy, unless by signal from me; so as to be unable to form a speedy junction with me; and the ships are to be kept in that order that the whole squadron may act as a single ship. When I make the

signal no 16, the commanders of divisions are to lead their separate squadrons, and they are to accompany the signal they may think proper to make with the appropriate triangular flag, viz. Sir James Saumarez will hoist the triangular flag, white with a red stripe, significant of the squadron under the commander in the second post; Captain Troubridge will hoist the triangular blue flag, significant of the rear squadron under the commander in the third post; and whenever I mean to address the centre squadron only I shall accompany the signal with a triangular red flag, significant of the centre squadron under the Commander-in-Chief.[105]

This memorandum makes nonsense of the idea that Nelson believed in leaving a wide discretion to his captains on every occasion. The discretion left to them over the disabling of prizes is one of seamanship rather than tactics. In other respects the memorandum might well have been the work of James Duke of York brought up to date.

Captain Berry adds that, had Nelson 'fallen in with the French fleet at sea,' in company with the convoy of Napoleon's troopships, a similar form of tactical control would have been employed. 'That he might make the best impression upon any part of it that might appear the most vulnerable, or the most eligible for attack,' Berry wrote, 'he divided his force into three sub-squadrons . . . Two of these sub-squadrons were to attack the ships of war, while the third was to pursue the transports, and to sink and destroy as many as it could.'[106]

The second Memorandum deals with action against the enemy at anchor and gives signals for attacking the enemy's rear, presumably when under sail.

Gen[eral] Mem[orandum] *Vanguard*, at sea, 8 June, 1798.

As the wind may probably blow along the shore when it is deemed necessary to anchor and engage the enemy at their anchorage, it is recommended to each line-of-battle ship of the squadron to prepare to anchor with the sheet-cable in abaft and springs, etc. – Vide Signal 54 and Instructions thereon, page 56, etc. Article 37 of the Instructions.[107]

Mem[orandum] PS. To be inserted in pencil in the Signal Book, at number 182. Being to windward of the enemy, to denote that I mean to attack the enemy's line from the rear towards the van, as far as thirteen ships, or whatever number of the British ships of the line may be present, that each ship may know his opponent in the enemy's line.

Number 183 I mean to press hard with the whole force on the enemy's rear.

The survival of a private signal book, with these signals actually inserted in pencil, confirms the authenticity of the Memorandum.[108]

---

103 *A Narrative of the Proceedings of His Majesty's Squadron under the command of Rear-Admiral Sir H Nelson KB from its sailing from Gibraltar to the conclusion of the Glorious Battle of the Nile, drawn from the Minutes by an Officer of Rank in the Squadron*, republished Quebec, 1799, pp13–6.

104 Printed in *Memoirs and Correspondence of Admiral Lord de Saumarez* by Sir John Ross, London, 1838, I, pp212–3.

105 Saumarez and Troubridge were the two senior captains. Signal Number 16 was 'For particular ships or divisions denoted, to attack the enemy on bearing indicated, or any number of their ships of war separated from the body of the fleet'.

106 pp18–9

107 Signal 54 is 'To prepare for battle; when it may be necessary to anchor with a bower or sheet cable in abaft and springs, etc.' Article 37 of the instructions is a fairly lengthy exposition of this. For the full text see *NMM*, KEI/S/11(c) (formerly Sp/22), Hood's Instructions of 25 December 1793, and Mead Collection, *NMM*, TUN/132 (formerly Mead/Ins/9), Jervis's Instructions of 19 November 1795.

108 *NMM*, Tunstall Collection, TUN/20 (formerly S/MS/R/5), manuscript Mediterranean Signal Book of 1796, has the signals 182 and 183 inserted in pencil in the exact words of the Memorandum.

When early in the afternoon of 1 August, the *Alexander* hoisted signal number 23 for 'Discovering a strange fleet,' Nelson's squadron was widely spread out because he had no frigates and was, therefore, using ships of the line as look-outs. The *Alexander* and the *Swiftsure* had been detached to look into Alexandria and were thus close enough to the coast to be able to spot Brueys's ships over the low Aboukir promontory.

After receiving private ships' signals that the enemy numbered sixteen of the line and were at anchor, Nelson made the following signals, using Jervis's Mediterranean Fleet arrangement:

No 9 (to *Alexander* and *Swiftsure*) leave off chase.
No 53 (general) prepare for battle.
No 54 (general) prepare for battle; 'when it may be necessary' to *anchor* with a bower or sheet cable in abaft, and springs, etc'.[109]
No 45 and No 46 [hoisted together] (general) engage enemy's centre, and engage enemy's van
No 31 (general) form line of battle ahead and astern of the Admiral as most convenient.
No 34 (general) Alter course one point to starboard in succession.
No 66 (general) make all sail (after lying by), the leading ship first.
No 5 [with red pendant] (general) engage the enemy closer.

No-one reading these signals as they stand would easily believe them to have been the prelude to one of the most overwhelming victories in maritime history.

Nelson had thirteen 74s and a 50 (*Leander*) and his squadron was widely spread. He decided to attack the French as quickly as possible. This meant rounding Aboukir Island

and the shoals surrounding it, which projected beyond the mainland promontory, and risking the hazards of the bay itself, of which no captain possessed a proper chart. By a combination of dash and caution which to the French suggested the possession of experienced pilots, the whole squadron rounded the island with only one casualty. The *Culloden* grounded by cutting in too fine.

Whatever happened now, Nelson was committed to the hazards of a night action against ships which had only to angle themselves correctly in order to exploit what might seem to be an artillerist's dream. Here were the British coming on in a rough, irregular line ahead inviting the French to rake their bows and shoot down their masts, spars and rigging. In the dark, the French ships must hold the advantage since being unable to hit each other they could assume that any ship opposed to their broadsides must be an enemy. Yet what happened? Instead of opening a careful and deliberate fire at long range to try and cripple the British at the start, and then opening with full broadsides at musket range, they allowed four British ships to pass right round the head of their line and a fifth to pass between their first and second ships. This not only showed their individual inefficiency in gunnery, but also Brueys's failure to put his leading ship right up against the shoals. All the British ships which passed round or through the French anchored by the stern and opened a heavy fire which was as good as a surprise attack, the French being unable to reply effectively with their larboard guns. Nelson and the two ships astern of him brought up on the outside of the French line opposite their third, fourth and fifth ships. All five of the French leading ships were silenced within two hours.

Apart from the general lack of fighting efficiency in the French fleet, it was Brueys's insistence on leaving such wide gaps between his ships that prevented them giving each other mutual support. This also prejudiced the ability of the ships farther along the line to bring their guns to bear on the leading British ships, that is supposing they had had springs on their cables and had actually used them. As it was, they had neither the range nor the arc of fire within which to intervene effectively and so had to wait for the British to come and attack them, individually.[110]

The remaining British ships, less the *Culloden*, now began to straggle into action, no regular line having been attempted or required. The *Bellerophon*, encountering the 120-gun *Orient*, was heavily repulsed, losing all her masts. The *Alexander* then passed astern and inshore of the *Orient*, which was also engaged on her starboard side by the *Swiftsure*. The *Peuple-Souverain*, with her cable cut by a shot, had drifted alongside the *Franklin*, her next astern, and the *Leander* (50) entered the wide gap now opened ahead of the *Franklin* and raked both her and the *Orient* simultaneously. With the *Defence* and *Swiftsure* on the outer side and the *Orion* and *Leander* all firing at the *Franklin*, and the *Leander*, *Swiftsure* and *Alexander* firing at the *Orient* as well, these two ships were obviously doomed. The *Orient* caught fire on her poop and eventually blew up with a terrific explosion. The ships astern of her were now under attack both from the British rear ships and from ships engaged earlier and now edging down. In the end only two of the line and two frigates survived to make their escape, out of the whole French fleet. Thus ended the greatest disaster suffered by the French Navy since Barfleur and La Hogue.

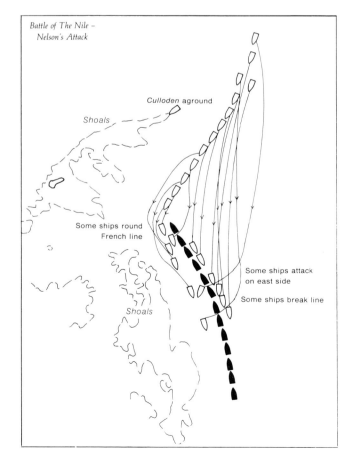

Battle of The Nile – Nelson's Attack

Culloden aground

Shoals

Some ships round French line

Some ships attack on east side

Some ships break line

Shoals

---

109 See second Memorandum above.

110 The plan of the battle drawn by Nelson with his left hand and now in the British Museum (Add. Ms. 18676) is too inaccurate to be of any significance.

Nelson's report to Lord Howe has since become famous:

I had the happiness to command a Band of Brothers: therefore, night was to my advantage. Each knew his duty, and I was sure each would feel for a French ship. By attacking the Enemy's van and centre, the wind blowing directly along their Line, I was enabled to throw what force I pleased on a few ships. This plan my friends readily conceived by the signals, (for which we are principally, if not entirely, indebted to your Lordship), and we always kept a superior force to the Enemy . . . Had it not pleased God that I had not been wounded [sic] and stone blind, there cannot be a doubt but that every ship would have been in our possession. But here let it not be supposed, that any Officer is to blame. No; on my honour, I am satisfied each did his very best.[111]

The use of the word 'friends' to describe his captains might well have outraged far less arrogant men than Rodney.

The first Admiralty signal books, 1799

With Howe's retirement and also Hood's, there was no senior admiral left capable of giving tactical and signalling direction to the service as a whole. Bridport, Howe's successor in the Channel command, was a frequent absentee and, in any case, lacked tactical and signalling initiative. Meanwhile there were men like St Vincent, Duncan, Lord Keith and Cornwallis, eager for tactical progress and ready, if left without direction, to begin carving out new systems for themselves. Luckily the Board of Admiralty realised that official action could no longer be postponed. The senior naval member of the Board was Captain James Gambier, who had been appointed in 1795 through Howe's influence after his meritorious conduct at the First of June. In 1798, therefore, Gambier was deputed to revise the whole tactical and signalling system. He began work in 1798 and finished early in the next year. Rear-Admiral Lord Hugh Seymour, an ex-colleague of his on the Board, wrote on 15 July 1798:

I am of the opinion that Lord Howe's signals which are now well understood, and much approved by the navy in general, should be as much followed as possible, and I am of opinion that you should not be too sparing of bunting, for, do what you will, every captain will contrive to steel enough of it to furnish himself with a set of flags, such as are now in fashion, and it is surely better economy, as well as more advantageous to the service, that each should have a perfect set allowed him.

The number of flags may certainly be reduced, and it is not necessary that captains should be furnished with divisional flags, and many others which are required by admirals, but I think that all signals when the fleet is formed in line of battle should be repeated by every ship, and they should be provided accordingly.' [112]

Captain Walter Young had made exactly the same plea in 1780. Sir John Laughton discusses Gambier in the *Dictionary of National Biography* as a mere Admiralty sailor who evaded sea service. This seems quite unjustified, the revision and codification of the signals and instructions being a task requiring special gifts not necessarily associated with seamanship.

When the Admiralty's work was revealed to the world it was seen that Howe's principle of separating signals and instructions was reversed, a new separation being made between day and night. Howe had always favoured putting all signals together in one book and all instructions in another, though he had somewhat compromised this princi-

Admiral James, Lord Gambier (1756–1833); engraving by F Flint after a painting by Sir William Beechey. BM

ple by issuing his separate volume of additional instructions. This apparent regression into the Admiralty errors of earlier years may well have caused some uneasiness to the service and so have contributed to the change now made.

The new Admiralty Signal-Book for the Ships of War, Howe's title being carefully retained, contained all the day and fog signals and all the day and fog instructions. The companion volume was entitled 'Night Signals and Instructions for the Conduct of the Ships of War', again retaining

SIGNAL-BOOK

FOR THE

SHIPS of WAR.

1799.

Left and overleaf: *The numerical code system employed in the Admiralty Day Signal Book for the Ships of War, 1799.* MOD

111 8 January 1799, Sir Harris Nicolas, *Dispatches and Letters of Lord Nelson*, London, 1844–46, seven vols. (hereafter cited *Nelson*), III, p230
112 Lady Chatterton, *Memorials of Lord Gambier*, I, p342

MOD NM/47 #3

24

N°	SIGNIFICATION.	Instruct. Pa.	Art.
31	Rake the enemy.		
32	Fire in succession upon the sternmost ships of the enemy, then tack or wear and take stations in the rear of the squadron or division specified (if a part of the fleet is so appointed) until otherwise directed.		
33	Engage the centre of the enemy.		
34	Engage the van, starboard or weather division of the enemy.		
35	Engage the rear, larboard or lee division of the enemy.		

N°	SIGNIFICATION.	Pa.	Art.
36	Engage the enemy on their starboard side if going before the wind, or to windward if by the wind.		
37	Engage the enemy on their larboard side, or to leeward if by the wind.		
38	Engage, in preference to others, as far as circumstances will admit, the ships of the nation whose jack is shewn with this signal.		
39	Discontinue the engagement.		
40	Are you ready to engage the enemy?		

If this be not addressed to particular ships every ship is to answer as her state may require.

MOD NM/47 #4

If an union jack be hoisted over a numeral signal, that signal will then represent the number of a ship on the list of the navy; and if distinguishing pendants be hoisted at the same time, they are to be the signal pendants of that ship, and she is to take the station in the line of battle to which those pendants belong. If, for example, an union jack be hoisted over the signal N° 19, it will represent the number of the Argonaut, which on the list of the navy is 19; and if the distinguishing pendants of the fourth ship in the first division of the van squadron be hoisted at the same time, they are to be the Argonaut's signal, and she is to take her station as the fourth ship in the first division of the van squadron. In disposing of these pendants to the ships of the fleet, several should be left in various parts of the line, that ships which join the fleet, may take stations, and have pendants assigned them, without interfering with those already appropriated.

Ships joining the fleet, or coming into a King's port, or passing the signal posts on the coast, are to make themselves known by shewing the signal expressive of their number as above directed.

If ships meeting at sea make this signal, after shewing the private signal to each other, it will serve, in some measure, as a corroboration of the same.

As private ships will not be furnished with the triangular flags, they are to make the distinguishing signals of other ships, by hoisting those of the van squadron at the fore-top-mast-head; those of the centre squadron at the main-top-mast-head; and those of the rear squadron at the mizen-top-mast-head; brigs are to hoist the signals of the rear squadron at the gaff-end.

MOD NM/47 #5

Numeral Flags.	Figures represented	REMARKS.
	1	The flags are intended to represent the figures placed opposite to them in the annexed table. A flag hoisted alone, or under another flag, is to represent units; when two flags are hoisted, the upper flag is to represent tens; when three are hoisted, the uppermost is to represent hundreds, the next tens, and the lowest units.
	2	
	3	When the substitute flag is hoisted under other flags, it is to represent the same figure as the flag *immediately* above it: when the substitute pendant is hoisted under two flags, it is to represent the same figure as the upper flag of the two. For example, to represent the N° 33, the substitute flag will be hoisted under the flag representing 3, and to represent the N° 303 the cypher flag will be hoisted under the flag representing 3, and the substitute pendant under both.
	4	
	5	
	6	In blowing weather, or when much sail is set, as it may be inconvenient to hoist three flags at the same place, the two upper flags may be hoisted at one part of the ship, and the lowest flag, or the substitute pendant, at any other part.
	7	If the Admiral should have reason to believe that the enemy has got possession of these signals, he will make the signal for changing the figures of the flags, and when that has been answered by every ship, he will hoist the numeral flags, two or three at a time, the uppermost flag of those first hoisted to represent 1; the next below it, 2; and so on till all the flags have been hoisted, the tenth flag representing the cypher, and the last being the substitute flag. To prevent mistakes, every ship is to hoist the same flags as the Admiral, and in the same order; the flag officers are to be particularly attentive to see this done, and to shew the distinguishing signal of any ship in which they observe a mistake. The figure which, by the new arrangement, each flag is to represent, is to be immediately entered in every ship's signal-book.
	8	
	9	
	0	
Substitute		The flags are always to represent the figures placed opposite to them in the annexed table, in signals made in port, or to the signal posts on shore; or by ships meeting accidentally at sea. But an Admiral may make any other arrangement for the use of the fleet he commands, while at sea.
Substitute		

Howe's wording. Each of the new volumes was issued 'By the Commissioners for executing the Office of Lord High Admiral . . . Given under our hands', the date of issue being inserted in handwriting. The book was usually signed by three members of the board and countersigned by the secretary. The original date of issue was 13 May 1799.[113]

The Day Book starts with a subject index to the signals under main and subheadings, itself thumb-indexed by letters of the alphabet in the margins, a refinement apparently attributable to John McArthur when acting for Hood. Then follows the index to signals from private ships, similarly treated.

The explanatory instructions and the explanatory observations on the use of triangular flags are much the same as Howe's; the triangular flags themselves and the accompanying pendants, for indicating squadrons, divisions, detachments, etc, being exactly the same, with the addition of red, white and blue wefts for indicating individual ships. The numbered flags 0–9, were the same as Howe's of 1790 for the Channel fleet and the First of June except that 1 was a yellow flag with a red horizontal stripe instead of plain red. The accompanying remarks include a procedure for changing the numbers of the flags, 'If the admiral should have reason to believe that the enemy has got possession of these signals'. There were only eleven extra flags for use with seven of the numerical flags for single flag signals. A very important innovation was that the Union hoisted over a numerical signal represented the number of a ship 'on the list of the navy', the list being given later in the book. Further innovations were a numbered table of pendants and wefts for the signal numbers of frigates, sloops, etc, and a numerical code of flags and pendants of any colour, to be used by ships 'not furnished with the signal flags' and for distance signalling, or in hazy weather.

The main admiral's signals, numbered 11–310, with 278–310 left blank, were thumb-indexed by numbers but had no 'Purport' or short-title reference. They included columns for page and article references to the instructions which follow. Unlike Howe's signal book, there was no guide to classification, but the battle signals were at the start. Breaking the enemy's line (number 27) was now worded, 'Break through the enemy's line in all parts where it is practicable, and engage on the other side. Page 160 Article 31: If a . . . is hoisted at the fore topmast head, break through the van: if at the main topmast head, break through the centre; if at the mizzen topmast head, break through the rear.' In later printings the signal was left blank and an insertion made in handwriting, presumably for security reasons.

In the Instructions themselves, Article 31 states,

If signal 27, to break the enemy's line, be made without a [red pendant] being hoisted, it is evident that to obey it the line of battle must be entirely broken; but if a [red pendant] be hoisted at either masthead, the fleet is to preserve the line of battle as it passes through the enemy's line, and to preserve it in very close order, that such of the enemy's ships as are cut off may not find an opportunity of passing through it to rejoin the fleet.

If a signal of number be made immediately after the signal, it will show the number of ships of the enemy's van or rear which the fleet is to endeavour to cut off; if the closing of the enemy's line should prevent the ships passing through the part pointed out, they are to pass through it as near to it as they can.

If any of the ships should find it impracticable, in either of the above cases, to pass thro' the enemy's line, they are to act in the best manner that circumstances will admit of for the destruction of the enemy.

Nothing is said about being to windward or leeward nor being either on the same or the opposite course to the enemy.

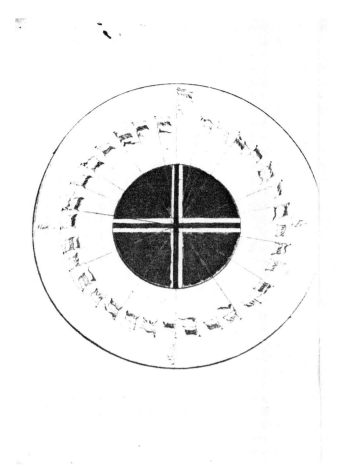

The remaining battle signals show nothing new, and the signals as a whole, through far in excess of those in Howe's book, are mainly extensions and variations of existing signals and with a stronger administrative element. Though following Howe's general arrangement, the book offers no means of telling where any particular type of signal begins. The signals from private ships (311–390) are well chosen and well arranged. The signals by pendants, wefts and jacks, and the compass signals by private ships are much the same as before.

A really important innovation was the numbered list of all ships in the navy, claimed incorrectly for Gambier as 'an invention of his own'.[114] With this list, the admiral could name ships not included in his fleet but which might be joining it or be encountered shortly. A total of 757 numbers were provided, and there were 112 blanks for handwritten insertions.

The 'General Instructions for the Conduct of the Fleet', consisted of thirty-four articles. Though composed and

---

113 Fourteen identifiable copies of the first printing of the Day Signals, dated 13 May, are in various collections and no doubt many more exist. The Corbett Collection copies of the Day & Night volumes (*NMM* TUN/18 and 19 (formerly S/P/R/2 and 3)) are richly bound in red morocco with gilt tooling and bear the bookplate of Lord Adolphus Fitzclarence superimposed on that of Prince William Henry, his father, later King William IV. They are dated 13 May 1799 in manuscript but are unsigned. The other copies are *NMM*, SIG/B/16, 74, 75 and 78 (formerly Sp/31, 32, 35, 36) HOL/51 (formerly S(H) 51), DUN/32 (formerly S(D) 32) and COR/66/1 (formerly Sp/120); *MOD*, Ec/53; NM/47 and 58 (formerly *RUSI* 47 and 58); Mead Collection; Royal Navy College Greenwich. Identified copies of the Night Signals issued 13 May 1799 are *NMM*, SIG/B/79 (formerly Sp/33) and DUN/33 (formerly S(D) 33), companion books to the Day Signals SIG/B/75 (formerly Sp/32) and DUN/32 (formerly S(D) 32).

114 Lady Chatterton, *Memorials of Lord Gambier*, I, p340. This is untrue; Howe thought of it long before. See *NMM*, Mx, 58/102 and ROW/7.

worded very much in the spirit of Howe, they were largely new, and aimed at attaining a still higher efficiency in station-keeping and in avoiding obstructions and collisions. They were drawn in very general terms and without special reference either to cruising or to battle. The seventeen Instructions Respecting the Orders of Sailing which follow were also modelled on Howe's, but were better worded and were illustrated by McArthur in eight plates, with three new ones. Provision was made for the order of sailing in two, three or six columns; the position of the admiral; opening and closing distances; tacking, wearing and bearing up in succession. The four 'Instructions Relating to the Line of Battle' were similar to those in Howe's Additionals. The eleven 'Instructions for Forming the Line of Battle from the Orders of Sailing', again illustrated with folding plates, were quite new. They provided for forming the order of battle in as many ways as practicable from either two, three or six columns, with signals.

The thirty-one 'Instructions for the Conduct of the Fleet, Preparatory to their Engaging, and when Engaged with an Enemy' (a more Howe-like wording even than Howe's), were conservative in tone. A new definition, however, was given to the purpose of a line of battle. Howe, in his Additional Instructions, said only that the line enabled ships to be 'as little as possible exposed to the fire of more than the particular ships corresponding in station with them in the Enemy's Line, or be subject to injury from each other'.

The new instruction (number 2) adopted a far more realistic attitude: 'that the ships may be able to assist and support each other in action; that they may not be exposed to the fire of the enemy's ships greater in number than themselves; and that every ship may be able to fire on the enemy without risk of firing into the ships of her own fleet'. A short table was given of the gradations of making and shortening sail for ships, 'suiting their rate of the sailing to that of the Admiral' (number 9).

When coming onto a parallel course with the enemy to engage, each ship was instructed to be careful to give her next astern 'room to haul up without running on board her' (number 14). The usual instruction was included that small numbers of the enemy beaten out of the line were not to be

pursued 'unless the commander-in-chief, or some other flag officer be among them'. The British ships concerned were instead to assist ships 'much pressed', 'and to continue their attack until the main body of the enemy be broken or disabled', unless specially signalled or instructed (number 15). Instruction number 18 dramatically stated that 'if there should be found a captain, so lost to all sense of honour, and the great duty he owes his country, as not to exert himself to the utmost to get into action with the enemy', he was to be suspended and replaced. Ships were to try to make a simultaneous attack by making or shortening sail, and be careful to avoid collisions (number 26). 'The leading ships must be very cautious not to suffer themselves to be drawn away so far from the main body of the fleet as to risk being surrounded and cut off' (number 27). There were twenty fog signals and nine simple fog instructions, similar to Howe's.

The 'Night Signals', a volume of 41 pages, was a far more elaborate and ambitious work than anything attempted before. There was a detailed index of signals and of 'Private Ships' Signals to the Admiral'. The method adopted was for one to four lights to represent the digits 1 to 4, for false fires or rockets (regardless of numbers) to represent the digit 5 and for guns to represent tens. Thus signal number 48 was represented by four guns, one or more false fires or rockets and three lights. There were seventy signals in all. Included were sailing instructions, battle instructions and 'Instructions for the Conduct of Ships Appointed to Watch the Motions of an Enemy's Fleet in the Night', all three of which were a revision of Howe's, many of the articles being word for word the same.[115]

It is worth asking just what it was that the new books were intended to supersede. Admiral Cuthbert Collingwood, at sea on 17 June 1799 and about to proceed from off Brest to reinforce Keith's fleet in the Mediterranean, issued his flag captain with day and fog signals with 'New Arrangement' printed on the cover and '1795' printed on the title page.[116] Clearly, there were many versions of Howe's, Hood's and St Vincent's signals and instructions in use and the Admiralty books were quite as necessary for securing standardisation as they were for their actual content.

With over 700 ships in the Royal Navy at this time and with continual changes of flag-officers and captains, the Signal Book had to be reprinted frequently during the next three years, even though the date, 1799, remained on the cover label and the short title. The differences were almost entirely typographical and extremely difficult to perceive, especially in the case of the night signals.[117] The Board of Admiralty, moreover, frequently omitted to fill in the date, 'Given under our hands the . . . of . . .', above the space left for the signatures. Second and later printings had twelve diagrammatic plates instead of eleven, and these were interspersed amongst the sailing instructions instead of being grouped together at the end. Handwritten insertions for the period 1799 to 1803 show no evidence of any important change in tactics.

In 1803 a British schooner was captured off Toulon. The lieutenant in command was not entitled to possess the Signal Book, but had evidently made a handwritten copy, which the French captured. The ship, flying the British signal to anchor, was afterwards used by the French in an attempted decoy operation. The Admiralty, therefore, on 4 November, issued a circular letter to all commanders-in-chief, ordering a change in the numbers assigned to the numerical flags and a change in the substitute flag.[118] They enclosed coloured printed slips for pasting over the flag and number columns in the Signal Book, as well as blank uncoloured slips.[119] Private copies of the signals were forbidden for the future and existing copies ordered to be handed in. The new issue of

115 Despite the standardisation now established, occasional anomalies appeared, as for instance *MOD, NM/45* (formerly *RUSI* 45) which consists of Howe's Instructions of 1793 'given on board the *Lancaster* at Spithead 1st September 1799, to Hon Capt Stopford, HMS *Excellent*, [signed Vice-Admiral Sir] Roger Curtis.' The signature has been verified.

116 *MOD, Ec/113,* issued to Thomas Seccombe, *Triumph,* at sea, 17 June 1793, but without Collingwood's signature.

117 *NMM, SIG/B/73* (formerly Sp/50), has typographical variants: SIG/B/76 (formerly Sp/102) bears the printer's name, William Winchester & Son, Strand (the flags are those of 1803 and it was issued after 2 May 1805 (Board signatures)); *NMM, TUN/60* (formerly S/P/R/7), also William Winchester, issued 5 October 1803; *MOD, Ec/54* issued 1804 (Board signatures). Of the Night Signals, *NMM, SIG/B/80* (formerly Sp/96) is undated and was issued 1804–05 (Board signatures). All three have typographical differences. DUC/49 (formerly Sp/34), undated, printed by Winchester & Son, was issued 1801–4 (Board signatures). *NMM,* Tunstall Collection, TUN/63 (formerly S/P/R/10) was issued 1805–6 (Board signatures), and *MOD, Ec/60(a)* was issued 29 October 1804.

118 *MOD, NM/47* (formerly *RUSI* 47), a copy of the 1799 Book issued 5 November 1803, with notes by Sir Julian Corbett inserted which in fact refer to *MOD, NM/58* (formerly *RUSI* 58).

119 The original number 5 flag, quarterly red and white, now became number 4. It is on the question of whether or not one or more of the coloured slips had the new number 4 painted quarterly white and red by mistake that the dispute about Nelson's Trafalgar Signal depends. See H P Mead, *Trafalgar Signals;* also Captain Holland, 'Number 4 Flag at Trafalgar', *The Mariner's Mirror,* vol 20 (October 1934) no 4, and D B Smith, review article on Mead's book, *The Mariner's Mirror,* vol 22 (1936) no 4

flags took effect in the Mediterranean Fleet on 16 January 1804.

By 1804, if not earlier, the Day Signal Book required expansion, which the Admiralty achieved by inserting an appendix between the day signals and the instructions, containing eighty-two new signals, in two sections.[120] The first section, entitled 'Signals from Senior to Junior Officers', gave printed form to the existing handwritten insertions in the original blanks of the Admiral's signals (ending at 310), and added a further group, numbered 500–506. In the second section, entitled 'Signals to be Added to Those Made From Private Ships, or from Junior to Senior Officers', existing handwritten insertions were again given printed form, and a new group added, numbered 391–417. Although certain tactical refinements were introduced, there was nothing of real importance. Rendezvous signals were added later. The appendix was issued as a separate publication; it was through this that Nelson made known his ideas for the attack at Trafalgar.

Another crisis arose in 1807 when the schooner *St Lucia* was captured in the West Indies and the French obtained 'the whole of the signals'.[121] Sir Alexander Cochrane, the commander-in-chief in the West Indies, changed the numbers and the substitute flag. This, however, does not seem to have gone beyond the West Indies station, as the flags remained the same even when the Admiralty published a completely new edition of the Signal Book in the following year. This edition of 1808 was substantially the same as the original except that the admiral's signals were extended from 310 to 360. This was to give room for expansion, since all signals from numbers 318–60 were left blank. The private ships had their quota increased by being given 175 signals (nos 361–535) instead of the previous eighty. Even so, the last sixty-eight were left blank.[122] There was no appendix, although the rendezvous signals were inserted separately.

It might have been expected that the impact of Nelson's tactics on the Navy as a whole would have resulted in some relevant additions to the Signal Book of 1808, as regards both the signals themselves and the battle instructions. No additions, however, in any way traceable to Nelson can be found. In a tactical sense, the war was as good as won and fleet actions were becoming less and less likely. The Admiralty was, no doubt, content to rely on traditional wisdom rather than promulgating new ideas – excellent when propounded and executed by Nelson but not to be encouraged as standardised procedure.

In 1811 the Admiralty issued a new printing of the day signals with a further alteration of the numerical flags, although no new flags were introduced. Nor was the book itself changed in any important respect. The night signals were not apparently reprinted.[123] The final consolidation by which the book itself and Popham's Telegraph became associated volumes did not take place until after the war, in 1816.

The Battle of Copenhagen

The Danish defence plan at Copenhagen in 1801 was entirely passive and aimed simply at protecting the city and harbour from sea bombardment and from assault by landing parties. At the northern end of the defence area, the difficult harbour approaches were covered by a small squadron of block ships and more effectively by the Three Crowns battery built out on piles above the mud flats. In front of the city itself and stretching well to the south of it was a long strip of mud flats over half a mile wide. Outside these mud flats was the navigable inner channel, about half a mile wide but very narrow at its southern entrance and made difficult at its

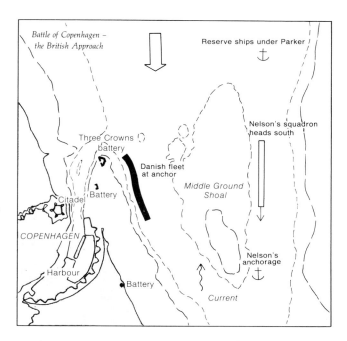

Battle of Copenhagen – the British Approach

northern and broader end by a small shoal, right in the middle, near the Three Crowns battery. Further out, beyond the navigable channel and severely limiting its usage, was the Middle Ground Shoal, with shallow-water extensions to the north of it. Instead of arranging movable block ships at both ends of the chancel, the Danes provided for covering the harbour approaches at the northern end, and strung out their defences to the south in a single line of warships and floating batteries ranged along the outer edge of the mud flats facing the city. They thus invited an attack by any force able to enter the channel at its narrow southern entrance and then steer north until opposed by the first floating battery or block ship, drawn up, not at right angles and across the bows of the advancing enemy, but merely parallel. In seeking to do no more than protect the very heart of their country, the capital and its port, the Danes adopted a system which denied every principle of true defence.

When first observed at a distance by the British fleet, the Danes were seen to have seven unrigged and dismasted ships-of-the-line and eleven floating batteries of various types ranged in line along the outer edge of the mud flats facing the city. There were two line ships at each end of the line, the other three being spread evenly between them, with the floating batteries filling the gaps. Nelson decided to make his attack from the southern end of the channel. For this, he needed a northerly breeze to carry his force through the narrow channel on the far and eastward side of the Middle Ground Shoal. He then needed a southerly breeze to enable him to round the southern tip of the Middle Ground Shoal, and sail straight up to engage the Danish defences, and if

120 *MOD*, NM/58 (formerly *RUSI* 58), originally issued not later than February 1801 (Board signatures) but may have had the Appendix bound in later. The flags of 1811 are pasted over the top. NM/76 (formerly *RUSI* 76) includes two copies of the Appendix as a separate publication, one issued by Admiral Rainier, Madras, East Indies, 1805 and the other by Cornwallis, probably in 1805.

121 General Memorandum to all captains, *Northumberland*, at sea, 20 April 1807. Typescript copy once in the Mead Collection, documentary source unknown.

122 *NMM*, SIG/B/88 and 89 (formerly 94, two copies) issued 13 July 1808; also SIG/B/91 (formerly Sp/108), issued 16 March 1812.

123 *MOD*, Ec/60 is a copy of the Night Signals printed by Winchester and dated 1806, issued 4 January 1811.

necessary to bombard the city and land troops. By nightfall on 1 April 1801, with the aid of a northerly breeze, he had achieved his first object, and was able to anchor at the southern end of the Middle Ground Shoal, scarcely a mile and a quarter from the nearest and most southerly Danish ship. He was, therefore, able to send Captain Hardy in a boat to take soundings ahead. His force was awkwardly placed in a narrow anchorage, and if, had the Danes been able even on the evening of 1 April to swing three of their ships or

batteries broadside on across the channel, they could have crossed the T of Nelson's advancing ships with a crushing broadside fire. Their attitude, however, was wholly passive, even neglecting night reconnaissance.

For Nelson, everything now depended on a favourable wind from the south, even though this would prevent Sir Hyde Parker with the rest of the fleet supporting him effectively by closing in on the harbour entrance from the north. Left unmolested by the Danes, although he was within

The Battle of Copenhagen, 2 April 1801 (detail), by Nicholas Pocock. NMM

gunshot, Nelson gave a dinner party for Rear-Admiral Thomas Graves[124] and several captains, showing himself 'in the highest spirits'. Then, at about 9.30pm he withdrew to the after cabin to draft his orders for the next day's attack,

that is if the wind was propitious. He asked two officers to assist him. One was Captain Thomas Foley (of the *Goliath* at the Nile) in whose ship, the *Elephant,* he now had his flag, his own flagship for the campaign, the *St George,* having too deep a draught for the shallows. The other was Captain Edward Riou whom Nelson had met, apparently, for the first time, on the previous day. Captain Hardy returned at 11pm, having actually sounded round the nearest Danish ship, the Danes apparently having failed to post sentries, let alone send

124 This is the same Thomas Graves who in the American War made the collection of signal books and fighting instructions once in the Royal United Service Institution and now divided between the National Maritime Museum and the Ministry of Defence Library, Whitehall.

out patrol boats after dark. For all his astonishing verve and ardour both in battle and in council, Nelson was an arch-planner when planning was required. It took him until 1am on 2 April before he was finished, having been persuaded to go on dictating from his cot, placed on the deck. Six clerks then began to copy out the orders and finished their work only by 6am. At 7.37 Nelson signalled for all captains so that, by 8 o'clock, each had received a copy of his own orders. The new Admiralty Day Signal Book of 1799 was now about to be used for the first time in a full-dress engagement. These facts are important, as they help to dispel the idea that Nelson acted solely on intuition. With the Danish defence both known and static, every British ship could be given a detailed role in the battle. Nelson's haste was due to the wind having changed during the night, just as required; it was now blowing fresh from SSE.

There was an agonizing moment when the pilots jibbed, their experience being mainly with merchant ships in the Baltic trade, smaller than those now in their care. But, at last, everything was ready. Nelson signalled to weigh, and the squadron began to advance in line ahead. The Danes were unable to do anything until the leading British ship drew level with the southernmost in their own line. Nelson's plan was, with certain special exceptions, for the ships to anchor by the stern in succession in line ahead to engage the Danes, each ship passing on the disengaged side of her next ahead, thus inverting the line according to the standard form. The exceptions were due to the special need for ensuring the early defeat of the two southernmost of the Danish ships-of-the-line which had first to be encountered. A frigate and some smaller ships were detailed to support the attack by going in close and taking their bows. All ships were ordered to have springs on their cables, and the end of the sheet cable taken in at the stern port. Under cover of Nelson's twelve of the line, the smaller vessels were to move still further ahead on the disengaged side. Seven bomb vessels were to anchor about half way along the line and as close to the Middle Ground Shoal as their shallow draught would allow. From there they could fire over the ships-of-the line at the Danish ships and also at the city and harbour. Flat-bottom boats were ready to embark troops for assaulting the Three Crowns battery, as a preliminary to a general bombardment of the city, dockyard and arsenal by the bomb vessels. The last two ships-of-the-line were to push on farthest north and attack the Three Crowns battery. Captain Riou, placed in command of a division of frigates and sloops, had been ordered to act as circumstances might require; he found himself committed to joining in the attack on the battery.

Several things went wrong, but Nelson had reckoned on this, and was prepared to take all risks. The *Agamemnon* grounded on the south-east corner of the Middle Ground Shoal and never entered the inner channel at all. The *Bellona* and *Russell,* going too far to starboard, also grounded, and, though able to fire on the Danes through gaps in the British line, at a range of only half a mile, they were out of position and completely immobilised. The average distance between the two lines during the battle was only a little over a quarter of a mile. Riou's attack on the Three Crowns battery and attendant block ships met with very strong resistance, especially as neither of the two ships-of-the-line ordered to open the attack reached her position, the *Russell* being already aground. Owing to an adverse current, the gun-brigs were unable to push forward to join in the bombardment. Riou

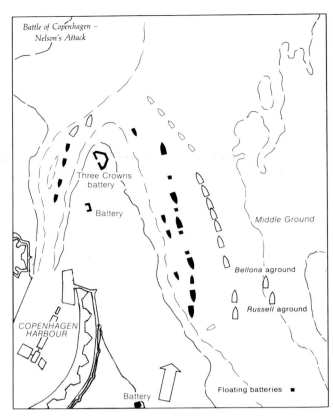

was killed while his ships were withdrawing in response to Sir Hyde Parker's signal 'To discontinue the engagement'.

Nelson continued his bombardment in spite of these reverses and in spite of Sir Hyde Parker's signal.[125] His great advantage lay in the fact that the Danes were stationary, so that some of his ships could shift the weight and direction of their attack. He was able, moreover, to make adjustments in his line by signal.

By about 2pm the battle was largely over. Many of the Danish ships and batteries were silenced. Desultory firing continued from the Three Crowns battery and their support-ing ships, both against what were now the three leading British ships and Sir Hyde Parker's advanced ships beating to windward from the north. Some of the Danish ships which had ceased fire, and had their flags shot away, opened irregular fire again when approached by boats sent to take possession of them. The Three Crowns battery was mean-while exchanging fire with the British across and through some of the Danish ships. It was this irregularity of conduct arising from ignorance of the customs of war (many of the Danes being volunteer landsmen) that caused Nelson to send his famous letter to the Crown Prince. When he eventually withdrew to the north, his own ship and another grounded within a mile of the Three Crowns, which showed how dangerous it would have been to have obeyed Sir Hyde Parker's signal and to have attempted to withdraw while still under fire.

British casualties were 1000 killed and wounded. The Danish total was more than twice that number; in addition, 3500 Danes were taken prisoner.

125 Number 39 in the Admiralty Signal Book

CHAPTER NINE

Tactical Development During the Napoleonic Wars

IN 1802 Audibert Ramatuelle published the most stimulating work in the French language on tactics under sail, *Cours Elémentaire de Tactique Navale, Dedié a Bonaparte*.[1] The author, an 'ancien officier de la Marine Militaire', was intent upon dealing with tactics in a practical manner. He regarded Morogues's obtuse angle formation for chasing and retreat as 'impracticable et désavantageuse dans tous les cas', and was critical of recent authors who had included a great

many evolutions in their books 'la plupart inutiles, dont le résultat est la complication et la combinaison d'une grande quantité de mouvements dont on ne peut faire aucun usage devant l'ennemi'. Meanwhile, the tactical developments of

---

1 *NMM*, Tunstall Collection, TUN/88 (formerly T/P/F/7), includes a volume of plates

A plate from Audibert Ramatuelle's Cours Elementaire de Tactique Navale, *Paris, 1802, showing his hopelessly impractical designs for signal flags. NMM*

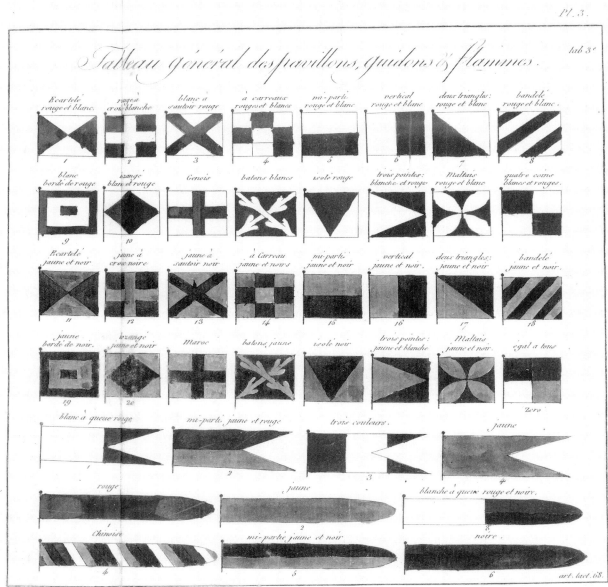

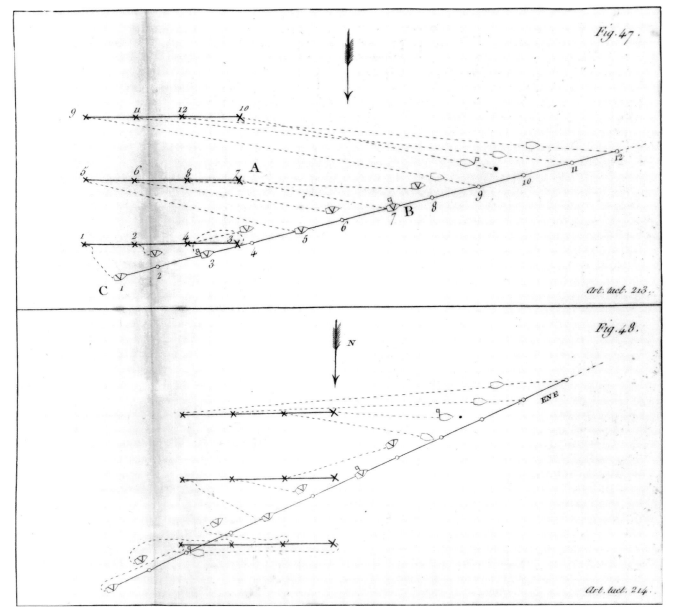

d'Orvilliers, de Guichen and Suffren had passed unnoticed. Few of Ramatuelle's own special ideas for attack may have been really feasible. Though lacking the simplicity and logic of John Clerk's work, however, no officer of the time could have read Ramatuelle's book without advantage.

In discussing battle tactics, Ramatuelle said that the fleet attacking from windward should come to the wind together when still at long range and dress their line carefully, ships taking station from the centre. He cited the example of Byng in describing the unwisdom of attacking from abeam, which obliged the attacking ships to run so much off the wind that they could not control their speed of advance by laying any of their sails aback. Unless a shift of wind was expected, Ramatuelle declared, it was better to attack from abaft the enemy's beam, with the ships lasking down in line of bearing, and thus well able to keep their station by laying sails aback, and able to use their broadsides all the time. In many respects the most interesting and original part of the book is an exposition of how to re-establish the order of sailing or battle after an unfavourable change of wind, and of how to take advantage of a favourable change of wind.[2]

He discussed in great detail the problem of doubling the enemy's van or rear from windward, and proposed various

defences against being doubled, including stationing the frigates in echelon ahead of the van to make it more difficult for the enemy's van to work round to windward. Doubling, he wrote, could also be executed by concentrating the heaviest ships in the centre and using the fastest sailers and frigates to double the enemy's rear. As an alternative, the fleet to windward might concentrate their fire against the weaker ships in the enemy's line, while leaving gaps in their own line arranged so that the enemy's heavy ships were left unopposed.

Another possibility suggested was that the windward fleet might refuse action and tack in succession. The fleets would then pass on opposite courses, but the windward fleet might defile in succession against the enemy's rear and then come to the wind again well astern so as not to lose the weather gage. Should the enemy wear together as a countermove, the van of the windward fleet, having already succeeded in their opening move, mighy tack in succession while the centre and rear formed astern of them and raked the enemy's van.

Further refinements included various ways of doubling the

2 Chapter XI

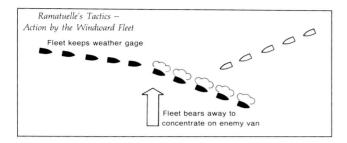

Ramatuelle's Tactics – Action by the Windward Fleet

Fleet keeps weather gage

Fleet bears away to concentrate on enemy van

enemy's new-formed van once they had worn. If the enemy wore, as previously, the windward fleet might divide, the van and centre passing to leeward and the rear immediately wearing so that they passed astern of the enemy now in reverse order. They might then rejoin their centre and van and concentrate against the enemy's rear. D'Orvilliers's plan at the Battle of Ushant for doubling Keppel's rear by ordering the van to wear failed only, Ramatuelle asserted, because his signal was not seen and acted on quickly enough.

Ramatuelle wrote very clearly of the kind of manoeuvre used by Admiral Pierre Martin against Hotham in 1795. The windward fleet, closing the enemy on opposite course, held just close enough to the wind to retain the weather gage. The fleets would converge practically head on. When fairly close to each other the windward fleet should defile to windward in succession, engaging the enemy's van with every ship of their own while passing out of range of the enemy's centre and rear. Should the enemy counter this by turning away together and running large, the windward fleet could do the same, first in line of bearing and then pursuing in line abreast.

Ramatuelle had some ingenious schemes for the fleet to leeward. If the two vans became extended ahead and the leeward fleet was more numerous, the centre might be able to cut through the enemy's centre. If the gap was wide enough, this could be done together rather than in succession. Similarly, if the leeward's rear could overlap the windward's rear they might be able to double them to windward. A far more ambitious manoeuvre was for the van of the leeward fleet to wear together and then come to the wind in succession and cut through the enemy's line. When through, they would defile along the windward side of the enemy's rear on the opposite course, having gained the windward position. Meanwhile, the centre and rear of the leeward fleet should bear away to leeward, and then take station in the wake of their van. An equally ambitious though simpler move was for the leeward fleet, acting in two divisions, to wear together, the leading division then attacking the enemy's rear while the stern division passed through the gap in the enemy's line. All these schemes would have required superior numbers, a considerable gap in the enemy's line and very quick changes of sail on the part of the leeward fleet. They also required a degree of complacency on the part of the enemy, without which all such plans are difficult to realise in practice.

Ramatuelle's variant of the traditional French method of meeting an attack from windward was so simple that it might

well have produced still better results against the traditional form of British attack. The fleets having begun to pass on opposite courses, the windward fleet was to bear up in order to attack. Beginning to form in reverse order, the new van was to form into line ahead parallel with the leeward fleet while the centre and rear ran large until they came within range. The leeward fleet could now haul to the wind together, coming into line of bearing, and then alter course again so as to resume their order in line of battle ahead. They thus would bring themselves more to windward, nearer the enemy, and thereby throw the deployment of the latter into confusion. The leeward fleet might then defile along their van, and rake their centre and rear as they approached. There might also be a chance for the van of the leeward fleet to cut through the enemy's van.

The experience of de Guichen and Rodney showed that, if both fleets were close-hauled on the same tack, and the centre of the windward fleet broke through a gap in the leeward fleet, there was no immediate or satisfactory countermove. If, however, both fleets had the wind abeam, the centre of the leeward fleet might be able to isolate the ships which broke through by themselves hauling to the wind and then turning away again, thus producing a small encircling movement.

Returning to the case of fleets passing each other on opposite courses, Ramatuelle showed how the leeward fleet, by crowding sail, might double the enemy's rear. Should the windward fleet wear together, reversing direction, the main body of the leeward would be able to haul to the wind and defile across their bows. Meanwhile, those ships of the leeward fleet that had doubled should also wear so that the new van of the windward fleet would still be kept between two fires. Supposing, however, the windward fleet held their course when the doubling movement began, the leeward fleet would then have a fine chance to cripple them. Those ships that had doubled would tack in succession as they rounded the enemy's rear, and get on the same course with them. The remainder, having passed the rear of the windward fleet on the leeward side, would wear together and crowd sail, to draw up level with them. The enemy's rear would thus be placed between two fires. Supposing the windward fleet tacked in succession, thus threatening to double on the doublers, the main body of the leeward fleet should wear together and disengage itself on the opposite course. The ships that had doubled, being now in some danger, should shorten sail, wear together, and form in the wake of their main body, which had just disengaged itself.

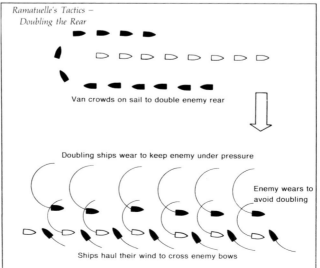

Ramatuelle's Tactics – Doubling the Rear

Van crowds on sail to double enemy rear

Doubling ships wear to keep enemy under pressure

Enemy wears to avoid doubling

Ships haul their wind to cross enemy bows

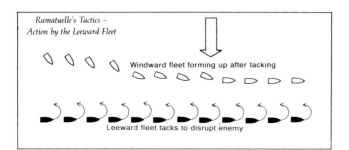

Ramatuelle's Tactics – Action by the Leeward Fleet

Windward fleet forming up after tacking

Leeward fleet tacks to disrupt enemy

Admiral Sir Robert Calder's action off Cape Finisterre, 23 July 1805, by William Anderson. (see pp246–7).
NMM

Almirante *Don Federico Gravina*
(1756–1806). MM

Ramatuelle's chapter on convoy escort reflects the strictness of French practice. He advocated the provision of signals by which the escort, if threatened by a superior force, could order the convoy to divide into two parts, which would try to escape in different directions.

Actions at anchor were illustrated by Linois's capture of the *Hannibal* in the first part of the Battle of Algeciras. Much depended on the co-ordination of the ships' broadsides with the shore batteries. Fireships could be an important consideration. Ramatuelle said that the least satisfactory way of attacking a fleet at anchor was to go in under sail, unless the wind especially favoured the attackers. Nelson's success at the Nile was a result of the double French mistake of allowing the British ships to get on the inside of their line, and in anchoring the ships too far apart. The second means of attacking a fleet at anchor was to anchor the attacking fleet close by. The third was to board the enemy, providing the enemy formation and the wind and depth of water made that possible. This option was advocated with great enthusiasm and ingenuity.

In a discourse on the line of battle *la prompte ligne* was shown as a line formed quickly 'without regard to the established order'. A line *par vitesse* was defined not as a line formed according to the relative speed of the ships, but in terms of the relative distances to be covered by the ships when entering a harbour hurriedly, especially if trying to seize an anchorage in anticipation of the enemy.[3] The term *la ligne de force* is employed to mean putting the heavy ships at the head of the line, and *la ligne de contre-force* means putting them at the tail.

Ramatuelle placed his tactical ideas in a strategic context, which suggests that he was not really seeking to develop more decisive tactics. He stated in an important footnote that holding conquests was more important though less spectacular than capturing enemy ships. What difference, he asked, would the loss of some ships make to the English? Their territories, their commercial wealth and their maritime power should be 'le point capital' of attack. In the American War the French held Grenada, captured St Kitts and reduced Yorktown where the English army surrendered. These events were the results 'de grand combats où l'on a laissé faire tranquillement sa retraite, pour ne pas s'exposer à lui laisser la faculté de jeter des secours dans les points attaqués'.[4] It might be answered that if d'Orvilliers, d'Estaing, de Grasse and de Guichen had invariably seized the opportunities before them, it might have been impossible for the British to have maintained fleets and squadrons of adequate strength, let alone to have given Rodney local superiority at the Saints. Certainly his *ancien régime* strategy was out of step with military developments of revolutionary warfare ashore.

Ramatuelle's work on signalling was undeveloped. He never established his own system clearly, being content with tentative proposals avowedly based on Pavillon. He was very concerned about the relative visibility of different colours and shapes of flags, of which he proposed twenty-six, all but one of which were combinations of red and white or yellow and black. His fog signals were made with drums and bells in accordance with an extremely formidable table. As an example, 'La Charge' played by drums to indicate the top line of the table, followed by bells 'isolée par trois' to indicate the side column of the table, was signal number 11, which gave the order to make sail on the larboard tack.

The 'Tratado de Señales', 1804

The Spanish ships which served under Don Federico Gravina with Villeneuve in his crossing and recrossing of the Atlantic, together with others present at Trafalgar, possessed one of the most sophisticated tactical and signalling treatises which has ever been produced. It was published in 1804 and is entitled *Tratado de Señales de Dia y Noche, e Hipótesis de Ataques y Defensas, dispuesto por el Retado Mayor de Marine para auxiliar la instruction de Este Ramo, Madrid en la Impresta Real 1804*. The usual preliminary matter appearing in tactical books was explained with great clarity and illustrated with excellent diagrams, but it was in analysis of battle tactics that it excelled. Eighteen separate forms of attack were described, with the appropriate methods of defence against them. This is a treatise which puts 'attack' first, with no hint of defensive thinking. The offensive spirit is particularly evident in the proposed means of meeting the various attacks, the majority taking the form of vigorous counterattacks. Knowing what we do about the Spanish navy's fighting efficiency at the time, these devices seem tragically unreal.

Summary of Spanish fighting instructions of 1804

1st Attack: The fleets are passing on opposite and converging courses. The windward fleet alters course to form parallel to the leeward fleet and then leads round the leeward fleet's rear in succession.

Defence: The leeward fleet turns away from the wind together and then turns to the wind again in line ahead.

2nd Attack: The windward fleet crosses the bows of the leeward fleet and passes them on the opposite tack. The fleet then tacks to come on the same course while sending a division to double the enemy's rear.

Defence: The leeward fleet turns away in succession, thus avoiding having their bows crossed. Alternatively, the leeward ships wear together and form line ahead on the opposite tack.

3rd Attack: The two fleets are parallel to each other on the same course, the windward slightly ahead. The leeward fleet tacks together and leads past the rear of the windward fleet on a converging course. It thus defiles against the windward rear so that the van gains the weather gage. The ships then wear in succession, passing the rear of the enemy's fleet to leeward at close range while the van doubles the enemy to windward.

Defence: The windward fleet wears together, crossing the T of the leeward fleet as it tacks and then defiles past them to leeward.

4th Attack: The fleets are on the same course. The leeward fleet doubles the windward fleet's van by crowding sail and tacking.

Defence: The van of the leeward fleet attacks the enemy's rear as they press ahead.

5th Attack: The fleets are on opposite courses. The windward fleet converges on the leeward's rear and doubles on it, while a division wears on the windward side to keep the enemy rear between two fires.

Defence: The leeward fleet turns away together, or else wears together, hauls to the wind in succession, and cuts across the head of the oncoming windward fleet.

3 pp244–5 (para 280)

4 pp363–4

6th Attack: The fleets are on opposite courses and the leeward fleet is about to tack together, while the windward fleet is wearing in succession to reach round the enemy van. Alternatively, the windward fleet is tacking together to reach round the ememy rear.

Defence: In the first case the leeward fleet wears together instead of tacking. In the second it forms line ahead on the opposite course to the windward fleet, and then wears together.

7th Attack: The fleets are converging on opposite courses. The windward fleet defiles against the van of the leeward fleet and then hauls to the wind in succession.
Alternatively, it passes the leeward fleet on the opposite course and then wears in succession to envelop the enemy rear.

Defence: The leeward fleet wears together and forms line on the opposite tack, then forces the windward fleet to withdraw to windward by hauling to the wind and then tacking with the leading ships.

8th Attack: The fleets are on the same course. The windward fleet is superior by 16 to 14. The leeward van in pressing ahead to hold off the windward van has left a gap between itself and their centre. The four van ships of the windward fleet press the leeward van hard. Numbers 5–8 cut through the gap in succession, while numbers 9–12 press hard on the leeward centre and prevent it closing the gap.

Defence: The leeward fleet turns away together, closes the gap and reforms on the original course. Alternatively, it wears together and reforms on the opposite course.

9th Attack: A similar situation, with the gap aft of the centre. The windward centre cuts through in succession, while the van holds off to windward *en echelon* and the rear presses the leeward rear.

Defence: The leeward fleet turns away as before and closes the gap.

10th Attack: There is a gap in the line of the leeward fleet and the windward fleet, though unformed, passes through it, forming line as they go, and then turns on the same course as the enemy van.

Defence: The leeward van shortens sail. The centre and rear crowd sail and turn away in succession, and then return to their original course in succession. The attackers thus find themselves between two fires. Alternatively, the leeward van turns away together and forms line ahead running large while the centre and van turn away in succession and run large on a parallel course. The attackers, or at any rate their van and centre, are thus boxed in between two lines of fire.

11th Attack: Similar to 10 except that the attackers, having passed through the gap, range along the centre and rear of the enemy on the opposite course, defiling against their rear.

Defence: The leeward fleet wears together and gets on the same course, so that the attackers are again caught between two fires, or else the fleet wears in succession, thus boxing the attackers between two lines ahead.

12th Attack: The windward fleet is in line ahead with a gap in the centre. The leeward fleet, unformed, breaks through on the opposite course. It then wears round the windward fleet's rear in succession, leaving its own rear squadron to wear together without passing round, thus placing the enemy's rear between two fires.
Alternatively, having cut through, the windward fleet tacks together and forms line ahead on the same course as the fleet attacked.

Defence: (a) The threatened fleet to windward tacks together and escapes.
(b) Alternatively, all the ships both ahead and astern of the gap wear together and then tack in succession, thus forming two close-hauled lines with the attackers boxed in between them.
(c) The same result, however, can be achieved more quickly by the van and centre wearing together and coming to the wind together while the rear division (astern of the gap) tacks in succession.
(d) Yet another way of achieving this is for the leading section to tack together and then haul to the wind in succession while the section astern of the gap tacks together, then wears together and finally tacks in succession.
(e) Here the van and centre, having tacked together, turn away together until the leading ship is within close range of the enemy. They then come to the wind in succession, thus forming in line parallel with the enemy. The rear section also tacks together, then wears together and hauls to the wind in succession when parallel with the enemy.
(f) A simpler defence, by which each section of the fleet tacks in succession.
(g) A more aggressive solution: the leading division wears together, getting on the opposite course, and then wears in succession. The rear division simply leads off the wind in succession. Both division are now enclosing the enemy's van and passing on the opposite course.

13th Attack: The fleets are equal and are on the same course and parallel. The rear half of the windward fleet turns away together and forms in line to rake the enemy's rear. It then hauls to the wind and resumes its original course. Meanwhile, the leading half of the windward fleet comes to close range. In a slightly different form, the leading half of the windward fleet holds off.

Defence: The windward fleet crowds sail. The rear thus escapes attack while the van tacks and threatens to cut off the leading half of the windward fleet from their rear. Alternatively, the windward fleet wears together on the opposite course and cuts through the ships attempting to attack the rear.

14th Attack: The windward fleet divides, the leading half enveloping the enemy van and the rear half the rear.

Defence: The leeward fleet closes up and turns away in succession. It thus blocks the attack on the van, while the windward fleet is too late to encircle the rear.
Alternatively, the fleet wears together and defiles in succession to windward across the head of the windward section attempting to encircle the rear.

15th Attack: The leeward fleet attacks in two separate divisions approaching on the opposite course. Having reached positions respectively opposite the rear and van of the windward fleet, both divisions wear in succession, come on the same course as the enemy and continue to engage.

Defence: The windward fleet leads to leeward through the gap between the enemy's two divisions. Alternatively, it wears together, the new van leading to leeward in succession round the head of the enemy's nearest division.

16th Attack: A similar situation; the leading division of the leeward fleet defiles against the rear of the windward fleet and then wears in succession and returns in wake of its own rear division, which has worn together on the same tack as the enemy but without becoming closely engaged.

Defence: The van of the windward fleet wears in succession and completely heads off the leeward attack.

17th Attack: The leeward fleet, being superior, doubles on the windward rear by causing its own rear to tack together. It then gets on the same course as the enemy, then tacks together again, thus passing astern of the enemy and then getting on the same course again, (the 'B' movement advocated earlier by the French). The van division of the leeward fleet also tacks together and gets on the same course as the enemy, pressing their van but not doubling. Meanwhile the leeward centre holds off.

Defence: The windward fleet tacks together and withdraws.

18th Attack: This article covers various forms of concentration when the fleets are equal, achieved by leaving gaps in different parts of the line. This can be answered by conforming movements or threats to cut through the gap or gaps.

The Spanish signalling system was just as efficient as that of the French, and in many respects more so. The general signals were made with a table of twenty-four flags, giving 576 signals in all. The first twenty-four were made with single flag hoists, and the remainder with double-flag hoists. Far more than 576 signals could in fact be made because many signals had numbered extensions, indicating details of application. For example, signal number 251, to attack the eleventh ship and those following, had the sub-headings: 1, to tack and get on the opposite course in order to make an attack from windward on the group cut off; 2 to 4, ordering one to three divisions of the rear to attack the group cut off from leeward; and 5, to tack to get on the opposite course. Another example was signal 537 to a particular ship, which had four numerical extensions indicating that rations should be reduced by a quarter, a third or a half, and (4) that a ration report should be returned. The flags employed to make the signals were unusual, and in some cases not very practical in design. The colours were either blue and white, or red, white and blue, no red and white combination being allowed unless with a narrow perpendicular blue stripe against the mast. Two separate tables provided 110 night signals for ships at anchor, and for those under sail, made by lights and a complicated system of gun signals at timed intervals.

The most intriguing feature of the book is a Spanish translation of practically the whole of Howe's Instructions for the Conduct of the Ships of War, only three of the night sailing articles being omitted. There was also a Spanish translation of large sections of Howe's 1790 Signal Book, with the colours of the numerical flags and flags for the single flag signals correctly described. A number of the main signals were given, and extracts were published of varying amounts from all the remaining sections of the book. The table for private ships was given in full, as well as all the compass signals and the signals with sails and guns combined. The translation of the Instructions, though sometimes summarised, was a careful piece of work. The extracts from the Signal Book, however, appear to have been made in a hurry, and possibly under some difficulty. Perhaps the translator had not had the book long enough in his hands. The book also includes summaries of Grenier's and d'Amblimont's works on signalling and tactics.

Sir Home Popham's telegraph, 1800–12

With the possible exception of Sir Sidney Smith, Sir Home Popham is the most controversial figure in the Royal Navy of Nelson's age. Everything about him was extraordinary. He was his parents' twenty-first child, a brilliant navigator and hydrographer. His private trading ventures in China and India were spectacular. His services in combined operations led to his successive promotions to commander and post captain, each time at the special request of the Duke of York and against the wishes of the Admiralty. His lawsuits, the Parliamentary enquiry into his financial conduct, his court-

Admiral Sir Home Popham (1762–1820); engraving by Anthony Carden after a painting by Mather Brown. BM

martial for attempting to liberate Buenos Aires from Napoleon, when he was actually concerned with the occupation of the Cape of Good Hope – all these combined to make him a dramatic and notorious figure, highly unpopular with men such as Hood and St Vincent. Prejudice outlived him, and in a substantial notice in the *Dictionary of National Biography*, Sir John Laughton dismisses his famous telegraphic and vocabulary signalling system in twenty-nine words. Nevertheless, it was through the unofficial adoption of this system by the navy (and at last in 1816 by the Admiralty) that Nelson had such a perfect liaison with the individual ships of his fleet during the Trafalgar campaign.

Popham, like many other officers, realised that the navy needed a much more detailed signalling system than that provided by the book of 1799. Communication by boat between the admiral and private ships was a waste of time and impossible in bad weather or with an unfavourable wind. Direct communication by speaking trumpet was only possible when two ships were very close together and the wind not too strong to drown voices. At the Battle of Ushant Keppel summoned frigates under his stern by flag signal and shouted orders which they in turn were to shout to Harland and Palliser; the loss of time was considerable, as well as the waste of a frigate which might otherwise have been more usefully engaged. A completely new system of signalling was needed if the latest requirements were to be met, and the system had to allow two-way communication. It was no longer a question of the admiral issuing orders, while the private ships sent oversimplified reports. A language of the sea was required.

A gradual change in the character of sea warfare in the Napoleonic era lay behind the need for more sophisticated signals. Trade protection, combined operations, colonial warfare, co-operation with allies by sea and land, and political and military relations with neutral states and their shipping, were expanding in such a way as to involve the Royal Navy in a world-wide web of politico-economic commitments and responsibilities. Fleet actions were becoming more rare, but general activity was increasing. Small detached squadrons abounded, under the command of the senior captain. To meet the needs of the time, a quick system of detailed communication was necessary between ship and ship, as well as between the admiral and the ships directly under his command.

Popham made his first experiment in what he termed telegraphic signalling when commanding the *Romney* off Copenhagen in 1800. He was then acting as liaison between Lord Whitworth, a British Minister on a special mission to the Danish Court, and Admiral Archibald Dickson, in command of the North Sea squadron off Elsinore.[5] The subject of the negotiation was Denmark's neutral rights, and Dickson had been instructed to apply pressure; an accurate, detailed and rapid means of communication between the ambassador and the admiral was highly desirable. Popham's solution was to make a numbered list of useful words which could be signalled in the ordinary way by a two- or three-flag hoist, using the Admiralty numerical flags 0 to 9, together with the usual preparative, affirmative, negative, annulling and substitute flags. A special preparative flag was used to indicate telegraphic signals. To begin with, he limited himself to 998 numbers so as to avoid the need for two substitute flags. For the same reason, he omitted all numbers having the same three figures: 333, 444, etc. He also began his numbers at 26; his first twenty-five numbers being used for indicating the letters of the alphabet, omitting J. By the use of these numbers, prefixed by the numerical pendant, to indicate alphabetical signalling, he could, if necessary, spell out letter by letter words not included in his list. In the list

itself, he used root forms, 'embark' for instance, also standing for 'embarks', 'embarked', 'embarking', 'embarkation' and 'embarkations'.

Popham chose his list of words with care, and this system, combined with his vocabulary spelling system, proved an immediate success. 'Its utility was in that instance so obvious and so generally allowed by the captains of the North Sea squadron that Sir Home Popham conceived it might be brought into more extensive practice'. His object, he stated, was not to interfere with the established signals. 'It frequently, however, happens that officers wish to make communications of very essential moment far beyond the capacity of the established signals and it is presumed that this vocabulary will remove such inconveniences.' It would save sending boats, could be useful for army co-operation and enable returns of stores, ammunition, equipment, etc, as well as detailed answers, to be asked for and sent.[6]

The exact nature and date of the original publication of his system is still obscure, but it was certainly printed in 1801, with a dedication to Lord Spencer, then First Lord, who had been favourably impressed and had encouraged publication. A Part II parallel list of words was added, numbered 1026 to 1998 with some intermediate omissions. Thus 903–6 in the first list, 'vocabulary', 'voyage', 'vegetable', are paralleled in the second list by 1903–6, 'visible-y-ion-ary', 'visit-ed-ing-ation' and 'violate-ation-ers'. The first list of words, moreover, is a revised version of the one sent to Keith. The thousand sign was a pendant or ball hoisted above the signal, wherever best seen.

The main edition of the *Telegraphic Signals or Marine Vocabulary* – the one adopted by Nelson – was published in 1803.[7] Parts I and II still appeared in parallel columns, but a Part III had been added with numbers from 2026 to 2998, the second thousand being indicated by a pendant or ball below the hoist. It consisted of simple phrases and sentences, names of the months and names of ports, anchorages and headlands throughout the world. Key words were dealt with in a more sophisticated manner than in Parts I and II, so as to avoid possible mistakes:

2210	dispatches, open
2211	do not open
2212	[blank]
2213	have you got
2214	I have got
2215	I have sent

The edition of 1803 appears to have been on sale to the public, but with the flags shown only in outline, presumably to safeguard security.

Popham's object was to supplement the Signal Book, not to supersede it. By doing so, he put immense new tactical powers into the admiral's hands. This can be seen by looking at the following excerpt from the signal log of the *Euryalus* on 19 October 1805:

550	notwithstanding		
458	little	591	outward(s)
960	wind	864	the
480	many	1719	rest
570	of	1261	except
249	enemy	578	on(e)

---

5 Not to be confused with Captain William Dickson, the signalling innovator in the American War.

6 Letter accompanying a draft sent to Admiral Lord Keith, 30 June 1801; *NMM*, KEI/S/2 (formerly Sm/165).

7 *NMM*, SIG/B/82 (formerly Sp/86); *MOD*, Ec/61 and NM/75 (formerly *RUSI* 75).

613	persevere	456	line
873	to	693	ready(,)
335	get	986	yards
		1374	hoisted[8]

The edition of 1803 came at an appropriate moment, with the renewal of the war. Small octavo printings were published in 1805, 1806 and 1809.[9] Three reasons may be given for the book's tremendous success: it was the work of a naval officer who well understood the needs of the service; it originated in an entirely *ad hoc* arrangement to fulfil an immediate local need; and it could be recommended because it had already been tried.

Popham showed himself to be a better lexicographer and philologist than McArthur, at least for this particular job. McArthur's claim that his 1799 scheme was 'susceptible of 250,000 mutations or signals, without the necessity of using a pendant; which is more than double the number of words in any dictionary of the English or other language, and exceeds by 170,000 the number of symbols or hieroglyphics in the Chinese language' revealed his inadequacy.[10] Popham was an egotist but he never made inaccurate or grandiose claims of this kind.

Popham had a collaborator, John Goodhew, a mysterious person about whom almost nothing is known. In 1806 he published *A General Code of Signals for the use of His Majesty's Navy*.[11] He used only twelve flags, all being reversible. These provided twenty-four letters of the alphabet (J and V omitted). No numbers were allotted to any of the flags, the signals being made by letters of the alphabet alone. Single letters were used for divisional signals and two-letter flags for nautical sentences; more elaborate signals were made by three letters, and a pendant was used as a substitute. He was much concerned with methods of altering the key for purposes of secrecy. The signals chosen were not very important, the innovation being his reliance on a completely alphabetical system. Although the French had introduced purely notional lettering as far back as Morogues and d'Orvilliers, they had never adopted letters as complete substitutes for direct signalling. Nor had Popham. Goodhew's system was given a trial by the Channel Fleet, where it seems to have met with approval, especially as it dispensed with the complications caused by combining letters and numbers. Popham, however, still held the advantage, first, because of the great range of signals he provided, and second, because of his practical experience and personal ingenuity in providing the kind of signals the navy required.[12]

In 1812 a completely new edition of Popham's *Telegraphic Signals or Marine Vocabulary* was published for sale to the public.[13] The vocabulary now contained nearly 6000 'primitive words, exclusive of the inflexions of verbs, etc, making in all upwards of thirty thousand real words'. There were nearly six thousand sentences. A table of 1500 syllables was included, mainly intended for spelling the names of officers; L32A, L263, for instance, spelled Pop-ham. A geographical list of about 1200 place names included Cambridge, Paris, Berlin and Rome. There were separate lists for provisions and stores and for technical words and terms used at sea, and also a spare list for local significations. The signals were denoted by numbers, numbers and letters, and letters alone.

Popham's own examples, given at the end of the book, included the following:

FA1	have you an idea
G647	a change of ministers is about to take place
52A	certainly
8BF	not
G647	ministers are gaining strength
—	
K678	I saw your wife on
8B9	Monday
K813	with your
6C3	family
KFE	left them all well

Popham's decision to represent his signals both by letters and numbers was due to purely technical reasons, which he explained in his introduction. It might be assumed from this that he would have been bound to introduce many new flags. All he needed, however, were the Admiralty flags 1 to 9, five cornets, or cornetts, (swallow-tail flags) designated A to E, five guidons (triangular flags) designated F to K (J omitted) and four pendants designated L to O. The Vocabulary of single words was made with two, three or four flags, etc. For this he used the Admiralty numerical flags 4 to 9 and cornets A and B both as the classification signs and the top flag of the hoist. In all cases, except the words under S in the vocabulary classified under cornet A, each classification symbol covered

Vice-Admiral Sir James de Saumarez (1757–1836); engraving by C Turner after a painting by Carboniero. BM

8 *Logs of the Great Sea Fights*, II, p163

9 *NMM*, GRE/18 (formerly Sp/99); former S(D) 58; and *MOD*, Ec/62

10 *Thoughts on Several Plans combining a system of Universal Signals*, 1799, p13

11 *MOD*, Ec/6

12 In 1808 Joseph Conolly published *A Treatise on Telegraphic Communication by Day & Night*, using six invertible flags and two pendants of poor design. His Numerical Dictionary provided 10,999 single words, no sentences. Figures 10 to 34 represent the alphabet less J. He placed too much reliance on substitutes and this made the code too complicated. *MOD*, Ec/31.

13 Many copies extant. *NMM*, DUC/16 (formerly Sm/69) is a manuscript draft of the Introduction and pp1–88 with a number of differences from the published text, not specially important.

two or more alphabetical sections of the vocabulary. Thus flag number 8 covered all words beginning with the letters M, N or O, while cornet B covered all words from T to Z. Meanwhile, all the letters represented in chief by the cornets, guidons and pendants were used in combination with the numbers. Hence cornet B followed by pendant L and the numerical figure 1 (BL1) stood for 'ungentleman-ly-like'.

In the sentences the classification flags were cornets C to E and guidons F to K. These represented the first letter of the key word of the sentence. Here again, a combination such as H39B was employed to indicate 'rudder [key word] is damaged'. The four pendants provided the classification signs for the remaining sections: L for syllables, M for geographical terms, N for provisions and stores, O for military phrases. The spelling out of words letter by letter, was done as before by using numbers. The great merit of the system was that signals could easily be read by noting the classification flag. They could be sent with equal ease, since each class of signal was arranged in alphabetical order, and so provided its own index. The chief disadvantage of the system was that it assigned classification letters to words and sentences other than the initial letters, and that the classification cornet letters were split in the middle, B representing vocabulary T to Z, and C representing sentences in which the key word began with A.

The Naval Library, Whitehall, possesses a manuscript copy of the 'Code of Signals Recommended by Sir Home Popham and Mr Goodhew', thus establishing Goodhew's collaboration.[14] The introduction states that John Goodhew had suggested the substitution of letters for numbers and the general alphabetical arrangement, though 'The vocabulary and sentences are entirely Sir Home Popham's but compressed'. Pendants and cornets were used towards the end of the alphabet. The flags representing the latters A to K were also used for the figures 0–9, these particular flags being almost the same as the Admiralty's of 1811.

Distance signals were made with black flags of various shapes flown at the main and foremasts. Although the document claims that 8778 words and sentences could be signalled as well as 924 names of places in the geographical table, only one double page of actual signals was shown. A loosely inserted sheet showed twenty coloured flags, eight pendants and four cornets, some of rather fanciful design. Nothing was known of the origin or usage of this code.

The publication of Popham's 1812 edition seems to have been the decisive act in establishing his system as a necessary extension of the official signal book. The edition of 1803 had been in general use in the service for a number of years and had no serious rival. Opposition had come mainly from those who failed to grasp his purpose, as for instance Admiral the Hon Sir George Cranfield Berkeley.[15]

Saumarez and Moreno

On 5 July 1801 Rear-Admiral Sir James de Saumarez learnt while blockading Cadiz that three French ships-of-the-line and a frigate, which had left Toulon for the Atlantic, were at Algeciras. He at once went to attack them with six of the line, but owing to an unfavourable wind could not get near enough to avoid the fire of the Spanish land battery. After five hours fighting he withdrew to Gibraltar, leaving the

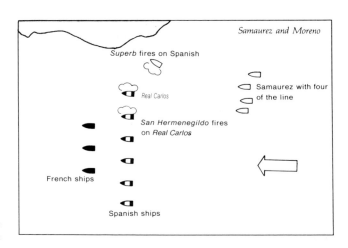

Samaurez and Moreno

Superb fires on Spanish

Real Carlos

Samaurez with four of the line

San Hermenegildo fires on *Real Carlos*

French ships

Spanish ships

Hannibal aground and captured after being very heavily damaged.

The French were now joined by six Spanish ships-of-the-line under Admiral Don Juan de Moreno. On 12 July all nine ships-of-the-line put to sea, accompanied by the frigates and a number of gunboats, steering west towards the Straits. Moreno placed his six Spaniards in line-abreast but sufficiently separated for the three French, also in line abreast but ahead of them, to fire through the gaps in the Spanish line with their stern chasers. If necessary, the whole force was to be prepared to form line ahead towards either the European or the African coast. With a strong following wind through the Straits all this seemed simple enough, though line-abreast is not always a good formation for a retreating force – which this was, even with two Spanish first rates included.

Saumarez had been reinforced by the *Superb* (74), commanded by Captain Richard Keats, who had been driven off by the Spaniards when they left Cadiz. He still could muster only five of the line and a frigate actually ready for action. Nevertheless, on the evening of 12 July, seeing the enemy across the bay putting to sea, he at once followed, but found it difficult to sight them in the dark. Hailing the *Superb*, he ordered Keats to crowd sail and attack the enemy's rear so as to delay them, keeping inshore on the European side. The *Superb*, once more a fast sailer after her terrible drag across the Atlantic and back with Nelson, went ahead at about 12 knots. Coming up with the *Real Carlos* (112), the *Superb* gave her a broadside. Some of the shots hit the *San Hermenegildo* (112), a quarter of a mile away. Thinking the *Real Carlos* to be British, she fired back and the *Real Carlos* fired at her as well as at the *Superb*. When the *Real Carlos* caught fire, the *San Hermenegildo* tried to capture her as an enemy. The flames caught her as well, and both ships burnt until they blew up with the loss of nearly all their men. 'A most awful sight', wrote Saumarez. Meanwhile, he and Keats captured the *St Antoine,* and the other three British ships drove the remaining six enemy ships into Cadiz. Next morning the *Venerable* nearly captured another but was prevented when her mainmast broke.

Sir John Laughton describes the action and Keats's part in particular as 'without a parallel in naval history'. No doubt Nelson took account of it when he planned his Trafalgar attack, especially as Keats became one of his closest friends.

Villeneuve and Calder

The battle fought by Sir Robert Calder on 23 July 1805, in an actual fog of war, perfectly illustrates the weather difficulties faced even by the ablest and best served tacticians.

With the wind at WNW, Villeneuve had the Combined

14 *MOD*, NM/85 (formerly *RUSI* 85). The date cannot be earlier than 1815, when Popham was knighted. Some pages are watermarked '1812'.

15 *NMM*, UNCAT (RACK 9) (formerly SN(P) 177), papers connected with the dispute and Popham's answer, 10 October 1811.

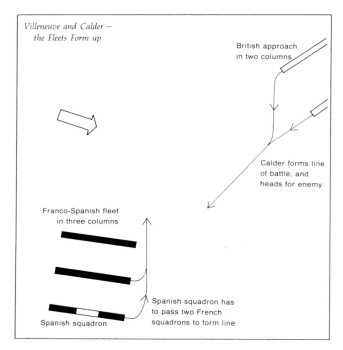

*Villeneuve and Calder —
the Fleets Form up*

British approach
in two columns

Calder forms line
of battle, and
heads for enemy

Franco-Spanish fleet
in three columns

Spanish squadron has
to pass two French
squadrons to form line

Spanish squadron

Fleet sailing large on a course east by south in three columns abreast with the Spanish squadron to starboard. His light squadron formed a fourth column away to larboard. On sighting the British to the north-east, Villeneuve deployed his fleet into a close-hauled line of battle ahead, steering north, with Gravina's starboard squadron leading. In this manner, Villeneuve risked losing the weather gage, since his leading ships, the Spaniards, had to parade right past the rest of the fleet (which had first to turn away to starboard to get into Gravina's water), then follow round astern of him to leeward. In other words, the deployment, though towards the British, was made from the column furthest removed from them. The light squadron, moreover, although nearest the British, had to take station astern of the whole fleet, which required it to run a good distance south before turning up again. Villeneuve presumably did this so as not to risk a shift of wind to the north, which would have enabled the British to steer straight for him.

On first report of the enemy being sighted to the south-west, Calder had turned his fleet towards them, signalling for two columns abreast in close order. Instead of trying to weather them, his intention was to interpose his squadron between Villeneuve and Ferrol and to engage him to leeward. He had two line-of-battle ships with his two frigates about six miles ahead and another 74 chasing a strange sail to leeward. The weather had been thick all day, but now it became worse. Although the two fleets were about sixteen miles apart when mutually sighted by their respective frigates, they were now approaching each other on opposite courses without any visibility at all. When the fog lifted, they were less than ten miles apart and Calder could see that he was outnumbered.

Sir Julian Corbett postulates that what followed shows that Calder shared 'Jervis's mistrust of formal tactics' and that 'He was not of the Howe–Kempenfelt School'.[16] This seems extremely hypothetical in view of what we now know of Jervis's theoretical, though not necessarily formalistic, approach to tactics. Conditions of weather and of relative strength encourage situational empiricism, and it seems unwise to postulate 'schools' when considering situations which in other ways are not comparable. Calder did not signal his intention to the fleet, presumably because he wished to leave himself with a free hand until the last moment. Unlike Rodney, moreover, he trusted his captains to act to the best advantage when the time came. At a distance of eight miles from the enemy, he recalled his advanced ships and signalled for the line of battle ahead to the SSW, the starboard column leading, and a few minutes later for turning another point to starboard. The fog now closed down again for a quarter of an hour, and when it lifted the two fleets were beginning to pass each other on opposite courses but still out of gunshot.

Calder evidently intended to attack the enemy's rear, or centre and rear, by doubling or otherwise, eschewing all idea of an attack on the van, with or without an attempt to seize the weather gage. Villeneuve divined this easily enough. He, therefore, made a preparative signal for the Combined Fleet to wear in succession as soon as the van heard firing in rear. Gravina, in the leading ship, spotted the signal being repeated by a French frigate and wore at once, followed by the rest of the fleet in good order. The effect of this was to turn the head of the Combined Fleet south, much closer to the British and roughly on a parallel course. As Gravina came down in line ahead, he interposed himself between the British and the rearmost ships of the Combined Fleet, which were still shuffling into station to form line ahead running north. Thus Calder's plan was frustrated in two senses. He could no longer bring off a remunerative attack on the enemy's rear nor obtain the extra benefit of catching them in confusion before they were properly formed. As each new ship appeared astern of Gravina, heading south, the rearmost ships of the Combined Fleet were drawing further away to the north, well covered.

For the moment the fog completely masked Gravina's action. Calder, seeing his leading ship roughly level with the French centre still heading north, he himself being eighth in the British line, signalled the British fleet to tack in succession. The effect of this should have been to bring his van within close range of the enemy, on the same course and about level with the tail ships of their centre. As soon as the leading ship began to tack, Calder signalled to engage the enemy's centre, and soon after for close order. His plan seemed to be going smoothly. However, just as the leading

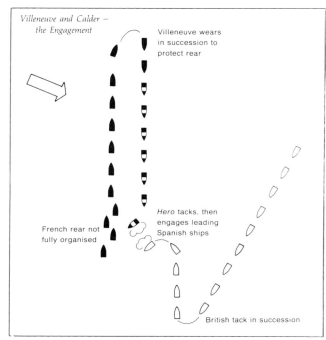

*Villeneuve and Calder —
the Engagement*

Villeneuve wears
in succession to
protect rear

French rear not
fully organised

Hero tacks, then
engages leading
Spanish ships

British tack in succession

16 *Campaign of Trafalgar*, pp197–8

British ship, the *Hero*, was coming within range, Gravina's flagship appeared out of the fog bearing down on the opposite course and at much closer range. Captain Alan Hyde Gardner, unable either to see or be seen by Calder, and, knowing the prejudice against fleets passing each other in action on opposite courses, tacked again to put himself on the same course as Gravina. As Gravina was leading the line to the south he had been hailed by Villeneuve in passing and ordered to steer WSW, so as to cover the rear of the Combined Fleet. Having fired a poorly aimed broadside at the *Hero*, still in stays, he bore away followed by the ship astern of him which, nevertheless, drove the *Hero*'s next astern out of the line in passing. The ships ahead of Calder all followed the lead given by the *Hero*, Calder himself signalling to tack in succession when his own turn came. Gravina was now in action with the fourth British ship, the *Barfleur*, and as the fog closed down again desultory firing developed all along Villeneuve's line as they headed south. The British rear ships, unable to judge where the enemy were or where the British van had gone, had no means of following the movement marked out by the two tacks in succession. They thus found themselves in action or otherwise on no real tactical basis. All, however, except the ship originally detached in chase, were engaged at various times, some very heavily.

In the Combined Fleet, equal if not greater confusion prevailed. The Spanish van squadron was heavily engaged; it alone met the British ships in any obvious and connected order, and its casualties amounted to nearly twice those of the whole British fleet. The two rearmost ships, fifth and sixth in the whole line, fell to leeward and were captured. The rear of the main Combined Fleet, including those ships which originally constituted the light squadron, was hardly in motion at all.

The Battle of Trafalgar

To understand what happened in the last and greatest battle between two fleets, each moving under sail, it is useful to look at the general intentions of the rival commanders-in-chief. In the Spanish naval base of Cadiz in early October 1805 lay the combined Franco-Spanish fleet under the supreme command of Vice-Admiral Villeneuve, under orders from the Emperor to carry a force of troops into the Mediterranean and land them in Neapolitan territory. The troops were in Cadiz ready for immediate embarkation. This was Villeneuve's first and paramount duty. He could leave Cadiz only if the wind was favourable, that is between northeast and south-east, and he knew that he would have to fight Nelson as soon as he got to sea. Nothing was said in Villeneuve's instructions about the absolute necessity for engaging and defeating the British fleet, so the threatened battle was relegated to an obstacle in the way of obeying unqualified orders to proceed from one place to another.

Villeneuve was a brave man, anxious to do his duty, but he had very natural misgivings about the quality of his ships and crews and their general lack of training. The crews of the Spanish ships were particularly poor. On balance, he knew that his ships were much less efficient and seaworthy than Nelson's. His misgivings were shared by the flag-officers and captains of the Combined Fleet, as was shown at his council of war held on 8 October. On the other hand, his relations with Admiral Gravina, and his Spanish allies generally, were excellent. There were no accusations and no intrigues.

Three considerations led Villeneuve to decide on what

Colonel Desbrière described succinctly as 'un acte de désespoir inutile'. The first was that Nelson had sent Rear-Admiral Sir Thomas Louis to Gibraltar with six of the line for water and victuals. Villeneuve was informed of this by his Gibraltar intelligence; contrary to what Nelson imagined, his main force, assumed to be out of sight, could often be counted from the tower of San Sebastian at Cadiz. Like all commanders of blockaded fleets, Villeneuve could choose his moment to emerge. Commanders of blockading fleets had to be constantly ready, and Nelson was now minus six of the line. Contrary to observed facts, Villeneuve represented himself to be numerically inferior to Nelson, but he realised that with Louis's absence he would never have a better chance of victory.

The second consideration was that the emperor had given secret orders for Villeneuve to be superseded by Vice-Admiral François Rosily, and Villeneuve knew or guessed this. The third was that the wind was favourable.

In Nelson's fleet the feeling was utterly different. Although his instruction from the Admiralty included responsibility for the reconstituted Mediterranean command, extending as far west as Cape St Vincent, every detail of their execution depended directly or indirectly on the defeat of the Combined Fleet which he was blockading. Nelson himself saw things in even more absolute terms. He aimed not only at victory but at the complete destruction of the enemy's fighting power. 'I should not doubt of spoiling any voyage they may attempt,' he wrote to Barham, 'yet I hope for the arrival of the ships from England, that as an enemy's fleet they may be annihilated.' [17] To Nelson the coming battle was to be not merely a fulfilment of all the efforts made since 1803 to defeat the French Toulon Fleet, now reinforced by the Spaniards, but a chance 'to take, sink and burn' every single ship in that fleet. 'Let the battle be when it may, it will never have been surpassed.' [18] This extraordinary boast and prophecy illustrates the level at which Nelson's spirit and mind were working. With the possible exception of Suffren, it is inconceivable that any other naval officer of the period

Amiral *Pierre Charles Jean-Baptiste Silvestre, Comte de Villeneuve (1763–1806).* MM

Admiral Horatio, Lord Nelson (1758–1805), first Viscount Nelson, by Lemuel Francis Abbott. NMM

17 Source unknown
18 Source unknown

should have used such words. Unlike Suffren, however, Nelson possessed that irresistible charm of manner and persuasive power which made even the stupid and the stubborn obey him. To Nelson the tactics necessary to ensure absolute victory must have been a constant preoccupation. The dream battle would be unlike anything that had happened in his own lifetime; unlike anything since Barfleur, and yet far more decisive. And so it proved to be.

During the seventeenth and eighteenth centuries, flag officers aged more quickly than their contemporaries ashore, amongst all of whom the expectation of life was considerably less than it is today. Rodney, Howe, St Vincent and Collingwood were all, in their various ways, men possessed of exceptional powers of endurance, but they all collapsed physically under the strain either of battle or prolonged command. Nelson, aged only just forty-seven when he died, was the youngest British commander-in-chief since Russell to win a major sea battle. It was this combination of supreme ardour and sufficient youth that enabled him to reject with scorn the accepted notion of a 'handsome victory'.

Nelson's fleet was a disparate lot brought together at different times. Only five out of his twenty-seven ships-of-the-line had been with the *Victory* in his original Mediterranean fleet. Only eight of his captains had served with him before. Only his flag captain, Hardy, and three other captains had been with him in his chase across the Atlantic and back again in pursuit of Villeneuve. Only five captains were commanding the same ships they had commanded in 1803. Only five had previously commanded a ship-of-the-line in battle. Of course they were able and experienced men and were assumed to be capable of efficient service in any fleet or squadron to which they might be suddenly appointed. Nevertheless, it was one thing to fit in with the well-known tactical requirements of a well organised fleet, but quite

another to help execute a new system of attack different from anything provided in any British fleet up to that time.

Tactical plans compared

Our understanding of Nelson's tactical plan comes from four sources.[19] The first was a plan of attack, probably written in 1803, and well worth quoting:[20]

> The business of a commander-in-chief being first to bring an enemy's fleet to battle on the most advantageous terms to himself (I mean that of laying his ships close on board the enemy, as expeditiously as possible, and secondly, to continue them there without separating until the business is decided), I am sensible beyond this object it is not necessary that I should say a word, being fully assured that the admirals and captains of the fleet I have the honour to command will, knowing my precise object, that of a close and decisive battle, supply any deficiency in my not making signals, which may, if extended beyond those objects, either be misunderstood, or if waited for very probably from various causes be impossible for the commander-in-chief to make. Therefore it will only be requisite for me to state in as few words as possible the various modes in which it may be necessary for me to obtain my object; on which depends not only the honour and glory of our country, but possibly its safety, and with it that of all Europe, from French tyranny and oppression.
>
> If the two fleets are both willing to fight, but little manoeuvring is necessary, the less the better. A day is soon lost in that business. Therefore I will only suppose that the enemy's fleet being to leeward standing close upon a wind, and that I am nearly ahead of them standing on the larboard tack. Of course I should weather them. The weather must be supposed to be moderate; for if it be a gale of wind the manoeuvring of both fleets is but of little avail, and probably no decisive action would take place with the whole fleet.
>
> Two modes present themselves; one to stand on just out of gun-shot, until the van ship of my line would be about the centre ship of the enemy; then make the signal to wear together; then bear up [and] engage with all our force the six or five van ships of the enemy, passing, certainly if opportunity offered, through their line. This would prevent their bearing up, and the action, from the known bravery and conduct of the admirals and captains, would certainly be decisive. The second or third rear ships of the enemy would act as they please, and our ships would give a good account of them, should they persist in mixing with our ships.

---

19 In his manuscript, Tunstall listed a fifth source of information, the famous and mysterious 'Nelson Touch'. Oliver Warner has revealed this for what it really was, the 'Nelson Trunk', to which quite naturally there was a key. See Oliver Warner, editor, *Nelson's Last Diary*, Kent, Ohio, 1971, p20.

20 Corbett, *Fighting Instructions*, pp313-6, with punctuation by Corbett from Clarke & McArthur, II, p427, and Nicolas, VI, p 443.

Nelson's Tactics

Nelson gains weather gage

Fleet wears together and concentrates on enemy van

Nelson bears away to head directly for enemy

Nelson declines weather gage

British ships wear, rearmost first

Breaks enemy line

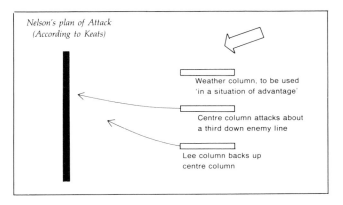

Nelson's plan of Attack (According to Keats)

Weather column, to be used 'in a situation of advantage'

Centre column attacks about a third down enemy line

Lee column backs up centre column

The other mode would be to stand under an easy but commanding sail directly for their headmost ship, so as to prevent the enemy from knowing whether I should pass to leeward or to windward of him. In that situation I would make the signal to engage the enemy to leeward, and cut through their fleet about the sixth ship from the van, passing very close. They being on a wind and you going large could cut their line when you please. The van ships of the enemy would, by the time our rear came abreast of the van ship, be severely cut up, and our van could not expect to escape damage. I would then have our *rear* ship and every ship in succession wear [and] continue the action with either the van ship or the second as it might appear most eligible from her crippled state; and this mode pursued I see nothing to prevent the capture of the five or six ships of the enemy's van. The two or three ships of the enemy's rear must either bear up or wear; and in either case, although they would be in a better plight probably than our two van ships (now the rear), yet they would be separated and at a distance to leeward, so as to give our ships time to refit. And by that time I believe the battle would, from the judgment of the admiral and captains, be over with the rest of them. Signals from these moments are useless when every man is disposed to do his duty. The great object is for us to support each other, and to keep close to the enemy and to leeward of him.

If the enemy are running away, then the only signals necessary will be to engage the enemy on arriving up with them; and the other ships to pass on for the second, third, &c., giving if possible a close fire into the enemy on passing, taking care to give our ships engaged notice of your intention.

The second source of information about the development of Nelson's plan of attack is Sir Richard Keats's account of a conversation with Nelson in late August or early September, 1805:

One morning, walking with Lord Nelson in the grounds of Merton, talking on naval matters, he said to me, 'No day can be long enough to arrange a couple of fleets and fight a decisive battle according to the old system. When *we* meet them,' (I was to have been with him), 'for meet them we shall, I'll tell you how I shall fight them. I shall form the Fleet into three Divisions in three Lines. One Division shall be composed of twelve or fourteen of the fastest two-decked Ships, which I shall keep always to windward or in a situation of advantage, and I shall put them under an Officer who, I am sure, will employ them in the manner I wish, if possible. I consider it will always be in my power to throw them into Battle in any part I may choose; but if circumstances prevent their being carried against the Enemy where I desire, I shall feel certain he will employ them effectually, and, perhaps, in a more advantageous manner than if he could have followed my orders.' (He never mentioned, or gave any hint by which I could understand who it was he intended for this distinguished service.)[21] He continued – 'With the remaining part of the Fleet formed in two Lines, I shall go at them at once, if I can, about one third of their Line from their leading Ship.' He then said, 'What do you think of it?' Such a question I felt required consideration. I paused. Seeing it he said, 'But I will tell you what *I* think of it. I think it will surprise and confound the Enemy. They won't know what I am about. It will bring forward a pell-mell battle, and that is what I want.'[22]

These two quotations are mainly valuable as evidence of preliminary thinking. Of more immediate importance was Nelson's memorandum written inboard *Victory*, off Cadiz, 9 October 1805. Again it is worth quoting at length:

Secret
Thinking it almost impossible to bring a Fleet of forty Sail of the Line into a Line of Battle in variable winds, thick weather, and other circumstances which must occur, without such a loss of time that the opportunity would probably be lost of bringing the Enemy to Battle in such a manner as to make the business decisive; I have therefore made up my mind to keep the Fleet in that position of sailing (with the exception of the First and Second in Command) that the Order of Sailing is to be the Order of Battle; placing the fleet in two Lines of sixteen Ships each, with an Advanced Squadron of eight of the fastest sailing Two-decked Ships *which* will always make, if wanted, a Line of twenty-four Sail on whichever Line the Commander-in-Chief may direct.

The Second in Command will, after my intentions are made known to him, have the entire direction of his Line to make the attack upon the Enemy, and to follow up the blow until they are Captured or destroy'd.

If the Enemy's fleet should be seen to Windward in Line of Battle and [in such a position] that the two Lines and the Advanced Squadron can fetch them, they will probably be so extended that their Van could not succour their Rear.

I should therefore probably make the 2nd in Command's signal to Lead through, about their Twelfth Ship from the Rear (or wherever he could fetch, if not able to get as far advanced); my Line would lead through about their Centre, and the Advanced Squadron to cut two, three or four Ships Ahead of their Centre, so far as to ensure getting at their Commander-in-Chief, on whom every effort must be made to capture.

---

21 Corbett's note: it was certainly not Keats himself, though afterwards Nelson meant to offer him command of the squadron he intended to detach into the Mediterranean. In the expected battle Keats, had he arrived in time, was to have been Nelson's second in the line; Nelson to Sir Alexander Hall, October 15 1805.

22 Nicolas, VII, p241

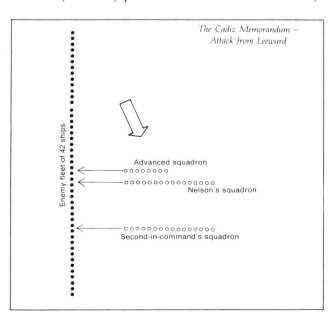

*The Cadiz Memorandum –
Attack from Leeward*

Enemy fleet of 42 ships

Advanced squadron
ooooooooo

Nelson's squadron
oooooooooooooooo

Second-in-command's squadron
oooooooooooooooo

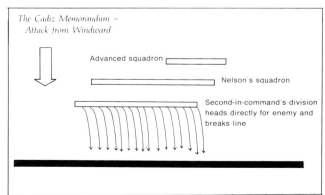

*The Cadiz Memorandum –
Attack from Windward*

Advanced squadron

Nelson's squadron

Second-in-command's division heads directly for enemy and breaks line

The whole impression of the British Fleet must be to overpower from two to three Ships ahead of their Commander-in-Chief, supposed to be in the Centre, to the Rear of their fleet. *I will suppose twenty Sail of the Enemy's Line to be untouched*; it must be some time before they could perform a manoeuvre to bring their force compact to attack any part of the British fleet engaged, or to succour their own Ships; which indeed would be impossible, without mixing with the ships engaged.[23]

Something must be left to chance; nothing is sure in a Sea Fight beyond all others. Shots will carry away the masts and yards of friends as well as foes; but I look with confidence to a Victory before the Van of the Enemy could succour their Rear and then that the British Fleet would most of them be ready to receive their twenty Sail of the Line, or to pursue them, should they endeavour to make off.

If the Van of the Enemy tacks, the Captured Ships must run to Leeward of the British fleet; if the Enemy wears, the British must place themselves between the Enemy and the Captured and disabled British Ships; and should the Enemy close, I have no fears as to the result.

The Second in Command will in all possible things direct the Movements of his Line, by keeping them as compact as the nature of the circumstances will admit. Captains are to look to their particular Line as their rallying point. But, in case Signals can neither be seen or perfectly understood, no Captain can do very wrong if he places his Ship alongside that of an Enemy.

The Divisions of the British fleet will be brought nearly within gun shot of the Enemy's Centre. The signal will most probably then be made for the Lee Line to bear up together, to set all their sails, even steering sails, in order to get as quickly as possible to the Enemy's Line and to Cut through, beginning from the twelfth Ship from the Enemy's Rear. Some Ships may not get through their exact place, but they will always be at hand to assist their friends; and if any are thrown round the Rear of the Enemy, they will effectually complete the business of Twelve Sail of the Enemy.

Should the Enemy wear together, or bear up and sail large, still the twelve Ships composing, in the first position, the Enemy's Rear, are to be *the* object of Commander-in-Chief; which is scarcely to be expected, as the entire management of the Lee Line, after the intentions of the Commander-in-Chief is [sic] signified, is intended to be left to the judgment of the Admiral Commanding that Line.

The Remainder of the Enemy's fleet, 34 Sail, are to be left to the Management of the Commander-in-Chief, who will endeavour to take care that the movements of the Second in Command are as little interrupted as possible.

The Memorandum introduces three major ideas: the separate role and responsibility of the second-in-command; the concentration of the attack on the enemy's rear, with their centre contained and thus prevented from interfering, and with their van unable to interfere until too late; concealment of the direction of attack until the very last moment. None of these ideas was new. A fourth and new idea is specifically mentioned: crowding sail, something entirely contrary both to contemporary means and traditional practice. Here a calculation is made of the risk of collision involved, and of the heavy sailers, not merely of dropping astern, but being out of alignment altogether. Against this is the dual advantage of presenting a quickly moving target to the enemy and of covering more quickly the space in which enemy fire must be endured without reply. Added to this is the association of speed with concealment of the point of attack. Not only will the enemy be kept guessing, but they will have less time in which to guess.

Yet a further idea, thought by Sir Julian Corbett to be implied in the Memorandum, when put alongside Nelson's latest known order of sailing, dated 10 October, is that he intended his three-deckers to lead the attack in the manner earlier suggested by Morogues. This, Sir Julian points out, was still valid even if Nelson and Collingwood, after leading their respective lines in the order of sailing, dropped back to second and third, respectively, when changing over to the order of battle. Nelson's conjectural order of battle shows:

Van or Weather squadron

First division
1 *Téméraire* 98
2 *Victory* 100
3 *Neptune* 98
4
5

Second division (six ships)
7 *Britannia* 100
8
9
10
11
12

Rear or Lee squadron

First division (eight ships)
1 *Prince* 98
2 *Mars* 74
3 *Royal Sovereign* 100
4 *Tonnant* 80
5 *Belleisle* 74
6
7
8

Second division (seven ships)
9 *Dreadnought* 98
10
11
12
13
14
15

In the battle itself, the order of engaging the enemy is still undetermined. We know for certain that in the weather line it was the *Victory, Téméraire, Neptune*, with the *Britannia* possibly sixth. For the lee line it was certainly the *Royal Sovereign* with the *Mars* and *Belleisle* either second and third or third and second. The *Dreadnought* was probably tenth and the *Prince* about fourteenth.

In Nelson's Memorandum of 9 October there is a note in the upper margin after the words 'The signal will most probably *then* [interlinear addition] be made for the Lee Line to bear up together and set all their sails even steering sails . . .'. It reads '*Vide* instructions for Signal Yellow with Blue fly, page 17, eighth Flag Signal Book, with reference to appendix'.

Sir Julian Corbett, when writing *The Campaign of Trafalgar*, suggested that since the signal did not appear in the

---

23 The words in italics were added by Mr Scott, Lord Nelson's Secretary. In the upper margin of the paper Lord Nelson wrote and Mr Scott added to it a reference, as marked in the text – 'the enemy's fleet is supposed to consist of 46 Sail of the Line, British fleet of 40 – if either is less, only a proportionate number of Enemy's ships are to be cut off; B to be ¼ superior to the E cut off.' Nicolas, VII, p89.

printed appendix it must have been added by Nelson himself.[24] His supposition was correct; the full text of the signal is in a copy of the Appendix in the Tunstall collection.[25] It reads:

Signals from Senior to Junior Officers by Lord Nelson.
Single Flag Yellow with Blue Fly. Cut through the Enemy's line and engage them on the other side.
NB: This Signal is to be repeated by all the Ships. The Number of the Enemy's ship under the Stern of which the Van Ship is to pass and engage will be pointed out by Signal counting from the Enemy's Rear. The Ships being prepared are to make all possible sail (keeping their relative Bearings and close order) so that the whole may pass thro' the Enemy's line as quick as possible and at the same time. It is recommended to cut away the Studding sails if set, to prevent confusion and fire. Each ship will of course pass under the Stern of the one she is to engage if circumstances permit, otherwise to refer to the Instructions Page 160 Article 31 [for breaking the enemy's line]. The Admiral will probably advance his Fleet to the van of theirs before he makes the Signal in order to deceive the Enemy by inducing them to suppose it is his intention to attack their Van.

The National Maritime Museum possesses a copy of the Signal Book of 1799, issued between 2 May 1805 and 10 February 1806,[26] marked 'Exd TMH' [Examined TM Hardy]. It was most probably used in Nelson's fleet at Trafalgar though attempts to identify it with the *Victory* appear optimistic rather than scholarly. It is the only one of many examined which has a yellow and blue flag for single flag signalling on page 17, though actually in the ninth and not the eighth position. In the blank signal space opposite, is the pencilled insertion: 'Cut through the Enemy's line and engage close on the other side . . . NB: This signal to be repeated by all the ships (*vide* Appendix)'. It would seem, therefore, that the signal inserted in the appendix must be accepted as having a very close relationship to Nelson's Memorandum, the reference in the signal quoted immediately above being taken as collateral evidence. In that case it was certainly valid twelve days before the battle, and must be accepted as the fourth source of information about Nelson's tactical plans.[27] One reason for the uncertainty which still persists about Nelson's actual tactics in the battle and the greater uncertainty still as to the actual text of the instructions under which it was fought, is that Nelson had no captain of the fleet. His personal staff consisted of a civilian and a divine: John Scott, his public or admiral's secretary, who was killed in the battle, and the Revd Alexander John Scott, the ship's chaplain, who acted as his private secretary, interpreter and intelligence officer. An efficient first captain would have taken care to preserve all copies of battle instructions and orders of battle and sailing, leaving the chaplain to attend, as indeed he did most faithfully, to Nelson's personal affairs. After the battle, Hardy of course was much too busy superintending the repair of the ship and keeping off a lee shore to collect and title the tactical papers. His own captain's clerk, equivalent of a secretary, was also killed in the battle.

On the day of the battle, Captain Henry Blackwood of the *Euryalus* acted as an *ad hoc* chief of staff. He was a captain of ten years' seniority with a great fighting record. Nelson had given him command of the whole of the inshore forces, Advance squadron and frigates included. He had even refused the offer of a ship-of-the-line in place of his frigates. On the morning of the battle he spent three hours with Nelson in the *Victory*, leaving only when the ship was already under fire. Nelson, he wrote, 'not only gave me the command of all the frigates [in the battle itself], for the purpose of assisting disabled ships, but he also gave me a latitude seldom or ever again, that of making any use I

pleased of his name in ordering any of the sternmost line of battle ships to do what struck me as best.' In using his frigates 'to complete the destruction of the enemy whether at anchor or not', Blackwood was ordered not to think about saving ships or men, destruction being the chief object, not prizes or prize money. This last command reveals the full measure of Nelson's demand for absolute victory.

Villeneuve's final instructions, dated 21 October, the actual morning of the battle, are practically a repetition of those he issued before leaving Toulon at the end of March. They are in very general form until the paragraph about probable British tactics is reached, in which we read:

The enemy will not confine himself to forming on a line of battle with our own and engaging us in an artillery duel, in which success is frequently with the more skilful but always with the more fortunate; he will endeavour to envelop our rear, to break through our line and to direct his ships in groups upon such of ours as he shall have cut off, so as to surround them and defeat them.

This is so startlingly accurate that one is at first inclined to think that the French Government must have been operating a very efficient intelligence service. It seems more likely, however, that this was no more than an appreciation made by Villeneuve himself, using his personal experience at the Battle of the Nile, the pocket manuscript signal book found in the captured schooner *Redbridge* during the previous year, and perhaps the 1804 edition of Clerk's *Essay on Naval Tactics*. A tactical study of the Battle of Camperdown would also have been useful. What seems equally surprising is that, having told his flag officers and captains what to expect, he vouchsafed nothing whatever about how to combat it. A final reference to fleet regulations and to the Specific Instructions for the Repetition of Signals, printed at the head of the Signal Book for the Use of the Fleet, sounds completely inadequate, especially when followed by the remark, 'They contain excellent maxims with which you must be fully conversant and dispositions which I am keeping in force to their full extent.'

In fact Villeneuve had nothing better to offer than he had had in the previous year. French tactics, so far as actual fighting was concerned, were completely barren. Spanish tactics, vigorous and resourceful on paper, were equally barren at sea.

Villeneuve's only contribution to executive tactics, his Squadron of Observation of twelve ships organised in two divisions and commanded by Gravina in the 112-gun *Principe de Asturias*, should have proved a decisive element once the

24 p449n

25 *NMM*, Tunstall Collection, TUN/61 (formerly S/P/R/8). This copy of the Appendix also contains a list of 'Additional Rendezvous by Lord Nelson' marked 'not in force' and two pages of 'Lord Nelson's Additional Rendezvous 26 September 1805' evidently written before joining the fleet off Cadiz on 29 September. This, however, does not mean that this copy of the Appendix was used in the *Victory*, though equally it does not preclude this.

26 *NMM*, SIG/B/76 (formerly Sp/102); Board signatures indicate date limits, otherwise undated. This is a later printing by William Winchester & Son, Strand.

27 Article 31 on p160 of this copy of the Signal Book has two separate pencilled additions: 'Signal 27 with a White Pendant is for the ships to break through the enemy wherever they can, filled in by Mr [?]'; and 'Signal 27 with a Dutch Pendant, according to the masthead it is hoisted'. Article 31 of the Fighting Instructions is for the execution of signal no 27, breaking the line. Curiously enough, this signal, which was printed in full in the original issue of 13 May 1799, is left blank in the later printing, the words being inserted in ink; this may have been a security measure. It is also curious that the variations of the signal noted above should appear as additions to the instruction and not to the signal itself.

battle was joined. They could have been used either to cover the gaps on the leeward side of the main line of battle and so deal severely with a breakthrough, or they could have been used from windward, to strike at the rear of British ships running down to attack. In fact the Squadron of Observation was no more than the calculated excess of ships which Villeneuve had imagined he possessed over Nelson. When he saw that Nelson had twenty-seven of the line instead of only twenty-one, as expected, he simply merged the Squadron of Observation in the main fleet, the result being confusion and ineffectiveness. Instead of fulfilling a detached and independent role, they merely became tied to the end of his main line of battle.

Preliminary movements

Ever since the blockade of Cadiz began, the British fleet had been keeping a close watch on the port. When Nelson arrived, he used four line of battle ships to supplement his four frigates, one schooner and a cutter. With these ships pushed comparatively close inshore, he was fed with continuous information about the enemy, who, boxed up in their own harbour waters, had to rely on what they could see from the town of San Sebastian. When the Combined Fleet at last put to sea they were kept under observation, so that, although there were various blank periods, Nelson had a good idea of where they were and what they were doing.

In all this, Nelson had the use, not only of an efficient day and night signal system, but of Popham's telegraph of 1803, as well. As his frigates gathered intelligence, they passed it back to the Advanced squadron, as well as to Nelson. There were always ships-of-the-line instantly ready and at short range to react to the enemy's movements, as well as to support the frigates if they were suddenly endangered. By contrast, Villeneuve gained no fresh information about Nelson's strength and movements from the time he left Cadiz until the morning of the battle.

Nothing better illustrates the contrast in attitude of the two commanders-in-chief than their reconaissance arrangements during 19–20 October. While Villeneuve was working his way out of Cadiz under some difficulty, made worse by the unhandiness of his ships, his frigates were continually reporting British ships within sight. Yet, except for one occasion when he made a show of ordering these shadowing ships to be driven away, he signalled 'general and unconditional recall' of his own frigates and advanced ships each night. Nothing marks so clearly the tactical inferiority of the Combined Fleet than this complete absence of a vigorous reconnaissance within sight of their own base, and that vulnerability to being spied upon at close quarters by the British advance ships.

The effective use of an advance squadron by the British dates from the occasion Howe employed it against the French rear at the start of the First of June operations. Since then, the idea had been adopted by Jervis for the Mediterranean fleet. The assumption apparently was that, provided the fleet was reasonably strong, it was worthwhile organising a special detachment to act either as an observation squadron and nothing more, or as a tactical *corps de réserve*. When Nelson took command off Cadiz he started building up his inshore squadron of fast two-deckers, amounting finally to eight. With these ships he could support his look-out frigates and at the same time use them in a battle-cruiser role, to pin down the enemy until the main fleet came up. In his Memorandum the 'Advance squadron of eight of the fastest sailing two-decked ships', were given a definite tactical role. In an attack from leeward they were 'to cut two, three, or four ships ahead of their [the enemy's] centre, so far as to

ensure getting at their commander-in-chief whom every effort must be made to capture'. Nothing is said about their role in a windward attack, but by an over-riding arrangement they could be signalled to reinforce either the admiral's squadron or that of his second in command. Had Nelson not been forced to send six ships to Gibraltar he could have had thirty-three in the battle, sufficient to allow for an advance squadron of eight, even supposing the enemy totalled forty-six, as he apparently assumed. After making the Gibraltar detachment, he had only twenty-seven left, and on the morning of 20 October he seems to have re-allotted all eight of his advance squadron to their respective original squadrons, while still retaining three to act as a temporary link between him and his frigates.

At dawn on 21 October the Combined fleet was in some disorder. It was not until 6am that Villeneuve saw the British for the first time, away to windward – and, to begin with at any rate, in somewhat of a jumble. At 6.20 he signalled the fleet to form in line of battle ahead on the starboard tack. Gravina then signalled the Squadron of Observation to take station ahead. Their course lay now between south and south-west by south. As, however, the wind was very light and the westerly swell heavy, the ships took a long time to carry out the evolution, and even then were not in their proper order.

At 7am Villeneuve received a signal from one of his lookout frigates that the enemy '26 of the line, 4 frigates and 3 corvettes or despatch craft . . . were advancing in groups without formation; they were bearing down on the rear division of our squadron [fleet].' Another frigate could see that the British were in two columns steering for the rear and centre of the Combined fleet. Villeneuve records that the British were 'standing in a body for my rear, with the double intention of attacking it advantageously and of cutting off the retreat of the Combined fleet to Cadiz. I made the signal to wear together and to form the line of battle on the larboard tack in reverse order; my sole object being to protect the rear from the projected attack of the whole enemy force.' In other words, Nelson's precipitate action in leading straight for the enemy without forming up had already achieved its purpose. True, he moved slowly enough, but having the wind well abaft the beam, he was moving very much faster than the Combined Fleet, which was hardly moving at all. To say that Nelson's action had achieved his purpose should not be understood as meaning that he wished to see the enemy go about and retreat towards Cadiz. Far from it; he was disappointed, having hoped to attack them as they passed Cape Trafalgar towards the Mediterranean. What he had achieved was to make Villeneuve put his fleet about when it was still not properly formed in its existing order.

Villeneuve was certainly in a most unenviable position. Weather conditions prevented him either from forming up properly or running. He was only twelve miles out from Cape Trafalgar on a lee shore and less still from the shoals running along the coast. If he tried a second run to the southeast, for the Straits, Nelson would be sure to concentrate on his rear, as well as being able to collect Louis's six of the line at Gibraltar. If a retreat to Cadiz was advisable, he had to act immediately, otherwise he would be cut off. He had already signalled the fleet to be at a cable's distance between ships, the standard close or fairly close formation, but his line was still irregular. Having made the signal to wear together, at about 8am, he kept it flying for a quarter of an hour before hauling it down, this being the executive signal to begin the movement. The method of execution was for the rearmost ship to wear first, followed by the rest in quick succession, that is, as quickly as they could start swinging without risk of collision. This generally required each ship to bear up a little more than

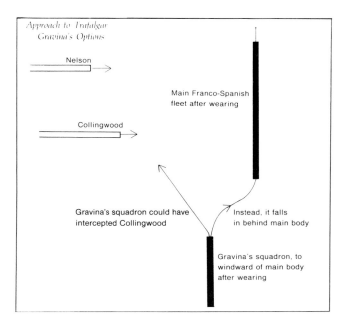

Approach to Trafalgar Gravina's Options

Nelson

Main Franco-Spanish fleet after wearing

Collingwood

Gravina's squadron could have intercepted Collingwood

Instead, it falls in behind main body

Gravina's squadron, to windward of main body after wearing

her future next ahead. Once round, it was the duty of the ship in the lead (previously the rearmost) to haul up to about 7 points from the wind, about north by east. The rest of the fleet, moving in succession, would then retrace their course, steering north-east by north about until the approximate point where the original rearmost ship had worn, and then haul to the wind in her wake.

Villeneuve now signalled his new leading ship to make more sail so as to clear some of the congestion astern, but as the fleet began working round, their confusion increased, and there was much bunching. At 10am Villeneuve signalled the leading ship to haul close to the wind, the rest to follow in succession. This was obeyed but the ships already fallen to leeward, fell off still more as the congestion of the line prevented them working up into station. In the centre the *Santa Ana* had to put about again and steer south to avoid a collision, so that, when she finally succeeded in wearing, there was a big gap ahead of her. With the fleet now in reverse order, the new van, having succeeded in hauling to the wind roughly in succession, left the ships of the *corps de bataille* in the centre somewhat to leeward. It was this that helped to bring about that sag in the Franco-Spanish line which was so much remarked on as an important feature of the battle.

All this time, Gravina was playing a very difficult role very badly. When the Combined Fleet was on its southerly course on the starboard tack, he had been ordered to take station ahead of the van. What his role really was has never been entirely clear. According to one of the current notions, the advance squadron or Squadron of Observation was composed of the number of ships by which the fleet exceeded that of the enemy. When the two fleets became engaged ship for ship, the Squadron of Observation, acting as a *corps de réserve*, was supposed to double on the enemy, assist in their discomfiture and prevent their escape. There was plenty of opportunity for this at Trafalgar but neither Villeneuve nor Gravina were prepared to try it.

When the Combined Fleet at first began to wear, Gravina was a little to windward of the original van, the fleet being then on a southerly course. Once round on the northerly course he seems to have thought that his duty was to take station in rear of what was now the rear squadron of the Combined Fleet, instead of acting independently and holding off to windward. He signalled his squadron to bear up together so as to fall into the wake of the rearmost ship of the

main body, thus sacrificing whatever advantage might have accrued from being slightly further to windward, instead of trailing along at the rear of the fleet. This was certainly not Villeneuve's intention and sometime between 11.30am and noon he signalled the Squadron of Observation to keep the wind; Gravina repeated the signal later. Had this signal been promptly and properly obeyed or had Gravina held his earlier windward position immediately after the whole fleet wore, he might have changed the course of the battle either by intercepting Collingwood's advance or doubling on his division later. As it was, the signal was never properly executed. Very soon the Squadron of Observation found themselves acting as if they were really the rear of the Combined Fleet and, therefore, the special object of attack by Collingwood's lee division. Admittedly it was very difficult for them in view of the light wind and heavy swell, both to come to the wind and to form in proper order. But here again it looks as if Collingwood's precipitate advance, in company with Nelson's, though slow enough in terms of absolute speed, was swift enough to catch Gravina's ships while still in disarray.

One result of Gravina's action, or lack of action, was that from the British ships it looked as if his squadron was the rear of the Combined Fleet, as in fact it had now become. Since, however, Villeneuve was stationed in the centre of the *corps de bataille* of the main fleet, it was difficult for the British to spot where he was.

The crowding and reduplication of the enemy's line, the result of general disorder and poor station keeping, was highly favourable for meeting Nelson's attack. The fact that Villeneuve had a good idea of the form the attack would take might suggest that he was actually trying to achieve this particular formation. This assumption, however, is mistaken. Though no doubt recognising, in theory, the value of a partially reduplicated line as a purely defensive formation, in

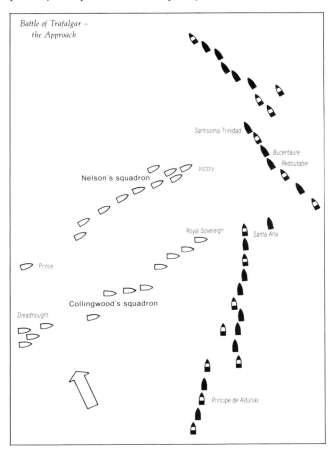

Battle of Trafalgar – the Approach

Santíssima Trinidad

Bucentaure
Redoutable

Nelson's squadron

Victory

Royal Sovereign

Santa Ana

Prince

Collingwood's squadron

Dreadnought

Principe de Asturias

practice he sought no more than a single line, albeit well closed up. As for initiating anything positive, he remained as calmly and sacrificially inactive as Villaret de Joyeuse had been at the First of June. Only when firing had already begun and it was almost too late, did he make any effort to improve his position. Being at last convinced that the *Victory* was heading either for the *Bucentaure* or the *Santissima Trinidad* and that his van was to be ignored, Villeneuve signalled 'every ship which by her present position was not engaging, to take any such steps as would bring her as promptly as possible into action'. This was a general signal. It was not addressed to any particular ship, division or squadron and it included no order as to how such steps should be taken. In Nelson's hands it would of course have expressed a requirement which his captains and flag officers both recognised and were able to meet. To address a signal of tactical sophistication to officers who were not able to obey it in a sophisticated manner was merely a self-justifying gesture. Villeneuve of course meant his signal to apply to Gravina's Squadron of Observation as well as to Dumanoir's van. Yet it seems extraordinary that, with the French tradition of relying on detailed signals as a means of ordering the battle, he should have failed to signal specific orders to particular flag officers. In the absence of signals to squadron and division commanders, captains of individual ships felt no personal obligation, and there was complete tactical paralysis.

By the time the battle began the Combined Fleet had assumed that crescent form so well known from the many plans of the battle. The van was hugging the wind and therefore slowing up. In the centre, the line showed a considerable sag to leeward, due partly to general disorder arising from the light wind and heavy swell, and partly from some ships having borne away slightly in trying to reach the wake of the ships ahead and so conform to the order of sailing arising from the wearing movement. The rear division by now was mixed up with Gravina's Squadron of Observation. Here, however, the ships were in better order, apart from three Spaniards well to leeward, but they were still steering east of north as against the van's west of north, thus

Vice-Admiral Cuthbert Collingwood (1750–1810), first Baron Collingwood, engraving by C Turner. NMM

completing the tail of the crescent. The rear ships also seemed to be trying to reach the wake of the ships ahead of them and were being slightly crowded by a local breeze from the south-west. The line as a whole with the distances between ships properly maintained, should have measured 6400 metres, but it is thought to have measured only 5 kilometres.

Nelson's attack

At 6am on the morning of 21 October, Nelson for the first time sighted the fleet he had been attempting to engage ever since he took command of the squadron watching Toulon in July 1803. In a very few hours, two years of planning was to be put to the test. He at once hoisted signal number 72, 'Form the order of sailing in two columns'. Almost immediately after he hoisted signal number 76, 'When lying to, or sailing by the wind, to bear up and sail large, on the course steered by the admiral, or that pointed out by signal', accompanied by the ENE compass signal. At that moment the fleet was scattered about in no particular order, but there were enough ships sufficiently close together for the columns to start forming.

Subtleties of interpretation based on the instructions (p101, Art 4, and p132, Art 14) dealing with these signals, seem entirely misplaced. To understand Nelson's tactics, one must understand not only the exact technical implications of his signals but also the immediate conventions governing their application. The general intention of the signals, as understood by the captains, is what really matters. Careful study seems to suggest that the problems of interpretation seen by some historians are illusory. Every captain knew that these signals indicated an immediate move towards the enemy. Even Captain Edward Codrington of the *Orion*, in Nelson's column, who still imagined himself to be part of a non-existent advance squadron and, therefore, assigned to the support of Collingwood's lee column, showed no hesitation in 'setting all sail' in conformity with Nelson. With the *Victory* under all sail, including 'steering sails' (studding or stunsails), how could it be expected that any careful alignment of ships could have been achieved, even supposing the ships took station 'as convenient'? Not only was there no time in which to form up in a regularly spaced line, but with all sail set, the heavy sailers found themselves at a still further disadvantage and with no means of catching up. Every captain sensed at once that Nelson was following the exact words of his Memorandum 'to set all their sails, even steering-sails in order to get as quickly as possible to the enemy's line'. Even so, and with a quartering wind, their progress was still slow.

By now the Combined Fleet could be seen in their original order, trying to form a line on the starboard tack, heading south. To do this, some ships had had to wear, which gave Nelson the temporary impression that the whole fleet was in the process of wearing to the north. After signalling to prepare for battle, Nelson again hoisted number 76 with the east course compass flag, though according to the Master's Log of the *Victory*, Nelson's own course was east by north. His plan, for Collingwood to overwhelm the enemy's rear while he himself cut through at about their thirteenth or fourteenth ship, counting from the van, could not possibly be carried out unless the two squadrons interchanged stations. Even so, there was no advance squadron left with which to reinforce either column. When, therefore, Villeneuve signalled the Combined Fleet to wear together and get on the larboard tack, heading roughly north, the difficulty solved itself. Nelson, without even referring to this convenient regularisation of his and Collingwood's respective roles, only showed concern lest the enemy should retire into Cadiz and escape him altogether. So the two British columns continued to advance as before.

Somewhere between 7am and 8am Nelson had signalled to the *Britannia*, in his own column, and the *Prince* and *Dreadnought* in Collingwood's, number 265, 'The ships may take such stations as are most convenient at the time, without regard to any established order of sailing'. At about 8.40 Nelson also made signal number 76 to the *Prince* only. With the exception of the *Africa*, which had missed the signal for the fleet to wear, made during the night, and was now a considerable distance to windward, Nelson's column was roughly in the last order of sailing which he is assumed to have issued. To conform with his assumed order of battle, the *Victory* should now have shortened sail to allow the *Témér-aire* to take station at the head of the column or line of battle ahead, as it was now becoming. Nelson, however, refused to give way. It would have delayed his attack.

Meanwhile Collingwood's division was proceeding on a course roughly parallel with Nelson's, in what the Admiralty Committee aptly described as an 'irregular line ahead'. Collingwood himself was leading it in the *Royal Sovereign*, with six or seven ships astern of his in reasonable order, though not in the prescribed order of sailing. The remainder were some distance in rear. At about 8.45 he signalled number 50 'For the ships of the fleet [in fact the so-called 'larboard division' [28]] to be kept on the larboard line of bearing from each other, though on the starboard tack.' This of course was a mistake as they were not on the starboard tack. Five minutes later it was replaced by number 42, 'Form the larboard line of bearing, steering the course indicated'. There has been intense controversy as regards the extent to which this order showed Collingwood executing or failing to execute the Memorandum in relation to the ships he intended to engage. Desbrière, moreover, raised the more subtle point as to what the larboard line of bearing, in this case, really meant. The Admiralty Committee carefully avoided discussing the point. The difficulty lies in interpreting the instructions, still reminiscent of Howe, related to the signals.[29]

Rear-Admiral Taylor, in his more recent account of the battle, shows with admirable clarity that, if and when this line of bearing was formed, the ships of Collingwood's division would still have been on their original easterly course, the same as those of the *Royal Sovereign* and the *Victory*.[30] They would have been so disposed that, if signalled to haul to the wind, they would have placed themselves in a line ahead, seven points off. With the wind then west, such a line of bearing would be north by west and south by east.

Why did Collingwood make this signal? The simple answer seems to be that he thought it would enable his ships to get quicker into action with Gravina's Squadron of Observation, which was sufficiently merged with the rear of the Combined Fleet to make it seem to be in fact the real rear. The signal meant that the ships following Collingwood should have steered away to starboard and crowded sail sufficiently to bring them up level with him on a line north by west and south by east, with their bows still pointing east. This was difficult even for the best sailers unless the *Royal Sovereign* shortened sail, which Collingwood had no intention of doing. Not only would they have had to increase their speed through the water very considerably, but each ship in

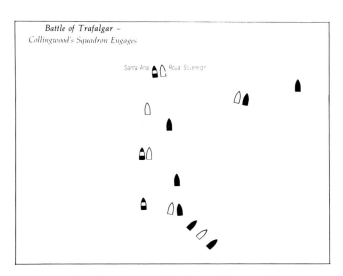

Battle of Trafalgar –
Collingwood's Squadron Engages

turn towards the rear would have had to sail increased distances, in order to make the double divergence necessary for reaching the prescribed station. This was always the difficulty which faced ships when forming on a line of bearing or changing the line of bearing while preserving the original course. It is true that Collingwood had not yet set his studding sails, but the amount of sail he already carried, as well as the lightness of the wind and the distance of his sternmost ships, contributed to making the signal almost impossible to execute. Nevertheless, its general intention, to direct the ships astern of him as quickly as possible towards the ships they were to engage, seems clear enough.

In order to try to achieve a more simultaneous impact, Collingwood followed up his line of bearing signal with number 88, 'Make more sail', and this was afterwards repeated to three particular ships. Owing to the length of the enemy's line once Gravina's squadron was joined with it, Collingwood could hardly use his fifteen ships to overwhelm the rearmost twelve, while at the same time leaving Nelson's twelve to deal with the remaining twenty-one. Perhaps also the sight of the 120-gun *Santa Ana*, now sixteenth from the rear in a very irregular line, tempted him to engage her, especially as the *Prince* and the *Dreadnought*, his only other three-deckers beside the *Royal Sovereign*, were right out of the picture. The *Principe de Asturias*, Gravina's 112-gun flagship, was rearmost ship but two of the whole fleet and had to be left to the rear ships of Collingwood's column.

As the morning advanced, Nelson and Collingwood urged on particular ships of their respective columns by appropriate signals, including number 307, 'Make all possible sail with safety to the masts'. By this time, however, Collingwood had set his studding sails, and there was no catching him. Just on noon, ships in the Combined Fleet opened fire on the *Royal Sovereign* and the three ships astern of her, at a range of about 1000 yards. Their progress was quite unchecked and Collingwood began the action on the British side by taking the *Royal Sovereign* through the enemy's line, then turning to larboard and engaging the *Santa Ana* with yardarms touching. Each successive ship in the lee division pushed through where she could, following the same tactics – that is, raking the bow and stern of the ships flanking the respective gaps, and then closing in to the nearest ship she could find on the lee side of the enemy's line. Two or three enemy ships were left unengaged at first, though they suffered later as the sternmost British ships arrived, nearly splitting their masts with the effort but at what today would appear to be an agonisingly slow speed of about two and a half knots. Gradually they built up a superior concentration of force against the enemy's extreme rear.

28 This misleading expression is one of status and not of navigation. The commander-in-chief was deemed to be in command of the starboard or weather division, his second-in-command having the larboard or lee division. These terms held despite reversals of wind and position.plates

29 Signal Book of 1799, pp149–50, Article 4plates

30 *The Mariner's Mirror*, vol 36 (1950), no 4plates

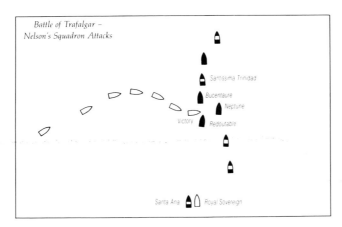

Battle of Trafalgar –
Nelson's Squadron Attacks

Santissima Trinidad
Bucentaure
Neptune
Victory   Redoutable

Santa Ana   Royal Sovereign

A little before 8am Nelson signalled for his four frigate captains to come on board the *Victory*. Meanwhile, he held on his course followed by his division, steering for what seemed to be roughly the centre of the whole Combined fleet, Gravina included. This was also the point of greatest concavity in the enemy line and where there was still a gap between the *Santa Ana* and the four ships ahead of her, all of which had fallen considerably to leeward. Ahead of these were the *Redoutable* (74), the *Bucentaure* (80), flying Villeneuve's flag, and the *Santissima Trinidad* (120), flying the flag of Rear-Admiral Don Báltasar Cisneros. The Combined Fleet had steerage way of about half a knot and if Nelson intended to cut through between the thirteenth and fourteenth ship, the place was between the *Redoutable* and her next astern, but actually he seemed to be steering for the *Santissima Trinidad*, the four-decker being the obvious big ship to attack in that part of the enemy's line. Apart from signals to individual ships to make more sail and get into station, even including one for the *Mars* to reach her station at the head of Collingwood's line (without, of course, hope of response), Nelson gave no further orders. At 10.50am he 'telegraphed' to Collingwood, 'I intend to pass or go through the enemy's line to prevent them getting into Cadiz.' This was purely informative but shows the way his mind was working.

As the *Victory* came within random gunshot, Nelson dismissed his frigate captains, though Blackwood remained a little longer. They were to hail the ships astern, telling them it was Nelson's intention 'to cut through the enemy's line about the thirteenth or fourteenth ship, and then to make sail on the larboard tack for their van'.

Nelson appears to have had three courses of action in mind. He could do as he told his frigate captains. This, however, would not necessarily enable him to attack and overwhelm the enemy's commander-in-chief. With as yet no flags showing, where could he be, now that Gravina's junction made the true centre of the enemy's line in doubt? Nelson could, on the other hand, attack the *Santissima Trinidad*, and so at least make certain of a flag officer. Still a little uncertain, he had the *Victory* turned slightly to larboard as if to attack the enemy's van. There seems no doubt about this turn. Was he looking for Villeneuve, or a good place to cut through, or was it a feint? Although the newly discovered signal with the yellow and blue flag was never made, we cannot today, in the light of this new evidence, entirely dismiss the possibility that some notion of a feint entered into his calculations. At this point the nearest ships of the Combined Fleet opened fire with full broadsides, all flags were hoisted, and the commander-in-chief was at last revealed in the *Bucentaure*, twelfth

in the line. Nelson at once had the *Victory* turned to starboard and ran back along the enemy's line under heavy fire. The wind had almost dropped and he could hardly have been doing two knots. Angling to pass astern of *Santissima Trinidad* and Villeneuve, he was foiled by the *Bucentaure* being brought up to the wind on the *Santissima Trinidad's* larboard quarter. At the same time, Captain Lucas brought the *Redoubtable* up astern of the *Bucentaure*. The *Victory* bashed through between them, but the immediate attempt of the following British ships to push through in succession was defeated. Ships became locked together and drifted to leeward, so that the original gap ahead of the *Santa Ana* opened wider. The remaining ships of Nelson's division were brought into action with great skill wherever there seemed a disengaged enemy to attack or a British ship to be relieved. Codrington was not far wrong in saying that they 'scrambled' into action as best they could. Nelson had succeeded in bringing the leading ships of both divisions to a very close action on the lee side of the enemy without having a single one disabled during the approach, the speed of which – even in case of the fastest sailer – never exceeded three knots. The pell-mell was now joined and just as Nelson guessed, planned and hoped, the enemy's van failed completely to help their centre and rear.

The prolonged controversy at the beginning of the present century about the tactics actually pursued by Nelson from dawn on 21 October until about 2.30pm, when most of the British ships were engaged, tended to centre on a question that had already been answered. The question was in what order did the two columns attack and how did this fit in with Nelson's Memorandum of 9 October? The answer had already been given by Colonel Edouard Desbrière in his monumental and exhaustive work, *La Campagne Maritime de 1805: Trafalgar*, published under the direction of the historical section of the French General Staff in 1907. Unfortunately, it was not read in England, though Sir Julian Corbett used it as his main authority for the French and Spanish side when writing *The Campaign of Trafalgar*, published in 1910. By 1912 the controversy was still so much alive that the Admiralty appointed a committee 'for the purpose of thoroughly examining and considering the whole of the evidence relative to the tactics employed by Nelson at the Battle of Trafalgar'. It took this committee fifteen months to produce their report, which provided the British counterpart of what Desbrière had already discovered and it in no way modified his main conclusions.[31] Their findings, moreover (though there was no such admission) confirmed the work of Sir Julian Corbett. The secretary to the committee, William Gordon Perrin, then Admiralty librarian, supplied a schedule of all the British signals made, just as he had done for Corbett's book three years previously.

The British fleet never formed properly in the technical sense of the word. Forming meant not only getting the ships into the right order but also at the standard distance from each other. To do this, it was necessary for ships already in their stations not to set so much sail that the ships out of station had no chance to gain it. Once formed, it was also necessary to avoid crowding sail so much that the heavy sailers were unable to keep up by crowding sail themselves. None of these considerations seemed to matter much at Trafalgar. All ships crowded sail. Ships at a distance had no chance to get into station, and of those the heavy sailers found themselves dropping further and further astern. As Nelson saw it, the most important thing was for all ships to reach the enemy as quickly as possible.

The British fleet triumphed because, once the pell-mell became general, the gunnery of the individual ships was better than that of the individual ships of the Combined Fleet.

31 Cd. 7120, 1913

In the pell-mell, moreover, the British captains were better at handling their ships so as to give mutual support, and in making such movements as were possible in order to bring superior fire to bear at angles from which their opponents could not so effectively reply. As far as bravery was concerned, there was little to choose between the two fleets. Apart from ships which sank with their crews still on board, the Combined Fleet suffered enormous casualties. The *Redoubtable* had 490 men killed, the *Bucentaure* 197, the *Santissima Trinidad* 216. Twenty-six flag-officers and captains were either killed or wounded. These are terrible figures. Many ships refused to surrender until either there were no men left capable of firing guns or no guns left capable of being fired. The *Victory* had fifty-seven men killed, the highest figure for any British ship.

The following propositions seem worth considering. If Nelson had realised that when Villeneuve wore the Combined Fleet towards Cadiz he was not intending to return to the port and take refuge, there would have been more time for the British to form. Furthermore, had both divisions of the British fleet been better formed, the impact on the Combined Fleet would have been greater and the victory more complete. It can be argued, however, that by launching his attack at once, slow though his advance was bound to be, Nelson forced the Combined Fleet out of their stride and paralysed Villeneuve's capacity for initiative. More particularly, Nelson's immediate advance prevented Villeneuve from taking preliminary steps to start an enveloping movement, either by tacking his van or by using Gravina's Squadron of Observation against Collingwood's lee division. It becomes a question, therefore, of how much was lost and how much was gained by Nelson's precipitate action. Opinions about this will always tend to reflect the personal temperament of those who give them.

Strachan and Dumanoir

In a brief action on 4 November 1805 Sir Richard Strachan, with an 80, three 74s and four frigates, captured Rear-Admiral Dumanoir's squadron of four French ships surviving from Trafalgar. The French squadron also, on paper, comprised an 80 and three 74s, but with no frigates. Dumanoir had been obliged to jettison twenty guns to keep the *Formidable* afloat. He was being slowly overhauled, though still fairly well ahead, when he was forced to haul to the wind and form in line of battle. This, surprising as it may seem, was because the two leading British frigates, ahead of the rest of the squadron, had gained enough distance to begin firing broadsides into his sternmost ship.

Strachan's fourth ship, the *Namur*, was still a good way astern, so he hailed the captains of the two ships already up with him to say that he intended to attack the enemy's rear. Strachan himself took the lead in the *Caesar* (80), and, being to windward, brought to opposite Dumanoir, who was second in the French line. Thus the main action began with these three British ships-of-the-line engaging the three rear French ships, the leading French ship being so far left alone. Meanwhile, all four British frigates, ignoring custom and tradition, came up on both quarters of the rearmost French ship and joined in the battle. Dumanoir signalled his squadron to tack in succession, so as to relieve the pressure on his rear and try to cut off the *Namur*, now approaching. It was a very bold manoeuvre and splendidly executed in the face of broadsides at close range. Strachan also tried to tack but had to wear, and this produced temporary confusion. With his three ships round, and with the *Namur* now into place as second in his reversed line, he soon began to get the upper hand. By this time, the frigates were engaging the

French on the leeward side of their line, thus putting them between two fires. After nearly four hours of fighting all four French ships surrendered, Dumanoir's *Formidable* being in danger of sinking. Only two masts out of a total of twelve remained standing in the French ships at the end. British casualties were twenty-four killed and 116 wounded.

Duckworth and Leissègues

After a remarkable series of strategic movements which took him from the post-Trafalgar blockade of Cadiz to the West Indies, Sir John Duckworth in the *Superb* (74), under Captain Richard Keats, found himself on the morning of 6 February 1806 in sight of a French squadron at San Domingo. He had besides an 80, four more 74s, a 64 and three frigates. His squadron included Rear-Admiral the Hon Alexander Cochrane and Rear-Admiral Thomas Louis. The French squadron, which was ready for sea, was commanded by Vice-Admiral Corentin Leissègues in the 120-gun *Impérial*; he had with him an 80, three 74s and three frigates. On sighting an enemy force to windward, Leissègues made sail to the west and formed in line ahead with the *Impérial* second and with the frigates in line ahead inshore of him. Duckworth immediately chased and with the wind almost directly astern of him, gained rapidly. Dividing his force as he went, he took four ships with him, including Cochrane's, against the French admiral and ordered Louis to attack the French rear with the remaining three ships.

Although the sailing powers of the British ships varied considerably, they were all in action in the end, despite the surprising statement that the French were running at 8 knots. As the *Superb* began to engage (with the band playing God Save the King on the poop), Keats hung a portrait of Nelson on the mizzen stay, where it remained untouched throughout the battle, though splattered with the blood and brains of a seaman killed close beside it. Smoke soon covered the scene, yet despite confusion and collisions with each other, the

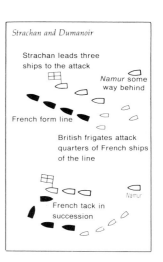

Admiral Sir John Thomas Duckworth (1748–1817), artist unknown, in the style of Sir William Beechey. NMM

Duckworth's action off San Domingo, 6 February 1806, by Nicholas Pocock (dated 1806). NMM

British ships fought with great energy and skill. Only the French frigates escaped; three of the line were taken and the *Impérial* and another wrecked.[32]

British tactical memoranda

Even before Trafalgar, the character of the war at sea had begun to change. Between the recommencement of the war in 1803 and its conclusion in 1814, Trafalgar was the only great naval battle. Apart from Strachan's defeat of Dumanoir, almost a part of Trafalgar itself, and Duckworth's overwhelming victory at San Domingo, tactical encounters by battle fleets were at an end. Shore bombardments, combined operations, attacks on ships at anchor and the constant watch on blockaded forces, absorbed the activities of substantial British fleets. However, the great efforts Napoleon made to acquire a new fleet with which to challenge the grip Britain had gained on the sea ensured that efficiency in battle tactics remained vital. There were at least three important British tactical documents issued for operations which in fact never took place.

Admiral Gambier, when proceeding to the Baltic in 1807, issued instructions for a fleet of thirty of the line. The orders of sailing and of battle were the same, with the fleet in three squadrons of ten ships each, and each squadron in two divisions of five ships each. Gambier place himself in the centre of the fleet, but the commanders of each subdivision of the van and rear were stationed in such a manner as to lead his own subdivision when his own squadron led the fleet. The effect was to strengthen the extremities, rather in the manner of the seventeenth century.

The commodores were to command their own subdivisions 'subject to the general direction' of their squadron commanders. The distance between ships was to be two cables, 'increasing that distance according to the state of the weather', no doubt a precaution for North Sea conditions. When sailing in two columns, which were referred to as grand divisions, the leading ship of the starboard (windward) division was to keep the admiral two points on her weather bow. The leading ship of the lee division was to keep the leader of the weather column two points before her beam, 'when sailing large, abreast of her'. These tactics were never really tested, because the Danes, as in 1801, declined a fleet action.

Collingwood's General Order of 23 March 1808 was much more detailed and, superficially at least, more reminiscent of the ideas of Nelson.

Should the enemy be found formed in order of battle with his whole force, I shall notwithstanding probably not make the signal to form the line of battle; but, keeping in the closest order, with the van squadron attack the van of the enemy, while the commander of the lee division takes the proper measures, and makes to the ships of his division the necessary signals for commencing the action with enemy's rear, as nearly as possible at the same time that the van begins. Of his signals, therefore, the captains of that division will be particularly careful.

If the squadron has to run to leeward to close with [the] enemy, the signal will be made to alter the course together, the van division keeping a point or two more away than the lee, the latter carrying less sail; and when the fleet draws near the enemy both columns are to preserve a line as nearly parallel to the hostile fleet as they can.

In standing up to the enemy from the leeward upon a contrary tack the lee line is to press sail, so that the leading ship of that line may be two or three points before the beam of the leading ship of the weather line, which will bring them to action nearly at the same period.

The leading ship of the weather column will endeavour to pass through the enemy's line, should the weather be such as to make that practicable, at one fourth from the van, whatever number of ships their line may be composed of. The lee division will pass through at a ship or two astern of their centre, and whenever a ship has weathered the enemy it will be found necessary to shorten sail as much as possible for her second astern to close her, and to keep away, steering in a line parallel to the enemy's and engaging them on their weather side.

A movement of this kind may be necessary, but, considering the difficulty of altering the position of the fleet during the time of combat, every endeavour will be made to commence battle with the enemy on the same tack they are; and I have only to recommend and direct that they be fought with at the nearest distance possible, in which getting on board of them may be avoided, which is always disadvantageous to us except when they are flying.

As Corbett pointed out, Gambier and Collingwood appear to have over-reacted to Nelson's bold tactics, which had counted upon the tactical incapacity of the Franco-Spanish fleet. The 'Nelson Touch' had come to be seen as a formula calling for swift attack, dual organisation and independent divisional control. No direction was given about concentration of force, or about breaking the line in one place or all places. There remained no vestige of the use of an advance squadron to contain part of the enemy force. The focus of attack on the enemy van contradicted centuries of experience that tactical necessity obliged the attacking fleet to seek to isolate the enemy rear. In the case of the attack from leeward, 'one cannot help wondering how far the leading ships after passing the line would have been able to lead down it [the enemy's line] before they were disabled'. However, 'the addition is interesting as the first known direction as to what was to be done after breaking the line in line ahead after Rodney's method.'[34]

A more constructive, if over-elaborate, use of the Nelson approach was made by the Hon Sir Alexander Cochrane who, while commander-in-chief on the Leeward Islands station, issued a Memorandum on 'Modes of Attack from the

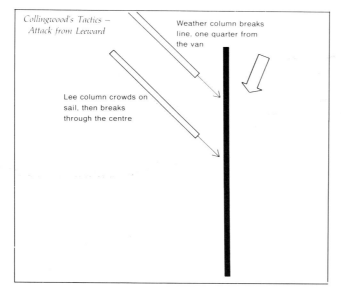

Collingwood's Tactics – Attack from Leeward

Weather column breaks line, one quarter from the van

Lee column crowds on sail, then breaks through the centre

32 A very detailed tactical plan of the action showing the track of all the British ships appears in the ship's log of the *Donegal*, which was sold at Hodgsons on 14 December 1962 (Lot 595).

33 G L Newnham, *A Selection from the Public and Private Correspondence of Vice-Admiral Lord Collingwood*, Collingwood, 1828, pp313–4

34 Corbett, *Fighting Instructions*, pp324–5

Windward, etc'.[35] Corbett saw in it the influence of Rodney, Clerk of Eldin and Morogues, although 'his excellent device for inverting the line after passing through the enemy's fleet' was Cochrane's own idea.

When an attack is intended to be made upon the enemy's rear, so as to endeavour to cut off a certain number of ships from that part of their fleet, the same will be made known by signal no 27,[36] and the numeral signal which accompanies it will point out the headmost of the enemy's ships that is to be attacked, counting always from the van, as stated in page 160, Article 31 (Instructions).[37] The signal will afterwards be made for the division intended to make the attack, or the same will be signalled by the ship's pennants, and the pennants of the ship in that division which is to begin the attack, with the number of the ship to be first attacked in the enemy's line. Should it be intended that the leading ship in the division is to attack the rear ship of the enemy, she must bear up, so as to get upon the weather quarter of that ship; the ships following her in the line will pass in succession on her weather quarter, giving their fire to the ship she is engaged with; and so on in succession until they have closed with the headmost ship intended to be attacked.

The ships in reserve, who have no opponents, will break through the enemy's line ahead of this ship, so as to cut off the ships engaged from the rest of the enemy's fleet.

When it is intended that the rear ship of the division shall attack the rear ship of the enemy's line, that ship's pennants will be shown; the rest of the ships in the division will invert their order, shortening sail until they can in succession follow the rear ship, giving their fire to the enemy's ships in like manner as above stated; and the reserve ships will cut through the enemy's line as already mentioned.

When this mode of attack is intended to be put in force, the other divisions of the fleet, whether in order of sailing or battle, will keep to windward just out of gunshot, so as to be ready to support the rear, and prevent the van and centre of the enemy from doubling upon them. This manoeuvre, if properly executed, may force the enemy to abandon the ships on his rear, or submit to be brought to action in equal terms, which is difficult to be obtained when the attack is made from windward.

When the fleet is to leeward, and the commanding officer intends to cut through the enemy's line, the number of the ship in their line where the attempt is to be made will be shown as already stated.

If the ships after passing the enemy's line are to tack, and double upon the enemy's ships ahead, the same will be made known by a blue pennant over the Signal 27;[38] if not, they are to bear up and run to the enemy's line to windward, engaging the ship they first meet with; each succeeding ship giving her fire, and passing on to the next in the rear. The ships destined to attack the enemy's rear will be pointed out by the number of the last ship in the line that is to make this movement, or the pennants of that ship will be shown; but, should no signal be made, it is to be understood that the number of ships to bear up is equal in number to the enemy's ships that have been cut off; the succeeding ships will attack and pursue the van of the enemy, or form, should it be necessary to prevent the enemy's van from passing round the rear of the fleet to relieve or join their cut-off ships.

If it is intended that the ships following those destined to engage the enemy's rear to windward shall bear up, and prevent the part of their rear which has been cut off from escaping to leeward, the same will be made known by a red pennant being hoisted over the Signal 21,[39] and the number of

Below and overleaf:
The title page and heavily corrected copy of Admiral Sir Home Popham's telegraphic signals, used when adapting his system for official use. MOD

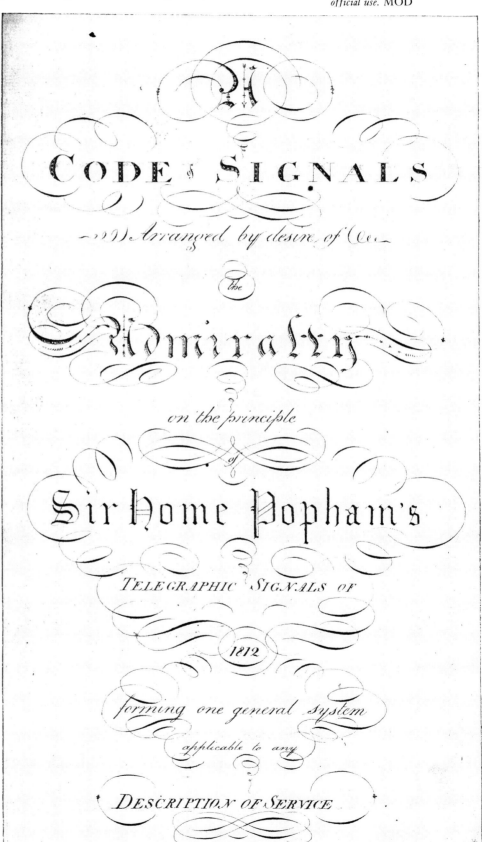

CODE of SIGNALS

Arranged by desire of

the

Admiralty

on the principle

of

Sir Home Popham's

TELEGRAPHIC SIGNALS OF

1812

forming one general system

applicable to any

DESCRIPTION OF SERVICE

---

35 Ekins, *Naval Battles*, pp394–7, and reprinted in Corbett, *Fighting Instructions*, pp330–4

36 'Break through the Enemy's line, in all parts, where it is practicable and engage on the other side . . .'

37 Variations of the attack with explanations.

38 Collingwood's own variation of the signal.

39 'Attack the enemy on the bearing indicated; or any number of their ships of war separated from the main Body of their fleet.'

ships so ordered will be shown by numeral signals or pennants. If from the centre division, a white pennant will be hoisted over the signal.

If the rear ships are to perform this service by bearing up, the same will be made known by a red pennant under. The numeral signal or pennants, counting always from the van, will show the headmost ship to proceed on this service. The ships not directed by those signals are to form in close order, to cover the ships engaged from the rest of the enemy's fleet.

When the enemy's ships are to be engaged by both van and centre, the rear will keep their wind, to cover upon which, each ship will luff up upon the weather quarter of her second

ahead, so as to leave no opening for the leading ship of the enemy to pass through: this movement will expose them to the collected fire of all that part of the fleet they intend to force.

What seems surprising today is that Collingwood and Cochrane, and even Nelson, could issue lengthy tactical memoranda in such rambling and chatty form. The fact that these documents deal with specific methods of attack, unsuitable, possibly, for issue as additional instructions, cannot absolve their respective authors from failure to give them businesslike form with numbered paragraphs, arranged in logical sequence. None seem to have had the help of an appointed captain of the fleet. Even Howe, with all his rambling obscurities, never issued such verbose and incomprehensible tactical directions as those drafted by some of his successors.

The Admiralty adopts Popham's system

On 23 April 1812 the Admiralty appointed a committee headed by Admiral Gambier, to examine Popham's telegraphic signals and see whether the book might officially be adopted. The other members of the committee were Admiral James Whitshed, Admiral Sir John Duckworth, Vice-Admiral George Martin, Rear-Admiral William Johnstone Hope, Rear-Admiral Lord Henry Paulet and Rear-Admiral Sir Henry Hotham. With the exception of Gambier and Duckworth, none was aged over fifty. Only Hope was at that moment a member of the Board. As a group, they represented a good cross-section of recent fighting and administrative experience. Gambier, it would seem, was appointed because of his work on the Signal Book of 1799.

The committee took only six days to make a favourable report.[41] They were all familiar with Popham's code, so that the whole business seems to have been a formality. As a result, a printed notice, dated 'Admiralty Office 13 September 1813', came to be pasted in the cover of Popham's *Telegraphic Signals* of 1812, stating that it was available for issue to captains and commanders, who were to treat it as a signal book.[42]

Popham had been careful to arrange his signal flags so that as much use as possible could be made of the flags in the Admiralty Signal Book of 1811 (1799 revised). Nevertheless, the official adoption of his telegraphic system involved the use of two sets of flags, counting the various guidons, cornet and pendants he employed. Furthermore it involved the use of two quite separate systems of signalling, Popham using both numerical and alphabetical flags quite apart from his system of spelling out words letter by letter. It had been comparatively easy for Nelson's fleet to acquire the flags necessary for Popham's telegraphic signals of 1803, but the official adoption of his far more elaborate book of 1812 faced the fleets at sea with a major signalling problem. A move had to be made to bring the two systems together, which could only mean applying Popham's telegraphic system as a whole to the Admiralty Signal Book. By this means it would be possible to have one system and one set of flags only. Apart from this, it would mean that the number of signals in the Admiralty book could be greatly increased without risk of

40 'Close nearer to the squadron division, or ships whose signal will be shewn after this signal has been answered.'

41 *MOD*, Ec/51

42 *MOD*, Ec/104 with the notice; *NMM*, Tunstall Collection, TUN/57 (formerly S/P/R/4) with the notice pasted over. *MOD*, Ec/145 contains another copy of the report together with an undated printed pamphlet by Popham entitled *Observations on Signal Flags and the best mode of applying them.*

confusion. In addition, by the hoisting of a single flag to indicate telegraphic signalling, the flow of naval speech could continue in exactly the same form, simply switching over to the larger vocabulary. Far from being an assimilation by the Admiralty, the adoption was a takeover by Popham.

The main step in the process was the drafting of 'A Code of Signals arranged by desire of the Admiralty on the principle of Sir Home Popham's Telegraphic Signals of 1812, forming one general system applicable to any description of service'.[43] The authorship of this important manuscript, now in the Naval Library, is uncertain. It may have been the work of the Admiralty committee, though internal evidence suggests otherwise. It can hardly have been Popham's, being far too long-winded. The battle signals read like the work of one or more persons intent on including under the bare title of the signal every scrap of explanatory and hortatory matter that could in any possible way be deemed relevant. The draft is heavily revised in pencil, that part left unerased or merely corrected forming word for word the contents of the new Admiralty Signal Book of 1816. The manuscript is thus proved to be the original draft and final revision of that work.

From what is included, and more particularly from what is excluded, it is possible to understand the kind of thinking that the eventual printed signal book represents. It would be unwise to treat the outright rejection of so many signals, and the exclusion of about three-quarters of the explanatory material of those actually adopted, as a sign of strongly opposed thinking. Possibly the author or authors of the draft were encouraged to write in full, so that the reviser or revisers would be saved the work of redrafting, needing only to accept or reject. The absence of all rewriting, apart from mere shortening and detailed correction, is one of the interesting features of the revision. One hundred and fifteen battle signals were proposed in the draft, of which sixty-three were finally adopted. Of these, only fifty appear in the battle signals section of the Signal Book of 1816, the others being introduced elsewhere; signals adopted from elsewhere in the draft are brought in to make a total of sixty-eight battle signals finally printed.

An auxiliary manuscript, loosely inserted, is entitled 'Subjects on which doubts have arisen during the revision and arrangement of the evolutionary signals for the decision of the Admiralty'. Those are mainly concerned with that old source of confusion, the line of bearing. The main draft includes a copy of the committee's letter of 29 April 1812, approving the adoption of Popham's Telegraphic Signals and a copy of a letter commending it, written by Vice-Admiral Sir Samuel Hood when commanding in the East Indies and dated 9 January 1814.[44] The draft also contains a note cautioning against the inclusion in a signal book of the sort of material found in Morogues's book, in which he proposed 106 evolutions for the exercise of fleets at sea, accompanied by 136 plates. It is a short course of naval mathematics, far exceeding the limits which ought to be prescribed to any signal book or from which any practical advantage can possibly be derived.'

Although the draft has much numbering and re-numbering both in ink and pencil, it is clear from the painted flags at the head of each page that the draft signals, when finally agreed, would be made according to Popham's system. The real task of the authors of the new signal book was to redraft the actual content of the signals and instructions – and it seemed that Popham himself had little in this. Sir Julian Corbett sensed Lord Keith's retrogressive influence at work, exercised through his ex-flag captain, Sir Graham Moore, who joined the Board on 24 May 1816,[45] but this seems unlikely, as the erasures were mainly of over-elaborate directions and speculative contingencies, rather than of a new ideas introduced by Nelson. On the question of lack of enterprise, it is clear that the original draft must have been begun before Sir Graham Moore took office.

The 1816 signal books

The final and printed form of the revision was entitled *A Book of Signals & Instructions for the use of His Majesty's Fleet: 1816*. With it, as a companion volume, was issued *Telegraphic Signals for the use of His Majesty's Fleet: 1816*. In each book the authorising statement by the Board of Admiralty begins: 'Whereas Rear-Admiral Sir Home Popham, KCB, has submitted to Us a system of Signals for the use of His Majesty's Fleet, which, with certain alterations, We have seen fit to adopt, you are hereby required & directed . . .'

The *Book of Signals* has only 216 pages, although it includes all the usual instructions as well as the fog signals and instructions, and the night signals and instructions. This is the first time that all types of signals, and all instructions bearing on them, appear between the covers of a single volume.[46] Yet there are far more signals in it than in the day and night books of 1799. The flags, guidons, cornets and pendants are the same as those in Popham's *Telegraphic Signals* of 1812, and in the same order and colours. In addition, there are three general-purpose flags and six pendants, as well as nine numbered 'distinguishing pendants'. No distinction is made between the admiral's signals and those made by private ships. The battle signals are at the beginning, using single flags, etc, for the first nineteen. A single-flag signal, using number 9 flag, is to 'Cut through the Enemy's Line'. The gloss explains how further signals will be made to indicate the point where the cut is to be made. Once the ships were through, unless previously directed by signal or in writing, they were 'to form in close order of battle on the other side [of the enemy's line]. The fleet is to push through with as much plain sail as possible, and in the closest and most compact order, so as effectually to separate their line, and render a rejunction of it impracticable.'

In the original draft the gloss was twice as long and introduced a good deal of complicated and hypothetical matter. Single flag B was to 'Break or cut through the enemy's line in all its parts and engage on the other side'. Ships should, if possible, pass under the stern of their respective opponents and engage on the other side; ships unable to conform were 'to act in the best manner circumstances will admit for the destruction of the enemy . . .' Lengthy explanatory matter in the original draft was deleted.

The tactics represented by the new set of signals were conservative, but Corbett was wrong to characterise them as retrograde. Nor should they be blamed for encouraging the reckless 'go at 'em' approach to tactics which undoubtedly grew up. Tactics breed countertactics. If the enemy are incapable of organising effective countertactics, irrespective of what they put down on paper, then the attackers will ignore skilful preparations and simply 'go at 'em'. Corbett

---

43 *MOD*, Ec/51. The whole document is in copperplate writing by the same hand throughout.

44 A cousin of Lord Hood's.

45 Corbett, *Fighting Instructions*, p336

46 This is not a rare work. The copy in the National Maritime Museum, Corbett Collection, TUN/54 (formerly S/P/R/1) is numbered in ink 'A2' and is inscribed 'For His Royal Highness the Admiral of the Fleet', the Duke of Clarence. It has the bookplate of Lord Adolphus Fitzclarence superimposed on that of his father, as Prince William Henry

quotes with approval the observation of an elderly French naval officer writing shortly after Trafalgar that Nelson's attack could have been defeated by a really firm double line 'with reserve squadrons on the wings stationed in such a manner as to bear down most easily upon the points too vigorously attacked'. He also approves of a Spanish tactician's observation that 'where two fleets are equally well trained, that which attacks in this [Nelson's] manner must be defeated'.[47] Given the incapacity of Napoleon's navies, the Admiralty were absolved from the need to work out counter-countertactics. Rightly, no doubt, they came to the conclusion that Nelsonian tactics were for Nelson only, and had no place in an official signal book.

The signal which particularly inspired Corbett's cricitism was class I signal 1.7, to 'cut through the enemy's line in the order of sailing in two columns'. From the detailed instruction and the accompanying diagram, it is easy to see that this was an attempt to regularise Nelson's attack at Trafalgar, and at the same time make it less risky. 'The Admiral will make known what number of ships [counting] from the rear at which the lee division is to break through their line.' The leading ship of each division, having broken through, would, unless otherwise directed, engage the ship astern of which she had just passed from leeward. The second ship in each division, and the rest in succession, would cut through one ahead of where their next ahead cut through, and engage from leeward the enemy ship astern of which they had just passed. Each ship was thus to break through one place ahead of its next ahead and so form an inverted line along the leeward side of the enemy. A special advantage of this procedure was that, as each fresh ship passed along the line seeking its place to break through, she could fire on those enemy ships already engaged from leeward by the ships of her own fleet which had already broken through. In this way the enemy were gradually and continuously doubled.

A great deal of attention was given to the preparation of this signal. An extremely long explanatory gloss in the draft had been entirely struck out, together with the diagram, and replaced with a new text and diagram. The working shows quite clearly that the attack was to be made on the enemy's centre or rear. The diagram, however, omitted to show the enemy van. In editing for the Navy Records Society his *Fighting Instructions*, Corbett thought this indicated an attack on the enemy's van and centre leaving their rear 'free to come into immediate action with an overwhelming concentration on the lee division'. He wrote, 'It is the last word of British sailing tactics, and surely nothing in their whole history, not even in the worst days of the old Fighting Instructions, so staggers us with its lack of tactical sense.' [48] Five years later he had seen his mistake and was able to take a more favourable view in his *Campaign of Trafalgar*.[49]

The original draft included a signal to 'cut through the enemy's line in the order of sailing on [sic] three columns', with a short explanation, the whole being erased.[50] Another original signal which did not survive, together with a long explanation and diagram, was to 'attack the enemy in the order of sailing on [sic] two columns to windward and to leeward'. 'This is a new evolution and it may in the first instance mislead the enemy as to its purpose.' The weather

division was to engage the enemy's van and centre by inverting the line along the enemy's windward side while the lee division did the same thing from the rear along the enemy's leeward side. No doubt it was rejected as being too complicated to bring off with any degree of certainty. Instead, the signal for Howe's attack on the enemy's rear by inverting the line was retained. His other device, of attacking the enemy's rear in succession, was also in the original draft, but was erased.

The experience of the Nile and possibly of Copenhagen is reflected in signals to 'anchor and engage the enemy', 'attack the enemy at anchor, passing between them and the land', 'attack the enemy at anchor, passing on the outside of them', 'anchor within the enemy and engage', and 'anchor on the outside of the enemy and engage'. Only the first of these was provided with a detailed explanation, representing about three-quarters of the original draft.

The ships are to have springs on their bower anchors, and the sheet cable taken through the stern port, to be prepared to anchor without winding, if they should go to the attack with the wind aft; the boats should be hoisted out, and hawsers coiled in the launches with the stream anchor ready to warp them into their stations or to assist other ships that may be in want of assistance.

No doubt this signal might be combined with one of the others.

Howe's signal, 'ships are to take stations for their mutual support and engage the enemy as they get up with them', was provided with a lengthy gloss which survived revision almost intact. It included the warning, 'but the leading ships must be very cautious not to suffer themselves to be drawn so far from the body of the fleet as to risk their being surrounded and cut off'.

A sign of the times was the group of signals dealing with the falling of 'shells'. There were also nineteen very good signals, clearly meant to be made by private ships reporting on various movements of the enemy. The thumb-indexing of the signal numbers included marginal notes on the subject matter. A scheme for distance signals in class 3 made use of a sixteen-sided table; the symbols to be flown included a ball, discovered by means of two further tables. Signals in the three main classes for which it was deemed necessary, had their distance signal numbers in a column next to the class number.

The instructions for the orders of sailing, with diagrams, were almost exactly the same as for 1799. The instructions for forming the line of battle omitted the diagrams and all the explanatory matter derived from the Hood-McArthur books. The battle instructions were reduced from thirty-one to nineteen, with no important differences. Ships' names in the list of the navy were signalled by the Union Jack over two other flags, etc. The fog signals and instructions were much the same as for 1799.

The night signals and instructions were also much the same as for 1799, though the night battle instructions were reduced from eleven to two. The real change was in the actual method of signalling, which was entirely new. There were sixty-three main signals, including three preparatory. The first fifteen were made with lights alone, up to four, hung in different patterns. The next fifteen were the same light patterns over again, but distinguished by 'one or more rockets or false fires'. The third fifteen were distinguished by 'one or more guns', and the last fifteen by 'one or more guns' and 'one or more rockets or false fires'. There were also nineteen signals at anchor. The scheme depended entirely on the visibility of the lights and was put forward in an effort to curtail the use of guns, rockets and false fires. The first fifteen

47 Corbett, *Fighting Instructions*, pp338-9; quotations from Mathieu-Dumas, *Précis des Evénements Militaires: Pièces Justificatives*, XIV, p408, and Fernandez Duro, *Armada Española*, VIII, p353.

48 Corbett, *Fighting Instructions*, p342

49 pp445-6

50 In the original draft, *MOD*, Ec/51, 'on' is also used for the attack with two columns.

signals with lights alone certainly included the most important ones: for altering course, tacking and wearing; forming line of battle and engaging; number 13, 'Engage as best able'; number 14, 'Being disabled in battle'; and number 15, 'Passed enemy without taking possession', which might be used after an enemy had surrendered.

Popham's *Telegraphic Signals for the use of His Majesty's Fleet: 1816* was based entirely on the scheme he first established in 1812. He employed nine flags, numbered 1 to 9; five cornets lettered A to E; five guidons (triangular flags) lettered F to K; and four pendants, lettered L to O. The shapes and colours were the same as those used in 1812. The main body of the signals consisted, as before, of words and sentences, arranged in alphabetical order. Each of them could be signalled with a three-flag hoist. Starting with flags 4 and 5, which represented A in the Telegraph, the remaining flags, guidons and pendants were used to cover all the words and

sentences. Thus all words and sentences beginning with the letter P were signalled by a three-flag hoist beginning with the character guidon G followed by two numbers. As before, syllables and parts of words were included, but these were now split up and placed at the end of the appropriate alphabetical sections. They included blanks for the insertion of words having 'local signification'. For these particular signals a four-flag hoist was necessary. As before, the lists of military and geographical terms were quite small. There was also a complete list of the navy with call numbers, exactly as in the Signal Book, and at the end of the book there was placed a system of semaphore signalling.

In order to spell out a word or words letter by letter, the 'orthographical pendant', white with a red cross, was first hoisted. The arrangements were then much as before. Character flag number 1 equalled A, and so on; number 9 equalled I and J; letter I equalled S and Z; L equalled U and V; and

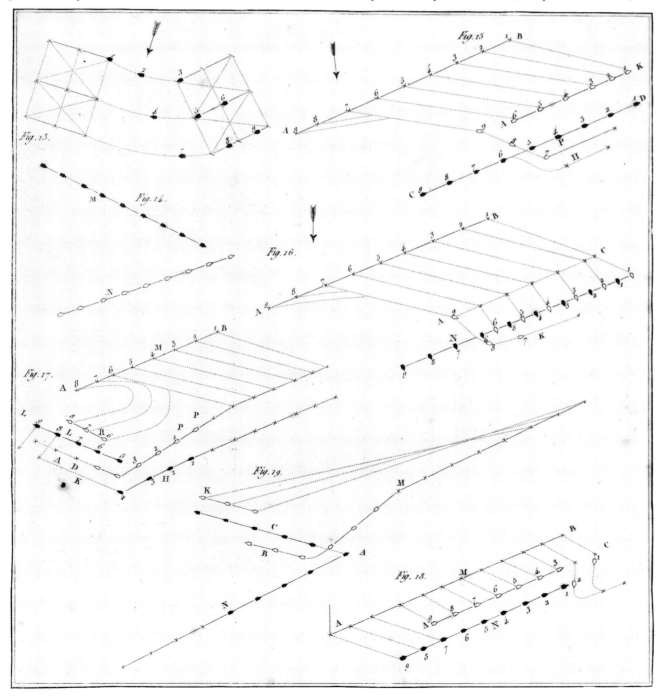

A plate from the Chevalier Delarouvraye's Traité sur l'art des Combats de Mer, Paris, 1815, *illustrating tactics for placing an enemy line of battle between two fires.*

M,N,O equalled W,X,Y respectively. To avoid confusion with the signals in the Signal Book, all of which started with a figure, all telegraphic signals were begun with a letter selected from the characters.

Something of the range and detail can be seen from the following examples: BA2 could mean 'fish', 'fishes', 'fished' or 'fishing'; BA3 meant 'cannot fish', BA4 meant 'plenty of fish'; BA5 meant 'no fish', BA6 meant 'for fishing'; BA7 meant 'after fishing'; BA8 meant 'fishing boats'; and BA9 meant 'fishing the anchor'.

In accepting Popham's telegraphic system as it stood, the Admiralty accepted responsibility for its main weakness – the confusion likely to arise from using flags possessing an alphabetical character to make a signal of which the key word began with quite a different letter of the alphabet. Against this, however, could be put the advantage of working with a much smaller number of flags than would have been needed to supply a complete alphabetical system with every letter having its own flag, guidon, cornet or pendant. Nor was there the same likelihood of confusion as must have frequently arisen in working the French system, where numbers given to flags could easily be confused with the actual signal numbers. In his signal book signals and in his telegraph, Popham was providing a wealth of orders, words, sentences, syllables and analogous matter undreamt of by those responsible for signalling arrangements in the French service.

French signals and tactics at the end of the war

In 1813 a code of vocabulary signals based on Popham, 'Signaux et Dictionnaire Telegraphique à l'Usage des Armées Navales', was issued to the French Navy, designed as an auxiliary to the official signal book.[51] The signalling was entirely with numerical flags, 0 to 9 with two pendants. The total of possible signals was 4998, consisting entirely of single words except for variations of *avoir*. The last forty-nine signal spaces were left blank. Other words could be signalled by using the numbers 57 to 82 to represent letters of the alphabet. There were separate lists for compass and horary signals. The code is simple and efficient, but its limitation to single words made it inferior even to Popham's code of 1803. This was essentially a vocabulary book, and was not concerned with ordinary executive signals.

Delarouvraye's 'Combats', 1815

When the war was over, Bonaparte dethroned and the king restored, a French naval officer at last produced the book for which the French navy had been waiting for over a hundred years. On the front cover of the *Traité sur l'Art des Combats de Mer*, under the name of the Chevalier Delarouvraye, *lieutenant de vaisseau*, was the inscription 'Delenda est Britannia'.[52] The book had only eighty-six pages, but a true offensive spirit is shown in the author's spirited and original approach to tactics. He stressed the individual responsibility of individual captains in a 'mélée navale', and wrote that in naval warfare, unlike land fighting, 'on ne peut abandoner la champ de bataille que quand il n'y a plus d'ennemis à combattre ou qu'on ne peut plus se battre'. An outnumbered fleet must fight on, and, 'avec une intrépidité sans égale, achèter une défaulte glorieux'.

Delarouvraye agreed with John Clerk's condemnation of attacking from windward, but none the less talked about the methods a windward fleet could employ to cut the enemy line. His proposed methods were somewhat ambitious and fanciful, and depended upon the assumption that the windward fleet did not itself fear being divided into two groups as a result of their attack. His whole scheme depended on the windward fleet being closed up to only half a cable distance between ships. The clear, though unstated, assumption was that the windward fleet possessed greater fighting efficiency. This applied not only to its seamanship, but to its morale and skill at arms. One of his schemes was for the windward fleet to bear down together, cut the leeward fleet at all points, and board the enemy. Each windward ship would grapple the bows of the approaching ship in the gap, and the frigates would join in from the stern.

Turning to the possibilities open to the leeward fleet, Delarouvraye showed a diagram for doubling on the windward fleet attacks in two columns in line ahead, the leeward fleet preventing their own line being cut by running aboard the enemy. If that failed, they were quickly to re-form their line, once the windward fleet had cut through, and then they would be in the windward position. Meanwhile their van, so far having been out of the fight, would drop to leeward and put the attacker between two fires. Superficially, this could have been a partial answer to Nelson's Trafalgar attack.

A delightful diagram shows how an inferior fleet, attacked from windward, might retreat by wearing together under cover of the smoke generated by the attackers' opening broadsides. Having re-formed on the opposite tack, the windward fleet would make a second attack on which the inferior fleet to leeward would wear back again to its original course. The superior fleet would thus be tired out, and unable to force action.

The French signal book of 1819

In 1819 the royal government published a completely new and official 'Livre des Signaux de Jour, à l'Usage des Vaisseaux de Guerre Français'. It was a perfect example, in the naval sphere, of the cruel gibe that the restored Bourbons had learnt nothing and forgotten nothing. On the first page of the main text we are told that Pavillon's signal system is still the best.

A comparison with the British Signal Book of 1816 suggests that each contains a far larger number of signals than were ever likely to be required or than could be seen and acted on in battle. However, the British book, using the numerical system in association with Popham's Telegraph, seems to radiate a great deal of tactical innovation and effervescence, whereas the French book gives the impression of being imprisoned in that period of the past when French arms were most successful. It took no account of post-Revolutionary experience. Service traditionalism seems combined with political sentiment to ignore the lessons of defeat.

The signals were divided into three classes, of which the first two covered compass points, numbers, latitude and longitude, soundings, recognition numbers, squadron and division numbers, van, rear, 'ships to windward', 'all frigates and corvettes', etc. No distinction was made between signals under sail and signals at anchor. As a piece of production, the signal book of 1819 was better than its predecessors. It was very clearly printed, and had a good index with main and sub-headings, referring directly to the signals, which were cross-referenced to the instructions. Signals were made by means of four complex tables. Twenty flags were employed, four guidons, which in this nomenclature indicated a

---

51 *NMM*, SIG/C/11 (formerly Sp/55).

52 Chevalier Delarouvraye, *Traité sur l'Art des Combats de Mer, Dédié ASAR Mgr Le Duc D'Angoulème, Grand Amiral de France*, Paris, 1815. (*NMM*, TUN/89).

swallow-tail flag, two triangular flags, and eight pendants. The first ten flags were a combination of red and white, and the second ten of blue and yellow.

The battle signals were scattered about, though most of them were amongst the first thirty, made with single-flag hoists. Signals numbers 35–586 were made by means of a twenty-four-sided table, and included in numbers 305–20 the admiral's advice of tactical intentions. Numbers 321–35 were for the *corps de réserve* and, though precise, were not very imaginative. Numbers 336–50 were further battle signals, including indications of the admiral's intentions. Numbers 351–74 were quasi-battle signals in the sense that they deal with the line, but without mentioning the enemy. Numbers 375–425 gave information about the enemy. Numbers 587–597 included some of the most important battle signals, which were made with a three-flag hoist and a complicated table. Numbers 590–2 were for cutting through the enemy's van, centre or rear as a gap offered, and number 593 was to cut the enemy's line by running aboard one of their ships if necessary.

The signals were followed by 331 general instructions, the numbers of which were printed in separate columns in the signal section, opposite the relevant signals. They were

A page from Henry Raper's manuscript Naval Signals, presented in 1827 or 1828 to the Duke of Clarence, the Lord High Admiral, showing his solution to the problem of signalling over extended distances.

mainly organisational, and for carrying out the usual evolutions in the order of sailing and changing into or out of the order of battle in line ahead. There were only five battle instructions of any interest. Number 233 was for the *corps de réserve* to be ready to attack and double on either the van or rear of the enemy, as directed. Nothing further was indicated, thus showing lack of any interest in the tactical possibilities of an advance squadron as conceived by Howe and Nelson. Number 234 indicated that, when the admiral made the signal for cutting the enemy's line, the ship which cut through first was to be followed by 'les autres', without specifying what group or division, their duty being to prevent the enemy ships cut off from rejoining their fleet. Number 235 instructed that, should the particular ship ordered to cut the enemy line be unable to do so, the first ship which found it possible should do so, and those astern should follow her. Number 236 indicated that the *corps de réserve* must be ready to intervene at the point where the enemy's line was cut. The rest of the fleet, on seeing the *corps de réserve* ready to support them, were to close the enemy and attempt to cut their line. Number 237 further advised that 'traversing' the enemy's line by cutting through the intervals, presumably between each enemy ship, was generally too dangerous. It could best be executed when the enemy were already doubled and between two fires, as for instance when the *corps de réserve* had doubled on them.

It is quite clear from the signals and instructions, and from the further tactical material which followed, that the lessons derived both from the more remote and the more immediate past admitted of only two kinds of attack. The first was doubling, with or without the intervention of the *corps de réserve*, and the second was cutting the enemy's line through a single gap or traversing by several ships together.

The book ended with a long chapter of 'Explications'. Section I discussed the history and principles of signals and established the value of tactics based on Hoste and Morogues. Section II discussed tactics, defined as 'l'art de ranger les vaisseaux de guerre sur une ou plusiers lignes, et de régler tellement leur mouvements, qu'ils puissent, avec promptitude, prendre la situation la plus favourable à l'attaque et à la défense'. Ships, it stated, were weakest at their bow and stern, and strongest on their broadsides, and various consequences were seen to flow from this, including the importance of concentration of fire. Section III expressed approval of Morogues's acceptance of 'l'ancien système des signaux', and rejection of the numerical system, including La Bourdonnais's method, after the vote taken at each of the naval bases.

Admiral Raper's new system of signals

Although by the time Admiral Henry Raper published in 1828 his *New System of Signals, by which colours may be wholly dispensed with* the Napoleonic wars were long over, his book is an interesting place at which to conclude this study because of the information in his preface, and other glimpses he provides of wartime practice. Raper (1767–1847) went to sea at the age of thirteen, was Lord Howe's signal lieutenant at the First of June, and was liaison officer with the Portuguese squadron subsequently attached to the Channel Fleet. He did not offer his system to the Admiralty until 1815 and, despite continued efforts, it was never accepted. Subsequent publication was presumably at the author's expense. A little earlier, possibly in 1827, he presented a somewhat different manuscript copy to the Duke of Clarence, then Lord High Admiral.[53]

Raper started from the proposition that colours had always been difficult to distinguish and hence a source of confusion in signalling. To get over this difficulty, he proposed a completely new set of colour symbols. He employed ten flags measuring 12ft by 10ft, a substitute flag of upright shape flown lengthwise to the mast, ten pendants, a triangular substitute flag, a short substitute pendant and a 27ft substitute pendant. Under ordinary conditions, these were used for ordinary numerical signalling with a full list of the navy and a telegraph vocabulary of words and sentences in addition. The only difference in principle from the Admiralty book was that the signals were placed in twelve classes. When colours could not be distinguished, it was possible to change over immediately to signalling by shapes. These consisted of:

1	Any flag	6	Ball
2	Upright flag	7	Irregular triangular flag
3	Triangular flag	8	ditto inverted
4	Pendant	9	Long pendant
5	Short pendant	0	Weft

He gives the following example of usage:

A ship having important intelligence rejoins the fleet and signals her number, – consisting of three pendants. The nearest ship to her is still too far off to distinguish the colours but can tell from the *shapes* used that the signal falls within Class XI, 'Ships Numbers'. She at once hoists a ball, to request Shapes, on which the stranger signals her number by *Shapes* alone. In the particular case illustrated, the number 932 is made with a long pendant (9), a triangular flag (3) and the upright flag (2).

To make the distant signal 'Pass through the [enemy's] line where the greatest interval appears', which is no 79 in Class III (Engaging and Evolutions), the signal will be an irregular triangular flag (7), over the short substitute pendant (for the class) and the long pendant for 9. The class pendant is thus flown between the two flags indicating the signal number. Telegraphic signals at a distance are made by signalling the page number and the number of the word or sentence on the page.

The feasibility of Raper's scheme depended on the shapes being clearly distinguishable when colours were not. The colours of his pendants were very confusing anyhow, though this was inevitable in view of the number employed. Raper failed to gain acceptance for his scheme mainly because he was too late, Popham having already secured the Admiralty's acceptance of his Telegraph. Besides, the Admiralty Signal Book of 1816 already contained a system for distant signalling which, if clumsier than Raper's, was designed to fit in with the main signalling system. Raper's handicap was that he wanted to reorganise the whole system in order to make it fit in with his distance arrangements, at the same time introducing a number of new and by no means desirable pendants. He did alter his scheme as time went on, in the light, no doubt, of useful criticisms. Nevertheless, it seems odd that after submitting it to the Admiralty for the last time in 1825 and offering it in similar form to the Duke of Clarence, two or three years later, he should have completely recast it before publication. He acknowledged, in detail, the complete change made in the distant signal shape, but said nothing about the changes made in the classes of signals, the signals themselves and the order of subjects and method of presentation.[54] In both versions he supplied diagrams and explanatory notes for his evolutionary signals. These also are changed in the published version. On the whole, the easier version in the form presented to the duke seems better arranged.

53 *NMM*, Tunstall Collection, TUN/29 (S/MS/Rap/1)
54 p92

Index